RaumFragen: Stadt – Region – Landschaft

Series Editors

Olaf Kühne, Forschungsbereich Geographie, Eberhard Karls Universität Tübingen, Tübingen, Germany

Sebastian Kinder, Forschungsbereich Geographie, Eberhard Karls Universität Tübingen, Tübingen, Germany

Olaf Schnur, Stadt- und Quartiersforschung, Berlin, Germany

RaumFragen: Stadt – Region – Landschaft | SpaceAffairs: City – Region – Landscape

Im Zuge des „spatial turns" der Sozial- und Geisteswissenschaften hat sich die Zahl der wissenschaftlichen Forschungen in diesem Bereich deutlich erhöht. Mit der Reihe „RaumFragen: Stadt – Region – Landschaft" wird Wissenschaftlerinnen und Wissenschaftlern ein Forum angeboten, innovative Ansätze der Anthropogeographie und sozialwissenschaftlichen Raumforschung zu präsentieren. Die Reihe orientiert sich an grundsätzlichen Fragen des gesellschaftlichen Raumverständnisses. Dabei ist es das Ziel, unterschiedliche Theorieansätze der anthropogeographischen und sozialwissenschaftlichen Stadt- und Regionalforschung zu integrieren. Räumliche Bezüge sollen dabei insbesondere auf mikro- und mesoskaliger Ebene liegen. Die Reihe umfasst theoretische sowie theoriegeleitete empirische Arbeiten. Dazu gehören Monographien und Sammelbände, aber auch Einführungen in Teilaspekte der stadt- und regionalbezogenen geographischen und sozialwissenschaftlichen Forschung. Ergänzend werden auch Tagungsbände und Qualifikationsarbeiten (Dissertationen, Habilitationsschriften) publiziert.

Herausgegeben von
Prof. Dr. Dr. Olaf Kühne, Universität Tübingen
Prof. Dr. Sebastian Kinder, Universität Tübingen
PD Dr. Olaf Schnur, Berlin

In the course of the "spatial turn" of the social sciences and humanities, the number of scientific researches in this field has increased significantly. With the series "RaumFragen: Stadt – Region – Landschaft" scientists are offered a forum to present innovative approaches in anthropogeography and social space research. The series focuses on fundamental questions of the social understanding of space. The aim is to integrate different theoretical approaches of anthropogeographical and social-scientific urban and regional research. Spatial references should be on a micro- and mesoscale level in particular. The series comprises theoretical and theory-based empirical work. These include monographs and anthologies, but also introductions to some aspects of urban and regional geographical and social science research. In addition, conference proceedings and qualification papers (dissertations, postdoctoral theses) are also published.

Edited by
Prof. Dr. Dr. Olaf Kühne, Universität Tübingen
Prof. Dr. Sebastian Kinder, Universität Tübingen
PD Dr. Olaf Schnur, Berlin

More information about this series at http://www.springer.com/series/10584

Dennis Edler · Corinna Jenal · Olaf Kühne
Editors

Modern Approaches to the Visualization of Landscapes

 Springer VS

Editors
Dennis Edler
Geographisches Institut
Ruhr-Universität Bochum
Bochum, Germany

Corinna Jenal
Eberhard Karls Universität Tübingen
Tübingen, Germany

Olaf Kühne
Eberhard Karls Universität Tübingen
Tübingen, Germany

ISSN 2625-6991   ISSN 2625-7009   (electronic)
RaumFragen: Stadt – Region – Landschaft
ISBN 978-3-658-30955-8   ISBN 978-3-658-30956-5   (eBook)
https://doi.org/10.1007/978-3-658-30956-5

Lektorat: Cori A. Mackrodt
This Springer VS imprint is published by the registered company Springer Fachmedien Wiesbaden GmbH part of Springer Nature.
The registered company address is: Abraham-Lincoln-Str. 46, 65189 Wiesbaden, Germany

Contents

Part I
Introduction

Modern Approaches to the Visualization of Landscapes—An Introduction

Dennis Edler, Olaf Kühne and Corinna Jenal

Abstract

Visualizations of landscapes have always played an important role in landscape research. Thanks to developments in the fields of cartography and related neighboring disciplines, modern approaches to the visualization of geographic space are constantly developing which includes, for example, new levels of image resolutions, interactivity, animation and three-dimensionality. New initiatives of (open) data availability, data creation and data exchange also help to improve these approaches and emerging visualizations. This anthology aims to gather different perspectives and approaches to the visualization of landscape, and to bring them together under what they all have in common—that is 'visualization' and 'landscape'. The present volume deals with the corresponding forms of representation in landscape analyses, as well as with landscape visualizations on the Internet and visual representations of multisensory landscapes. The collection is complemented by references to modern virtual landscapes and modern urban societies, landscape visualizations in the field of education, but also landscape and its visualization in conflicts.

D. Edler (✉)
Ruhr-Universität Bochum, Bochum, Germany
e-mail: dennis.edler@ruhr-uni-bochum.de

O. Kühne · C. Jenal
Eberhard Karls Universität Tübingen, Tübingen, Germany
e-mail: olaf.kuehne@uni-tuebingen.de

C. Jenal
e-mail: corinna.jenal@uni-tuebingen.de

© Springer Fachmedien Wiesbaden GmbH, part of Springer Nature 2020   3
D. Edler et al. (eds.), *Modern Approaches to the Visualization of Landscapes*,
RaumFragen: Stadt – Region – Landschaft, https://doi.org/10.1007/978-3-658-30956-5_1

Keywords

Visualization · Landscape · Landscape Visualization · Landscape Research · Cartography

1   Introductory Remarks

Visualizations of landscapes have always played an important role in landscape research. Thanks to developments in the fields of cartography and related neighboring disciplines, modern approaches to the visualization of geographic space are constantly developing. This includes, for example, new levels of image resolutions, interactivity, animation and three-dimensionality (c.f. Clarke et al. 2019; Dickmann 2018; Griffin et al. 2017; Virrantaus et al. 2009). New initiatives of (open) data availability, data creation and data exchange also help to improve these approaches and emerging visualizations. These visualizations serve as media to unite and communicate relevant spatial information. Moreover, they enable researchers to access represented landscapes through different epistemological and empirical perspectives. Both positivist and constructivist perspectives on landscapes may benefit from modern approaches to the visualization of landscapes (Edler et al. 2018).

For example, visual stimuli have a prominent role in the social and individual construction of landscape (amongst many: Crandell 1993; Daniel 2001; Ode et al. 2008). Daniel Cosgrove referred to landscape as a way of seeing the world (Cosgrove 1984). Seeing the world as a landscape is only one side of the visual relationship to landscape. The other side is the visualization of landscape. This is of outstanding importance for the social and individual construction of landscape, as we provide interpretations and patterns for the interpretation of the world through visualizations. Without the visual representation of the world in the form of landscape paintings in the Renaissance, we would not have learned that we can develop landscape interpretations referring to physical spaces that appear rather profane (amongst others: Büttner 2006; 2019; Eberle 1980; Kühne 2018a; Schneider 2009; see also contribution Schenk 2020).

In this respect, the visualization of landscape has a constitutive importance for its construction. In the landscape-related socialization process, people have to learn which elements (including their characteristics and spatial arrangements) can be socially accepted as 'landscape', without having to fear the loss of social recognition (Kühne 2008, 2019a; Stotten 2013; Tillmann 2007; Visscher and Bouverne-De Bie 2008; Tessin 2008). The patterns of interpretation and evaluation of physical spaces as landscapes are socially (e.g. age, place of residence, gender and ideological-political orientation) and culturally quite differentiated (among many others: Bruns et al. 2015; Bruns and Kühne 2013, 2015; Buijs et al. 2009; Gehring and Kohsaka 2007; Hunziker et al. 2008; Jenal 2019; Kühne 2018b). A further social differentiation of the construction addresses the question whether the individual acting, in relation to landscape, has acquired 'expert special knowledge' (Kühne 2019b, 2020b). This refers to specific patterns of interpretation and evaluation which have been acquired during a course of study (Hokema 2015; Kühne 2013; Stemmer et al. 2019; Ticino 2008; Wojtkiewicz 2015).

Visualizations of what we—quite socially and culturally differentiated—call 'landscape' are not limited to depicting physical geographic spaces (or a deliberate pictorial composition of objects), such as a painted image or photographed (and often edited) images or videos. Thanks to corresponding databases and portals on the Internet (e.g. YouTube, Instagram, Facebook), photos and videos today enjoy a potentially high degree of dissemination. They include specific and individual information and they can satisfy aesthetic needs. In other words, they are used be used to generate social recognition (in this context e.g. Burgess and Green 2018; Garrett 2011; Kühne and Jenal 2020a; Kühne and Weber 2015; Linke 2019; Münderlein et al. 2019).

Thanks to technological innovations in the software and hardware industries, the creation of virtual landscapes has been severely improved in the last decade (see also Kersten and Edler 2020). Companies which produced commercial software, especially in the gaming industry, released their 'engines' which can be used to develop individual applications. These applications can be experienced in real-time and from a first-person perspective. Body movements are 'virtually translated' into the movements of an avatar in the virtual 3D landscape (see also Edler et al. 2018; Hruby et al. 2019; Jerald 2016). As virtual reality (VR) applications are closely related to modern video and computer games, they also include aspects of a gamified experience of a represented environment (see also Fontaine 2017; Kersten et al. 2018; Vetter 2019). The same applies to augmented reality (AR) applications, which can also be created with freely released game engines (c.f. Keil et al. 2019; 2020; Lindner et al. 2019). Social constructivist approaches have only hardly been combined with augmented reality and immersive virtual environments so far (but see Edler et al. 2018).

Several cartographers take the view that modern virtual reality applications are extensions of traditional cartographic approaches, and other cartographic media, such as 2D maps, are often used as (temporarily implemented) information source to locate 3D objects in virtual environments (e.g. Edler et al. 2019; Hruby et al. 2020). Other VR applications are derived from remotely sensed point cloud data (Lütjens et al. 2019; Zhao et al. 2019). First examples also exist that expand virtual environments with auditory elements (Berger and Bill 2019; Edler et al. 2019b; Hruby 2019). The implemented 3D sound files are used to simulate an audiorealistic impression of soundscapes, an important information layer of multisensory landscapes (see Dodt et al. 2017; Edler 2020; Edler and Kühne 2019; Kühne and Edler 2018; Schafer 1977; Winkler 2005). This can be regarded as an extension of audiovisual cartography (Krygier 1994). In these kinds of multimedia maps, sound files are usually bound to 2D graphical base maps (e.g. Edler and Vetter 2019; Edler et al. 2015; Laakso and Sarjakoski 2010; Lammert-Siepmann et al. 2017; Siepmann et al. 2020a).

Aside from the creation of multisensory virtual environments with immersive technologies, modern software and hardware developments as well as web technologies have broadened the opportunities to discuss, create, edit and share various kinds of media representing landscapes. This includes images, sound and videos recorded with smartphones, tablets and other mobile devices. Moreover, the increase of open initiatives has also promoted the release of open source software. This publicly available software often convinces through its simplicity. Images can be edited and improved, sounds and videos can be cut or extended, and digital maps can be created. Today's straightforward opportunities to create visual or audiovisual landscape representations has differentiated the dichotomy of expert (professionals) and non-experts ('home brewers').

This is a differentiation which has also been made regarding specific expert knowledge on the topic of landscape: Much information, including scientific texts, is available online and can be accessed without any restrictions. Selecting this information depends on individual interests. These interests are often related to a resistance to certain physical manifestations of social interests, such as the energy revolution or the extraction of raw materials (Gobert 2016; Jenal 2018; Kühne and Weber 2018 [online first 2017]; Schweiger et al. 2018; Weber et al. 2017; Weber et al. 2018).

So, not only the increasing opportunities to create landscape visualizations, but also the increasing differentiation within the creators of these visualizations are topics of modern landscape research. This is where this volume comes into play:

The concept of this volume is to cover a broad range of theoretical, empirical and practical approaches of landscape visualization. This includes data and its sources, modern digital visualization techniques and application scenarios in education and transforming urban societies. The volume comprises 30 articles. It is divided into seven chapters, beginning with a first chapter dedicated to the versatility of landscape analysis and methods of data processing. The second chapter is focused on the Internet as modern source of landscape visualizations. In Chap. 3, the contributors leave the visual dimension of landscapes and consider non-visual stimuli, including smellscapes and soundscapes. The three-dimensional presentation of soundscapes is also taken up in Chap. 4. This chapter highlights current practical opportunities and theoretical foundations of creating virtual landscapes. A specific focus is given to computer games and immersive virtual reality (VR). Potentials of modern VR applications are also explored in present studies on education. Landscape visualizations in educational scenarios are addressed in Chap. 5. Examples of modern approaches to the visualization of landscapes are not only applied in education. They shape modern urban societies. Visualizations are not only used to represent these urban landscapes. Visualizations may occur as parts of urban areas. Examples can be found in protest movements or in the design of neighborhoods or a "fabulous" city. Chapter 6 puts emphasis on landscape visualizations in conflicts. Chapter 7 offers more examples of landscape visualizations and modern urban societies.

2   Chapters and Contributions—A Brief Overview

The first chapter emphasizes the many available sources of landscape visualizations used in quantitative and qualitative empirical investigations of landscapes. It begins with a contribution by Winfried Schenk, who highlights three (cultural) landscape dimensions—resource, work and social structuring—in historical landscape paintings (Schenk 2020). Based on modern remotely sensed data, M. Fabian Meyer-Hess presents a method of detecting "forgotten" historical landscapes (Meyer-Hess 2020). Volker Hochschild, Andreas Braun, Christian Sommer, Gebhard Warth and Adel Omran give an overview of landscape visualization using a selection of different and modern geospatial techniques (Hochschild et al. 2020). The complexity of landscapes is addressed by Fivos Papadimitriou, who presents a novel quantitative method (Papadimitriou 2020a). Mohammed Al-Khanbashi suggests an approach to the organization and representation of qualitative data in social constructivist landscape research (Al-Khanbashi 2020).

The second chapter highlights the Internet as modern media source for landscape visualization. It begins with an introductory article by Olaf Kühne, who addresses theoretical foundations of the social construction of landscapes in online available videos (Kühne 2020a). Simone Linke puts emphasis on a selection of online available images (Linke 2020). Online images are also used by Mirella Loda, Olaf Kühne and Matteo Puttilli, who refer their image selection to the social construction of Tuscany, Italy (Loda et al. 2020). Based on online available open spatial data provided by the OpenSeaMap community and modern cartographic web-based visualization techniques, Alexander Kleber, Dennis Edler and Frank Dickmann present a JavaScript-based application which was programmed to support sea navigation and the management of maritime shipping in the Western Scheldt, Netherlands (Kleber et al. 2020).

The third chapter includes contributions showing the relevance of different sensory dimensions in landscape research. Karsten Berr deals with visuality (Berr 2020), aesthetic and landscape from a constructivist perspective. Kate McLean conducted a sensory walk in Kiev, Ukraine, and visualizes her empirical findings on the smellscape in maps (McLean 2020). Nils Siepmann, Dennis Edler and Olaf Kühne provide an overview of 2D and 3D cartographic representations of soundscapes (Siepmann et al. 2020b).

The fourth chapter is dedicated to modern virtual landscapes. Dominique Fontaine opens the chapter with theoretical considerations of virtuality and landscapes (Fontaine 2020a). Recent methodological opportunities of landscape visualization with virtual reality (VR) equipment and software are presented by Mark Vetter (2020). Dennis Edler, Julian Keil and Frank Dickmann also address modern VR visualization methods, with a focus on the potentials of open spatial data shared by official surveying departments and the web community of gamers (Edler et al. 2020). The chapter is closed by Dominique Fontaine, who deals with virtual landscapes as dynamic sociocultural products in selected computer games (Fontaine 2020b).

The fifth chapter has a focus on educational applications of modern examples of landscape visualization. Christopher Prisille and Marko Ellerbrake take up modern VR visualizations and discuss the potentials of visualizations made with 360° cameras for geography education in secondary schools (Prisille and Ellerbrake 2020). Maximiliam Stintzing, Stephan Pietsch and Ute Wardenga address the teaching of landscapes using a gamified approach and an Augmented Reality (AR) application (Stintzing et al. 2020). Fivos Papadimitriou also broadens the media spectrum and brings together future landscapes, postmodern cinema and geography education (Papadimitriou 2020b). Petér Bagoly-Simó expands the chapter by emphasizing landscape representations in geography textbooks (Bagoly-Simó 2020).

The sixth chapter is about landscape visualization and conflicts (see also Walsh et al. 2020). In this context, Olaf Kühne and Corinna Jenal refer to the threefold landscape change in further differentiating societies and the related conflicts in social negotiation processes concerning 'landscape'. The authors emphasize the potentials of virtual landscape research for productive conflict regulation (Kühne and Jenal 2020b). Erik Aschenbrand and Thomas Michler (2020) examine the conflicts of interpretation and attribution concerning bark beetle infestation and windthrow in the Bavarian Forest. The authors address how this is reflected in the visualizations of the Bavarian Forest by the various participants in the discourse. The conflicts about landscapes have become particularly evident in Germany, in the context of the implementation of the so-called energy transition. Using the example of the expansion of the power grid, Corinna Jenal shows how visual and interpretative patterns are performed and structured in protest movements through landscape visualizations. Physical objects are explicitly included or excluded in the display of landscape (Jenal 2020). The contribution by Peter Martin Thomas provides an outlook on the long-term effects of digital transformation. In view of its irreversibility, the author pleads for an active shaping of the digital transformation in one's own living environment (Thomas 2020).

The final seventh chapter focuses on landscape visualization and modern urban societies. In his contribution, Florian Weber (2020) addresses landscape visualizations in the desert city of Las Vegas (USA). He shows how specific landscapes are simulated here, both externally and internally, and how utopian and real spaces merge into artificial worlds. In a post-structuralist discourse and based on a Google image analysis, Albert Roßmeier (2020) examines the media representations of the neighborhoods East Village and Barrio Logan in San Diego (USA). He works out differences and similarities between the various neighborhoods. Judith Stratmann, Alina Ristea, Michael Leitner and Gernot Paulus apply different spatial analysis methods to create urban blightscapes of Baton Rouge, LA (USA). It is shown that spatial videos and specific GIS tools are suitable technologies to visualize urban blightscapes (Stratmann et al. 2020). The article by Alenka Poplin et al. (2020) examines tangible and intangible features of places and focuses on collecting features, emotions, memories and stories related to self-chosen evocative locations in a city (Poplin et al. 2020). The anthology concludes with the contribution by Andrea Bellini and Laura Leonardi, who examine various forms of hybridization and stereotyping in the context of the textile crisis and immigration. The spatial case study is the Italian industrial district of Prato (Bellini and Leonardi 2020).

This volume brings together recent technological and methodological developments of spatial visualization and current topic of landscape research. The editors hope that you will enjoy reading all research articles compiled in this volume.

Dennis Edler, Olaf Kühne and Corinna Jenal

References

Al-Khanbashi, M. (2020). Using matrix as a qualitative data display for landscape research and a reflection based on the social constructivist perspective. In D. Edler, C. Jenal, & O. Kühne (Eds.), *Modern approaches to the visualization of landscapes* (pp. 103–118). Wiesbaden: Springer VS.

Aschenbrand, E., & Michler, T. (2020). Linking socio-scientific landscape research with the ecosystem service approach to analyze conflicts about protected area management – The case of the Bavarian Forest National Park. In D. Edler, C. Jenal, & O. Kühne (Eds.), *Modern approaches to the visualization of landscapes* (pp. 403–425). Wiesbaden: Springer VS.

Bagoly-Simó, P. M. (2020). Landscape in geography textbooks. In D. Edler, C. Jenal, & O. Kühne (Eds.), *Modern approaches to the visualization of landscapes* (pp. 371–285). Wiesbaden: Springer VS.

Bellini, A., & Leonardi, L. (2020). Prato: The social construction of an industrial city facing processes of cultural hybridization. In D. Edler, C. Jenal, & O. Kühne (Eds.), *Modern approaches to the visualization of landscapes* (pp. 549–572). Wiesbaden: Springer VS.

Berger, M., & Bill, R. (2019). Combining VR visualization and sonification for immersive exploration of urban noise standards. *Multimodal technologies and interaction, 3*(2), 34. https://doi.org/10.3390/mti3020034.

Berr, K. (2020). Visuality, Aesthetics and Landscape. For the enlightenment and self-enlightenment of constructivist landscape research. In D. Edler, C. Jenal, & O. Kühne (Eds.), *Modern approaches to the visualization of landscapes* (pp. 189–215). Wiesbaden: Springer VS.

Bruns, D., & Kühne, O. (2015). Zur kulturell differenzierten Konstruktion von Räumen und Landschaften als Herausforderungen für die räumliche Planung im Kontext von Globalisierung. In B. Nienaber, & U. Roos (Eds.), *Internationalisierung der Gesellschaft und die Auswirkungen auf die Raumentwicklung. Beispiele aus Hessen, Rheinland-Pfalz und dem Saarland* (Arbeitsberichte der ARL, vol. 13, pp. 18–29). Hannover: Selbstverlag. https://shop.arl-net.de/media/direct/pdf/ab/ab_013/ab_013_02.pdf. Accessed: 26 November 2018.

Bruns, D., & Kühne, O. (Eds.). (2013). *Landschaften: Theorie, Praxis und internationale Bezüge. Impulse zum Landschaftsbegriff mit seinen ästhetischen, ökonomischen, sozialen und philosophischen Bezügen mit dem Ziel, die Verbindung von Theorie und Planungspraxis zu stärken.* Schwerin: Oceano Verlag.

Bruns, D., Kühne, O., Schönwald, A., & Theile, S. (Eds.). (2015). *Landscape culture – Culturing landscapes. The differentiated construction of landscapes.* Wiesbaden: Springer VS.

Buijs, A. E., Elands, B. H. M., & Langers, F. (2009). No wilderness for immigrants: Cultural differences in images of nature and landscape preferences. *Landscape and Urban Planning, 91*(3), 113–123. https://doi.org/10.1016/j.landurbplan.2008.12.003.

Burgess, J., & Green, J. (2018). *YouTube. Online video and participatory culture* (Digital Media and Society Series (2nd ed.)). Cambridge: Polity Press.

Büttner, N. (2006). *Geschichte der Landschaftsmalerei.* München: Hirmer.

Büttner, N. (2019). Landschaftsmalerei. In O. Kühne, F. Weber, K. Berr, & C. Jenal (Eds.), *Handbuch Landschaft* (pp. 577–584). Wiesbaden: Springer VS.

Clarke, K. C., Johnson, J. M., & Trainor, T. (2019). Contemporary American cartographic research: A review and prospective. *Cartography and Geographic Information Science, 46*(3), 196–209. https://doi.org/10.1080/15230406.2019.1571441.

Cosgrove, D. E. (1984). *Social formation and symbolic landscape.* London: University of Wisconsin Press.

Crandell, G. (1993). *Nature pictorialized: "the view" in landscape history.* Baltimore: Johns Hopkins University Press.

Daniel, T. C. (2001). Whither scenic beauty? Visual landscape quality assessment in the 21st century. *Landscape and Urban Planning, 54*(1–4), 267–281. https://doi.org/10.1016/s0169-2046(01)00141-4.

Dickmann, F. (2018). *Kartographie*. Braunschweig.

Dodt, J., Bestgen, A.-K., & Edler, D. (2017). Ansätze der Erfassung und kartographischen Präsentation der olfaktorischen Dimension. *KN – Journal of Cartography and Geographic Information, 67*(5), 245–256. https://doi.org/10.1007/BF03545321.

Eberle, M. (1980). *Individuum und Landschaft. Zur Entstehung und Entwicklung der Landschaftsmalerei*. Gießen: Anabas-Verlag.

Edler, D. (2020). Where spatial visualization meets landscape research and "Pinballology": Examples of landscape construction in pinball games. *KN – Journal of Cartography and Geographic Information*. https://doi.org/10.1007/s42489-020-00044-1.

Edler, D., & Kühne, O. (2019). Nicht-visuelle Landschaften. In O. Kühne, F. Weber, K. Berr, & C. Jenal (Eds.), *Handbuch Landschaft* (pp. 599–612). Wiesbaden: Springer VS.

Edler, D., & Vetter, M. (2019). The simplicity of modern audiovisual web cartography: An example with the open source javascript library leaflet.js. *KN – Journal of Cartography and Geographic Information, 69*(1), 51–62. https://doi.org/10.1007/s42489-019-00006-2.

Edler, D., Husar, A., Keil, J., Vetter, M., & Dickmann, F. (2018b). Virtual Reality (VR) and open source software: A workflow for constructing an interactive cartographic VR environment to explore urban landscapes. *KN - Journal of Cartography and Geographic Information, 68*(1), 3–11. https://doi.org/10.1007/BF03545339.

Edler, D., Keil, J., & Dickmann, F. (2020). From Na Pali to Earth—An 'Unreal' engine for modern geodata? In D. Edler, C. Jenal, & O. Kühne (Eds.), *Modern approaches to the visualization of landscapes* (pp. 279–291). Wiesbaden: Springer VS.

Edler, D., Jebbink, K., & Dickmann, F. (2015). Einsatz audio-visueller Karten in der Schule – Eine Unterrichtsidee zum Strukturwandel im Ruhrgebiet. *KN - Journal of Cartography and Geographic Information, 65*(5), 259–265. https://doi.org/10.1007/BF03545162

Edler, D., Keil, J., Wiedenlübbert, T., Sossna, M., Kühne, O., & Dickmann, F. (2019). Immersive VR experience of redeveloped post-industrial sites: The example of "Zeche Holland" in Bochum-Wattenscheid. *KN – Journal of Cartography and Geographic Information, 38*(3), 1–18. https://doi.org/10.1007/s42489-019-00030-2.

Edler, D., Kühne, O., Jenal, C., Vetter, M., & Dickmann, F. (2018a). Potenziale der Raumvisualisierung in Virtual Reality (VR) für die sozialkonstruktivistische Landschaftsforschung. *Kartographische Nachrichten, 68*(5), 245–254.

Edler, D., Kühne, O., Keil, J., & Dickmann, F. (2019b). Audiovisual cartography: Established and new multimedia approaches to represent soundscapes. *KN – Journal of Cartography and Geographic Information, 69*(1), 5–17. https://doi.org/10.1007/s42489-019-00004-4

Fontaine, D. (2017). *Simulierte Landschaften in der Postmoderne. Reflexionen und Befunde zu Disneyland, Wolfersheim und GTA V*. Wiesbaden: Springer VS.

Fontaine, D. (2020a). Landscape in computer games—The examples of GTA V and Watch Dogs 2. In D. Edler, C. Jenal, & O. Kühne (Eds.), *Modern approaches to the visualization of landscapes, Modern approaches to the visualization of landscapes* (pp. 293–306). Wiesbaden: Springer VS.

Fontaine, D. (2020b). Virtuality and landscape. In D. Edler, C. Jenal, & O. Kühne (Eds.), *Modern approaches to the visualization of landscapes* (pp. 267–278). Wiesbaden: Springer VS.

Garrett, B. L. (2011). Videographic geographies: Using digital video for geographic research. *Progress in Human Geography, 35*(4), 521–541.

Gehring, K., & Kohsaka, R. (2007). ,Landscape' in the Japanese language: Conceptual differences and implications for landscape research. *Landscape Research, 32*(2), 273–283. https://doi.org/10.1080/01426390701231887.

Gobert, J. (2016). *Widerstand gegen Großprojekte. Rahmenbedingungen, Akteure und Konfliktverläufe* (Essentials). Wiesbaden: Springer VS.

Griffin, A. L., Robinson, A. C., & Roth, R. E. (2017). Envisioning the future of cartographic research. *International Journal of Cartography, 3*(S1), 1–8. https://doi.org/10.1080/23729333.2017.1316466.

Hochschild, V., Braun, A., Sommer, C., Warth, G., & Omran, A. (2020). Visualizing landscapes by geospatial techniques. In D. Edler, C. Jenal, & O. Kühne (Eds.), *Modern approaches to the visualization of landscapes* (pp. 47–78). Wiesbaden: Springer VS.

Hokema, D. (2015). Landscape is everywhere. The construction of landscape by US-American Laypersons. *Geographische Zeitschrift, 103*(3) 151–170.

Hruby, F. (2019). The sound of being there: Audiovisual cartography with immersive virtual environments. *KN – Journal of Cartography and Geographic Information, 69*(1), 19–28. https://doi.org/10.1007/s42489-019-00003-5.

Hruby, F., Ressl, R., & de La Borbolla del Valle, G. (2019). Geovisualization with immersive virtual environments in theory and practice. *International Journal of Digital Earth, 12*(2), 123–136. https://doi.org/10.1080/17538947.2018.1501106.

Hruby, F., Sánchez, L. F. Á., Ressl, R., & Escobar-Briones, E. G. (2020). An empirical study on spatial presence in immersive geo-environments. *PFG – Journal of Photogrammetry, Remote Sensing and Geoinformation Science*. https://doi.org/10.1007/s41064-020-00107-y.

Hunziker, M., Felber, P., Gehring, K., Buchecker, M., Bauer, N., & Kienast, F. (2008). Evaluation of landscape change by different social groups. Results of two empirical studies in Switzerland. *Mountain Research and Development, 28*(2) 140–147. https://doi.org/10.1659/mrd.0952.

Jenal, C. (2018). Ikonologie des Protests – Der Stromnetzausbau im Darstellungsmodus seiner Kritiker(innen). In O. Kühne & F. Weber (Eds.), *Bausteine der Energiewende* (pp. 469–487). Wiesbaden: Springer VS.

Jenal, C. (2019). *„Das ist kein Wald, Ihr Pappnasen!" – Zur sozialen Konstruktion von Wald. Perspektiven von Landschaftstheorie und Landschaftspraxis.* Wiesbaden: Springer VS.

Jenal, C. (2020). Visualizations of ‚landscape' in protest movements. On exclusive and inclusive patterns of vision and interpretation using the example of resistance to the expansion of the electricity grid in Germany. In D. Edler, C. Jenal, & O. Kühne (Eds.), *Modern Approaches to the Visualization of Landscapes* (pp. 427–445). Wiesbaden: Springer VS.

Jerald, J. (2016). *The VR book.* San Rafael, CA: Human-centered design for virtual reality.

Keil, J., Edler, D. & Dickmann, F. (2019). Preparing the holoLens for user studies: an augmented reality interface for the spatial adjustment of holographic objects in 3D indoor environments. *KN – Journal of Cartography and Geographic Information, 69*(3) 205–215. https://doi.org/10.1007/s42489-019-00025-z.

Keil, J., Korte, A., Ratmer, A., Edler, D., & Dickmann, F. (2020). Augmented Reality (AR) and Spatial cognition: Effects of holographic grids on distance estimation and location memory in a 3D indoor scenario. *PFG – Journal of Photogrammetry, Remote Sensing and Geoinformation Science*. https://doi.org/10.1007/s41064-020-00104-1.

Kersten, T. P., Deggim, S., Tschirschwitz, F., Lindstaedt, M., & Hinrichsen, N. (2018). Segeberg 1600 – Eine Stadtrekonstruktion in Virtual Reality. *KN – Journal of Cartography and Geographic Information, 68*(4), 183–191. https://doi.org/10.1007/BF03545360.

Kersten, T. P, & Edler, D. (2020). Special issue "Methods and Applications of Virtual and Augmented Reality in Geo-Information Sciences". *PFG – Journal of Photogrammetry, Remote Sensing and Geoinformation Science*. https://doi.org/10.1007/s41064-020-00109-w.

Kleber, A., Edler, D., & Dickmann, F. (2020). Cartography and the sea: A JavaScript-based web mapping application for managing maritime shipping. In D. Edler, C. Jenal, & O. Kühne (Eds.), *Modern approaches to the visualization of landscapes* (pp. 173–186). Wiesbaden: Springer VS.

Krygier, J. B. (1994). Sound and geographic visualization. In A. M. MacEachren & D. R. F. Taylor (Eds.), *Visualization in modern cartography* (pp. 149–166). Oxford: Elsevier.

Kühne, O. (2008). Die Sozialisation von Landschaft – Sozialkonstruktivistische Überlegungen, empirische Befunde und Konsequenzen für den Umgang mit dem Thema Landschaft in Geographie und räumlicher Planung. *Geographische Zeitschrift, 96,*(4) 189–206.

Kühne, O. (2013). Macht und Landschaft: Annäherungen an die Konstruktion von Experten und Laien. In M. Leibenath, S. Heiland, H. Kilper, & S. Tzschaschel (Eds.), Wie werden Landschaften gemacht? *Sozialwissenschaftliche Perspektiven auf die Konstituierung von Kulturlandschaften* (pp. 237–271). Bielefeld: transcript.

Kühne, O. (2018a). *Landscape and power in geographical space as a social-aesthetic construct.* Dordrecht: Springer International Publishing.

Kühne, O. (2018b). *Landschaft und Wandel. Zur Veränderlichkeit von Wahrnehmungen.* Wiesbaden: Springer VS.

Kühne, O. (2019a). Die Sozialisation von Landschaft. In O. Kühne, F. Weber, K. Berr, & C. Jenal (Eds.), *Handbuch Landschaft* (pp. 301–312). Wiesbaden: Springer VS.

Kühne, O. (2019b). *Landscape theories. A brief introduction.* Wiesbaden: Springer VS.

Kühne, O. (2020a). The social construction of space and landscape in internet videos. In D. Edler, C. Jenal, & O. Kühne (Eds.), *Modern approaches to the visualization of landscapes* (pp. 121–137). Wiesbaden: Springer VS.

Kühne, O. (2020b). Landscape conflicts—a theoretical approach based on the three worlds theory of Karl Popper and the conflict theory of Ralf Dahrendorf, illustrated by the example of the energy system transformation in Germany. *Sustainability, 12*(17), 6772.

Kühne, O., & Edler, D. (2018). Multisensorische Landschaften – Die Bedeutung des Nicht-Visuellen bei der sozialen und individuellen Konstruktion von Landschaft und Herausforderungen für ihre Erfassung und Wiedergabe. *Berichte. Geographie und Landeskunde, 92*(1), 27–45.

Kühne, O., & Jenal, C. (2020a). *Baton Rouge – The multivillage metropolis. A neopragmatic landscape biographical approach on spatial pastiches, hybridization, and differentiation.* Wiesbaden: Springer VS.

Kühne, O., & Jenal, C. (2020b). The threefold landscape dynamics—Basic considerations, conflicts and potentials of virtual landscape research. In D. Edler, C. Jenal, & O. Kühne (Eds.), *Modern approaches to the visualization of landscapes* (pp. 389–402). Wiesbaden: Springer VS.

Kühne, O., & Weber, F. (2015). Der Energienetzausbau in Internetvideos – Eine quantitativ ausgerichtete diskurstheoretisch orientierte Analyse. In S. Kost & A. Schönwald (Eds.), *Landschaftswandel – Wandel von Machtstrukturen* (pp. 113–126). Wiesbaden: Springer VS.

Kühne, O., & Weber, F. (2018). Conflicts and negotiation processes in the course of power grid extension in Germany. *Landscape Research, 43*(4), 529–541. https://doi.org/10.1080/01426397. 2017.1300639 (Online first 2017).

Laakso, M., & Sarjakoski, L. T. (2010). Sonic maps for hiking – Use of sound in enhancing the map use experience. *The Cartographic Journal, 47*(4), 300–307. https://doi.org/10.1179/00087 0410X12911298276237.

Lammert-Siepmann, N., Bestgen, A.-K., Edler, D., Kuchinke, L., & Dickmann, F. (2017). Audiovisual communication of object-names improves the spatial accuracy of recalled object-locations in topographic maps. *PLOS ONE, 12*(10), e0186065. https://doi.org/10.1371/journal. pone.0186065.

Lindner, C., Rienow, A., & Jürgens, C. (2019). Augmented reality applications as digital experiments for education – An example in the Earth-Moon system. *Acta Astronautica, 161,* 66–74. https://doi.org/10.1016/j.actaastro.2019.05.025.

Linke, S. (2020). Landscape in internet pictures. In D. Edler, C. Jenal, & O. Kühne (Eds.), *Modern approaches to the visualization of landscapes* (pp. 139–156). Wiesbaden: Springer VS.

Linke, S. I. (2019). *Die Ästhetik medialer Landschaftskonstrukte. Theoretische Reflexionen und empirische Befunde*. Wiesbaden: Springer VS.

Loda, M., Kühne, O., & Puttilli, M. (2020). The social construction of Tuscany in the German and english speaking world – Presented by the analysis of internet images. In D. Edler, C. Jenal, & O. Kühne (Eds.), *Modern approaches to the visualization of landscapes* (pp. 157–171). Wiesbaden: Springer VS.

Lütjens, M., Kersten, T., Dorschel, B., & Tschirschwitz, F. (2019). Virtual reality in cartography: Immersive 3D visualization of the Arctic Clyde Inlet (Canada) using digital elevation models and bathymetric data. *Multimodal Technologies and Interaction, 3*(1), 9. https://doi.org/10.3390/mti3010009.

McLean, K. (2020). Temporalities of the smellscape: Creative mapping as visual representation. In D. Edler, C. Jenal, & O. Kühne (Eds.), *Modern approaches to the visualization of landscapes* (pp. 217–246). Wiesbaden: Springer VS.

Meyer-Heß, F. (2020). Discovering forgotten landscapes. In D. Edler, C. Jenal, & O. Kühne (Eds.), *Modern approaches to the visualization of landscapes* (pp. 33–45). Wiesbaden: Springer VS.

Münderlein, D., Kühne, O., & Weber, F. (2019). Mobile Methoden und fotobasierte Forschung zur Rekonstruktion von Landschaft(sbiographien). In O. Kühne, F. Weber, K. Berr, & C. Jenal (Eds.), *Handbuch Landschaft* (pp. 517–534). Wiesbaden: Springer VS.

Ode, Å., Tveit, M. S., & Fry, G. (2008). Capturing landscape visual character using indicators: Touching base with landscape aesthetic theory. *Landscape researchn 33*(1), 89–117.

Papadimitriou, F. (2020a). Modelling and visualization of landscape complexity with braid topology. In D. Edler, C. Jenal, & O. Kühne (Eds.), *Modern approaches to the visualization of landscapes* (pp. 79–101). Wiesbaden: Springer VS.

Papadimitriou, F. (2020b). Visualization of future landscapes, postmodern cinema and geographical education. In D. Edler, C. Jenal, & O. Kühne (Eds.), *Modern approaches to the visualization of landscapes* (pp. 351–369). Wiesbaden: Springer VS.

Poplin, A., de Andrade, B., & Mahmud, S. (2020). Exploring tangible and intangible landscapes of evocative places: Case study of the city of Vitória in Brazil. In D. Edler, C. Jenal, & O. Kühne (Eds.), *Modern approaches to the visualization of landscapes* (pp. 519–547). Wiesbaden: Springer VS.

Prisille, C., & Ellerbrake, M. (2020). Virtual Reality (VR) and Geography education: Potentials of 360° 'Experiences' in Secondary Schools. In D. Edler, C. Jenal, & O. Kühne (Eds.), *Modern approaches to the visualization of landscapes* (pp. 321–332). Wiesbaden: Springer VS.

Roßmeier, A. (2020). Urban/Rural hybridity in pictures—The creation of neighborhood images using the example of San Diego's urbanizing Inner-Ring suburbs East Village and Barrio Logan. In D. Edler, C. Jenal, & O. Kühne (Eds.), *Modern approaches to the visualization of landscapes* (pp. 479–547). Wiesbaden: Springer VS.

Schafer, R.M. (1977). *The soundscape. Our sonic environment and the tuning of the world*. Rochester: Destiny Books.

Schenk, W. (2020). Visualization of the fundamental dimensions of "landscape" in landscape paintings around 1500 A.D. In D. Edler, C. Jenal, & O. Kühne (Eds.), *Modern approaches to the visualization of landscapes* (pp. 19–349). Wiesbaden: Springer VS.

Schneider, N. (2009). *Geschichte der Landschaftsmalerei. Vom Spätmittelalter bis zur Romantik*. Darmstadt: WBG.

Schweiger, S., Kamlage, J.-H., & Engler, S. (2018). Ästhetik und Akzeptanz. Welche Geschichten könnten Energielandschaften erzählen? In O. Kühne, & F. Weber (Eds.), *Bausteine der Energiewende* (pp. 431–445). Wiesbaden: Springer VS.

Siepmann, N., Edler, D., Keil, J., Kuchinke, L., & Dickmann, F. (2020a). The position of sound in audiovisual maps: An experimental study of performance in spatial memory. *Cartographica, 55*(2), 136–147. https://doi.org/10.3138/cart-2019-0008.

Siepmann, N., Edler, D., & Kühne, O. (2020b). Soundscapes in cartographic media. In D. Edler, C. Jenal, & O. Kühne (Eds.), *Modern approaches to the visualization of landscapes* (pp. 247–263). Wiesbaden: Springer VS.

Stemmer, B., Philipper, S., Moczek, N., & Röttger, J. (2019). Die Sicht von Landschaftsexperten und Laien auf ausgewählte Kulturlandschaften in Deutschland – Entwicklung eines Antizipativ-Iterativen Geo-Indikatoren-Landschaftspräferenzmodells (AIGILaP). In K. Berr & C. Jenal (Eds.), *Landschaftskonflikte* (pp. 507–534). Wiesbaden: Springer VS.

Stintzing, M., Pietsch, S., & Wardenga, U. (2020). How to Teach "Landscape" through games? In D. Edler, C. Jenal, & O. Kühne (Eds.), *Modern approaches to the visualization of landscapes* (pp. 333–349). Wiesbaden: Springer VS.

Stotten, R. (2013). Kulturlandschaft gemeinsam verstehen – Praktische Beispiele der Landschaftssozialisation aus dem Schweizer Alpenraum. *Geographica Helvetica, 68*(2), 117–127. https://doi.org/10.5194/gh-68-117-2013.

Stratmann, J., Ristea, A., Leitner, M., & Paulus, G. (2020). Exploring urban "Blightscapes" applying spatial video technology and geographic information system: A case study from Baton Rouge, USA. In D. Edler, C. Jenal, & O. Kühne (Eds.), *Modern approaches to the visualization of landscapes* (pp. 499–517). Wiesbaden: Springer VS.

Tessin, W. (2008). *Ästhetik des Angenehmen. Städtische Freiräume zwischen professioneller Ästhetik und Laiengeschmack.* Wiesbaden: VS Verlag.

Thomas, P. M. (2020). The digitalizing society—Transformations and challenges. In D. Edler, C. Jenal, & O. Kühne (Eds.), *Modern approaches to the visualization of landscapes* (pp. 447–457). Wiesbaden: Springer VS.

Tillmann, K.-J. (2007). *Sozialisationstheorien. Eine Einführung in den Zusammenhang von Gesellschaft, Institution und Subjektwerdung* (Rororo Rowohlts Enzyklopädie, vol. 55476, 15. Auflage). Reinbek bei Hamburg: Rowohlt.

Vetter, M. (2020). Technical potentials for the visualization in virtual reality. In D. Edler, C. Jenal, & O. Kühne (Eds.), *Modern approaches to the visualization of landscapes* (pp. 307–317). Wiesbaden: Springer VS.

Vetter, M. (2019). 3D-Visualisierung von Landschaft – Ein Ausblick auf zukünftige Entwicklungen. In O. Kühne, F. Weber, K. Berr, & C. Jenal (Eds.), *Handbuch Landschaft* (pp. 559–573). Wiesbaden: Springer VS.

Virrantaus, K., Fairbairn, D., & Kraak, M.-J. (2009). ICA research agenda on Cartography and GI Science. *The Cartographic Journal, 46*(2), 63–75. https://doi.org/10.1559/152304009788188772.

Visscher, S. De, & Bouverne-De Bie, M. (2008). Recognizing urban public space as a Co-Educator: Children's socialization in Ghent. *International Journal of Urban and Regional Research, 32*(3), 604–616. https://doi.org/10.1111/j.1468-2427.2008.00798.x.

Walsh, C., Kangler, G., & Schaffert, M. (Eds.) (2020). *Landschaftsbilder und Landschaftsverständnisse in Politik und Praxis.* Wiesbaden: Springer VS.

Weber, F. (2020). Blurring the boundaries of landscape visualization: Welcome to Fabulous Las Vegas. In D. Edler, C. Jenal, & O. Kühne (Eds.), *Modern Approaches to the Visualization of Landscapes* (pp. 461–457). Wiesbaden: Springer VS.

Weber, F., Jenal, C., Roßmeier, A., & Kühne, O. (2017). Conflicts around Germany's Energiewende: Discourse patterns of citizens' initiatives. *Quaestiones Geographicae, 36*(4), 117–130. https://doi.org/10.1515/quageo-2017-0040.

Weber, F., Kühne, O., Jenal, C., Aschenbrand, E., & Artuković, A. (2018). *Sand im Getriebe. Aushandlungsprozesse um die Gewinnung mineralischer Rohstoffe aus konflikttheoretischer Perspektive nach Ralf Dahrendorf.* Wiesbaden: Springer VS.

Winkler, J. (2005). Raumzeitphänomen Klanglandschaften. In: Denzer, V., Hasse, J., Kleefeld, K.-D. und Recker, U. (Eds.), *Kulturlandschaft. Wahrnehmung – Inventarisation – Regionale Beispiele,* (pp. 77–88). Bonn.

Wojtkiewicz, W. (2015). *Sinn – Bild – Landschaft. Landschaftsverständnisse in der Landschaftsplanung: eine Untersuchung von Idealvorstellungen und Bedeutungszuweisungen.* Berlin: Technische Universität Berlin.

Zhao, J., Wallgrün, J. O., LaFemina, P. C., Normandeau, J., & Klippel, A. (2019). Harnessing the power of immersive virtual reality – Visualization and analysis of 3D earth science data sets. *Geo-spatial Information Science, 22*(4). https://doi.org/10.1080/10095020.2019.1621544.

Dennis Edler is a Senior Lecturer in the Institute of Geography at the Ruhr-University Bochum. In his research and teaching activities, he deals with open spatial data and modern approaches to visualizing multimedia and 3D environments. He also chairs the joint commission "Virtual and Augmented Reality" of the German Cartographic Society and German Society for Photogrammetry, Remote Sensing and Geoinformation. He is also very interested in landscape research.

Olaf Kühne studied geography, modern history, economics and geology at Saarland University and received his doctorate in geography and sociology there and at the Open University of Hagen. After working in various Saarland state authorities and at Saarland University, he was Professor of Rural Development/Regional Management at Weihenstephan-Triesdorf University of Applied Sciences from 2013 to autumn 2016 and Associate Professor of Geography at Saarland University in Saarbrücken. Since autumn 2016, he has been a professor in the Department of Geography at the Chair of Urban and Regional Development at the Eberhard Karls University of Tübingen. His research interests include landscape and discourse theory, social acceptance of landscape change, sustainable development, transformation processes in Southern California and the Southern States of the USA, regional development, and urban and landscape ecology.

Corinna Jenal studied German language and literature, political science and philosophy at the University of Trier and completed the "Sustainability Certificate" at Saarland University at the Endowed Chair for Sustainable Development. At Saarland University and the Weihenstephan-Triesdorf University of Applied Sciences she worked on various research projects. Since autumn 2016 and summer 2019, respectively, she has been working as a research assistant or academic councillor in the research area of geography at the Chair of Urban and Regional Development at the Eberhard Karls University of Tübingen, where she received her doctorate in 2019 on the social construction of forests. Her research focuses on landscape research, energy system transformation, urban-rural hybrids, old industry, and social construction and negotiation processes of nature and forest as their associated part.

Part II

Landscape Analysis. From Traditional to Digital Data Sources

Visualization of the Fundamental Dimensions of "landscape" in Landscape Paintings Around 1500 A.D.

Winfried Schenk

Abstract

Following Ipsen (Ipsen D (1999) Landschaft als Raum nachhaltigen Handelns. In: Friedrichs J, Hollaender K (Eds.) Stadtökologische Forschung. Analytica, Berlin, pp. 217–226) three dimensions of (cultural)landscape can be distinguished: (nature-) resource, labor and social structuring. Labor describes the energetic input, which is necessary to create or change physical objects from the potential within nature. This can only work effectively on a wider scale thanks to a social structuring of work in social systems. "Culture" is to be understood as a specific combination of these dimensions and the product is "the cultural landscape". In preindustrial times the concept of culture was driven by agricultural logics, to be reflected in specific forms of agricultural landscape. This article tries to verify, that landscape paintings visualize these three aspects of landscape around the turn of the 16th century.

Keywords

Cultural landscape · Early modern times · Energy system · Social metabolism · Landscape paintings

W. Schenk (✉)
Universität Bonn, Bonn, Germany
e-mail: winfried.schenk@giub.uni-bonn.de

© Springer Fachmedien Wiesbaden GmbH, part of Springer Nature 2020
D. Edler et al. (eds.), *Modern Approaches to the Visualization of Landscapes*,
RaumFragen: Stadt – Region – Landschaft, https://doi.org/10.1007/978-3-658-30956-5_2

1   The Three Fundamental Dimensions of Landscape: Resource, Work and Social Structuring

From the historical linguistics' point of view landscape is a "tinted" (Schenk 2020, p. 7) or "composed" (Ipsen et al. 2003, p. 13) term. Since its first documentation in 830 A.D. a whole variety of meanings has accumulated. Especially two threads of meanings can be traced down (Schenk 2013, 2019), in which the word itself in semantic terms partially got overlaid with new meanings without losing the older aspects (for the conceptual history of landscape please refer to Berr and Schenk 2019):

1. In the Middle Ages the meaning of "landscape" was transmitted from the designation of self-organized local dwellers of a clearly defined area, who are able to act politically to an area (politically defined or natural area), which was inhabited by these groups of people.
   In this sense landscape is to be understood as a territorial and legal policy concept similar to the Latin word "regio". Hence regionalizations do become possibilities of defined, mostly medium-sized segments of space.
2. From Early Modern Age onwards started the reification (put down on canvas) of a painted segment of space influenced by an aesthetic attitude while being created by overlaying physically tangible elements with aesthetical meaning. Hence one can interpret "landscape" as a "picture" and a "symbol of soul", whereby it became possible by the turn of the 20th century to understand "landscape" in the German-speaking parts of Europe as a positive metaphor of rural life in contrast with urban life. This is the reason why in this context landscape meant the space in rural areas.
3. Finally, with recourse to customs in geography and other disciplines linked to landscape, landscape can be seen as a physical entity, as an area, which is filled with physical objects. These objects can be described, and their origins can be explained by analysing their individual functions.

Historical Geography deals with this kind of (cultural) landscape history (among other subjects) and the paper takes precisely this perspective (Schenk 2011).

From a conceptual history's point of view, Kühne's (2018) stance is coherent, whereby landscape is formed by synopsizing physical objects based on social conventions. These conventions are conveyed during the process of socialization. Landscape is therefore a social construct which is per se processual on the material as well as on the perceptive and attributive level (see in this volume: Bellini and Leonardi 2020; Kühne 2020; Kühne and Jenal 2020). Therefore, we can notice in every respect a dual change of landscape, whereas the change on the physical and on the perceptual level determine each other. Therefore, every generation has a different perception of landscape (for

Fig. 1 Dimensions of (cultural) landscape according to Ipsen (1999)

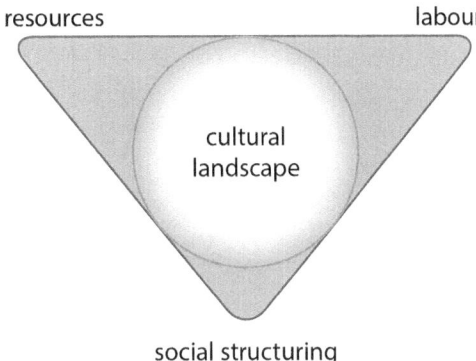

resources   labour

cultural
landscape

social structuring

studies on the methodological and technological transfer of landscape perceptions into modern cartographic visualizations, see in this volume: Edler et al. 2020a; Siepmann et al. 2020; Stratmann et al. 2020).

Following Ipsen (1999) on a more abstract level, three fundamental dimensions of (cultural)landscape can be distinguished: (nature-)resource, labour and social structuring (see illustration 1). Labour describes the energetic input, which is necessary to create physical objects from the potential within nature or to change – according to the concept of social metabolism, it can be understood as an act of colonizing nature (Fischer-Kowalski 1997). This can only work effectively on a wider scale thanks to a social structuring of work in social systems. In turn, values will emerge and with these values one can mentally construct "landscape".

I would like "culture" to be known as a specific combination of these dimensions and the product is "the cultural landscape". We can presume that in preindustrial times the concept of culture was driven by agricultural logics, to be reflected in forms of agricultural stamped landscape (see Fig. 1).

This article tries to verify that landscape paintings visualize these three aspects of landscape around the turn of the 16th century. This specific time is characterized by the limitations of the solar energy system (see Fig. 2) and by feudal hegemonic structures (a structure where landowners exert power over their country and the people living on their land), which influences the perception of landscape.

Looking at the interpretation of Thomas Gainsboroughs picture "Mr. and Mrs Andrews", dated 1784 (see Fig. 3), Hugh Prince depicts such a landowner of the late 17th-century, who is shown in a distant relationship to his wife on one hand, and, on the other hand, he is shown, most notably, in a distant relationship to the people who cultivate his land. They are not shown at all—only the result of their hard work, i.e. the cultivated land, has been painted (Prince 1998, p. 104).

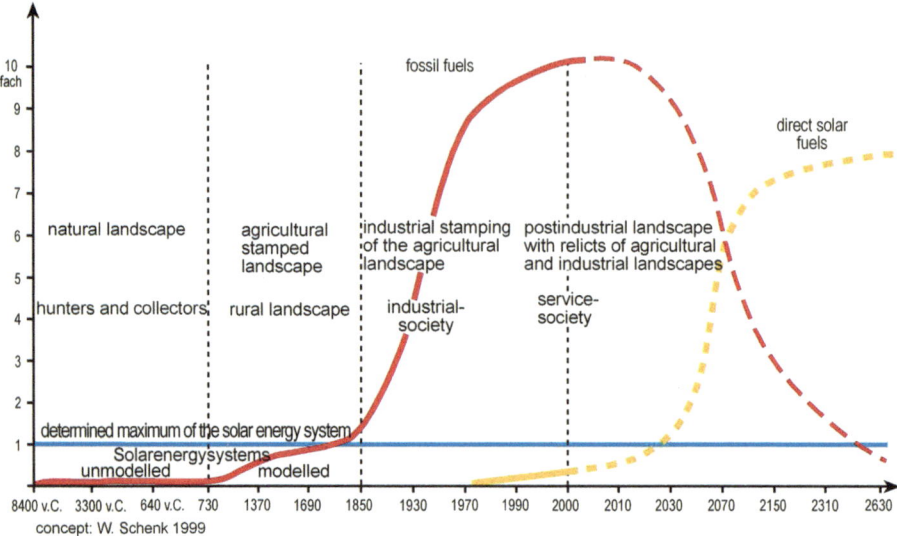

Fig. 2 Energy systems in the history of mankind (after Schenk 2005, p. 231)

Fig. 3 Thomas Gainsborough: Mr. and Mrs. Andrews (1784; from https://www.wikiart.org/de/thomas-gainsborough/mr-und-mrs-andrews-1749)

2 Fundamental thoughts about the interpretation of historical landscape paintings

Landscape paintings are based on the poetic inspiration of the painter, which means that every scenic element and structure shown is based on the subjective reality of the painter (Tolkemitt and Wohlfeil 1991). One has to bear in mind that many painters, who e.g. painted the river Rhine as a sample for a Central European ideal of an early 19th-century landscape (Dix 2002), had never visited the area by themselves. They worked from templates of graphics and prints with rural themes and chose any available themes. The river, one can conclude, was not scenic reality for them. The painters did not aim for a realistic reproduction of the river itself and its surroundings. Hence, they did not draw inspiration or regulated the composition of their work from images of the river seen by themselves. Büttner (2019), Steingräber (1985) and Schneider (1999) provide general overviews of the European landscape painting. In contrast to Dutch landscape paintings of the early 16th century, one can notice the tendency to capture precisely the special atmosphere of a rural landscape and the topographic situation of the time.

One example for this type of genre paintings is the famous painting by Pieter Bruegel, the Elder called "The gloomy day", 1565 (see Fig. 4; as well as Makowski and Buderath 1983, p. 99).

From the series "The Seasons", "The gloomy day" depicts in the front highly realistic images of the intensive agricultural use and its effects – the damaged state of the forests in the vicinity of the villages in the Early Modern Age. Men are shown in the act

Fig. 4 ‚The gloomy day' of Pieter Bruegel the Elder, 1565 (from Makowski and Buderath 1983, p. 99)

of cutting rods from the willows which were used for housebuilding or to weave baskets (Braun and Konold 1998).

Here, we undoubtedly see a scene of a landscape portrayed as the result of the use of a natural resource through physical labour. The social structuring is the village we see in the middle of the picture, which constitutes a community. The background shows a rugged and untamed mountain scene, which seems to grow out of the stormy ocean. It symbolizes that nature constantly threatens or oppresses the work of man. Notwithstanding the undoubtedly allegorical character of the composition, more and more contemporary voices are heard who complain that, on these genre paintings, a tree is only "a tree". Heroes and heroic scenes of battle, symbolism and glorification are gone. However, exactly this constituted the landscape paintings up until the Romantic period.

The fact that artists tried to express crypticness as well as mysticism and mythological information in themes of their time is the reason why one rarely finds an exact reproduction of the painters' surroundings in historical landscape paintings.

If one looks at these kinds of drawings in the light of Ipsen's definition of the three dimensions of landscape, one needs to know, first of all, the agricultural practices and the social and economic circumstances of that time. Historical Geography and Agrarian History provide this type of knowledge; second one needs to keep in mind that one's perspective, especially the ones taken on rural areas, is influenced by a common attitude of distancing oneself from ordinary, so to speak, innocent landscape paintings – this being a reaction to the latent reproach of the paintings being tacky (Eisel 2001, p. 160).

In Germany landscape paintings are linked automatically to "being tacky". The term "landscape" carries automatically a romantic aesthetic notion and meaning, as it has been used so often and frequently in the field of painting in general. It also bears the mark of "being tacky", because of the reason given above (Linke 2019; see in this volume: Linke 2020).

In the context of art or artisan craftwork, if the term is explicitly used to insinuate mood, and mood only, the connection to tackiness (= "Kitsch") is easily made, especially when symbols are used which are commonly understood by men throughout the world (Linke 2019; see in this volume: Linke 2020; Weber 2020). Quoting Franz Kafka, who gets to the heart of the special relationship the Germans have with the terminus "landscape: "The landscape is interfering with my thoughts …. It is pretty and demands to be looked at." (quoting Schwarze 1996, p. 413).

3   Resource, Labor and Social Structuring Shown on a Certain Landscape Painting Around 1500 A.D.

I would like to show on three examples that landscape paintings visualize the dimensions of landscape namely "resource", "labor" and "social structuring", despite their "poetic character" and Germans' liking to link them to tackiness. They were painted around 1500 A.D., therefore they belong to the early landscape paintings of Central Europe and show environmental conditions at the turn of the Middle Age to Early Modern Age. The pictures illustrate the agricultural landscape under the strings of Feudalism. One can see the functional interdependencies between production on the farm- and pastureland, in the garden and orchard as well as in the forest. Hence the chosen pictures break tradition with the set-up of paintings, which usually show only a small clip of the landscape, bend to fit into the desired whole composition, which was common practice at those times (Steingräber 1985).

3.1 Albrecht Dürer, "The Wire-Drawing Mill at the river Pegnitz" (1496–1500)—showing an overexploitation of an economic landscape at the end of the Middle Ages

The shown landscape (see Fig. 5) includes all signs of a stressed economy landscape of the "solar energy time"- model (see Fig. 2), where people try to control the flow of their only source of energy, the sun, hence attempting to optimize agricultural production.

From a vast range of species, people of that time chose those species as crop plant or livestock, which, in the ideal case, should solely inhabit this space. Agricultural economy can therefore be understood as a concentration of species. Competitors of the favourite crop or livestock, which also used sunlight, water and nutrition, were called "weeds" and "pest"—therefore, they lost the right of existence and got eliminated, or people tried at least to protect their land from them. Moreover, people changed plants and animals

Fig. 5 The wire-drawing mill at the Pegnitz river by Albrecht Dürer, around 1496/1500 (from Makowski and Buderath 1983, p. 51)

through cultivation and breeding. In this sense, the preindustrial agricultural economy is understood as a social system used to redirect solar energy for human purposes as shown in the already mentioned concept of social metabolism (Winiwarter and Sonnlechner 2001). Crop plants and livestock are "artificial bio-convertors" in this sense and should therefore be rated as forms of technic like pick or plough. The result of this kind of energetic manipulation is "agrarian–" or "agricultural landscape" (Radkau 2000) of preindustrial times. It was shaped by dimensions, which were guided by the limited power of men and animals; looking at the area of agrarian transportation, it is estimated that more than 50% of the total amount of workload, put out by humans' and animals' physical strength, is used for this kind of transportation (Sieglerschmidt 1999).

Having said this, the appearance of Dürer's landscape is mainly defined by space- and fodder requirements—even though the livestock is not visible. The landscape is traversed by "living fences" (hedges) and enclosures, which shall keep the animals away from orchards, pastures and fields. In the case of hedges being permeable by nature (they have holes), they will be repaired by setting up paling fences. In areas to which animals have open access to, one will not find any natural growth anymore. The remaining few trees are debranched and defoliated up until the height the animals are able to reach. The communal land is threatened by soil erosion due to its intensive usage. The bare earth stands out painted as a bright area around the top house of the mill.

The settlement is threatened by tides as it is located in the meadows, which always represent a precarious place to live in. Nevertheless, dwellers are willing to take chances, as they are able to raise taxes when travelers ford the river (Schenk 2001). Besides, water is the only source of energy which is able to drive the mill wheel continuously. In contrast to the unprotected milling settlement, a promontory village or town rises up in the background of the picture protected from floods.

Art history can prove that the spectator's view drifts across the riversides of the Pegnitz, across the large and small mill pastures towards Nuremberg (Koschatzky 1971). It catches sight of the impressive Spittlertor gate tower and the town walls. At the horizon the Schwabach, mountains arise as part of the Franconian Jura. The high altitudes of the Jura are partially deforested, because the towns Nuremberg and Schwabach close by needed a lot of wood for their various trades (Radkau 1997). All buildings of the mill are timber frameworks or they are completely made of wood, e.g. the barns.

3.2   Two Calendar Pictures of Hans Wertinger (Around 1525/30)—Manifold Forms to Use Resources Under Feudalistic Conditions

Looking at the not widely known calendar pictures of Hans Wertinger, born in Landshut (Lutze and Wiegand 1937; Ehret 1976), dated 1525/30, it is possible to get a glance at a landscape at the turn from the European Middle Age to Early New Age (see "Haus der Bayerischen Geschichte" 1992). Even though the paintings are drawn in the tradition of the calendar-month-pictures, they do concentrate on the various forms of farm work in the context of feudalistic social structuring by repositioning the pleasures at court.

The images are idealized, indeed, which is obvious, as the labor is always done in good weather conditions and the peasantry seemed to be in best mood. The pictures do not show people's hardship, e.g. fear for one's existence, if the harvest was bad and the landowners still insisted on their tributes (natural produce, money and manpower), which they received in turn allowing the peasantry to work on the allotted land. Moreover, one finds hints to the gender role.

The first picture shows work done in the month of June, hay harvest (see Fig. 6). Men cut hay by using scythe. Women rake it up and drive the hay wagon to the barn in the background. Some of the men lie in the grass to have a break and eat. Please note—it is a mowed pasture used to harvest winter fodder. The pasture is separated by wattle fences and hereby protected from trespassers (humans and animals). In this area, the grass is high, whereas behind the fence the ways are trodden down so badly that one can see the bare earth. In the distance, the grass is low, probably eaten up by animals.

Wattlework, made of rods taken from branches, separates the pond from the pasture. Thus, it protects its fringe from subsurface erosion (denting by erosion) done by the stream close by which is shown in the front of the picture. The water surface suggests a

Fig. 6 Monthly picture June by Hans Wertinger, approx. 1525/30 (from Haus der Bayerischen Geschichte, Bauern in Bayern 1992, p. 110)

high ground-water-level; therefore, another use of the area other than using them as pastures does not seem possible.

The realistic reproduction of the houses (made in the way of the log building method) on the right-hand side is indicative for Wertinger's precise observation skills. It seems to be the house of a peasant because of its simplicity and convenience. Three beehives are displayed which are placed underneath the porch. On the left-hand side, there is a two-story stone building with a pediment and an oriel (= a projecting alcove), probably the domain of the landowner. The stately houses shown in these series are built from stone. Hence the picture indicates the social differences by means of the architecture.

The second picture probably addresses either the month September or October (see Fig. 7). Two agricultural production areas are shown: first farmland is cultivated for winter; in the front the plough is shown. Behind it, one can see the seeds for sowing and the chain harrows.

To show everything at once, Wertinger spreads the depiction of labor (which needs to be done at this time of the year) on various, very short acres rather than reflect the conventional larger patches of the time. The plough drawn by three-horses and a menial sitting on top of one horse indicates a seigniorial estate. The stone buildings (including tile roofs) make up the estate, where the landlord himself or his administrator live. The horses wear a collar, a rigid leather harness which enables the landlord to use the animals for ploughing. Horses were rarely seen in peasant holdings; they usually used oxen or

Fig. 7 Monthly picture of September or October by Hans Wertinger, approx. 1525/30 (from Haus der Bayerischen Geschichte, Bauern in Bayern 1992, p. 111)

cows to draw the plough. With the power of even three horses, the earth was ploughed very deeply retrieving clay minerals hidden in the lower levels of the earth. The plough itself already has a moldboard, which enables it to overturn the soil to furrows. With this equipment, people were able to work even on land with heavy soil. Furthermore, the plough also has wheels, a Saxon invention of the 10th-century. Even though the picture provides so many details of obvious signs of wealth, the iron parts on the plough are minimized. Iron was very expensive compared to wood. The punishment for steeling iron was five times higher than for steeling wood.

The harvest still was not very high, even though the technical standard of working the land was. In an average year of the time, a large estate harvested only five times the amount of the seeds sown, but they yielded the double number of what peasants harvested on their fields (Glaser and Schenk 1988; Hüßner 2019).

In the background, fruit was being picked. If these were apples or pears, they could be kept fresh for months if stored in a cool and dry place. Or fruit was dried, as was done with prunes, in the afterglow of the baking oven, or distilled into fruit schnapps. The choice of locating the fruit trees outside of the fields, in fenced orchards and in proximity of the houses reflected the real landscape of this time, even though the size of the individual fields was inaccurate.

Meadow orchards amid the farmland emerged in the 19th-century. Then sections of the common agricultural meadows and pastures used by all were transferred to individual users, who were able to work the land by themselves.

4   Conclusion: Historical Landscape Paintings Being a Visualization of the Fundamental Dimensions of Landscape

Looking at the principles of art-historical source verification combined with a variety of methods used in Historical Geography (Jäger 1994, Schenk 2005), it seems that historical landscape paintings are fit to visualize historical-ecological processes, as well as past landscape situations of pre-industrial times. They show how human society colonizes nature in terms of the ecological metabolism by labor. Nature was regarded as a resource. One can learn to "read" landscapes using this medium. Thus, landscape paintings are an important media representing spatial details and they enrich the multitude of modern digital media (c.f. Edler et al. 2020b). Moreover, it is possible to gain an impression of landscape in pre-industrial times, how it used to look and how it functioned, dominated by feudalistic structures (Rösener 1997) and determined by the shortage of energy from solar energy systems.

References

Bellini, A., & Leonardi, L. (2020). Prato: The social construction of an industrial city facing processes of cultural hybridization. In D. Edler, C. Jenal, & O. Kühne (Eds.), *Modern approaches to the visualization of landscapes* (pp. 549–572). Wiesbaden: Springer VS.

Berr, K., & Schenk, W. (2019). Begriffsgeschichte [Landschaft]. In O. Kühne, F. Weber, K. Berr, & C. Jenal (Eds.), *Handbuch Landschaft, RaumFragen: Stadt – Region – Landschaft* (pp. 23–38). Wiesbaden: Springer VS.

Braun, B., & Konold, W. (1998). *Kulturgeschichte und Bedeutung der Kopfweiden in Südwestdeutschland.* Ubstadt-Weiher: Verlag regionalkultur.

Dix, A. (2002). Das Mittelrheintal – Wahrnehmung und Veränderung einer symbolischen Landschaft des 19. Jahrhunderts. *Petermanns Geographische Mitteilungen, 146,* 44–53.

Edler, D., Jenal, C., & Kühne, O. (2020a). Modern approaches to the visualization of landscapes—An introduction. In D. Edler, C. Jenal, & O. Kühne (Eds.), *Modern approaches to the visualization of landscapes* (pp. 3–15). Wiesbaden: Springer VS.

Edler, D., Keil, J., & Dickmann, F. (2020b). From Na Pali to Earth—An 'Unreal' Engine for Modern Geodata? In D. Edler, C. Jenal, & O. Kühne (Eds.), *Modern approaches to the visualization of landscapes* (pp. 279–306). Wiesbaden: Springer VS.

Ehret, G. (1976). *Hans Wertinger. Ein Landshuter Maler an Wende der Spätgotik zur Renaissance.* München: Tuduv-Verlagsgesellschaft.

Eisel, U. (2001). Angst vor der Landschaft? *Erdkunde, 55,* 159–171.

Fischer-Kowalski, M. (Ed.). (1997). *Gesellschaftlicher Stoffwechsel und Kolonisierung der Natur. Ein Versuch in Sozialer Ökologie.* Amsterdam: G+B Fakultas Verlag.

Glaser, R., & Schenk, W. (1988). Einflussgrößen auf die Anbau- und Ertragsverhältnisse des Ackerlandes im frühneuzeitlichen Mainfranken - Forschungsstand, Ergebnisse und offene Fragen. *Mainfränkisches Jahrbuch, 40,* 43–69.

Hüßner, R. (2019) „den Zehenten allhier in Münchsundheim belangent". Eine Untersuchung zum ländlichen Abgabewesen am Beispiel Mönchsondheim. In *Franken unter einem Dach, Fränkisches Freilandmuseum Bad Windsheim* p. 35–50.

Haus der Bayerischen Geschichte (Eds.). (1992). *Bauern in Bayern von der Römerzeit bis zur Gegenwart.* München: Haus der Bayerischen Geschichte.

Ipsen, D. (1999). Landschaft als Raum nachhaltigen Handelns. In J. Friedrichs & K. Hollaender (Eds.), *Stadtökologische Forschung* (pp. 217–226). Berlin: Analytica.

Ipsen, D., Reichhardt, U., Schuster, S., Wehrle, A., & Weichler, H. (Eds.). (2003). *Zukunft Landschaft. Bürgerszenarien zur Landschaftsentwicklung.* Kassel: Universitätsverlag.

Jäger, H. (1994). *Einführung in die Umweltgeschichte.* Darmstadt: Wissenschaftliche Buchgesellschaft.

Koschatzky, W. (1971). *Albrecht Dürer. Die Landschaftsaquarelle. Örtlichkeit – Datierung Stilkritik.* Wien: Jugend & Volk.

Kühne, O. (2018). Der doppelte Landschaftswandel. Physische Räume, soziale Deutungen, Bewertungen. *ARL-Nachrichten, 1,* 14–17.

Kühne, O. (2020). The social construction of space and landscape in internet videos. In D. Edler, C. Jenal, & O. Kühne (Eds.), *Modern approaches to the visualization of landscapes* (pp. 121–137). Wiesbaden: Springer VS.

Kühne, O., & Jenal, C. (2020). The threefold landscape dynamics—Basic considerations, conflicts and potentials of virtual landscape research. In D. Edler, C. Jenal, & O. Kühne (Eds.), *Modern approaches to the visualization of landscapes* (pp. 389–402). Wiesbaden: Springer VS.

Linke, S. (2019). Landschaftsästhetik. In O. Kühne, F. Weber, K. Berr, & C. Jenal (Eds.), *Handbuch Landschaft, RaumFragen: Stadt – Region – Landschaft* (pp. 441–452). Wiesbaden: Springer VS.

Linke, S. (2020). Landscape in internet pictures. In D. Edler, C. Jenal, & O. Kühne (Eds.), *Modern approaches to the visualization of landscapes* (pp. 139–156). Wiesbaden: Springer VS.

Lutze, E., & Wiegand, E. (1937). *Die Gemälde des 13.-16. Jahrhunderts.* Leipzig: Kataloge des Germanischen Nationalmuseums zu Nürnberg.

Makowski, H., & Buderath, B. (1983). *Die Natur dem Menschen untertan. Ökologie im Spiegel der Landschaftsmalerei.* München: dtv Verlagsgesellschaft.

Prince, H. (1988). Art and Agrarian change, 1710–1815. In D. Cosgrove & S. Daniels (Eds.), *The iconography of landscape* (pp. 98–119). Cambridge: Cambridge University Press.

Radkau, J. (1997). Das Rätsel der städtischen Brennholzversorgung im „hölzernen Zeitalter". In D. Schott (Eds.), *Energie und Stadt in Europa. Von der vorindustriellen ‚Holznot' bis zur Ölkrise der 1970er Jahre,* VSWG Beihefte 135, (pp. 43–75). Stuttgart.

Radkau, J. (2000). *Natur und Macht. Eine Weltgeschichte der Umwelt.* München: C. H. Beck.

Rösener, W. (1997). *Einführung in die Agrargeschichte.* Darmstadt: Wissenschaftliche Buchgesellschaft.

Schenk, W. (1997). Der Blick auf die Landschaft. Die Geschichte unserer Umwelt im Spiegel von Landschaftsgemälden. *Praxis Geschichte, 11* (4), 40–42.

Schenk, W. (2001). Auen als Siedlungs- und Wirtschaftsräume vor den ingenieurtechnischen Veränderungen des 19. Jahrhunderts – Das Mittelmaingebiet als Beispiel. *Zeitschrift für Geomorphologie* N.F. Supp.-Bd., T124, 55–67.

Schenk, W. (2002). „Landschaft" und „Kulturlandschaft" – „getönte" Leitbegriffe für aktuelle Konzepte geographischer Forschung und räumlicher Planung. *Petermanns Geographische Mitteilungen, 146*(6), 6–13.

Schenk, W. (2005). Historische Geographie. In W. Schenk & K. Schliephake (Eds.), *Allgemeine anthropogeographie* (pp. 215–264). Gotha: Klett-Perthes.

Schenk, W. (2011). *Historische Geographie.* Darmstadt: Wissenschaftiche Buchgesellschaft.

Schenk, W. (2013). Landschaft als zweifache sekundäre Bildung. Historische Aspekte im aktuellen Gebrauch von Landschaft im deutschsprachigen Raum, namentlich in der Geographie. In D. Bruns & O. Kühne (Eds.), *Landschaften: Theorie, Praxis und internationale Bezüge* (pp. 23–34). Schwerin: Oceano Verlag.

Schenk, W. (2018). Defining and exploring cultural landscape from the perspective of historical geography in German speaking Countries. In V. G. Sychev & L. Mueller (Eds.), *Novel methods and results of landscape research in Europe, Central Asia and Siberia, Band 1, Landscapes in the 21th Century: Status analyses, basic processes and research, concepts* (pp. 52–56). Moskau: Russian Academy of Sciences.

Schenk, W. (2019). Landscape. In L. Kühnhardt & T. Meyer (Eds.), *The Bonn handbook of globality* (Vol. 1, pp. 621–633). Wiesbaden: Springer.

Schenk, W. (2020): Der Wert von Kulturlandschaften für die Umweltbildung in Deutschland aus kulturgeografischer und ideengeschichtlicher Perspektive. In O. Kühne, T. Strobel, R. Traba, & M. Wiatr (Hrsg.), *Kulturlandschaften in Deutschland und Polen.* (pp. 63–73). Göttingen.

Schneider, N. (1999). *Geschichte der Landschaftsmalerei. Vom Spätmittelalter bis zur Romantik.* Darmstadt: Wissenschaftliche Buchgesellschaft.

Schwarze, T. (1996). Landschaft und Regionalbewusstsein – Zur Entstehung und Fortdauer einer territorialbezogenen Reminiszenz. *Berichte zur deutschen Landeskunde, 70*(2), 413–433.

Sieferle, R. P. (2001). *The subterranean forest: Energy systems and the industrial revolution.* Cambridge: The White Horse Press.

Sieglerschmidt, J. (1999). Wandlungen des Energieeinsatzes in Mitteleuropa in der Frühneuzeit. In W. Schenk (Eds.), *Aufbau und Auswertung „Langer Reihen" zur Erforschung von historischen Waldzuständen und Waldentwicklungen, Tübinger Geographische Studien 125* (pp. 47–89). Tübingen.

Siepmann, N., Edler, D., & Kühne, O. (2020). Soundscapes in Cartographic media. In D. Edler, C. Jenal, & O. Kühne (Eds.), *Modern approaches to the visualization of landscapes* (pp. 247–263). Wiesbaden: Springer VS.

Steingräber, E. (1985). *Zweitausend Jahre europäische Landschaftsmalerei*. München: Hirmer.

Stockhammer, A. (2009). *Landschaftsmalerei als Spiegel gesellschaftlichen Naturverständnisses*. Dissertation, Universität Wien.

Stratmann, J., Ristea, A., Leitner, M., & Paulus, G. (2020). Exploring urban "Blightscapes" applying spatial video technology and geographic information system: A case study from Baton Rouge, USA. In D. Edler, C. Jenal, & O. Kühne (Eds.), *Modern approaches to the visualization of landscapes* (pp. 499–517). Wiesbaden: Springer VS.

Tolkemitt, B., & Wohlfeil, R. (Eds.). (1991). Historische Bildkunde. Probleme – Wege – Beispiele. *Zeitschrift für Historische Forschung,* Beiheft 12.

Weber, F. (2020). Blurring the boundaries of landscape visualization: Welcome to Fabulous Las Vegas. In D. Edler, C. Jenal, & O. Kühne (Eds.), *Modern approaches to the visualization of landscapes* (pp. 461–478). Wiesbaden: Springer VS.

Winiwarter, V., & Sonnlechner, C. (2001). *Der soziale Metabolismus der vorindustriellen Landwirtschaft in Europa. Der europäische Sonderweg 2*. Breuninger Stiftung. Stuttgart.

Dr. Winfried Schenk is Professor of Historical Geography at the University of Bonn. He received his doctorate in geography (on the Shaping of landscapes by the monastery Ebrach of the Cistercian Order in Early modern Times in Franconia) at the University of Würzburg. His habilitation, also in Würzburg, dealt with the interface of regional development in selected regions of Germany with respect to the usage of forests in preindustrial times. Today, his focus lies on historical landscape research in various approaches (esp. ecological, social, perceptional approach) and applied historical geography in Central Europe.

Discovering Forgotten Landscapes

M. Fabian Meyer-Heß

Abstract

Human activities associated with agriculture and settlement have been a relief-shaping element in landscape history for thousands of years. Different stages of these processes can be visualized using laser-derived Digital Terrain Models (DTM) that revolutionized exploration of landscapes especially in archaeological research. Conventional models support understanding of landscapes general character. For detailed analysis, however, these are too coarse. That is what complex micro-relief visualizations are specialized in. These allow for detailed investigations and documentations of relief features, e.g. historical field systems and their transformation over time. Furthermore, they can be used to automatically detect different types of relief structures, which is the most recent research topic.

Keywords

Landscape Visualization · LiDAR · Digital Terrain Model · Automated Detection

1   Introduction

Today's landscapes are results of both natural and anthropogenic processes. Reconstructions of past conditions help to understand their development and eventually to project it into the future (related in this volume: Poplin et al. 2020; Schenk 2020;

M. Fabian Meyer-Heß (✉)
Ruhr-Universität Bochum, Bochum, Germany
e-mail: matthias.meyer@rub.de

Stratmann et al. 2020). One aspect of the physical landscape component is agriculture as a relief shaping activity over thousands of years. In this context, this chapter defines landscape as a combination of geomorphological features of different scales (Kühne 2019; in this volume: Kühne and Jenal 2020).

LiDAR (Light Detection And Ranging) is amongst the most suitable techniques for investigating the geomorphological appearance of landscapes. For about two decades, lasers have been used to examine landscapes without their vegetation canopy, revealing geomorphological features that were hidden for up to thousands of years. These can be both natural and anthropogenic, e.g. fossil river terraces, remains of ancient field systems or trading routes. Especially archaeology and historical geography are interested in these structures, as they allow for conclusions about (pre-)historical settlement pattern.

Initial studies revealed the enormous potential of LiDAR-derived Digital Terrain Models (DTM) for reconstructions of historical landscapes and archaeological prospection. Using the new technique, it was possible to see the landscape without vegetation, revealing numerous new insights even in old investigation sites (e.g. Holden 2001; Motkin 2001; Holden et al. 2002; Sittler 2004; Devereux et al. 2005; Crutchley 2006; Bofinger et al. 2007; Doneus et al. 2008).

Research soon pointed out that simple visualizations such as Shaded Relief have certain disadvantages regarding visibility of linear structures that align with the direction of illumination (e.g. Doneus and Briese 2006; Devereux et al. 2008). As a consequence, a variety of visualizations were developed that are optimized towards highlighting or extracting the micro relief (e.g. Hesse 2010; Kokalj et al. 2011; Zakšek et al. 2011; Doneus 2013). In times of increasing computing power and Open Geodata, proven visualizations are used for automated detection approaches, tackling the constantly growing amounts of data. The techniques are manifold, such as Template Matching (e.g. de Boer 2007; Schneider et al. 2015; Trier et al. 2015), Object-based Image Analysis (OBIA) (e.g. Freeland et al. 2016; Sevara et al. 2016; Cerrillo-Cuenca 2017; Meyer et al. 2019), combinations or other techniques (e.g. Heinzel and Sittler 2010; Davis et al. 2019). Most recently, research is investigating Machine Learning as well (e.g. Trier et al. 2018; Verschoof-van der Vaart and Lambers 2019).

This chapter illustrates research challenges and achievements of the last two decades regarding archaeological feature visualization and extraction. At first, the issue of choosing an adequate visualization is introduced. Secondly, it is demonstrated, how certain terrain models allow retracing relief imprints of agricultural field systems of different epochs. Thirdly, a brief outlook points out the role of visualizations not only for visibility but for automated detection approaches as well.

2   Choosing an Adequate Terrain Visualization

In the era of INSPIRE (European Commission 2007) and Open Geodata, access to LiDAR data is easier than ever, e.g. in the federal state of North Rhine-Westphalia (Bezirksregierung Köln 2020). Its government provides a variety of products, including various LiDAR-derived DTM as well as a Digital Landscape Model (Edler and Dickmann 2019), free of charge (related in this volume: Edler et al. 2020; Hochschild et al. 2020; Siepmann et al. 2020; Vetter 2020).

However, visualizing surface structures using LiDAR is not as simple as it seems. Results of measurement flights initially are long lists of three-dimensional coordinates, each of which represents a reflection point of the laser beam. These point clouds are classified at first, e.g. in order to remove vegetation and extract ground points, from which DTM can be calculated.

DTM are usually visualized as grayscale images, so that every tone represents a specified height. In these kinds of simple models, the large-scale character of a landscape can already be determined. If a landscape is mountainous, however, contrast is often insufficient for detecting subtle small-scale relief anomalies. Therefore, artificial shadow marks, known from aerial archaeology and cartography, are generated additionally (shading, Hake et al. 2002). This is done using a digital sun, which usually shines from top left (northwest) over the terrain increasing the local contrast significantly (Fig. 1, see e.g. Song et al. 2019 for a recent methodological overview of aerial archaeology).

A disadvantage of the *Shaded Relief* visualization is that the visibility of structures depends on illumination in two ways. Firstly, a deviation from the familiar illumination leads to the effect of relief inversion, turning depressions into hills and removing valleys. Secondly, linear structures such as ditches, embankments or pathways does not cast shadows if they run parallel to the light beams. Although both real and digital sun change their position either during the course of a day or by a mouse click, seeing all features at once is not possible. Therefore, a wide variety of visualizations were developed that are

Fig. 1 Conventional Shaded Relief of a forest near Westerkappeln, North-Rhine Westphalia, Germany. The DTM reveals hollow ways on the left, although more than these are present. *Data source* Land NRW 2020—dl-de/ by-2-0

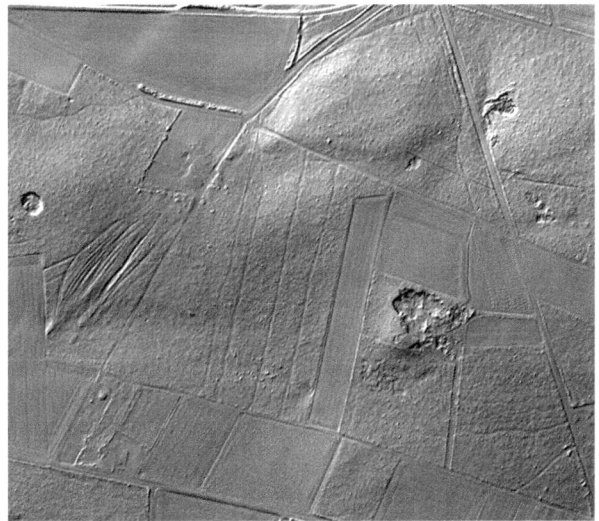

independent of illumination, further enhance contrast and isolate the micro-relief in different ways.

The visualizations are accessible via tool collections like the *Lidar Visualization Toolbox* (LiVT) or the *Relief Visualization Toolbox* (RVT), the latter of which was used to produce the figures presented here. See Kokalj and Hesse (2017) for guide and overview of these tools and available visualizations. For presentation purposes, the figures were slightly contrast enhanced.

The Process of generating a *Local Relief Model* (LRM) discriminates between macro- and micro- relief. The first of which is produced by applying a lowpass filter to the original DTM, preserving only plains, hills and valleys. These are then subtracted to produce a *Difference Map*. This model only contains the micro-relief and could already be used, as its grid values represent the relative height above ground level. However, ground level is not represented perfectly by the smoothed DTM as the micro-relief is rather smoothed than eliminated. Better results are generated by cutting out micro-relief features completely. This is done using the Difference Map for extraction of zero contour lines which are then transferred to the original DTM and used to cut out micro-relief features. Afterwards, the resulting gaps are interpolated using a Triangulated Irregular Network. Subtracting this from the original DTM finally produces the LRM (Fig. 2). The process parameters, such as the size of the low pass filter, decide whether large or small features will be exposed. On the one hand, this model better represents relative heights and increases visibility significantly, on the other hand, artefacts are produced at locations with strong slope changes (Hesse 2010).

Significant contrast enhancements are achieved using the *Sky-View Factor* (SVF) as well (Fig. 3). This visualization is based on diffuse illumination from a hemisphere that

Fig. 2 LRM of a forest near Westerkappeln, North-Rhine Westphalia, Germany. The LRM reveals additional hollow ways, running from the lower right corner to the upper left. *Data source* Land NRW 2020—dl-de/by-2-0

Fig. 3 SVF visualization of
a forest near Westerkappeln,
North-Rhine Westphalia,
Germany. SVF reveals
additional hollow ways,
running from the lower right
corner to the upper left. *Data
source* Land NRW 2020—
dl-de/by-2-0

is centered on the current point to calculate. For every point, the amount of light coming
from that hemisphere is calculated, which is high at exposed locations and low at depressions. This measurement is equal to the sky that is visible from the current point (Kokalj
et al. 2011; Zakšek et al. 2011).

The idea of *Openness* visualization is quite similar to SVF, which is why the models are similar. Openness indirectly describes relief illumination by calculating the
maximum zenith angle of profiles running from the currently observed point to a given
number of cardinal directions. The mean of these zenith angles is finally equal to the
Positive Openness (Fig. 4). *Negative Openness* models are calculated by measuring nadir
angles instead. In contrast to SVF, angles can be greater than 90° for exposed locations,
as this calculation is not based on a hemisphere. (Yokoyama et al. 2002; Doneus 2013).

Micro-relief visualizations are beneficial on different scales. Delimited structures,
such as castles or fortifications, become clearly visible and even allow better assessment
of locations from distance than with conventional DTM or a field survey. Planning projects, e.g. archaeological excavations, are thus simplified.

The same applies to relief features on medium scales, e.g. ancient road systems that
are cut by modern land use and whose connection to the surrounding landscape is no
longer obvious. As shown in Fig. 1 to 4, adequate terrain visualization results in a more
complete and detailed picture of the historical landscape condition.

Fig. 4 Positive Openness
visualization of a forest near
Westerkappeln, North-Rhine
Westphalia, Germany. Positive
Openness reveals additional
hollow ways, running from the
lower right corner to the upper
left. *Data source* Land NRW
2020—dl-de/by-2-0

3   Visualization of Historical Field Systems

The introduction of agriculture and sedentism during the Neolithic Revolution in the
6. Century B.C. (for Central Europe) increased anthropogenic impact on landscapes
and started turning natural landscapes, mostly forests, into cultural landscapes (Schenk
2011). Different stages of this process left relief imprints such as historic field systems.
These can be visualized using micro-relief terrain models, connecting the geomorphological findings in the field with knowledge gained in history science.

Irregular Bank-Sink Field Systems (Arnold 2011) and the related *Celtic Fields* are
remnants of early agricultural activities. These are rectangular ridges that date from the
Neolithic period up to the Roman Iron Age. They define rectangular fields, which are the
results of ards – ploughs that were merely scratching the soil instead of turning it (Küster
2010). In some cases, the ridges are still preserved today and become visible in terrain
models (Fig. 5, Pfeffer 2017).

The invention of soil turning mould-board ploughs, presumably in the pre-Roman
Iron Age, is associated with a change of field systems. These ploughs were able to turn
soil only to one side, which made returning the whole plough at the end of a field necessary. In order to have to turn as seldom as possible, fields became long and narrow, often
hundreds of meters long, but only a few meters wide. As soil was turned to one side
repeatedly, the centers of the fields were lifted over time, while the borders sank. This
reduced erosion and improved drainage at the same time, which again optimized water
balance (Küster 2010; Schenk 2011). As usually several fields were aligned, the corresponding field pattern is called *Ridge and Furrow*, which is a frequent record in forests
today (Fig. 6).

Fig. 5 LRM of an Irregular Bank-Sink Field System, probably Celtic Fields, in Hopsten, North Rhine-Westphalia, Germany. The ridges are represented by white shadows. *Data source* Land NRW 2020—dl-de/by-2-0

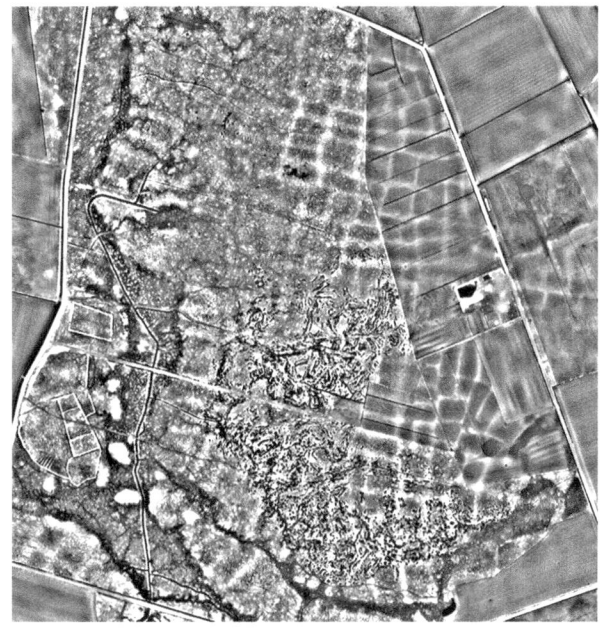

In the late Middle Ages, parts of the poor, rural population migrated to the flourishing cities, hoping for a better life or being forced by plagues or expulsion while many villages became abandoned. Wealthy landlords remained in the countryside and annexed their properties. They reforested parts of the abandoned farmland (which is one reason, why Ridge and Furrow areas are frequently found in forests today) and adapted their agricultural activities to the needs of the constantly growing cities. This process already resembled first modern land consolidations and led to large, rather square fields and pastures. These were typically enclosed by hedge banks, e.g. to allow cattle to graze unattended, and were merged even further in later centuries (Küster 2010; Schenk 2011). Today, these temporary field boundaries can still be observed frequently in terrain models (Fig. 7).

4 Automated Detection of Medieval Field Systems

On a regional scale, certain combinations of visualization and classification algorithms allow for automated detections of diverse micro-relief structures, whose spatial distribution can be used to reconstruct settlement structures. This especially applies to structures that are as similar as possible towards each other but are, in their entirety, as different as possible from other types of relief features.

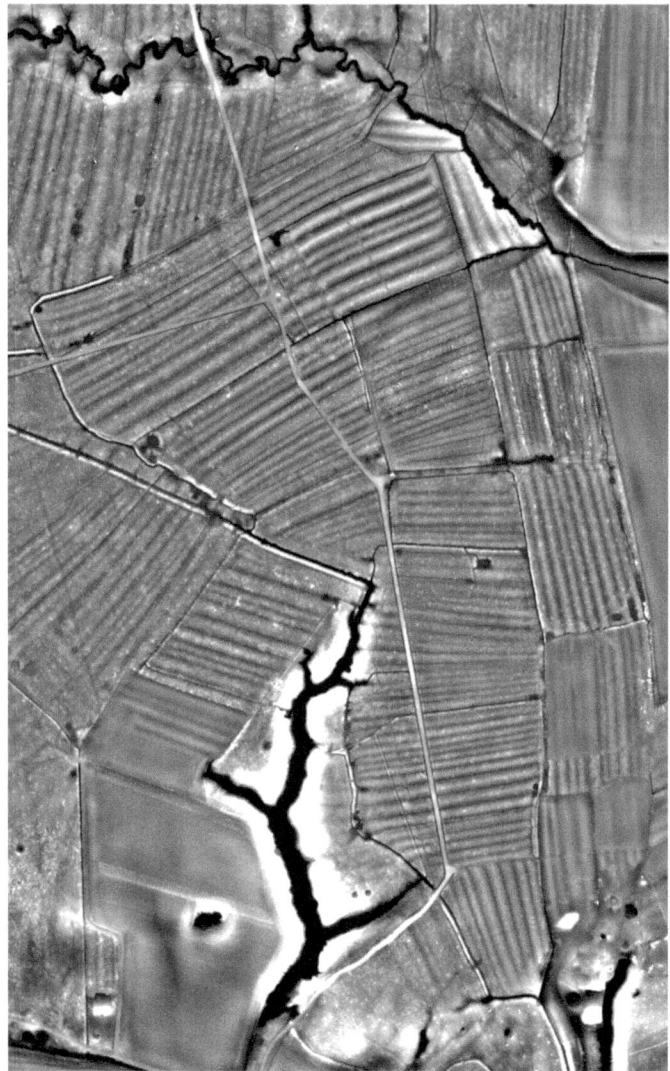

Fig. 6 LRM of an extensive Ridge and Furrow area near Schloss Cappenberg, North-Rhine Westphalia, Germany. *Data source* Land NRW 2020—dl-de/by-2-0

One possible approach is the use of OBIA for detection of Ridge and Furrow areas in Difference Maps or LRMs. Finding single fields is not reasonable as the risk of confusions with other ridges is too high. However, taking the surroundings into account, detecting ridges that are part of an accumulation of such is more promising. This way, Ridge and Furrow areas are identified by finding some kind of indicating ridges.

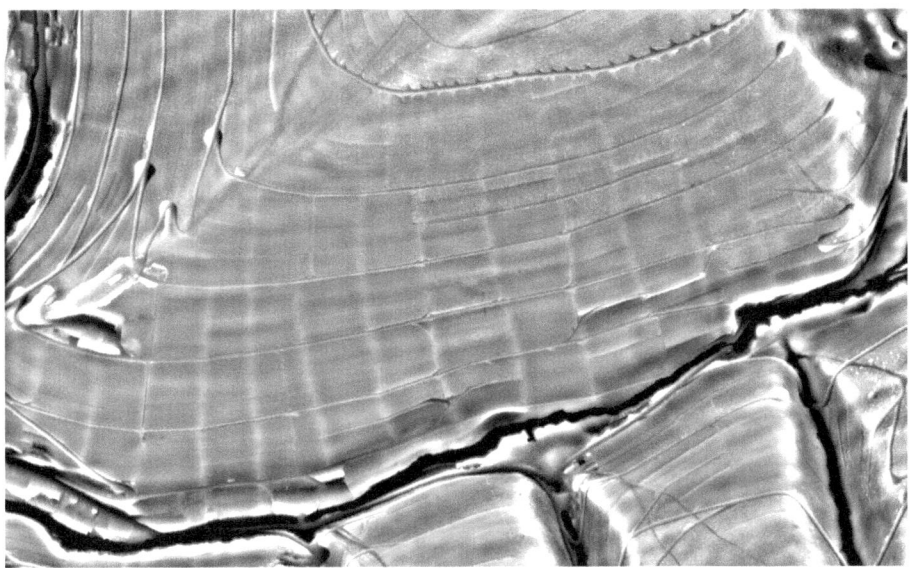

Fig. 7 LRM of probable former enclosed pastures in Girkhausen, North-Rhine Westphalia, Germany. The ridges of former boundaries run down the hill and are displayed as straight white shadows. *Data source* Land NRW 2020—dl-de/by-2-0

On a technical level, this approach is quickly implemented, as individual fields stand out as homogeneous areas with positive grid cell values, while furrows are represented by low or even negative values in between. These homogeneous areas can be separated from each other as unique objects. Each of which represents an elongated local maximum or minimum and is thus distinguishable from other relief anomalies. Finally, each object is checked regarding the number of other possible ridge or furrow objects in its vicinity. If a threshold is exceeded, an object is presumably representing a part of a Ridge and Furrow area. The more of such objects are found next to each other, the higher is the probability that a Ridge and Furrow area was detected (Meyer et al. 2019). Other approaches for Ridge and Furrow detection were successfully carried out by Heinzel and Sittler (2010), Brouwer (2015) or Noack (2019).

5   Conclusion and Outlook

This chapter gave an insight into the field of landscape visualization using terrain models. Different approaches require certain visualizations, some of which are useful to get an overview of a landscape while being intuitive and easy to read. Others extract micro-relief features but need specific software to be calculated and training to be fully exploited. The correct choice is essential for optimal visibility and correct reconstruction of past landscape conditions. Certain visualizations allow for automated detection approaches, which is a current research topic due to constantly growing amounts of data.

The combination of visualization and automation finally allows for extensive mapping surveys. Research still is in a methodological stage, but automated prospection might exceed the borders of distinct investigation areas in the near feature, aiming at detecting and preserving unknown field monuments that are almost lost. However, it is common ground that results of automated detections will probably still need interpretation and confirmation by archaeologists, which is why automation should only serve as an assistance, e.g. by highlighting structures. It will be interesting to see how the various methods perform, how they deal with features whose recognition is difficult to train and how well they can be integrated into the archaeologist's daily work routine.

References

Arnold, V. (2011). Celtic Fields und andere urgeschichtliche Ackersysteme in historisch alten Waldstandorten Schleswig-Holsteins aus Laserscan-Daten. *Archäologisches Korrespondenzblatt, 41*(3), 439–455.

Bezirksregierung Köln. (2020). Digitales Basis-Landschaftsmodell (Basis-DLM). https://www.bezreg-koeln.nrw.de/brk_internet/geobasis/opendata/index.html. Accessed 17.4.2020.

Bofinger, J., Kurz, S., & Schmidt, S. (2007). Hightech aus der Luft für Bodendenkmale – Airborne Laserscanning (LIDAR) und Archäologie. *Denkmalpflege in Baden-Württemberg – Nachrichtenblatt der Landesdenkmalpflege, 36*(3), 153–158.

Brouwer, S. (2015). *Ermittlung von Wölbäckern aus dem Digitalen Geländemodell mit ArcGIS. Eine Reliefanalyse zur Rekonstruktion historischer Landnutzung in NRW.* Unpublished Master's Thesis, Hochschule Osnabrück and Universität Osnabrück.

Cerrillo-Cuenca, E. (2017). An approach to the automatic surveying of prehistoric barrows through LiDAR. *Quat. Inter., 2017*(435), 135–145.

Crutchley, S. (2006). Light Detection and Ranging (lidar) in the Witham Valley, Lincolnshire – An Assessment of New Remote Sensing Techniques. *Archaeological Prospection, 13,* 251–257.

Davis, D., Sanger, M., & Lipo, C. (2019). Automated mound detection using lidar and object-based image analysis in Beaufort County, South Carolina. *Southeastern Archaeology, 38*(1), 23–37.

De Boer, A. (2007). Using pattern recognition to search LIDAR data for archaeological sites. In: A. Figueiredo & G. Velho (Eds.), *The world is in your eyes – CAA2005 – Computer applications and quantitative Methods in archaeology – Proceedings of the 33rd conference, Tomar, March 2005* (pp. 245–254). Tomar: CAA Portugal.

Devereux, B., Amable, G., Crow, P., & Cliff, A. (2005). The potential of airborne lidar for detection of archaeological features under woodland canopies. *Antiquity, 79,* 648–660.

Devereux, B., Amable, G., & Crow, P. (2008). Visualisation of LiDAR terrain models for archaeological feature detection. *Antiquity, 82,* 470–479.

Doneus, M. (2013). Openness as visualization technique for interpretative mapping of Airborne LiDAR derived digital terrain models. *Remote Sensing, 5,* 6427–6442.

Doneus, M., & Briese, C. (2006). Full-Waveform Airborne laser scanning as a tool for archaeological reconnaissance. In S. Campana & M. Forte (Eds.), *From space to place: 2. International conference on remote sensing in archaeology; Proceedings of the 2. International workshop,*

CNR, Rome, Italy, December 2–4, 2006 (BAR International Series, vol. 1568, pp. 99–106). Oxford: Archaeopress.

Doneus, M., Briese, C., Fera, M., & Janner, M. (2008). Archaeological prospection of forested areas using full-waveform airborne laser scanning. *Journal of Archaeological Science, 35,* 882–893.

Edler, D., Keil, J., & Dickmann, F. (2020). From Na Pali to Earth—An 'Unreal' Engine for Modern Geodata? In D. Edler, C. Jenal, & O. Kühne (Eds.), *Modern approaches to the visualization of landscapes* (pp. 279–291). Wiesbaden: Springer VS.

Edler, D., & Dickmann, F. (2019). Landschaft im amtlichen Geoinformationswesen. In O. Kühne, F. Weber, K. Berr, & C. Jenal (Eds.), *Handbuch Landschaft* (pp. 507–515). Wiesbaden: Springer VS.

European Commission. (2007). Directive 2007/2/EC of the European Parliament and the of the council of 14 March 2007 establishing and Infrastructure for Spatial Information in the European Community (INSPIRE). https://eur-lex.europa.eu/legal-content/EN/TXT/PDF/?uri=CELEX:32007L0002&qid=1581958705103&from=EN. Accessed: 17.4.2020.

Hake, G., Grünreich, D., & Meng, L. (2002). *Kartographie. Visualisierung raum-zeitlicher Informationen.* Berlin: De Gruyter.

Heinzel, J., & Sittler, B. (2010). LiDAR surveys of ancient landscapes in SW Germany. Assessment of archaeological features under forests and attempts for automatic pattern recognition. In S. Campana, M. Forte, & C. Liuzza (Eds.), *Space, Time, Place. Third International Conference on Remote Sensing in Archaeology* (BAR International Series, vol. 2118, pp. 113–121). Oxford: Archaeopress.

Hesse, R. (2010). LiDAR-derived local relief models. A new tool for archaeological prospection. *Archaeological Prospection, 17,* 67–72.

Hochschild, V., Braun, A., Sommer, C., Warth, G., & Omran, A. (2020). Visualizing landscapes by geospatial techniques. In D. Edler, C. Jenal, & O. Kühne (Eds.), *Modern approaches to the visualization of landscapes* (pp. 47–78). Wiesbaden: Springer VS.

Holden, N. (2001). Digital airborne Remote Sensing – The Principles of LIDAR and CASI. *AARGnews, 22,* 23–24.

Holden, N., Horne, P., & Bewley, R. (2002). High-resolution digital airborne mapping and archaeology. In R. Bewley & M. Raczkowski (Eds.), *Aerial archaeology – Developing future practice* (pp. 173–180). Amsterdam: IOS Press.

Freeland, T., Heung, B., Burley, D., Clark, G., & Knudby, A. (2016). Automated feature extraction for prospection and analysis of monumental earthworks from aerial LiDAR in the Kingdom of Tonga. *Journal of Archaeological Science, 69,* 64–74.

Kühne, O. (2019). *Landscape theories. A brief introduction.* Wiesbaden: Springer VS.

Kühne, O., & Jenal, C. (2020). The threefold landscape dynamics—Basic considerations, conflicts and potentials of virtual landscape research. In D. Edler, C. Jenal, & O. Kühne (Eds.), *Modern approaches to the visualization of landscapes* (pp. 389–402). Wiesbaden: Springer VS.

Küster, H. (2010). *Geschichte der Landschaft in Mitteleuropa.* München: C.H. Beck.

Kokalj, Ž., Zakšek, K., & Oštir, K. (2011). Application of sky-view factor for the visualization of historic landscape features in Lidar-Derived relief models. *Antiquity, 85*(327), 263–273.

Kokalj, Ž., & Hesse, R. (2017). *Airborne laser scanning raster data visualization: A guide to good practice.* Ljubljana: Založba ZRC.

Land NRW. (2020). OpenGeodata.NRW. Data licence germany: dl-de/by-2-0 (www.govdata.de/dl-de/by-2-0). https://www.opengeodata.nrw.de/produkte/geobasis. Accessed: 5.4.2020.

Meyer, M. F., Pfeffer, I., & Jürgens, C. (2019). Automated Detection of Field Monuments in Digital Terrain Models of Westphalia Using OBIA. *Geosciences, 2019*(9, 3), 109.

Motkin, D. (2001). An assessment of LIDAR for archaeological use. *AARGnews, 22,* 24–25.

Noack, D. (2019). *GIS-gestützte Analyse zum Wölbackervorkommen in der Prignitz*. Unpublished Master's Thesis, Universität Salzburg.

Pfeffer, I. (2017). Celtic Fields. Neu entdeckte eisenzeitliche Ackersysteme in Westfalen. *Archäologie in Westfalen Lippe, 2016*, 207–211.

Poplin, A., de Andrade, B., & Mahmud, S. (2020). Exploring tangible and intangible landscapes of evocative places: Case study of the city of Vitória in Brazil. In D. Edler, C. Jenal, & O. Kühne (Eds.), *Modern approaches to the visualization of landscapes* (pp. 519–547). Wiesbaden: Springer VS.

Schenk, W. (2020). Visualisation of the fundamental dimensions of "landscape" in landscape paintings around 1500 A.D. In D. Edler, C. Jenal, & O. Kühne (Eds.), *Modern approaches to the visualization of landscapes* (pp. 19–32). Wiesbaden: Springer VS.

Schenk, W. (2011). *Historische Geographie*. Darmstadt: WBG.

Schneider, A., Takla, M., Nicolay, A., Raab, A., & Raab, T. (2015). A template-matching approach combining morphometric variables for automated mapping of charcoal kiln sites. *Archaeological Prospection, 22*, 45–62.

Sevara, C., Pregesbauer, M., Doneus, M., Verhoeven, G., & Trinks, I. (2016). Pixel versus object – A comparison of strategies for the semi-automated mapping of archaeological features using airborne laser scanning data. *Journal of Archaeological Science: Reports, 5*, 485–498.

Siepmann, N., Edler, D., & Kühne, O. (2020). Soundscapes in Cartographic Media. In D. Edler, C. Jenal, & O. Kühne (Eds.), *Modern approaches to the visualization of landscapes* (pp. 247–263). Wiesbaden: Springer VS.

Sittler, B. (2004). Revealing historical landscapes by using airborne laser scanning – A 3-D modell [sic!] of ridge and furrow in forests near Rastatt (Germany). In M. Thies, B. Koch, H. Spiecker, & H. Weinacker (Eds.), *Laser-Scanners for Forests and Landscape Assessment* (pp. 258–261). Freiburg: Institute for Forest Growth, Department of Remote Sensing.

Song, B., Leidorf, K., & Heller, E. (2019). *Luftbildarchäologie. Archäologische Spurensuche aus der Luft. Methoden und Techniken. Klassisch und virtuell*. Darmstadt: wbg Theiss.

Stratmann, J., Ristea, A., Leitner, M., & Paulus, G. (2020). Exploring Urban "Blightscapes" applying spatial video technology and geographic information system: A case study from Baton Rouge, USA. In D. Edler, C. Jenal, & O. Kühne (Eds.), *Modern approaches to the visualization of landscapes* (pp. 499–517). Wiesbaden: Springer VS.

Trier, Ø., Zortea, M., & Tonning, C. (2015). Automatic detection of mound structures in airborne laser scanning data. *Journal of Archaeological Science: Reports, 2*, 69–79.

Trier, Ø., Cowley, D., & Waldeland, A. (2018). Using deep neural networks on airborne laser scanning data – Results from a case study of semi-automatic mapping of archaeological topography on Arran, Scotland. *Archaeological Prospection, 26*(2), 1–11.

Verschoof-van der Vaart, W., & Lambers, K. (2019). Learning to look at LiDAR. The use of R-CNN in the automated detection of archaeological objects in LiDAR data from the Netherlands. *Journal of Computer Applications in Archaeology, 2*(1), 31–40.

Vetter, M. (2020). Technical potentials for the visualization in virtual reality. In D. Edler, C. Jenal, & O. Kühne (Eds.), *Modern approaches to the visualization of landscapes* (pp. 307–317). Wiesbaden: Springer VS.

Yokoyama, R., Shirasawa, M., & Pike, R. J. (2002). Visualizing topography by openness. A new application of image processing to digital elevation models. *Photogrammetric engineering and remote sensing, 68*, 251–266.

Zakšek, K., Oštir, K., & Kokalj, Ž. (2011). Sky-View factor as a relief visualization technique. *Remote Sensing, 3*, 398–415.

M. Fabian Meyer-Heß studied history and geography at the Ruhr University Bochum. His master's thesis dealt with automated detection approaches of different types of field monuments in terrain models of Westphalia. Since 2017, he is a research assistant at the Department of Geography of the Ruhr University Bochum, doing further research in the field of automation and open geodata utilization for the purpose of archaeological feature detection.

Visualizing Landscapes by Geospatial Techniques

Volker Hochschild, Andreas Braun, Christian Sommer, Gebhard Warth and Adel Omran

Summary

This chapter will provide an overview on the manifold approaches to the visualization of landscapes. There are many techniques on the market nowadays, but we first start with a historical view on the development of the different presentation methods of landscape models. In the beginning, there have been historical maps, followed by physical 3D models of cities and landscapes and then entering the digital world of GIS-systems, integrating nadir viewing remote sensing data with equidistant topographical maps including contour lines and 2.5 dimensional shaded relief maps of Digital Terrain Models. The second paragraph is dedicated to recent hardware technology and image processing techniques of remote sensing data in order to derive height information of the landscape. Section 2.3 summarizes the Digital Landscape Models, the basic geodata infrastructure available through the German surveying agencies. These multiscale national topographical datasets lead to the harmonization within the Infrastructure for Spatial Information in Europe (INSPIRE). The last paragraph gives some examples of

V. Hochschild (✉) · A. Braun · C. Sommer · G. Warth · A. Omran
Geographisches Institut, Universität Tübingen, Tübingen, Germany
e-mail: volker.hochschild@uni-tuebingen.de

A. Braun
e-mail: an.braun@uni-tuebingen.de

C. Sommer
e-mail: christian.sommer@uni-tuebingen.de

G. Warth
e-mail: gebhard.warth@uni-tuebingen.de

A. Omran
e-mail: adel.omran@uni-tuebingen.de

© Springer Fachmedien Wiesbaden GmbH, part of Springer Nature 2020   47
D. Edler et al. (eds.), *Modern Approaches to the Visualization of Landscapes*,
RaumFragen: Stadt – Region – Landschaft, https://doi.org/10.1007/978-3-658-30956-5_4

applications in geomorphology, landscape evolution models and animated maps as well as mixed reality, artificial intelligence and landscape architecture and planning. Finally, we not only try to provide a comprehensive overview of the published literature, we also give a comprehensive overview on the available techniques in Fig. 8, where we plotted the "amount of virtuality" against the user involvement. The digital visualization of landscapes as well as the spatial modelling of landscapes under predefined border conditions are a prerequisite for future sustainable decision making.

Keywords

3D-Visualization · Animated maps · Digital landscape visualization techniques · Digital elevation models · Image processing · Landscape evolution models · Virtual reality · Visualization hardware

1   Introduction

1.1   Background and Structure

This chapter will provide an overview on the current state of landscape visualization from a Geographers point of view. In recent decades, the understanding of what is referred to as landscape has become more differentiated in the various scientific disciplines and also in geography. In addition to the classical understanding of landscape as an object, constructivist understandings have developed and, most recently, approaches that attempt to abolish the subject-object separation (Cosgrove 1984; Greider and Garkovich 1994; in summary: Howard et al. 2019; Kühne 2019). The present contribution focuses on the possibilities of how the material world can be visualized, especially through the use of computer technologies. It does not focus on the constructive character of these visualizations. In this respect, we follow the classical understanding of landscape by Karl Sauer (1925): computerized approaches today (Tiede and Lang 2010) enable analytical 3D views of landscape elements through draped earth observation images on digital terrain models, or through the cartographic representation of a third dimension either by contour lines or an artificial light source, providing a 2.5 dimensional view of a shaded relief.

Landscape visualization can be seen from many different perceptions. Urban planners, civil engineers, landscape architects or geographers have different approaches to landscapes in general. Meanwhile landscape architects primarily see landscapes as an addition of infrastructural elements to keep the urban systems running (Bélanger 2017), some basic work was presented by Geier et al. 2001; Lange 2002; Cantrell and Michaels 2010; Bélanger 2013a. Architects are more dedicated to single building environments (Frampton 1999, Amoroso 2019). The Geographers point of view is closely connected with the development of Geographical Information Systems within the last 40 years, enabling not only computerized digital cartography, but providing also information on the characteristics of landscape objects and providing last but not least the ability to carry out spatial modelling capabilities, an irreplaceable prerequisite for spatial decision support.

Landscape visualization has of course also to do with hardware issues. There are many different media approaches to simulate landscapes in three dimensions, sometimes with the possibility to move through the landscape. Whether there are mirror screens with beamsplitter technology, virtual reality 3D domain glasses, CAVE surroundings or red-green polarizing glasses, all of them provide the opportunity to get a three-dimensional impression of landscapes, not to forget the 3Dprinter-technology or the projection of GIS maps on white, physical 3D-Modells (Coucelo et al. 2013). Many authors have published various approaches to visualize realistic landscapes through different modelling environments (Berry et al. 1998; Lange 1998; Lange and Bishop 2001; Blaschke et al. 2004; Tiede and Blaschke 2005; Faulkner 2006; Harmon 2006; Martínez-Graña and Valdés Rodriguez 2016). Also the applications are manifold, they reach from urban studies (Meyboom 2009, Bélanger 2013b, Waldheim 2016) over climate change (Sheppard 2005) up to tourism (Hartmann 2019).

In this regard the following paragraphs will draw a landscape image based on the historical development of landscape visualization, the necessary hardware to do so as well as image processing techniques to gain information from remote sensing data (Sect. 2) and to use this data in Digital Landscape Models and Landscape Evolution Models (Sect. 3.3). Finally, we present some applications of landscape visualizations through animated maps, the visualization of geomorphological processes, mixed reality, landscapes and artificial intelligence as well as landscape architecture and planning.

1.2 Landscape Visualization in the Past and Today

Before the dawning of the digital age, the visualization of landscapes was mainly addressed within the domain of cartography, which originates from the Greek words χάρτης, which can be translated as "papyrus" or "sheet of paper", and γράφειν, which stands for "to write". It was defined by the Portuguese scholars Manuel Francisco de Barros e Sousa Santarém, who wanted to find a term for the study of old maps (de Santarem 1852). Yet, mapping the environment goes back to prehistoric times when depictions of landscapes were found etched in Neolithic and Bronze Age rocks, so-called Petroglyphs, to mark important reference or gathering points (Frachetti and Chippindale 2001). These maps give us insights on how people perceived their surroundings. For example, the *Imago Mundi*, a map carved into a clay tablet, shows the Babylonian view of the world in the 6th century, where Mesopotamia is divided by the Euphrates River and entirely surrounded by the ocean. The city of Babylon lies in the center of the land and is framed by southern marshes (represented by parallel lines) and the Zagros Mountains in the North (indicated as a curved line, Horowitz 1988).

Over the centuries, the quality of the media on which the maps were sketched became more advanced: Beginning from papyrus as it was developed by the ancient cultures and later turning into more persistent paper, while more wealthy people could afford to utilize even materials of higher grade, such as leather for nautical charts, for example

(Bagrow 2017). Today, analogous outdoor maps are made of plastic or coated by other waterproof substances (Senda-Cook 2013). They also differ regarding topic, scale and generalization but all of them have in common, that they are mostly restricted two a two-dimensional representation of the environment (McGranaghan 1993).

The challenge of visualizing the third dimension, expressed as the altitude of the landscape or the portrayed objects, had been tackled differently throughout history. In early times, mountainous regions were depicted by triangles or molehills, often connected to long ridges, as displayed in Fig. 1. While this works well for elongated mountains, such as the Alps or the Pyrenees, it meets its limits for landforms of more complex shapes and maps at a higher level of detail.

Besides the inefficiency of many forms of terrain signatures, there was no semblance of standardization until the 18th century, which led to a variety of systems, such as hachures and broken lines until the French topographical engineer Jean Louis Dupain-Triel published a map of France suggesting horizontal equidistant curves indicating areas of the same elevation (Dupain-Triel 1791). It is considered to be the beginning of the concept of contour lines, which has been consistently adapted since then, because it was applicable independently from the scale and the orientation of a map. As another advantage of this technique, cartographers were now able to include further thematic contents in their maps, for example different types of forest, because their signatures were

Fig. 1 Portrayal of topography as triangles (left) and molehills (right) in early historical maps. *Left* World map from 1410 by Pirrus de Noha (detail of Europe), *Right* Map of Europe from 1493 by Hieronymus Münzer (detail of Central Europe). Source: Public Domain

no longer overlapping with the underlying topographical features. An example is given in Fig. 2 (middle), where contour lines and a thematic layer are combined. To give the maps an even more three-dimensional look, the concept of hillshading was introduced by the German cartographer Johann Georg Lehmann, which was implemented as line hachures drawn normal to the adjacent contour and of variable thickness corresponding to the slope (Lehmann 1799). It is based on the assumption of an artificial light source, mostly coming from the top left corner, which casts shadows at the shape of the underlying landscape. This pattern is intuitively understood by the human eye and became a paradigm in the visualization of landscapes, because it described the shape and steepness of the relief complementary to its height delineated by contour lines. It was soon adopted by the Swiss general and cartographer Guillaume Henri Dufour, who created the first official maps of entire Switzerland at a scale of 1:100 000 (Dufour and Flamsteed 1833). As one of the first comprehensive works on this topic, Eduard Imhof published a set of rules and standards on the measurement of topography and its depiction in maps using contour lines, shading techniques, colors and other signatures in the context of scale, orientation, vertical exaggeration and generalization (Imhof 1965). His initial book on terrain and maps contained a set of eight anaglyph images illustrating typical landforms, which could be visualized by a pair of spectacles with red and blue lenses to give the reader a three-dimensional impression (Imhof 1950). An example is given in Fig. 2 (bottom), where the shaded relief is superimposed by an anaglyph color coding.

However, the aspect of three-dimensional landscape visualization, as it is described in Chap. 2, is not an invention of the digital age. During the 18th and 19th, over 250 models of French cities were constructed, most of them under reign of Louis XIV of France for military purposes (Rothrock 1969). These socalled plan-reliefs were made of cork, pasteboard, wood and silk and show miniatures of cities, their fortifications and their surrounding landscapes at a scale between 1:75 and 1:600 (Buisseret 1998) and can reach several tens of square meters. Their initial purpose was to prepare the defense of the city in times of war, but as they were too large and fragile to be used on the battlefield, they were also used for urban planning and as a prestige object shown to visiting dignitaries (Ellis 2018). As they were built for selected cities and their specific needs, they largely differed regarding size, scale and level of detail. However, all shared the essential characteristic of a traditional map that they were created from the plan view perspective, but with the additional information on the height of the surrounding landscape and the buildings of the city, which gave their users a clearer perspective for planning, strategic decisions and future visions for the area. Today, only about one hundred of these models are preserved, but they regained new attention in the past years due to various reasons. Many of them are used in museums to give the visitors insights about the cities historic appearance and its scenic embedding in earlier times. The plan relief in Fig. 3 is used in combination with coordinated light effects to draw the visitors' attention on different parts of the model. Its presentation is furthermore enriched by cinematic holograms, which blend over the city to merge both traditional and modern techniques. Besides the aspect of visualization and historic education, these models are also being surveyed with

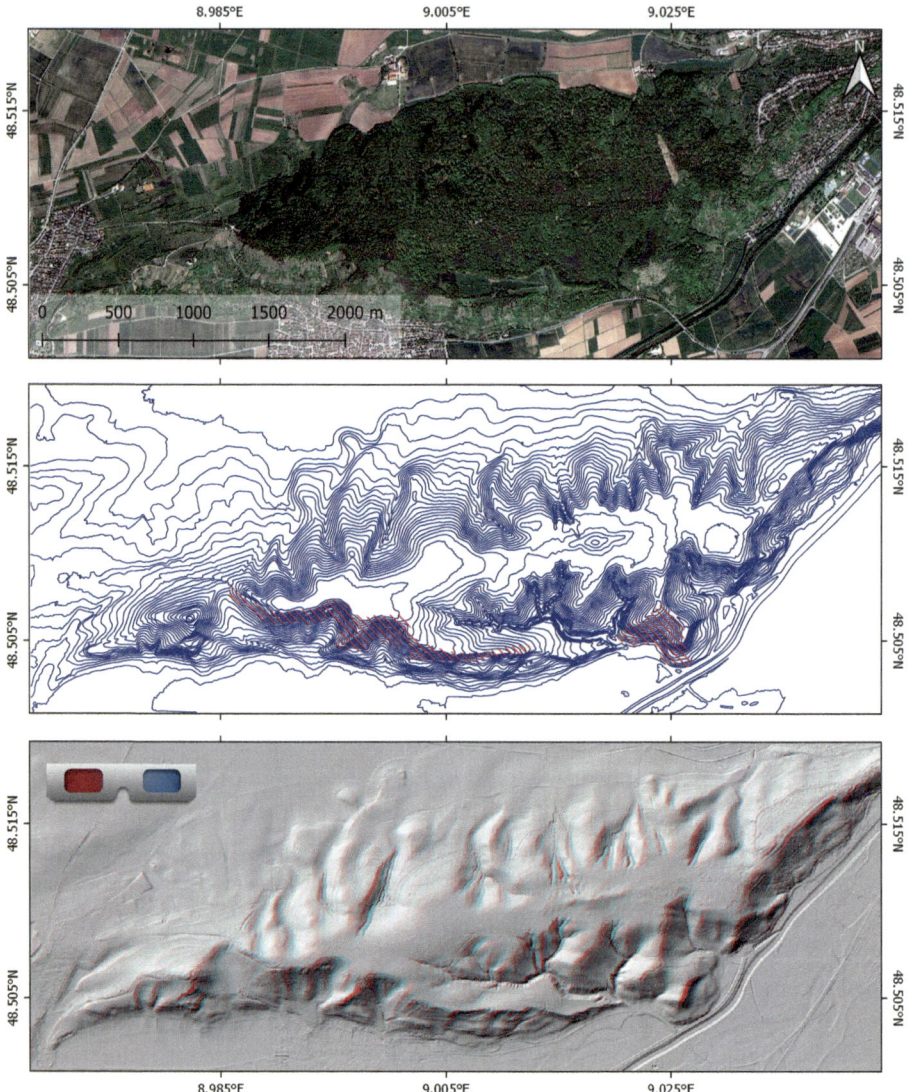

Fig. 2 Spitzberg near Tübingen (Germany): *Top* Satellite image. *Middle* Contour lines (blue, equidistant 5 meters) and nature preservation area (red haches). *Bottom* Anaglyph hillshading based on a digital elevation model (topography enhanced, requires speculars with lenses of red [left] and blue [right]). Source: own presentation

photogrammetric methods to create digital and semantically enriched 3D models for the later use in Geographic Information Systems (Macher et al. 2017).

These kinds of three-dimensional models currently experience a renaissance, as their value has been acknowledged in various domains. Administrations have 3D models

Fig. 3 Plan-relief of the city of Strasbourg in the Musées de la ville de Strasbourg. (Source: Rama 2010, under the license CC BY-SA 2.0 FR)

of their cities printed to communicate the locations of landmarks or tourist attractions (Shiode 2000), or to assist the mobility and spatial orientation of visually impaired persons (Gual et al. 2012). In their digital form, 3D city models are used especially for urban planning, for example in visibility analyses of landmarks (Delikostidis et al. 2013), facades or billboards with respect to marketing (Albrecht et al. 2013) or the ideal placement of surveillance cameras (Ming et al. 2003), but also for estimates on solar irradiation or energy flows (Perez et al. 2013).

But also outside cities, landscape visualization by solid 3D models gains popularity. Brazen topographic models are often placed along hiking trails as a complementary representation of the surrounding landscape and its landforms, or to highlight the direction and distance of surrounding summits (Fig. 4, left). And also laser engravings are used to etch blocks of glass for three dimensional depictions of large mountain ridges (Buchroithner and Knust 2013). Lastly, technological advancements allow to include the dimension of time in landscape visualization: Variable map contents are projected by installed illuminants, and interact with the topographical features of three-dimensional surfaces, for example representing historic developments or future scenarios of a certain area (Fig. 4, right). This gives the audience an additional impression of temporal dynamics and spatial variability of the observed phenomena. More integrative systems allow users to modify and re-shape the displayed surface to actively change the thematic information displayed by the projected colors (Mitasova et al. 2006).

Fig. 4 Usage of printed 3D models for tourism (Source left: Pflug/Gomaringen) and the visualization of complex scientific topics (Source right: Phi-Experience, Frascati)

2   Systems

2.1   Hardware

Head-Mounted Displays, red-green- or red-blue-glasses, polarizing glasses as well as shutter-systems from the gaming industry, they all use the parallaxe difference between the left and the right eye of the observer. Viewing an object from a slightly different viewing angle from your left and your right eye generates the virtual stereoscopic view. This is implemented either by two different colors (red-green, red-blue) or two different polarizations (vertical, horizontal) for the left or the right eye. In the case of the shutter-systems there is a flap opening and closing very fast per single eye so that the observer also gets a 3D impression.

Another possibility to visualize landscapes are Cave Automatic Virtual Environments (CAVE), available since the 1990ies determining a 3D room with immersive virtual reality in spatial environments. You need a cube with six projected areas, sometimes also a glass floor is projected from down under and it may be realized also with LED walls or LC displays. In opposite to Head-Mounted Displays, the CAVE enables more than one person to experience the virtual landscape.

Projecting GIS data on solid 3D terrain models also provides the opportunity for groups to jointly view on virtual landscapes. All types of thematic or topographical maps may be projected, but the majority is on urban issues (moving traffic, infrastructure projects, etc.) as well as for the study of geological or geomorphological phenomena (Coucelo et al. 2013).

One of the actual most sophisticated technologies is to use passive 3D Stereo-Monitors for stereoscopic image evaluation. These innovative stereo-photogrammetric monitors use due to the beamsplitter technology the full resolution per eye up to 4 K/UHD. Together with polarizing glasses this enables professional 3D GIS mapping applications or the visualization of 3D city models (Fig. 5).

Fig. 5 Full 3D stereoscopic visualization of high-resolution remote sensing data. (Source: 3DpluraView)

2.2   Landscape Visualization Through Image Processing Techniques

To describe landscape, the relief is one of the elementary bases. Traditional approaches to capture elevation information as key parameter to characterize relief, terrestrial approaches have been used traditionally. The invention of the theodolite in the 16th century made large-area measurements possible. Triangulations based on point measurements enable precise results. Modern geodetic techniques are still based on point measurements, which can be performed using high precision theodolites or GPS data to locate the measuring points. To perform areal measurements, elevation information has to be captured using a defined regular sampling. For this purpose, the workload for individual measurements is not efficient. Therefore, the opportunities of remote sensing based approaches offer valuable opportunities due to the largescale acquisition techniques. In this chapter, the most important remote sensing based approaches for measuring digital elevation models (DEM) are explained (for another remote sensing study using DEM data in this volume, see: Meyer-Heß 2020).

2.2.1 Radar Interferometry

The process of elevation generation by radar interferometry is based on the phase information of two radar images. Radar remote sensing uses a Synthetic Aperture Radar (SAR) technique to spatially visualize surface reflections on earth sent by a space- or airborne radar antenna. Compared to passive remote sensing techniques, such as optical systems, by sending radar impulses to the surface it is exposed by the antenna itself and therefore independent from daylight. Besides this advantage of active remote sensing systems, microwave radiation with frequencies between 9.65 GHz (X-Band), 5.3 GHz (C-Band) and 1.2 GHz (L-Band) is able to penetrate almost all atmospheric conditions and thus is operable in cloudy conditions (Woodhouse 2005).

To derive elevation by SAR processing techniques a SAR image pair is required. This image pair needs to fulfill the most important prerequisites of identical radar frequencies, acquisition positions of the carrier vehicle within 40–300 m and stable surface conditions in the covered scene to avoid decorrelation of the phase (Gens and Van Genderen 1996).

By forming an interferogram through relating the phase information of both acquisitions of the image pair, an intermediate product is generated. After unwrapping the interferogram a dataset is generated, containing phase information directly related with surface heights. The interferometric baseline and the radar frequency define the factor to convert the unwrapped interferogram information to elevation heights (Cloude and Papathanassiou 1998).

The interferometric processing of SAR imagery (InSAR) allows generating very large area covering elevation datasets with high spatial resolution and very high precision of the elevation data. For example, the Shuttle Radar Topography Mission (SRTM) DEM and the TanDEM-X global DEM were generated by interferometrically processing SAR data (Farr et al. 2007; Wessel 2018). Nevertheless there are some technical and radiation caused problems which limit the use of SAR interferometry elevation data. Besides the ability to penetrate clouds, radar radiation has the ability to penetrate surfaces accordingly. Depending on the wavelength, the radiation has different penetration capabilities. As an example, X-Band radiation usually is scattered at canopies, whereas L-Band radiation has the ability to penetrate the foliage and might be scattered at the solid structures of the trees or the ground surface. In many cases depths of the center of reflection is not clear, as on dense vegetation, snow and sand. Steep terrain causes errors due to shadow and foreshortening effects. There are actually many operational systems available: TanDEM-X (X-Band), Sentinel-1 and Radarsat-2 (C-Band) and ALOS-2 (L-Band), see Table 1, Fig. 6.

2.2.2 LiDAR/Laser Altimetry

Laser altimetry is an active approach to generate height information. A laser sensor sends out a laser pulse and measures the reflections from the surface. The signal propagation is the key parameter to be acquired as the laser pulse travels at the speed of light and therefore the distance from the sensor to the reflector is linked to the signal propagation (Bufton 1989). The knowledge of the exact position of the sensor allows the determination of surface heights. The laser altimetry approach is often referred to as light detection and ranging (LiDAR).

Laser altimetry missions are mainly operated as airborne missions. Airborne systems have the advantage that individual specifications can be considered. In addition, the measurement density and thus the spatial resolution can be varied by means of adapted

Table 1 Comparison between SRTM-C-Band DEM and TanDEM-X Global DEM (after Farr et al. 2007; Wessel 2018)

	SRTM C-Band DEM	TanDEM-X Global DEM
Acquisition period	February 2000	2010–2014
Radar frequency, wavelgt.	5.3 GHz, 5.6 cm	9.65 GHz, 3.1 cm
Spatial resolution	1 arc second ~ 30 m	0.4 arc seconds ~ 12 m
Coverage	60 °N–56 °S	global

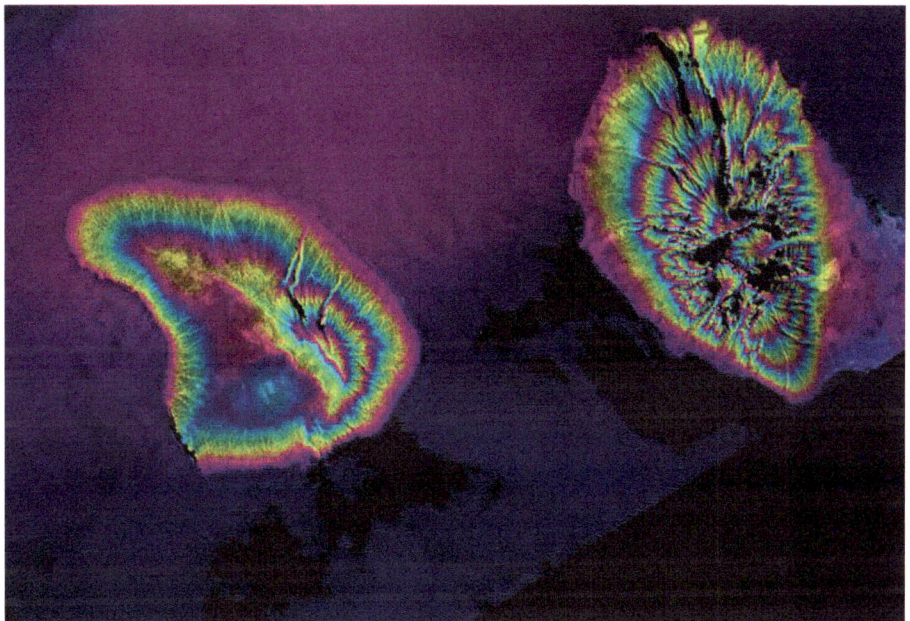

Fig. 6 Hawaiian islands Lanai and West Maui represented as unwrapped interferogram generated by interferometric processing of SRTM data. Each colored fringes cycle indicate 400 m/1300 feet of elevation difference. (Source: NASA 2020)

flight heights. By mounting rotating mirrors in front of the laser sensor, the laser pulses can be dispersed covering wide areas. Few satellite-based missions have installed laser altimeters, to perform point measurements for areas, which are covered by the satellite tracks. As an example, the ICESat satellites are to be mentioned here, whose laser altimeters measure the changes in the global ice caps and ice shields.

Recent developments in LiDAR techniques enabled the reduction of the size and the weight of the altimeters with the possibility to operate the sensors on unmanned aerial vehicles (UAV) (Lin et al. 2010). This enables individual and small-scale analyses with

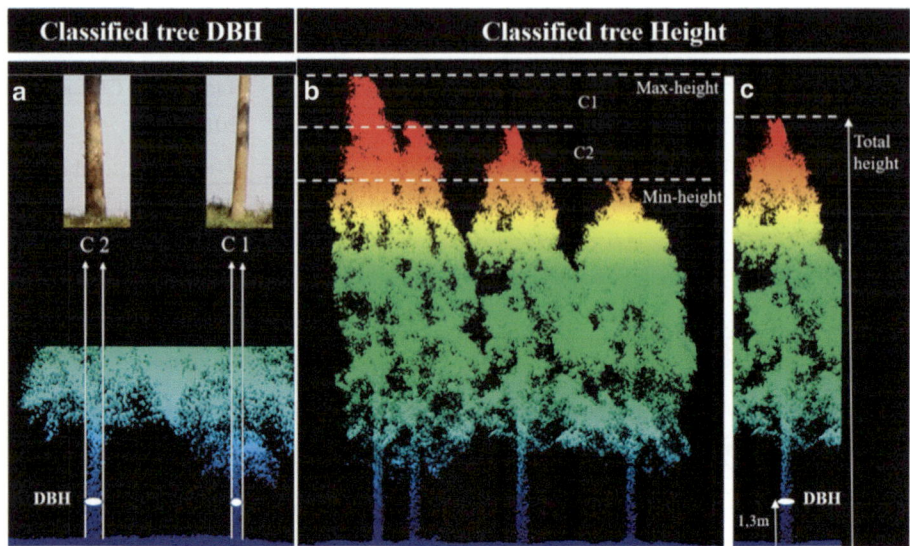

Fig. 7 Information content of lidar data applied on forest stock. First and intermediate reflections reveal volumetric tree stock information. (Source: Dalla Corte et al. 2020)

the highest spatial resolution, which might be carried out with less financial, time and organizational effort.

Similar to long wave radar radiation, the radiation of the laser is usually able to penetrate vegetation. The laser beam does not penetrate the vegetation at full intensity, parts are always reflected. This leads to different reflections until the laser beam reaches the ground, where there remaining beam intensity is fully reflected (Blair et al. 1999). The first reflection, which is also known as "first pulse", represents the surface including all objects of the measured area. The last reflection measured, also known as "last pulse", represents in vegetation stocks the height of the ground. Besides first and last pulse, in the raw data set a height value is generated from every reflection, therefore multi-reflectability of LiDAR measurements can reveal the vertical structure of vegetation stocks (Lim et al. 2003).

Due to the very high spatial resolution, the first pulse and last pulse information and the possibility to measure volumetric information about vegetation, laser-altimetry measurements are one of the most valuable elevation data because of its height resolution and precision (see Fig. 7).

2.2.3 Photogrammetry

Photogrammetric approaches allow the generation of elevation information on the base of optical remote sensing imagery. Both space-borne and airborne acquisition systems are used for this approach. Main prerequisite is the multiple coverage by images of the surveyed area from different acquisition positions. Similar to the human cognition, the photogrammetric approach uses panoramic difference in the image to create a three dimensional space.

When taking at least two images with a fixed camera, objects that are close to the sensor appear strongly offset when these images are overlaid. Objects far away from the camera appear less offset. This so-called parallax effect can be used to estimate relative distances from the camera to the objects. Objects with a large parallax show a small distance to the sensor, objects with a small parallax are further away. Based on inner and outer camera orientation, the relative distances can be converted into metric distances. Thus, the distance between camera and earth's surface can be calculated by photogrammetric methods and elevation models can be generated with known flight altitude of the carrier system (Mikhail et al. 2001). The interior orientation parameter contains information on metric characteristics of cameras to evaluate the relation between the camera hardware and the image coordinates. Exterior orientation parameters deliver precise knowledge on the camera location by the time of the image acquisition. Satellite positioning sensors and inertial measurement units (IMU) measure precisely horizontal and vertical positions as well as lateral and transverse accelerations.

Depending of the covered area of the imaging system or the number of images included in the process, large areas are covered through photogrammetric processing. However, not all satellite systems may be utilized. The Landsat satellites or Sentinel-2 have nadir looking imaging geometries, which prevent the essential panoramic effect in the acquisitions. On the other hand, a growing number of very high-resolution optical satellites have specially for the purpose of photogrammetric use installed cameras. To create stereo or tri-stereo image pairs, these satellites have cameras tilted in flight direction and against flight direction. Besides these recent systems, there is a growing number of historical imagery that is getting declassified by intelligence authorities. These images are very valuable sources to reconstruct past states of environments and can be used for photogrammetric processing.

Closely related to the photogrammetric processing is the approach of structure from motion (SfM, Westoby et al. 2012). The SfM approach is applied for imagery resulting from unstructured image acquisition plans. This means that an overlapping ratio between the images has to be ensured, but the images do not necessarily have to be acquired in defined positions. Besides this, the cameras do not have to be fixed frame cameras, the SfM processing works with simple digital compact cameras as well. On the one hand the quality of the elevation results depends on the quality of the images, but on the other hand the number of provided images have an effect on the result as well.

For smaller scale height data processing, unmanned aerial vehicles (UAV) are ideal camera carriers and therefore can be cost-effective alternative to satellite imagery. Multi-copter UAV systems are ideal for acquisition campaigns covering some hectares, whereas fixed wing systems allow higher flight velocities and therefore are suitable for larger areas.

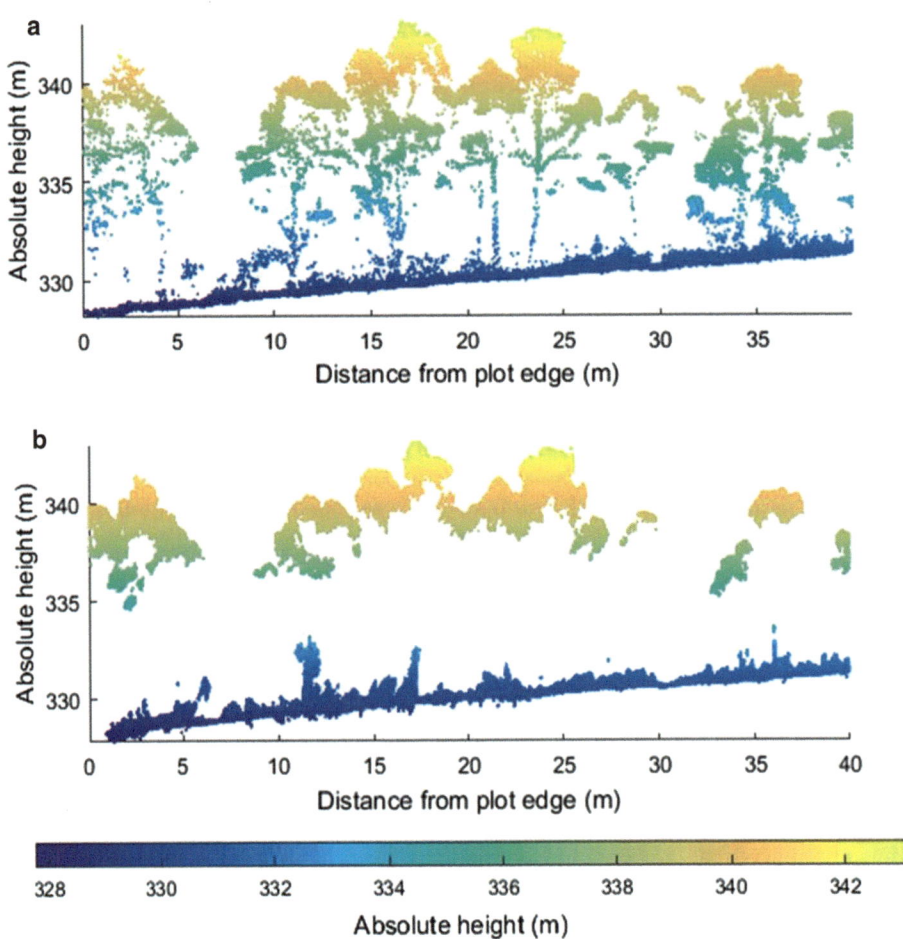

Fig. 8 *a* Side-view of a transect through a forest area generated from LiDAR data and *b* the same area produced from SfM processing of RGB photography. (Source: Wallace et al. 2016)

The barriers of SfM approaches are very low. Standard compact cameras can be used for image acquisition, but up to date UAVs often are equipped with high-quality cameras. Additionally, the SfM processing is implemented very user friendly in software, that is designed for users without any processing experience of spatial data, though the processing itself demands increased hardware requirements. The computational effort for the SfM processing is high, therefore standard components for computing, working memory, graphic processing are not sufficient.

Meanwhile, more and more software providers are entering the market for SfM processing software (see Fig. 8). Commercial software offer a very comfortable and user-friendly implementation of the processing in the software. Agisoft (Agisoft 2020) is

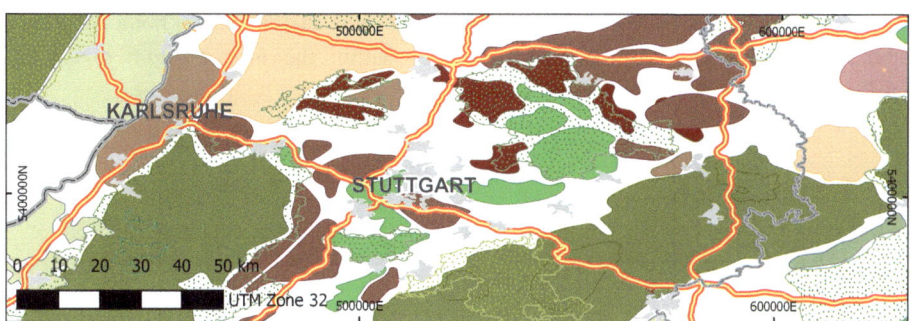

Fig. 9 Example of the Digital Landscape Model 1:1 Mio. (Source: GeoBasis-DE/BKG 2020, licensed under dl-de/by-2)

present on the market since the appearance of the SfM topic and Pix4D (Pix4D 2020) has a major position as well, just to name a few, but there is also a large community, that works on implementing the SfM approach in open source software. In these software developments, users usually have the chance to intervene in data processing to adjust parameters. Established open source software are VisualSFM (Wu 2015), Colmap (Schönberger and Frahm 2016) or MicMac (Rupnik et al. 2017).

2.3 Digital Landscape Models

"Digital Landscape Model" (DLM) is a term for a geospatial data infrastructure, which contains primary data of all relevant topographic objects of the landscape in a vector format (Fig. 9). A Base DLM is often provided by national cadaster authorities and displays the most detailed and accurate representation of surveyed geo data. According to Kohlstock (2018), a Base DLM meets the following properties:

- A common coordinate reference system, geodetic datum and elevation reference for all data.
- Coded information comprising object type, attributes and name follow an object catalog, thus enriching the spatial information.
- Maximal spatial accuracy of the surveyed data without any cartographic generalization.

A Base-DLM is usually available in scales of 1:10.000 or 1:25.000 (e.g. German *Basis-DLM* as part of *ATKIS/ALKIS*, Dutch *TOP10NL*) and has a high update frequency. Secondary DLMs of lower resolution (commonly 1:100.000, 1:250.000, 1:1 Mio.) are derived from the Base DLM through semantic model generalization, whereas a Digital Cartographic Model Is produced through further cartographic generalization (João 1998, Hardy et al. 2004, Stoter et al. 2010).

An ongoing major undertaking of the 2000s and 2010s is the homogenization, publication and visualization of national data within the *Infrastructure for Spatial InfoRmation in Europe* (INSPIRE).

---

3   Application

3.1   Animated Maps

Early geomorphological maps have often been produced for military and engineering purposes and are designed for the use in the field (Klimaszewski 1982). A morphological view of landscapes and the description of their physiography at the turn of the 20th century, landform analysis and morphological description became new purposes for geomorphological maps. Therefore, maps were more than just a medium for visualization but represented a research tool for providing a generalized inventory of landforms, surface structures, geomorphological processes, surface and subsurface materials, and genetic information. The applications of geomorphological maps range from simple descriptions of a field site, to land system analyses (Bennett et al. 2010), land surveys, land management or natural hazard assessment (Seijmonsbergen and de Graaff 2006).

Maps have continued as the main geomorphological output through to the 1980s as they afforded both 2D visualization and an effective data storage paradigm for spatial data. Because of a preoccupation with cartographic symbolization and a move to field scale experimentation, geomorphological mapping subsequently has been declined. Since the 1990s, widely available remotely sensed data and the improvement of information systems as example in ESRI 2003 have allowed the combination of field scale and regional approaches, causing improvements in mapping. In particular, the development of GIS (Geographic Information Systems) through the field of geomatics has added a digital tool through which disparate datasets can be stored, manipulated, analyzed and presented. This provides an impressive tool to combine and integrate diverse data such as GPS, satellite imagery, postcodes and historic sources.

Interactive maps have become a standard component on many websites (see in this volume: Kleber et al. 2020). They are currently considered a state-of-the-art method for the propagation of geomorphological data. Virtual globes provide the 3D visualization of satellite imagery and topographic data of the Earth and are used to locate any kind of information from photographs to scientific data. A wide range of GIS applications and tools have been developed to generate applications of web cartography, either for the visualization of research outcomes or to enhance the search for, or distribution of, scientific data (for other GIS-based projects in this volume see: Poplin et al. 2020; Stratmann et al. 2020). Additionally, animated maps have become increasingly popular in recent years because they congruently represent the passage of time with changing graphic displays (or maps). Animated maps, sometimes called movie maps or change maps, are primarily used to depict geographic change and processes, while static maps present all of their information simultaneously; animated maps present information over time (Peterson 1995).

However, this technology goes beyond just virtually exploring the surface of the Earth: expert users of animated maps may now explore 'attribute space' using the same kind of techniques, developed to explore 'geographic space', for example, when cycling through different themes of a geographic dataset, or by moving along a timeline that represents ordered data values of a certain variable of interest. With Google Earth or similar virtual-globe interfaces that can render rich, immersive three-dimensional maps in real-time, the dynamic change of one's point of view or geometric perspective on a three-dimensional landscape has become as easy as one mouse click.

One of the primary reasons for making animated maps is to show spatial processes. As Dorling and Openshaw (1992, p. 643) already note:

'It is self-evident that two-dimensional still images are a very good way (if not the only way) of showing two-dimensional still information. However, when the underlying patterns (and processes) start to change dynamically, these images rapidly begin to fail to show the changes taking place'.

In cartography, two basic animation types are known: temporal animation and non-temporal animation (Dransch 1997). Temporal animation deals with the depiction of dynamic events in chronological order and depicts actual passage of time in the world. In a temporal animation, 'world time' (e.g. days, centuries) is typically proportionally scaled to 'animation time' (e.g. typically seconds). Examples of temporal animations are population growth, diffusion processes of diseases, commodities and the like or wild fire spreads and glacier movements. Meanwhile, non-temporal animations use animation time to show attribute changes of a dynamic phenomenon. The morphing technique is a good example of a non-temporal animation. For instance, animation time is used to show a phenomenon's transformation from an orthographic two-dimensional map depiction (e.g. 'god's eye' view) into a perspective three-dimensional view. Other very popular non-temporal animation examples are fly-bys or fly-through of three-dimensional terrain, where the viewer's perspective changes over time (e.g. animation of camera motion).

3.2   Visualization of Geomorphological Processes

There might be a communication problem when geoscientists develop interpretive and interactive products explaining geomorphologic processes presented to non-specialists. However, some research on methods for geo-interpretation and the visualization of the geomorphological processes are not widespread (for example Cayla et al. 2010; Megerle 2008). Visualization tools can be used for teaching and learning at universities and high school. Compared to learning via a conventional instructional method, students learning with Google Earth or other learning animation models do not have different geomorphological development concepts because both settings enable students to learn with similar static representation. However, students learning with Google Earth or visualization tools improve topographic map skills significantly compared to the conventional instructional method. Many different websites give options as interactive tools for teaching geomorphological processes with different visualization tools such as animation and videos.

Therefore, Martin (1994) has published a media interactive tool for the interpretation of geomorphological features through the use of interactive visual media. It focuses on the use of interactive functions to go beyond the design or cartographic limits of classic visual media with the aid of interactivity, multimedia, and animation. Interpretive web applications were developed on selected geomorphological features from three sites in the Swiss Hautes Alpes Calcaires in the Helvetic domain. These sites are classified as landscapes of national significance and as geo sites of regional or national significance. They have been selected because they are well known as specific places by many tourists, which are connected with Google Earth. These web sites include nine interactive applications for the interpretation of geomorphological features. Each application is connected to the fundamentals of geomorphology: (1) causality and landform/process relations, (2) erosional/depositional processes and (3) evolution of landscapes (in this case, especially related to glaciers).

The modeling of channel erosion as Gully erosion has been achieved outside of a GIS because most of the required data, as Digital Elevation Model (DEMs), did not have a detailed enough resolution to establish the geometry of the gully. The Gully Erosion requires a high resolution DEM, daily discharge data, topographical data and soil parameters. Through these data, gully and channel erosion models can be applied using GIS-based approaches. There are several models to predict stable gully morphology (Mirtskhulava 1988), the conditions of ephemeral gully initiation (Poesen and Govards 1990), the rate of gully head growth and gully longitudinal profile transformation (Sidorchuk 1996). Sidorchuk (1999) explained the mechanism of the gully development dynamically and statically. The main Gully Erosion Model idea was specifically developed to estimate soil loss as volume and change areas by erosion process.

The actual erosion/deposition rates will be affected on the change in the elevation surface only marginally and the morphology of the gully system can be observed only after many events. Gully erosion, however, has an immediate effect on topography and the elevation change must be taken into account during the simulation. The model of dynamical erosion will illustrate the changes in topography through time which is called visualization of the erosion effect. To visualize the gully formation, it is assumed that erosion starts before water flow reaches steady state, the sediment transport is close to detachment-limited conditions, and all eroded material is transported outside the gully. A new set of topographic parameters are then derived from the updated DEM to reflect new flow gradients used by the water flow and soil erosion model in the next iteration (Fig. 10).

Fig. 10 Simulated Gully bottom evolution over 50 years – 2D-3D view. Source: own presentation

3.3 Landscape Evolution Models

Landscape Evolution Models (LEM) allow one to simulate, analyze and visualize the change of the earth's surface over timescales that exceed human observability. The landscape, as we see it today, is the result of natural (and to an increasing amount human) forces and processes. Di Pietro (2018) distinguishes between 'forcing agents' and five related 'mechanisms', that act upon the landscape. Thereafter, plate tectonics are a primary agent that builds a landscape through the mechanisms of uplift, subsidence and volcanism. For example, mountain ranges like the Himalaya are folded and their elevation and relief is increased. The competing primary agent is climate, which in turn triggers the complementary pair of erosion and deposition. These mechanisms equalize the landscape by loosening material through physical and chemical weathering in exposed areas and subsequent accumulation thereof in low-energy areas. In the high lying Himalayas, glaciers and melt water streams dislocate enormous amounts of rock, transport it through rivers like the Ganges into the low lying foreland and deposits it in the Ganges-Brahmaputra-Delta. However, in most cases these agents and mechanisms do not act in a constant manner. They rather emerge in pulses, phases (e.g. tectonics) and cycles (e.g. climate), so that mechanisms superimpose each other and dominate different stages of the lifecycle of a landscape. The mindful observer may recognize some relics of this changeful past in landforms, such as planation surfaces, river terraces, meander migration, etc.

In order to reflect the coaction of all these processes, LEM incorporate different processes in terms of numerical equations, which together form a numerical model. A digital landscape in these models is either represented by a continuous raster surface with equally spaced pixels, or as a triangular irregular network (TIN), where surface points are connected through a mesh. Independent from the type of surface representation, the data contain at least elevation information for each pixel or node, but according to the complexity of the LEM, additional information is provided, e.g. slope, soil properties, bedrock type etc. Upon this initial surface, several sub-processes calculate iteratively the surface change of a given time step and modify the initial landscape accordingly. The type and implementation of included sub-processes differs from model to model, however the majority comprises these elements: (1) a tectonic process, (2) a sediment flux process incorporating erosion, transport and deposition and (3) a hillslope process (Temme et al. 2017). These sub-processes may require additional information on varying climate, uplift rates, vegetation changes, etc.

Therefore, LEMs can be considered as laboratories, where scientists simulate and reproduce landscape development under varying border conditions. This allows on the one hand the testing of geomorphological hypotheses and the evaluation of potential mechanisms in idealized landscapes under controllable conditions. On the other hand, these models help to reconstruct actual landscapes and disentangle the multiple factors that were involved in their creation.

3.4   Mixed Reality

Mixed reality is an umbrella term for a variety of applications, that range between the visualization of the real world and a completely virtual environment (Milgram and Colquhoun 1999). Augmented Reality (AR) is a combination of the view of the physical world with a virtual component, which enriches the perspective with digital objects or information (see in this volume Stintzing et al. 2020). This is implemented through a virtual layer that projects texts, videos, objects etc. onto the real-world layer. A true Virtual Reality (VR) in contrast is a completely virtual environment, where all objects are computer generated (for other studies addressing virtual reality in this volume see: Edler et al. 2020; Fontaine 2020; Kühne and Jenal 2020; Prisille and Ellerbrake 2020; Siepmann et al. 2020; Vetter 2020). Augmented Virtuality (AV) is positioned between those technologies and enriches a virtual environment with real-world information. However, a distinction between the different grades of reality-virtuality is hardly feasible and combinations are legion (Carmigniani et al. 2011).

While earlier definitions of these techniques included specific hardware components (Azuma et al. 2001), recent developments in display technologies widened this field and rendered constraint unnecessary (Carmigniani et al. 2011; Kipper and Rampolla 2012). The hardware comprises head-mounted displays (HMD), binoculars, handheld smartphones or tablet computers, projectors, see-through devices or even 3D holograms. Interaction with virtual objects can be executed through touch displays, hardware controllers, gloves and sensors that detect hand gestures (Mistry et al. 2009) and artificial skins, that are able to provide haptic feedback (Yu et al. 2019). The AR Sandbox developed at UC Davis (https://arsandbox.ucdavis.edu) allows to manipulate a physical medium (here sand), while the AR reacts by displaying mountain reliefs or water bodies. Such a projector-based AR allows multiple users to experience the same visualization, while other technologies, like headsets, create a distinct AR for each individual user.

Independently from the type of display, the overlay generated by an AR onto the real world relies on tracking technology, which matches the digital information with the view of the physical world (Papagiannakis et al. 2008). For example, this could be a combination of the device's GPS position, pointing direction and predefined markers, which cause the AR device to render the names and elevations of mountain peaks onto the view field. Using geo-located AR, the physical landscape becomes an augmented place, a concept proposed by Oleksy and Wnuk (2016), which enables to merge the experience of the real world with new levels of interaction and depth of information. The application of AR was found to intensify the experience of a place by enhancing elements such as attention, involvement, engagement and immersion (Ermi and Mäyrä 2005; Scott and Le 2017; Han et al. 2019).

AR has been widely adapted in a conservational and educational context of cultural heritage sites and tourism, where 3D models of ancient buildings or historical images can be draped over the modern landscape (Seo et al. 2010; Han et al. 2013, 2019; Chang et al. 2015) or parts of non-existing buildings could be reconstructed virtually, giving the observer an imaginary view of historical sites. Such an AR enriched experience showed to provide a deeper understanding and emotional involvement for the users, but also improved visitor satisfaction and a positive image for providers (Oleksy and Wnuk, 2016; Moorhouse et al. 2019).

3.5   Landscapes and Artificial Intelligence

Artificial Intelligence (AI) is well known as the working horse of the Age of Information. It's no wonder, that these techniques found their way into the generation of synthetic landscapes and the human-like interpretation of landscapes. According to current research, Landscape is a phenomenon, which emerges in the mind of a recipient. It is more than the bare combination of landscape elements, but rather an amalgamation of their physical existence with added information and emotions interlaced by the observing individual. Within the framework of Machine Learning, a subset of AI, the Artificial Neural Network (ANN) is a technique that is inspired by biological neural circuits and its functionality comes closest to the way, the human brain works. A learning algorithm is trained to fulfil a classification task, like the recognition of landscape elements from a photography, by providing it with sample images and preliminary assigned classes. Such a supervised learner would take an input image and compare it to the desired output pair, which, for example, says the landscape is composed of a city and a surrounding forest. Informed with this information, a set of rules is created that would probably classify the forest based on its color or the texture of the leaves or any other feature. In fact, it is quite difficult to understand, which features exactly are regarded as important by the model that is why neural networks are often described as a black-box function. However, this procedure is performed with millions of input-output-pairs and later applied to additional test-pairs, in order to evaluate and improve the rule set. A typical application of such a classifier is the categorization of large image datasets in social networks, which allow users to find their holiday images from the seaside or the mountain hike in the vast haystack of self-portrayals.

PlaNet is an approach developed by Google (Weyand et al. 2016), to not only recognize the landscape depicted in a photography, but rather to estimate the most likely geolocation of the image. Based on a subtype of ANNs, a convolutional neural network (CNN), it is able to detect landmarks and other visual clues in images and create a heat map of positions, where other images with similar information were taken. Other methods in the field of computer vision try to detect the geolocation by matching unfocalized images with GPS-tagged counterparts from a database (Hays and Efros 2008, 2015; Quack et al. 2008; Avrithis et al. 2010) or by comparing "urban canyons" with 3D city models (Ramalingam et al. 2010). These techniques work best in areas, where distinct landmarks are easily recognizable, like Paris with the Eiffel Tower, and validations show, that they are still not ready for operational use. However, tests showed that the models already outperform humans given the same task.

The project GauGAN, developed by Nvidia, aims at the visualization of artificial landscapes, generated from human sketches (Park et al. 2019). A simple drawing, for instance a tree behind a lake under a sunny sky, is turned into a photorealistic image based on a generative adversarial network (GAN), which learned from millions of sample photographs, how these objects look like in nature. So far, the algorithm is even capable to create reflections in the water or render the scene in different seasons of the year.

Some of these AI applications may seem a bit playful at their current stage of development and are of course far from the complexity of human perception of landscapes, especially when it comes to personal experiences or emotions related to landscapes. Yet the results are impressive and show future trends, how to bring real landscapes into the digital world, and furthermore visualize digital landscapes.

3.6   Landscape Architecture and Planning

Landscape architecture is a profession and a field of research, addressing the design, planning and management of the landscape (Nijhuis 2015). The landscape is regarded under different aspects, such as ecology, social and behavioral aspects as well as aesthetics (Jellicoe 1987). This multilayer approach towards landscape offers many opportunities to facilitate GIS, which is distinguished to model and process multi-dimensional data of various disciplines in a structured way. Although several studies have demonstrated, that the potential of GIS in the field is still not tapped in the whole range (Drummond and French 2008; Göçmen and Ventura 2010; Nijhuis 2016), best-practice examples do exist.

The analysis of viewsheds within a landscape plays an important role in planning. The use of Digital Surface Models (DSM) allows to map the area, which can be seen from a specific point in the landscape, or vice versa, the area from which a certain place is visible. This is applied to help planners detect aesthetically appealing views, or even construct such (Domingo-Santos et al. 2011) e.g. to illustrate the "visual pollution" of wind generators. Chamberlain and Meitner (2013), for example, estimated the visual relevance of a tourist route through a UNESCO Reserve in British Columbia. In another application, the method has been used to support the development of a hotel complex with a scenic view on the Niagara Falls and it allowed the simulation of the negative impact of a high-rising hotel tower on the touristic environment. Accordingly, this method is also used to quantify and mitigate the effect of man-made structures. Palmer (2019) describes a Visual Impact Assessment (VIA), which helped decision makers to evaluate two proposed alternatives of a high-voltage transmission line. De Vries et al. (2012) demonstrate to which extent planning of vegetation can ease the disturbing view on industrial parks, barns or wind turbines. Especially the latter are an object of dispute (in Central Europe) and require particular consideration in planning (Bishop and Miller 2007; Torres Sibille et al. 2009; Möller 2010).

The example of wind parks shows, that the public reacts increasingly sensitive to landscape changes and it is therefore crucial for the acceptance of an undertaking, to integrate citizens early through participation. Beside classic approaches to visualize planned infrastructure, like rendered images or edited photographs, Augmented Reality (AR) applications can inform residents by overlaying the artificial objects, such as wind turbines, over the actual landscape (Wagner et al. 2009; Bishop 2015; Kerr and Lawson 2019). Through interactive web applications, planners can propose different scenarios and inquire public preferences (Smith et al. 2012).

Ideally, local stakeholders are engaged through collaborative planning and participation into the design process as early as possible. Digital collaborative environments, like SIEVE, visualize GIS data as 3D virtual reality and enable participants to make changes, implement different ideas and get feedback from other users (Bishop and Stock 2010). Recent projects combine immersive virtual reality techniques and locomotion tools established in video and computer gaming with planning scenarios (Jamei et al. 2017; Edler et al. 2019; Ma et al. 2020).

3.7 Comprehensive Overview

Finally, we would like to give a quick summary overview of the variety of visualization possibilities mentioned in this chapter (Fig. 11). Milgram's reality-virtuality continuum, which has already been explained in Sect. 3.4, shall be used as a basis for this. It comprises the domains of physical reality, mixed reality and full virtual reality, which do not have to be separated but can also merge into each other. In addition, we have added another dimension to the model that is very relevant for landscape visualization, namely user involvement. On the one hand, this includes a rather static area where the user can passively consume a pre-produced visualization. This is the case, for example, for tangible scale models and printed maps (Sect. 1.2), edited images and prerendered computer graphics (Sect. 3.6). They were all created by a designer according to his ideas, but do not allow direct feedback from the user. With 360° videos, web maps (Sect. 3.1)

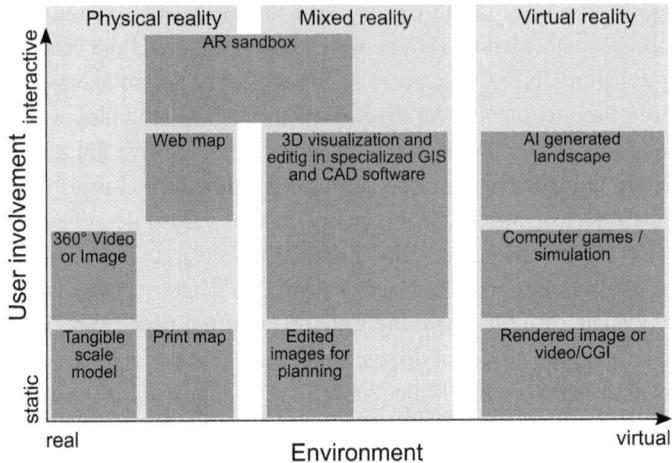

Fig. 11 An approach to structure the manifold of landscape visualization techniques using Milgram's reality-virtuality continuum and the additional dimension of user involvement. Source: own presentation

or computer simulations, the user can intervene in what is happening to varying degrees, for example by simply changing his perspective, dynamically inserting layers to generate new knowledge from superimposed information, or making direct changes to the virtual reality. An example of the latter is the AI generated landscape, in which the user can create his own landscapes with the support of artificial intelligence (Sect. 3.5). GI and CAD systems play a key role in merging the view of the physical landscape with auxiliary information, such as the potential threats through erosion (Sect. 3.2) and the development of the landscape (Sect. 3.3). Furthermore, user interaction with GIS allows to compute potential outcomes of different user-driven scenarios and thus foster immersion. Augmented Reality applications, such as the educational AR sandbox (Sect. 3.4) bridge the gap between a physical landscape and its virtual representation.

GIS as a design and visualization tool is just one application among others in this continuum. However, none of the others would be imaginable without the underlying geoinformation provided through modern image processing techniques (Sect. 2.2) as well as the ability to store large amounts of location information in accurate geodatabases with sophisticated scalable data models (Sect. 2.3).

References

Agisoft. (2020). Agisoft Metashape. https://www.agisoft.com/. Accessed: 15.05.2020.

Albrecht, F., Moser, J., & Hijazi, I. (2013). Assessing façade visibility in 3D city models for city marketing. In U. Isikdag (Ed.), *Proceedings of the ISPRS 8th 3D GeoInfo Conference & WG II/2 Workshop* (pp. 1 – 5). Istanbul.

Amoroso, N. (2019). *Representing landscapes analogue*. London.

Avrithis, Y., Kalantidis, Y., Tolias, G., & Spyrou, E. (2010). Retrieving landmark and non-landmark images from community photo collections. *Proceedings of the 18th ACM international conference on Multimedia*. Firenze.

Azuma, R., Baillot, Y., Behringer, R., Feiner, S., Julier, S., & MacIntyre, B. (2001). Recent advances in augmented reality. *IEEE Computer Graphics and Applications, 21*(6), 34–47. https://doi.org/10.1109/38.963459.

Bagrow, L. (2017). *History of cartography*. New York.

Bélanger, P. (2013a). The new geographic landscape. *Landscape Architecture Frontiers, 1*(1), 42–55.

Bélanger, P. (2013b). *Landscape infrastructure: Urbanism beyond engineering*. Wageningen University.

Bélanger, P. (2017). *Landscape as infrastructure*. New York.

Bennett, G. L., Evans, D. J. A., Carbonneau, P., & Twigg, D. R. (2010). Evolution of a debris-charged glacier landsystem, Kviarjokull, Iceland. *Journal of Maps*, 40–67.

Berry, J., Buckley, D., & Ulbricht, C. (1998). Visualize realistic landscapes. 3D modeling helps users envision natural resources. *GIS World, 11*(8), 42–27.

Bishop, I. D. (2015). Location based information to support understanding of landscape futures. *Landscape and Urban Planning, 142,* 120–131. https://doi.org/10.1016/j.landurbplan.2014.06.001.

Bishop, I. D., & Miller, D. R. (2007). Visual assessment of off-shore wind turbines: The influence of distance, contrast, movement and social variables. *Renewable Energy, 32*(5), 814–831. https://doi.org/10.1016/j.renene.2006.03.009.

Bishop, I. D., & Stock, C. (2010). Using collaborative virtual environments to plan wind energy installations. *Renewable Energy, 35*(10), 2348–2355.

Blair, J. B., David, Rabine, D. L., & Hofton, M. A. (1999). The laser vegetation imaging sensor: A medium-altitude, digitisation-only, airborne laser altimeter for mapping vegetation and topography. *ISPRS Journal of Photogrammetry and Remote Sensing, 54*(2–3), 115–122.

Blaschke, T., Tiede, D., & Heurich, M. (2004). 3D landscape metrics to modelling forest structure and diversity based on laser scanning data. *International Archives of the Photogrammetry, Remote Sensing and Spatial Information Sciences, 36*(8/W2), 129–132.

Brown, L. A. (1950). *The story of maps*. Boston.

Buchroithner, M. F., & Knust, C. (2013). True-3D in cartography. Current hard- and softcopy developments. In: A. Moore & I. Drecki (Eds.), *Geospatial visualization* (pp. 41–65). Berlin.

Bufton, J. L. (1989). Laser altimetry measurements from aircraft and spacecraft. *Proceedings of the IEEE, 77*(3), 463–477.

Buisseret, D. (1998). Modeling cities in early modern Europe. In D. Buisseret (Ed.), *Envisioning the city: Six studies in urban cartography* (124–143). Chicago.

Cantrell, B., & Michaels, W. (2010). *Digital drawing for landscape architecture: Contemporary techniques and tools for digital representation in site design*. London.

Carmigniani, J., Furht, B., Anisetti, M., Ceravolo, P., Damiani, E., & Ivkovic, M. (2011). Augmented reality technologies, systems and applications. *Multimedia Tools and Applications, 51*(1), 341–377. https://doi.org/10.1007/s11042-010-0660-6.

Cayla, N., Hobléa, F., & Gasquet, D. (2010). Guide des bonnes pratiques de médiation des géosciences sur le terrain. *Géologie de la France, 1,* 47–55.

Chamberlain, B. C., & Meitner, M. J. (2013). A route-based visibility analysis for landscape management. *Landscape and Urban Planning, 111,* 13–24. https://doi.org/10.1016/j.landurbplan.2012.12.004.

Chang, Y. L., Hou, H. T., Pan, C. Y., Sung, Y. T., & Chang, K. E. (2015). Apply an augmented real-ity in a mobile guidance to increase sense of place for heritage places. *Journal of Educational Technology & Society, 18*(2), 166–178.

Cloude, S. R., & Papathanassiou, K. P. (1998). Polarimetric SAR interferometry. *IEEE Transactions on Geoscience and Remote Sensing, 36*(5), 1551–1565. https://doi.org/10.1109/36.718859.

Cosgrove, D. E. (1984). *Social formation and symbolic landscape*. London: University of Wisconsin Press.

Coucelo, C., Duarte, P., & Crespo, R. (2013). gison3dmap – Efficient geographic communication with GIS data projection on solid terrain models. In H. Kremers (Ed.), *Proceedings CEGeoIC 2013, International Conference on Environmental Information and Communication*, Bogotá, Feb. 6–8.

Dalla Corte, A. P., Rex, F. E., Almeida, D. R. A., Sanquetta, C. R., Silva, C. A., Moura, M. M., et al. (2020). Measuring individual tree diameter and height using GatorEye High-Density UAV-Lidar in an integrated crop-livestock-forest system. *Remote Sensing, 12,* 863.

de Santarem, M. F. (1852). *Essai sur l'histoire de la cosmographie et de la cartographie pendant le moyen-age, et sur les progrès de la géographie après les grandes découvertes du XVe siècle: pour servir d'introduction et d'explication à l'atlas composé de mappemondes et de portulans, et d'autres monuments géographiques, depuis le VIe siècle de notre ère jusquau XVIIe* (Vol. 3). Paris.

de Vries, S., de Groot, M., & Boers, J. (2012). Eyesores in sight: Quantifying the impact of man-made elements on the scenic beauty of Dutch landscapes. *Landscape and Urban Planning, 105*(1), 118–127. https://doi.org/10.1016/j.landurbplan.2011.12.005.

Di Pietro, J. A. (2018). Geology and Landscape Evolution. General Principles Applied to the United States.

Delikostidis, I., Engel, J., Retsios, B., Van Elzakker, C. P., Kraak, M. J., & Döllner, J. (2013). Increasing the usability of pedestrian navigation interfaces by means of landmark visibility analysis. *The Journal of Navigation, 66*(4), 523–537.

Domingo-Santos, J. M., de Villarán, R. F., Rapp-Arrarás, Í., & de Provens, E. C.-P. (2011). The visual exposure in forest and rural landscapes: An algorithm and a GIS tool. *Landscape and Urban Planning, 101*(1), 52–58. https://doi.org/10.1016/j.landurbplan.2010.11.018.

Dorling, D., & Openshaw, S. (1992). Using computer animation to visualize space-time patterns. *Environment and Planning B: Planning and Design, 19,* 639–650.

Dransch, D. (1997). *Computer-Animation in der Kartographie: Theorie und Praxis*. Heidelberg.

Dufour, G. H., & Flamsteed, J. (1833). *Topographische Karte der Schweiz*. Service Topographique Fédéral.

Dupain-Triel, J. L. (1791). *La France considérée dans les différentes hauteurs de ses plaines: ouvrage spécialement destiné al'instruction de la jeunesse*.

Drummond, W. J., & French, S. P. (2008). The future of GIS in planning: Converging technologies and diverging interests. *Journal of the American Planning Association, 74*(2), 161–174. https://doi.org/10.1080/01944360801982146.

Edler, D., Keil, J., & Dickmann, F. (2020). From Na Pali to Earth—An 'Unreal' Engine for Modern Geodata? In D. Edler, C. Jenal, & O. Kühne (Eds.), *Modern Approaches to the Visualization of Landscapes* (pp. 279–291). Wiesbaden: Springer VS.

Edler, D., Keil, J., Wiedenlübbert, T., Sossna, M., Kühne, O., & Dickmann, F. (2019). Immersive VR experience of redeveloped post-industrial sites: The example of "Zeche Holland" in Bochum-Wattenscheid. *KN - Journal of Cartography and Geographic Information, 69*(4), 267–284. https://doi.org/10.1007/s42489-019-00030-2.

Ellis, P. (2018). The Panstereorama: City models in the balloon era. *Imago Mundi, 70*(1), 79–93.

Ermi, L., & Mäyrä, F. (2005). Fundamental Components of the Gameplay Experience: Analysing Immersion. *DiGRA - International Conference: Changing Views: Worlds in Play.*

Farr, T. G., Rosen, P. A., Caro, E., Crippen, R., Duren, R., Hensley, S., Kobrick, M., Paller, M., Rodriguez, E., Roth, L., Seal, D., Shaffer, S., Shimada, J., Umland, J., Werner, M., Oskin, M., Burbank, D., & Alsdorfet, D. (2007). The shuttle radar topography mission. *Reviews of geophysics, 45*(2).

Faulkner, L. (2006). Physical terrain modeling in a digital age. *Simulation series, 38*(1), 373.

Fontaine, D. (2020). Virtuality and landscape. In D. Edler, C. Jenal, & O. Kühne (Eds.), *Modern approaches to the visualization of landscapes* (pp. 267–278). Wiesbaden: Springer VS.

Frachetti, M., & Chippindale, C. (2001). Alpine imagery, alpine space, alpine time, and prehistoric human experience. In G. Nash & C. Chippindale (Eds.), *European landscapes of rock-art* (116–143). London.

Frampton, K. (1999). Megaform as urban landscape. University of Michigan, A. Alfred Taubman College of Architecture + Urban Planning.

Geier, B., Egger, K., & Muhar, A. (2001). Integrierte 3D-Visualisierungs-Systeme für die Landschaftsplanung: Konzepte und Marktrealität. In M. Schrenk (Ed.), *CORP Geo-Multimedia'01* (231–236). Vienna.

Gens, R., & Van Genderen, J. L. (1996). SAR interferometry—issues, techniques, applications. *International Journal of Remote Sensing, 17*(10), 1803–1835.

Göçmen, Z. A., & Ventura, S. J. (2010). Barriers to GIS use in planning. *Journal of the American Planning Association, 76*(2), 172–183. https://doi.org/10.1080/01944360903585060.

Greider, T., & Garkovich, L. (1994). Landscapes: The social construction of nature and the environment. *Rural Sociology, 59* (1), 1–24. https://doi.org/10.1111/j.1549-0831.1994.tb00519.x.

Gual, J., Puyuelo, M., Lloverás, J., & Merino, L. (2012). Visual Impairment and urban orientation. Pilot study with tactile maps produced through 3D Printing. *Psyecology, 3*(2), 239–250.

Han, D. I., Jung, T., & Gibson, A. (2013). Dublin AR: Implementing augmented reality in tourism. *Information and Communication Technologies in Tourism.* Cham.

Han, D. I., Weber, J., Bastiaansen, M., Mitas, O., & Lub, X. (2019). Virtual and augmented reality technologies to enhance the visitor experience in cultural tourism. In M. C. tom Dieck & T. Jung (Eds.), *Augmented reality and virtual reality: The power of AR and VR for business* (113–128). Cham: Springer International Publishing.

Hardy, P., Briat, M.-O., Eicher, C., & Kressmann, T. (2004). Database-driven cartography from a digital landscape model, with multiple representations and human overrides. *ICA Workshop on 'Generalisation and Multiple Representation'.* Leicester.

Harmon, R. S. (2006). *Real-Time Landscape Model Interaction Using a Tangible Geospatial Modeling Environment.*

Hartmann, R. (2019). 12 Virtualities in the new tourism landscape. The case of the Anne Frank house virtual tour and of the visualizations of the Berlin Wall in the Cold War context. Tourism Fictions, Simulacra and Virtualities.

Hays, J., & Efros, A. A. (2008). IM2GPS: Estimating geographic information from a single image. *IEEE Conference on Computer Vision and Pattern Recognition.*

Hays, J., & Efros, A. A. (2015). Large-Scale Image Geolocalization. In J. Choi & G. Friedlad (Eds.), *Multimodal location estimation of videos and images.*

Horowitz, W. (1988). The Babylonian map of the world. *Iraq, 50,* 147–165.

Howard, P., Thompson, I., Waterton, E., & Atha, M. (Eds.). (2019). *The Routledge companion to landscape studies* (2nd ed.). London: Routledge.

Imhof, E. (1950). *Gelände und Karte.* Erlenbach-Zürich.

Imhof, E. (1965). *Kartographische Geländedarstellung.* Berlin.

Jamei, E., Mortimer, M., Seyedmahmoudian, M., Horan, B., & Stojcevski, A. (2017). Investigating the role of virtual reality in planning for sustainable smart cities. *Sustainability, 9*(11), 2006. https://doi.org/10.3390/su9112006.

Jellicoe, S. (1987). *The landscape of man shaping the environment from prehistory to the present day*. London.

João, E. M. (1998). *Causes and Consequences of Map Generalisation*. London.

Kerr, J., & Lawson, G. (2019). Augmented reality in design education: Landscape architecture studies as AR experience. *International Journal of Art & Design Education*. https://doi.org/10.1111/jade.12227.

Kleber, A., Edler, D., & Dickmann, F. (2020). Cartography and the sea: A JavaScript-based web mapping application for managing maritime shipping. In D. Edler, C. Jenal, & O. Kühne (Eds.), *Modern approaches to the visualization of landscapes* (pp. 173–186). Wiesbaden: Springer VS.

Klimaszewski, M. (1982). Detailed geomorphological maps. *ITC Journal, 3*, 265–271.

Kohlstock, P. (2018). *Kartographie*. Stuttgart.

Kipper, G., & Rampolla, J. (2012). *Augmented reality: An emerging technologies Guide to AR*.

Kühne, O. (2019). *Landscape theories. A brief introduction*. Wiesbaden: Springer VS.

Kühne, O., & Jenal, C. (2020). The threefold landscape dynamics—Basic considerations, conflicts and potentials of virtual landscape research. In D. Edler, C. Jenal, & O. Kühne (Eds.), *Modern approaches to the visualization of landscapes* (p. 389–402). Wiesbaden: Springer VS.

Lange, E. (1998). *Realität und computergestützte visuelle Simulation: eine empirische Untersuchung über den Realitätsgrad virtueller Landschaften am Beispiel des Talraums Brunnen/Schwyz*. Doctoral dissertation, ETH Zurich.

Lange, E., & Bishop, I. (2001). Our Visual Landscape: Analysis, Modelling, Visualization and Protection. *Landscape and Urban Planning, 54*, 1–4.

Lange, E. (2002). Visualization in landscape Architecture and Planning – Where we have been, where we are now and where we might go from here. Trends in GIS and virtualization in environmental planning and design. *Proceedings at Anhalt University of Applied Sciences*, 8–18.

Lehmann, J. G. (1799). *Darstellung einer neuen Theorie der Bezeichnung der schiefen Flächen im Grundriß oder Situationszeichnung der Berge*. Leipzig.

Lim, K., Treitz, P., Wulder, M., St-Onge, B., & Flood, M. (2003). LiDAR remote sensing of forest structure. *Progress in Physical Geography, 27*(1), 88–106.

Lin, Y., Hyyppa, J., & Jaakkola, A. (2010). Mini-UAV-borne LIDAR for fine-scale mapping. *IEEE Geoscience and Remote Sensing Letters, 8*(3), 426–430.

Ma, Y., Wright, J., Gopal, S., & Phillips, N. (2020). Seeing the invisible: From imagined to virtual urban landscapes. *Cities, 98*, 102559. https://doi.org/10.1016/j.cities.2019.102559.

Macher, H., Grussenmeyer, P., Landes, T., Halin, G., Chevrier, C., & Huyghe, O. (2017). Photogrammetric recording and reconstruction of town scale models – The case of the plan-relief of Strasbourg. *The International Archives of the Photogrammetry, Remote Sensing and Spatial Information Sciences*, XLII-2/W5: 489–495.

Martin, S. (1994). Interactive visual media for geomorphological heritage interpretation. *Theoretical Approach and Examples. Geoheritage, 6*, 149–157. https://doi.org/10.1007/s12371-014-0107-y.

Martínez-Graña, A., & Valdés Rodríguez, V. (2016). Remote sensing and GIS applied to the landscape for the environmental restoration of urbanizations by means of 3D virtual reconstruction and visualization (Salamanca, Spain). *ISPRS International Journal of Geo-Information, 5*(1), 2.

McGranaghan, M. (1993). A cartographic view of spatial data quality. *Cartographica: The International Journal for Geographic Information and Geovisualization, 30*(2–3), 8–19.

Megerle, H. (2008). *Geotourismus.: Innovative Ansätze zur touristischen Inwertsetzung und nachhaltigen Regionalentwicklung*. Nürnberg.

Meyboom, A. (2009). Infrastructure as practice. *Journal of Architectural Education, 62*(4), 72–81.

Meyer-Heß, F. (2020). Discovering forgotten landscapes. In D. Edler, C. Jenal, & O. Kühne (Eds.), *Modern approaches to the visualization of landscapes* (pp. 33–46). Wiesbaden: Springer VS.

Mikhail, E. M., Bethel, J. S., & Chris McGlone, J. S. (2001). *Introduction to modern photogrammetry.* New York.

Milgram, P., & Colquhoun, H. J. (1999). A taxonomy of real and virtual world display integration. In: Y. Ohta & H. Tamura (Eds.), *Mixed reality. Merging real and virtual worlds* (pp. 1–26). Berlin.

Ming, Y., Jiang, J., & Bian, F. (2002). 3D-City Model supporting for CCTV monitoring system. *International Archives of Photogrammetry Remote Sensing and Spatial Information Sciences, 34*(4), 456–459.

Mirtskhulava, T. Y. (1988). *Osnovy Fiziki i Mekhaniki Erozii Rusel (Principles of physics and mechanics of channel erosion).* Leningrad: Gidrometeoizdat. (in Russian).

Mistry, P., Maes, P., & Chang, L. (2009). WUW – wear Ur world: a wearable gestural interface. *CHI '09 extended abstracts on human factors in computing systems.* Boston.

Mitasova, H., Mitas, L., Ratti, C., Ishii, H., Alonso, J., & Harmon, R. S. (2006). Real-time landscape model interaction using a tangible geospatial modeling environment. *IEEE Computer Graphics and Applications, 26*(4), 55–63.

Möller, B. (2010). Spatial analyses of emerging and fading wind energy landscapes in Denmark. *Land Use Policy, 27*(2), 233–241. https://doi.org/10.1016/j.landusepol.2009.06.001.

Moorhouse, N., Jung, T., & tom Dieck, M. C. (2019). Tourism marketers perspectives on enriching visitors city experience with augmented reality: An exploratory study. In M.C. tom Dieck & T. Jung (Eds.), *Augmented reality and virtual reality: The power of AR and VR for business* (pp. 129–144). Cham.

NASA. (2020). Space Images | Radar Image, Wrapped Color as Height, Lanai and West Maui, Hawaii. https://www.jpl.nasa.gov/spaceimages/details.php?id=PIA02723. Accessed: 18.05.2020.

Nijhuis, S. (2015). *GIS-based landscape design research.* Ph.D.: Delft University of Technology, Delft.

Nijhuis, S. (2016). Applications of GIS in landscape design research. *Research in Urbanism Series, 44,* 43–56. https://doi.org/10.7480/rius.4.1367.

Oleksy, T., & Wnuk, A. (2016). Augmented places: An impact of embodied historical experience on attitudes towards places. *Computers in Human Behavior, 57,* 11–16.

Palmer, J. F. (2019). The contribution of a GIS-based landscape assessment model to a scientifically rigorous approach to visual impact assessment. *Landscape and Urban Planning, 189,* 80–90. https://doi.org/10.1016/j.landurbplan.2019.03.005.

Park, T., Liu, M.-Y., Wang, T.-C., & Zhu, J.-Y. (2019). GauGAN: semantic image synthesis with spatially adaptive normalization. *ACM SIGGRAPH 2019 Real-Time Live!.* Los Angeles.

Papagiannakis, G., Singh, G., & Magnenat-Thalmann, N. (2008). A survey of mobile and wireless technologies for augmented reality systems. *Computer Animation and Virtual Worlds, 19*(1), 3–22. https://doi.org/10.1002/cav.221.

Perez, D., Kämpf, J. H., & Scartezzini, J. L. (2013). Urban area energy flow microsimulation for planning support: A calibration and verification study. *International Journal on Advances in Systems and Measurements, 6*(3–4), 260–271.

Peterson, M. P. (1995). *Interactive and animated cartography.* NJ and Englewood Cliffs: Prentice Hall.

Pix4D (2020). Professional photogrammetry and drone mapping software | Pix4D. https://www.pix4d.com/. Accessed: 15.05.2020.

Poesen, J., & Govers, G. (1990). Gully erosion in the loam belt of Belgium: typology and control measures. In J. Boardmann, I. D. L. Foster, & J. A. Dearing (Eds.), *Soil erosion on agriculture land* (pp. 513–530). UK: Chichster.

Poplin, A., de Andrade, B., & Mahmud, S. (2020). Exploring tangible and intangible landscapes of evocative places: Case study of the city of Vitória in Brazil. In D. Edler, C. Jenal, & O. Kühne (Eds.), *Modern approaches to the visualization of landscapes* (pp. 519–547). Wiesbaden: Springer VS.

Prisille, C., & Ellerbrake, M. (2020). Virtual Reality (VR) and Geography Education: Potentials of 360° 'Experiences' in Secondary Schools. In D. Edler, C. Jenal, & O. Kühne (Eds.), *Modern approaches to the visualization of landscapes* (pp. 321–332). Wiesbaden: Springer VS.

Quack, T., Leibe, B. & Gool, L. V. (2008). World-scale mining of objects and events from community photo collections. *Proceedings of the 2008 international conference on Content-based image and video retrieval*. Niagara Falls.

Ramalingam, S., Bouaziz, S., Sturm, P., & Brand, M. (2010). SKYLINE2GPS: Localization in urban canyons using omni-skylines. *2010 IEEE/RSJ International Conference on Intelligent Robots and Systems*.

Rothrock, G. A. (1969). The Musee des plans-reliefs. *French Historical Studies, 6*(2), 253–256.

Rupnik, E., Daakir, M., & Pierrot Deseilligny, M. (2017). MicMac – A free, open-source solution for photogrammetry. *Open Geospatial Data, Software and Standards, 2*, 1–9.

Sauer, C. (1925). The Morphology of Landscape.

Scott, N. & Le, D. (2017). Tourism experience: A review. In *CABI*, 30–49. Wallingford.

Schönberger, J.F., & Frahm, J.-M. (2016). Structure-from-Motion Revisited. *2016 Conference on Computer Vision and Pattern Recognition (CVPR)*.

Seijmonsbergen, A. C., & de Graaff, L. W. S. (2006). Geomorphological mapping and geophysical profiling for the evaluation of natural hazards in an alpine catchment. *Natural Hazards and Earth System Science, 6*, 185–193.

Seo, B.-K., Kim, K., Park, J., & Park, J.-I. (2010). A tracking framework for augmented reality tours on cultural heritage sites. *Proceedings of the 9th ACM SIGGRAPH Conference on Virtual-Reality Continuum and its Applications in Industry*. Seoul, South Korea.

Senda-Cook, S. (2013). Materializing tensions: How maps and trails mediate nature. *Environmental Communication: A Journal of Nature and Culture, 7*(3), 355–371.

Sheppard, S. R. (2005). Landscape visualisation and climate change: the potential for influencing perceptions and behaviour. *Environmental Science & Policy, 8*(6), 637–654.

Shiode, N. (2000). 3D urban models: Recent developments in the digital modelling of urban environments in three-dimensions. *GeoJournal, 52*(3), 263–269.

Sidorchuk, A. (1996). Gully erosion and thermo-erosion on the Yamal Peninsula. In O. Slaymaker (Ed.), *Geomorphic Hazards* (153–168). New York.

Sidorchuk, A. (1999). Dynamic and static models of gully erosion. *CATENA, 37*(3–4), 401–414.

Siepmann, N., Edler, D., & Kühne, O. (2020). Soundscapes in cartographic media. In D. Edler, C. Jenal, & O. Kühne (Eds.), *Modern approaches to the visualization of landscapes* (pp. 247–263). Wiesbaden: Springer VS.

Smith, E. L., Bishop, I. D., Williams, K. J. H., & Ford, R. M. (2012). Scenario Chooser: An interactive approach to eliciting public landscape preferences. *Landscape and urban planning, 106*(3), 230–243. https://doi.org/10.1016/j.landurbplan.2012.03.013.

Stintzing, M., Pietsch, S., & Wardenga, U. (2020). How to Teach "Landscape" through Games? In D. Edler, C. Jenal, & O. Kühne (Eds.), *Modern approaches to the visualization of landscapes*. Wiesbaden: Springer VS.

Stoter, J. E., Meijers, B. M., Van Oosterom, P. J. M., Grunreich, D. & Kraak, M. J. (2010). Applying DLM and DCM concepts in a multi-scale data environment. *GDI 2010, a symposium on Generalization and Data Integration*. Boulder.

Stratmann, J., Ristea, A., Leitner, M., & Paulus, G. (2020). Exploring urban "Blightscapes" applying spatial video technology and geographic information system: A case study from Baton Rouge, USA. In D. Edler, C. Jenal, & O. Kühne (Eds.), *Modern approaches to the visualization of landscapes* (pp. 499–517). Wiesbaden: Springer VS.

Temme, A. J. A. M., Armitage, J., Attal, M., Gorp, W., Coulthard, T. J., & Schoorl, J. M. (2017). Developing, choosing and using landscape evolution models to inform field-based landscape reconstruction studies. *Earth Surface Processes and Landforms, 42*(13), 2167–2183. https://doi.org/10.1002/esp.4162.

Tiede, D., & Blaschke, T. (2005). A two-way workflow for integrating CAD, 3D visualization and spatial analysis in a GIS environment. In *The 6th international Conference for Information Technologies in Landscape Architecture: Real-Time Visualization and Participation, Visualization in Landscape Architecture*, 26–28.

Tiede, D., & Lang, S. (2010). Analytical 3D views and virtual globes – Scientific results in a familiar spatial context. *ISPRS Journal of Photogrammetry and Remote Sensing, 65*(3), 300–307.

Torres Sibille, A. d. C., Cloquell-Ballester, V.-A., Cloquell-Ballester, V.-A. & Darton, R. 2009: Development and validation of a multicriteria indicator for the assessment of objective aesthetic impact of wind farms. *Renewable and Sustainable Energy Reviews, 13*(1), 40–66. doi:https://doi.org/10.1016/j.rser.2007.05.002.

Vetter, M. (2020). Technical potentials for the visualization in virtual reality. In D. Edler, C. Jenal, & O. Kühne (Eds.), *Modern approaches to the visualization of landscapes* (pp. 307–317). Wiesbaden: Springer VS.

Wagner, I., Basile, M., Ehrenstrasser, L., Maquil, V., Terrin, J. J., & Wagner, M. (2009). Supporting community engagement in the city: urban planning in the MR-tent. *Proceedings of the fourth international conference on Communities and technologies*. University Park.

Waldheim, C. (2016). *Landscape as urbanism: A general theory*. Princeton University Press.

Wallace, L., Lucieer, A., Malenovský, Z., Turner, D., & Vopěnka, P. (2016). Assessment of forest structure using two UAV techniques: A comparison of airborne laser scanning and Structure from Motion (SfM) point clouds. *Forests, 7*, 62.

Weyand, T., Kostrikov, I., & Philbin, J. (2016). PlaNet – Photo geolocation with convolutional neural networks. *Computer Vision – ECCV 2016*. Cham.

Wessel, B. (2018). TanDEM-X ground segment–DEM products specification document.

Westoby, M. J., Brasington, J., Glasser, N. F., Hambrey, M. J., & Reynolds, J. M. (2012). 'Structure-from-Motion' photogrammetry: A low-cost, effective tool for geoscience applications. *Geomorphology, 179*, 300–314.

Woodhouse, I. H. (2005). Introduction to microwave remote sensing.

Wu, C. (2015). Structure from Motion using Structure-less Resection. *ICCV 2015*

Yu, X., Xie, Z., Yu, Y., Lee, J., Vazquez-Guardado, A., Luan, H., et al. (2019). Skin-integrated wireless haptic interfaces for virtual and augmented reality. *Nature, 575*(7783), 473–479. https://doi.org/10.1038/s41586-019-1687-0.

Yoëli, P. (1959). *Relief shading. Surveying and mapping, 19*(2), 229–232.

Volker Hochschild has been Professor of Geoinformatics at the Geographical Institute of Tuebingen University since 2004. His major research subjects are multisensoral remote sensing, web-based spatial data bases, qualitative compilation of environmental risks and vulnerabilities using integrative assessment methods and indices as well as GIS-based management recommendations for natural resources. He has published 76 refereed papers (>170 in total) and headed several national and international research projects (DFG, BMBF, BMWi, EU, State of Baden-Wuerttemberg). Within his lectures on geospatial visualization techniques he always presents the newest state of the art technology to his students as well as for the commercial continuing education course on "geo data management". He is reviewer for funding agencies like BMBF, DAAD, VW-Stiftung, Landesstiftung Baden-Württemberg, GACR (Czech Republic), PRIN (Italy) as well as for several scientific journals and international geoscientific space programmes.

Andreas Braun studied geography at the University of Tuebingen (B.Sc. 2010, Master 2013) with a strong focus on geospatial landscape analysis and human-environment interactions. He received his PhD in 2019 for his work on "Radar satellite imagery for humanitarian response. Bridging the gap between technology and application". His research deals with the assistance of humanitarian organizations by means of earth observation and radar remote sensing, the monitoring of ecosystems and the contribution of very high-resolution imagery for urban planning. He furthermore received the Baden-Württemberg certificate of university didactics and is a lecturer of several undergraduate and graduate courses on spatial data analysis (geographic information systems, remote sensing, statistics and cartography). In 2020, his dissertation was awarded the German young researcher award (Förderpreis Geoinformatik).

Christian Sommer is a Geographer, who uses spatial data and quantitative analytical tools to study the formation of landscapes, their morphology, processes and the diverse human interactions with these. His interests range from the implications of landscapes on our earliest ancestors (ROCEEH: The Role of Culture in Early Expansions of Humans) to the landscape-shaping impact of modern humans throughout the Anthropocene.

Gebhard Warth is a Geographer, who works on remote sensing methods and spatial data analysis for environmental monitoring and to support urban planning. Within his research focus he identifies and evaluates parameters, based on very high-resolution elevation information derived by interferometric and photogrammetric processes, to map and to characterize environmental and urban change processes.

Adel Omran is working as Post doctorial researcher at the Geographical institute at Tuebingen University. In 2013, he completed his PhD in the field of GIS and remote sensing from Tuebingen University, Germany. His research areas of interest include Mineral mapping, soil erosion modelling, GIS Programming tools and hydrological models. He has published several research papers in international journals and conferences. He held lectureships at diverse universities in Europe and Africa. He is membership of different geological societies in Europe and Africa, especially in Egypt.

Modelling and Visualization of Landscape Complexity with Braid Topology

Fivos Papadimitriou

Abstract

A new method is presented here for temporal modelling and visualization of the spatial complexity changes of landscape boundaries. The method is based on topological braid theory. It is shown how braids can serve as spatio-temporal models of changes in landscape boundaries through time. A braid consists of a set of interwoven strands which can represent temporal changes of geographical boundaries in landscapes. Braids can also be used to model various alternative isotopic simplifications of landscape changes. Isotopic simplifications can be considered as scenarios for possible future developments of landscapes. So using braids instead of graphs might be a useful approach for modelling landscape complexity qualitatively and at different instants in time. Further, braids can be used to calculate the spatial complexity of landscapes by means of algebraic methods, by making use of known matrix representations of braid forms, therefore also offering the possibility to treat landscape changes also algebraically. Other standard topics in braid theory, such as Artin words and braid closures are also applied to geography for first time with this paper.

Keywords

Landscape   Visualization · Landscape   Complexity · Braids · Topology · Boundaries · Synoriology · Spatiotemporal Modelling · Landscape Modelling · Mathematical Modelling

F. Papadimitriou (✉)
Eberhard Karls Universität Tübingen, Tübingen, Germany
e-mail: fivos.papadimitriou@mnf.uni-tuebingen.de

1   Introduction

The importance of boundaries emerges by considering the multitude of studies in geography, geopolitics, human geography, and, moreover, in landscape ecology. For this reason, it has been suggested that a special discipline dealing exclusively with borders and boundaries of all kinds should be established (Papadimitriou 2015). Boundaries are important in ecostructures (Kent et al. 1997; Boniolo et al. 2009; Gruninger 2011), in determining species mobility patterns and flows of biomass (Merckx et al. 2003; Fiege 2005; Li et al. 2007) and are widely regarded as significant determinants of species richness and diversity (Martin et al. 2006; Aavil et al. 2008; Aavik and Liira 2010; Bassa et al. 2011) etc. For these reasons, several researchers have been concerned with changes of boundaries over a landscape, or "boundary dynamics" (i.e. Wiens et al. 1985; Chang et al. 2004), or even the complexity of boundaries (Metzger and Muller 1996; Du et al. 2008), while changing landscape boundaries is also significant for practical applications of spatial planning (Boothby 2004; Karstens et al. 2007; Huber et al. 2010).

Yet, there is a growing body of scientific literature on "landscape complexity", increasing over the last years (i.e. Papadimitriou 2002; Batty 2005; Bolliger et al. 2005; Gabriel et al. 2005; Herzon and O'Hara 2006; Cadenasso et al. 2006; Prouix and Parrott 2008, 2009, 2010a, b). As there are various ways by which landscape complexity can be considered, certain researchers have studied the spatial/structural aspects of landscape complexity (Papadimitriou 2009; 2012a; 2020), while others have focused on the functional context (Yodzis 1988; Maurer 1999; Kolasa 2005; Werner 1999; Fonstad 2006; Murray and Fonstad 2007; Papadimitriou 2012c, 2013; Keil et al. 2020) or even on the semantic and representational aspects of landscape complexity (Papadimitriou 2012b).

Issues of complexity have emerged in land use planning, which have been addressed in a number of approaches, most commonly by using multi-agent systems (Ligtenberg et al. 2004). However, we still do not possess a mathematical model that would enable us to describe and visualize the complexity of boundary changes, despite the fact that we need to monitor landscape changes for biodiversity planning, as well as for land use planning (Haines-Young et al. 2003). More boundaries mean higher fragmentation, which is a significant parameter in planning for sustainable development (Jaeger et al. 2008). Thus, we need to monitor changes in landscape patch boundaries, but also field boundaries (Petit et al. 2003) in a spatiotemporal context, that would also aid in the visualization of changes in landscape boundaries. This may also have repercussions for cadastral management (Nan et al. 2006), as well as for valuating ecosystem services (Martin-Lopez et al. 2009).

Yet, the topological theories of landscape complexity available to date relate to the boundaries among landscape units (Papadimitriou 2002; 2012a) or to non-archimedean topologies relating landscape units to landscape functions (Papadimitriou 2013). But beyond these approaches, other topological methods can be applied, which have been hitherto disregarded by geographers and ecologists. One such is "braid theory".

There are several classic introductory texts to braid theory (Artin 1947; Markoff 1945; Crowell and Fox 1963; Gordon 1978; Kauffman 1987; Hemion 1992), which is a field of topology with many applications in physics (Atiyah 1990; Drinfeld 1985), particularly in relativity (Mielke 1977), quantum mechanics (Kauffman 1987; Baez and Muniain 1994; Blanchet et al 1995), statistical mechanics (Jimbo 1989), and, moreover, string theory.

Simply put, an interwoven set of strings is called a "braid". But a more rigorous definition is given by considering the sets of points of plane $A_i = (i, 0,0)$ and of a parallel plane $B_i = (i, 0,1)$ in the space R^3, where $i = 1, 2, ..., n$. Then, a curve or a polygonal line joining one of the points A_i with one of the points B_j is called "ascending" if moving from A_i to B_j along the line its z-coordinate increases monotonously. Hence, a braid that consists in n strands is defined as a set of pairwise nonintersecting ascending lines (the strands) joining the points of the one plane $(A_1,..., A_n)$ to the points of the parallel plane $(B_1,..., B_n)$.

A braid diagram of n strands is a set, which is a union of n topological intervals called "strands" of this set. At each intersection point of two intersecting strands, one of the strands is going "under" the other strand while the other is going "over". The point where there is one strand going under and another going over, is called a "crossing".

Among the many interesting properties of braids, for the purpose of this study it suffices to recall that braids can be multiplied and give a "product" and can also be simplified to give "isotopic" braids. These two actions have rigorous definitions. Two braids X and X' with n strands are isotopic if X can be continuously deformed into X'.

The connections between topology and complexity theory have been established over the last decades by several studies (Makowsky and Marino 2003; Diao and Ernst 1998; Kholodenko and Rolfsen 1996; Barenghi et al 2001; Orlandini et al. 2005). The complexity of braids in particular (Hamidi-Tehrani 2000; Nechaev et al. 1996; Bangert et al. 2002; Garber et al. 2002; Fiedler and Rocha 1999) is a characteristic example and various measures of braid complexity have been proposed, such as the "braiding exponent" (Thiffeault 2005), the "minimum braid energy" (Bangert et al 2002), the amount of "distortions" per braid (Dynnikov and West 2007). Nechaev et al (1996) suggested the use of the Lyapunov exponents of generators of braid groups in order to assess the complexity of braids, while Bangert et al (2002) defined the problem of braid complexity as equivalent to the minimal crossing number of braids. Yet, other researchers (Kholodenko and Rolfsen 1996) defined braid complexity in the context of crossings. All in all, it can be said that the minimum number of crossings of a braid can been considered as the most widely accepted measure of braid complexity (Raymer and Smith 2007; Simsek et al. 2003; Berger 1994; Bangert et al. 2002) and it will therefore also be adopted here.

To date, braid theory has never been applied to geographical or landscape analysis and hence the aim of this paper is to show that braid theory can be a useful spatio-temporal model for visualizing the complexity of changes in landscape boundaries.

2   Methods and Data

2.1   Model Formulation and Test Site

The topological model of landscape complexity is developed to describe the landscape dynamics of an area near Athens, Greece. The landscape of North Attica (the piece of land stretching north of Athens) was studied at a scale 1/80.000. The landscape transformations in this area have been measured from maps of land use from aerial photographs of the year 1967, Landsat.TM satellite imagery of 1988 and imagery of 2009 and after field verification (for other studies considering remotely sensed imagery in this volume, see Hochschild et al. 2020; Meyer-Heß 2020; Stintzing et al. 2020). From these maps, several landscape transformations were identified but were selected only a few, in order to apply the model. The main land uses in the landscapes selected are forest (F), agriculture (A), shrubland (S) and bare ground (B). The boundaries between forest, agriculture and semi-natural vegetation have been considered as the most important ones in several landscape ecological studies (Le Coeur et al. 2002; Croissant 2004), particularly in Mediterranean landscapes (Papadimitriou and Mairota 1996; Christodoulou and Nakos 1990). The forests (F) of North Attica are mainly characterised by pine trees (*Pinus halepensis, Pinus brutia, Pinus pinea* L.), the shrublands (S) consist of maquis species (*Quercus coccifera, Erica arborea, Pistacia lentiscus etc*), as well as degraded vegetation (*Sarcopoterium spinosum, Calycotome vilosa* etc*), which expand to bare ground areas (B). Some of the bare ground areas have been denuded due to forest fires during the last years. The agricultural (A) lands are mainly viticulture and olive tree plantations (*Olea europaea*).

A representative map showing landscape changes in North Attica as a landscape L is shown in Fig. 1. The boundary changes between these land uses are the following seven landscape transformations (symbolised with the code numbers 1 to 7), which have taken place from time 1967 to 2009 in this order:

1 Shrubs (S) expand on abandoned wheat cultivations (A)
2 Shrubs (S) expand in abandoned barley cultivations (A)
3 Shrubs (S) expand in olive groves of (A)
4 Shrubs (S) expand in citrus tree crops (A)
5 New arable crops (A) expand on formerly forested areas (F)
   i. after the forest was burnt
6 New tree plantations (A) extend on sparsely forested areas (F)
7 Certain forest species (F) expand on bare ground areas (B)

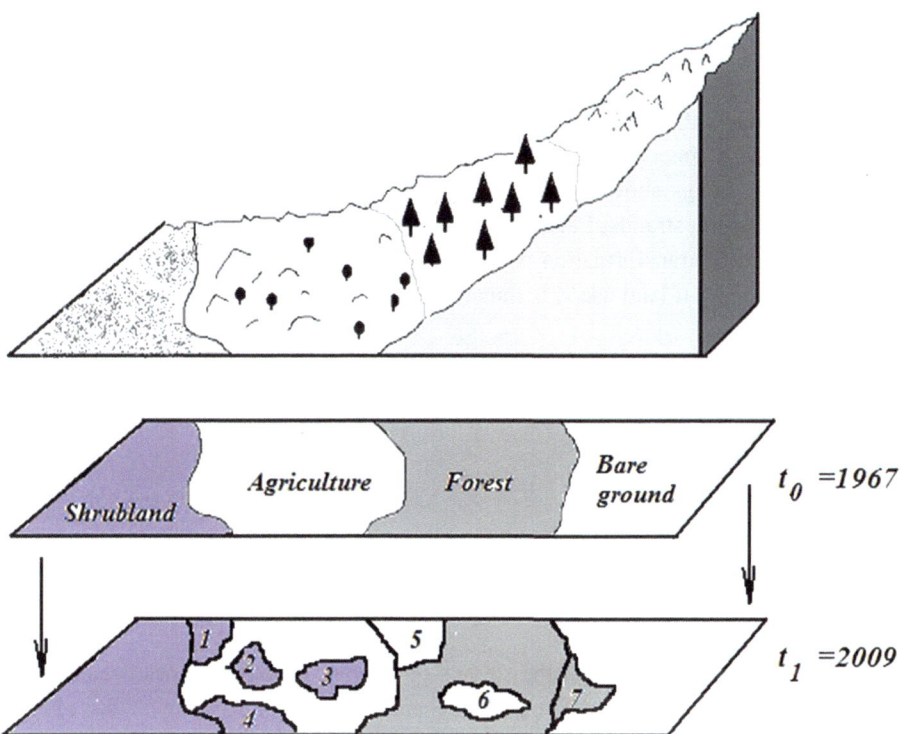

Fig. 1 The landscape L representative of changes in landscape boundaries in North Attica, with its four landuses: S, A, F, B and transformations 1 to 7. (Source: own presentation)

2.2   Topological Model and Visualization

Landscape changes can be visualized by a braid diagram by using:

a) a layer at time $t = 0$ on which the nodes (landscape units) are allocated,
b) an opposite layer where the nodes end up (not necessarily with the same order) representing the state of the landscape at time $t = 1$
c) the strings (called "strands") or bands in between the two layers, connecting the nodes, representing the course of boundary changes from $t = 0$ to $t = 1$.

Each strand may have none or one or several crossings with any other strand. The higher the number of crossings, the higher the complexity of the braid. The complexity of landscape transformations over time can thus be modelled by the (high or low) number of interlaces (crossings) of their respective strands from time 0 to time 1.

The topological model of the landscape L shows how landscape changes can be visu-
alizaed by braids. The braid complexity will be measured in the sense of Kholodenko
and Rolfsen (1996), as the cardinality of crossings among the braid's strands.

Braid R_1 models the landscape transformations 1 to 7 between the land use types of
the landscape L as interwoven interactions involving four types of boundary changes
(Fig. 2). The crossings show the complexity of transformations: there are seven interlaces between all the strands. Each strand is either above or under the other, depending
on which landscape transformation occurs (on whether the landuse X is transformed to
Y or the reverse). So if land use X is transformed to Y, the strand of Y will appear "over"
the strand of X.

The matrix of landscape transformations (code-numbered from 1 to 7) is given below
(row: transformation from, column: transformation to):

$$
\begin{bmatrix}
 & S & A & F & B \\
S & & - & - & - \\
A & 1,2,3,4 & - & - & - \\
F & - & 5,6 & & - \\
B & - & & - & 7
\end{bmatrix}
$$

So the number of crossings of the strands of the four landscape types is equal to the
number of boundary changes: the strands of S and A have four crossings, those of A and
F have two and the strands of F and B have one.

A further application of the topological model of landscape complexity is developed
to describe the landscape dynamics of North Attica in more detail, by making use of the
landscape map shown in Fig. 3.

This map shows how two small landscapes of North Attica have transformed from
1967 to 1988. There is an upper region (landscape L_1) and a lower region (landscape L_2).
Parts of these regions have remained the same over the time period 1967-1988: these are
the areas of landscape types A and S (agriculture and shrublands respectively). But other
areas have changed, according to the coding of transformations shown in Table 1.

Fig. 2 aid R_1, modelling the
transformations of boundaries
S, A, F, B in landscape L
shown on Fig. 1. (Source: own
presentation)

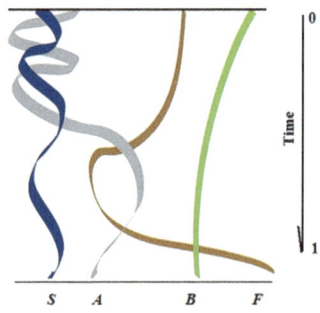

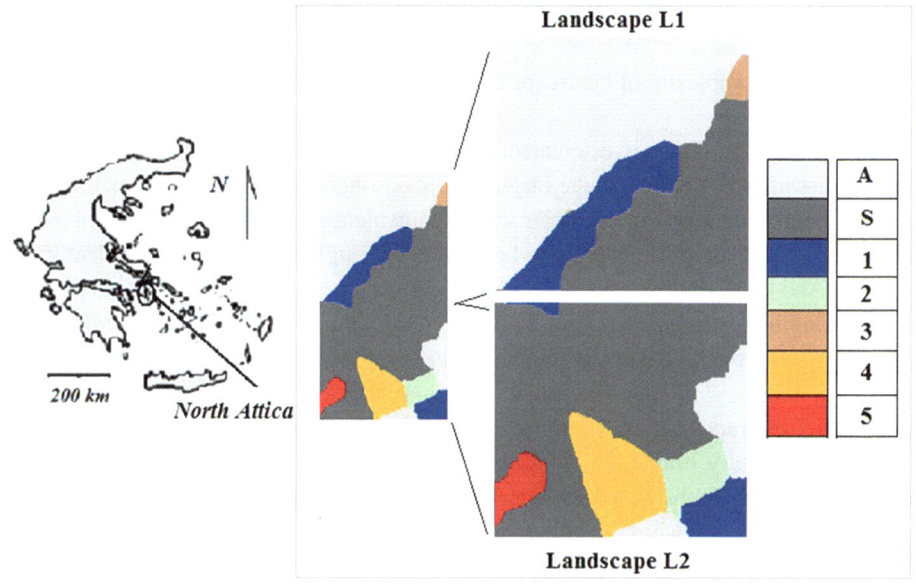

Fig. 3 Location of the two landscapes L_1 and L_2 in the landscape transformations map of North Attica, Greece (1967–1988). Landscape change (1967–1988) in two adjacent landscapes (L2 and L3). Symbols A, S stand for the areas of agriculture and shrublands which have remained unaltered from 1967 to 1988. The numbers 1, 2, 3, 4, 5 stand for codes of land use transformations over the same time period (see Table 1). (Source: own presentation)

	Code	Landscape Transformation (1967–1988)
Table 1 Code numbers of landscape transformations in East Attica (1967–1988). A = agriculture, B = bare ground, F = forest, S = Shrubland. *Source* own presentation	1	From A to S
	2	From A to B
	3	From A to F
	4	From S to B
	5	From S to F

3   Results

3.1   Braid Complexity of the Model Landscape

As stated earlier, the topological modelling of the braid's complexity $C(R_1)$ consists in measuring the total number of crossings in a braid: the higher the number of crossings, the more complex the braid is, The number of times the some strand of braid R_1 crosses another strand is seven:

$$C(R_1) = \text{card}\{\text{crossings}\} = 7$$

Changes in the complexity of landscape dynamics can be measured by using the concept of "braid isotopy".

Once an isotopic braid is calculated, the initial braid can be represented by a simpler braid, which is isotopic to the original. Untangling a braid so as to visualize these landscape transformations in a simpler way is equivalent to the mathematical problem of determining the "isotope" of a braid (a braid with topological properties equivalent to the original). And this isotopy represents a state of reduced complexity of spatial boundaries.

Following braid theory, the braid R_1 can be reduced to a simpler braid R_2, which is a braid "isotopic" to R_1. To calculate braid R_2, we need to break braid R_1 into its "elementary braids" b_1, b_2, b_3. They are called elementary because they model the cases of a braid with four strands and with one crossing at most each time (Fig. 4).

Every elementary braid has its inverse (e.g. b_1^{-1}, b_2^{-1}), so the topology of a braid can be turned into algebra and each braid can be represented by an algebraic polynomial or "word". Hence, the word, or polynomial, for braid R_1 can be written, by reading the braid from time 0 to time 1 (Fig. 5):

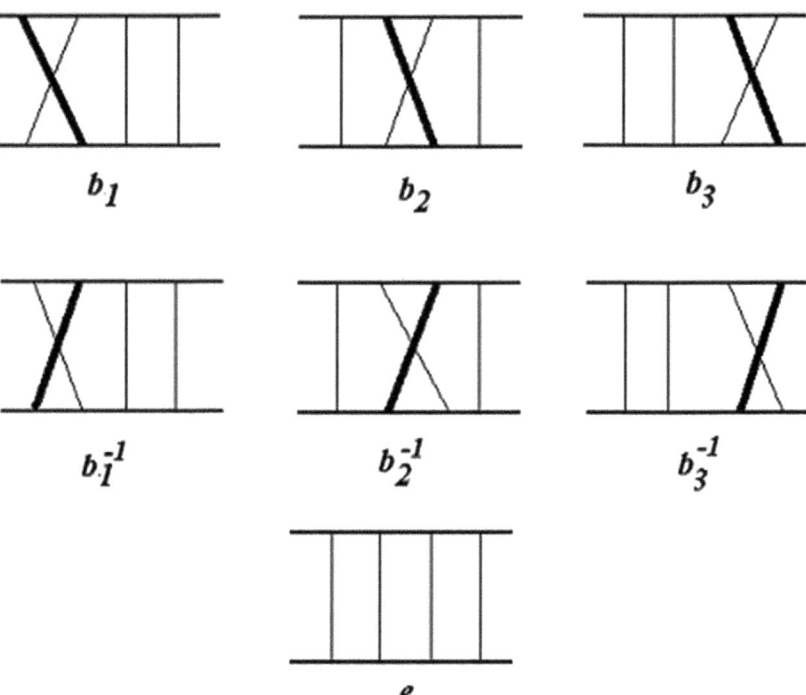

Fig. 4 Elementary braids with their inverses and the trivial braid e. Strands in bold are "upper" (frontal) with respect to the non-bold ones, which are "under". (Source: own presentation)

Fig. 5 Breaking down braid R_1 into a sequence of elementary braids. (Source: own presentation)

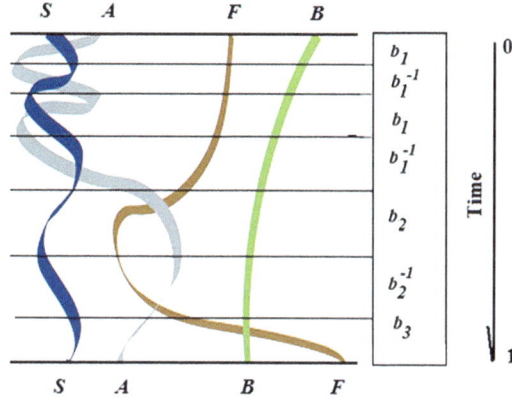

Fig. 6 Braid R_2: a braid which is isotopic to braid R_1, with $C(R_2) = 1$. (Source: own presentation)

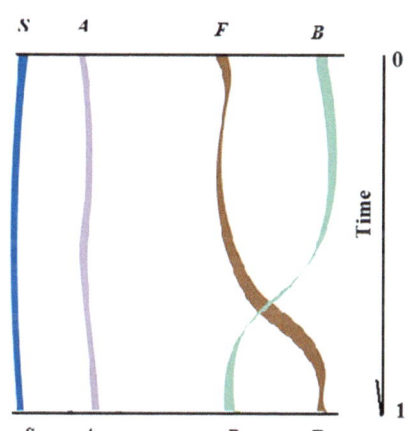

$$b_1 \, b_1^{-1} \, b_1 \, b_1^{-1} \, b_2 \, b_2^{-1} \, b_3$$

Each braid, when multiplied by its inverse, is reduced to the "trivial braid" e (Fig. 4):

$$b_i \, b_i^{-1} = e$$

Associativity applies to braids, so the polynomial of braid R_1 can be reduced to the elementary braid $b3$ and thus the entire braid R_1 is isotopic to the elementary braid $b3$.

Consequently, a braid which is isotopic to R_1 and simpler than R_1 is braid R_2 (Fig. 6).

This braid has only one crossing, it is nothing but the elementary braid $b3$ and its complexity is equal to 1:

$$C(R_2) = 1.$$

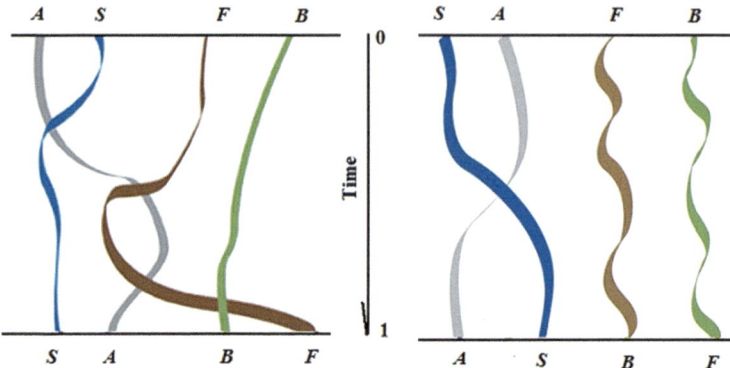

Fig. 7 Two different braids, both isotopic to braid R_1, but corresponding to different landscape transformation senarios. Braid R_3, (on the left) is isotopic to braid R_1 with $C(R_3) = 4$, representing a case of simpler landscape transformations. Braid R_4 (on the right side) is isotopic to braid R_1, with $C(R_4) = 1$, representing an almost unchanged landscape, apart from one transformation only that involves the landscape types S and A. (Source: own presentation)

Braid R_2 is simpler than R_1, modelling a new and less complex condition among the four landscape types, because in braid R_2 the landscape types A and F no longer interact, while S and A have less interlaces than in braid R_1 (which is equivalent to saying that they have attained a less complex spatial interaction).

Hence, the relationships among the four landscape types are simpler in braid R_2, although the whole structure of relationships is still topologically "the same" as in braid R_1.

This procedure shows how braids we can reduce complex, interlaced braids to simpler ones by applying the notion of braid isotopy to such visualizations of landscape change.

In the same way, alternative landscape futures can be derived from the basic braid R_1.

When the encroachment of S in A will have finished, the only impact remaining that S will have on A will be the expansion of shrubs in olive groves of A. All other relationships between A, F and B will remain unchanged (Fig. 7, left). We would therefore have another visualization, that of braid R_3, with a lesser number of crossings and therefore with lower complexity value: $C(R_3) = 4$.

Later, once the F and B will have stabilized, the visual appearance of the landscape will be less complex, as modelled by braid R_4, with $C(R_4) = 1$ (Fig. 7, right).

3.2   Visualization of Landscape Dynamics with Braids

In braid theory, it is well known that braids can be treated algebraically: they can be added, subtracted, multiplied and divided pair wise. Multiplication of braids in particular, can be useful to represent changes in relationships among landscape types with time. Consider, for instance, a situation at time $t = 1$, whereby the landscape types of the model landscape interact in some ways, as modelled by braid R_5. If the boundaries between land units S to A develop at time $t = 2$ to a situation, where landscape transformations have increased in complexity with F interchanging strands only with B and S only with A, this condition is modelled by braid R_6, shown in Fig. 8.

The landscape change from time t = 1 to time t = 2 can be modelled topologically as the product $R_6 * R_5$

This product is visualized in Fig. 9: the landscape begins simpler with S and A interchanging strands and F and B only once (during the time 0 to 1) and end up more complex for all landscape types involved at time 2.

Fig. 8 Braid R_6, modelling the complexity of changing landscape boundaries, at time t = 2. (Source: own presentation)

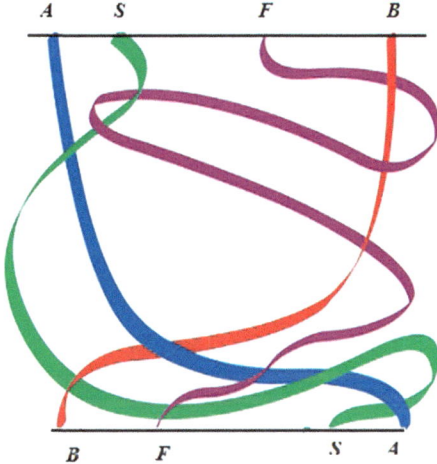

Fig. 9 The braid $R_6 * R_5$ modelling the changing relationships between landuses, from t = 0 to 1 (braid R_5) to t = 1 to 2 (braid R_6). (Source: own presentation)

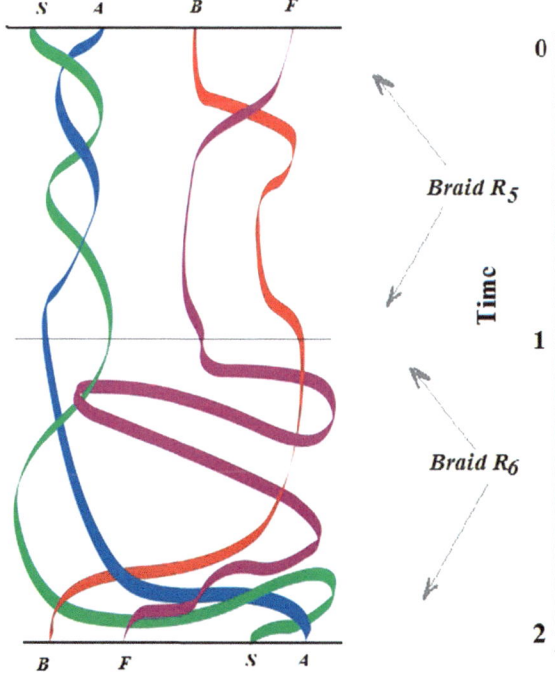

In this case, the change in the complexity of the landscape transformations is:

$$C(R_6 * R_5) = C(R_6) - C(R_5) = 12 - 4 = 8.$$

It can therefore be deduced that, as time lapsed from time 0 to time 2, the complexity of the landscape increased from 4 to 12, that is an increase by 66%.

3.3   Algebraic Calculation of Landscape Complexity from Braids

As each elementary braid corresponds to a square matrix, the complexity of a braid may also be measured by the maximum eigenvalue of its corresponding square matrix (Boyland et al. 2000; Boyland et al. 2003; Vikhansky 2004; Thiffeault 2005; Thiffeault and Finn 2006; Nechaev and Voituriez 2005).

The eigenvalues of each one of the elementary braids of the braid group *B3* can be calculated from the following algebraic equivalences to the elementary braids, which are named "Burau matrices" (Nechaev and Voituriez 2005):

$$b_1 = \begin{pmatrix} 1 & 0 \\ -1 & 1 \end{pmatrix}, b_1^{-1} = \begin{pmatrix} 1 & 0 \\ 1 & 1 \end{pmatrix},$$

$$b_2 = \begin{pmatrix} 1 & 1 \\ 0 & 1 \end{pmatrix}, b_2^{-1} = \begin{pmatrix} 1 & -1 \\ 0 & 1 \end{pmatrix}$$

From these relationships, the braid complexity can be evaluated for the entire landscape L_1 (which is modelled by braid R_1) as:

$$C(R_1) = 1 + \frac{\sqrt{3}}{2}$$

For each triplet of strands of the braid R_I the calculations yield:

$$C(R_1(S, A, F)) = C(e) = 1$$

$$C(R_1(S, A, B)) = C(b_1 b_1^{-1} b_1 b_1^{-1}) = C(e) = 1$$

$$C(R_1(S, F, B)) = C(b_2) = 1$$

$$C(R_1(A, F, B)) = C(b_1 b_1^{-1} b_2) = C(b_2) = 1$$

From these results, it can be concluded that

- there is no triplet of landscape types entailing a complexity higher than any other.
- all triplets of landscape types have the same complexity.
- each triplet has a complexity that is lower than that of the landscape as a whole.

3.4   Visualizing alternative possible Landscape Transformations

With the convention that lower (behind) strands correspond to land use in 1967 and upper (front) strands to land use in 1988, the braid model of the upper landscape L_1 is given in Fig. 10. We observe that there can be two alternative braids, depending on the time ordering by which we consider the landscape transformations to have occurred. It was not possible to verify from aerial photographs and satellite data which landscape transformation took place earlier, so we retain both alternative braid models. In both alternatives however, the complexity of the braid model is (as the number of crossings):

$$C(L_1) = 2.$$

Similarly, in the lower landscape L_2, there is a number of possible braids, but after field observations, it was decided that the correct representation is the one shown in Fig. 11 (that is the model which reflects the temporal order that the transformations have taken place in the landscape), so its complexity is

$$C(L_3) = 4.$$

The braid of landscape L_1 has either one of the two words:

$$b_1^{-1}b_2^{-1} \quad \text{or} \quad b_2 b_1$$

(depending, as previously stated, on the time by which each land transformation has taken place) while L_2 has a longer (and non-reducible word):

$$b_2 b_1 b_2^{-1} b_3^{-1}$$

and hence it is more complex than landscape L_1.

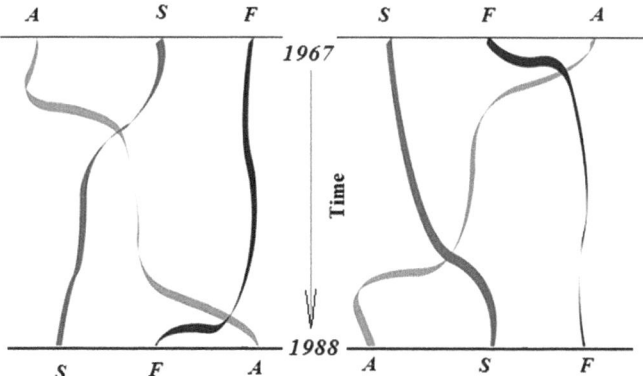

Fig. 10 Two alternative braid models of landscape L_1. As it is unknown which one of the landscape transformations 2 and 5 took place earlier and which one later, we are forced to adopt anyone of two alternative braid models (left and right), which nevertheless have the same complexity: $C(L_1) = 2$. (Source: own presentation)

Fig. 11 Braid model of
landscape L_3. The complexity
of landscape transformations
of this landscape is $C(L_2) = 2$.
(Source: own presentation)

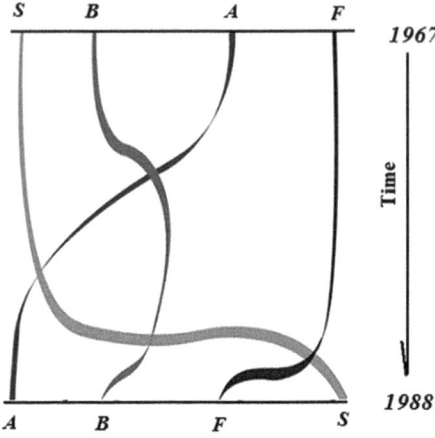

4   Discussion

Clearly, we need to address landscape-related problems in the context of transdisciplinarity (Berr 2018), which would ideally extend to combinations of quantitative with qualitative approaches to landscape analysis, as well as other potentially diverging approaches to landscape analysis, i.e. essentialist and positivist (Kühne et al. 2018). This is particularly true in the context of landscape complexity, whereas we need to combine qualitative methods intertwined with quantitative ones. With the application of braids theory, topology offers the possibility to extend quantitative descriptions of landscape changes to qualitative analyses. It is thus one step ahead towards a synthesis of qualitative with quantitative approaches to landscape research.

Evidently, braids analysis can be extended to cross-border analyses at international level, which may have a landscape-related context; for instance in cases of Interreg projects that have been carried out to resolve conflicts over landscape-related issues. In such studies of cross-border analyses (i.e. Berr and Jenal 2019), strands might be used to model particular landscape parameters instead of landscape types while braids would be useful to visualize the conflictual situations among these landscape parameters. Further, braids theory may be implemented to visualize the degree of hybridization of a landscape, and this applies to cross-border areas also. In example, processes of cross-border cultural hybridization (i.e. as documented by Kühne and Schönwald 2018) may be visualized by means of braids, where each strand would correspond to a different social, cultural, aesthetic characteristic that changes spatially along a transect from one side of the border to the other, in the context of a hybridization process that entails changing shades of hybridity (i.e. hybridization processes as defined by Kühne 2012).

The need to visualize both real and subjectively constructed landscapes is on the rise and this has prompted the use of advanced technologies offering augmented experiences of landscape settings (Edler et al. 2018; Keil et al.2019; Lütjens et al.2019; Edler et al.

2019; see in this volume: Edler et al. 2020, Fontaine 2020; Kleber et al. 2020; Kühne and Jenal 2020; Prisille and Ellerbrake 2020; Siepmann et al. 2020; Stintzing et al. 2020; Vetter 2020). The theoretical background of braid theory can become a source of interesting applications in modelling and visualization of the complexity of landscape transformations, for the following reasons:

i. Braids can be used to model and visualize the topology of changes in boundaries within a landscape, regardless of area change, boundary shape or boundary length.

ii. Braid theory presents an advantage compared with usual graph models, in that the analyses made with it for landscapes could not be carried out with graph models. For instance, instead of braid R_1, we would simply have the graph: S-A-F-B (four nodes and three connections among them). Evidently, none of the topological analyses and classifications shown here with the use of braids would have been possible with such a graph.

iii. Further, it has to be noted that graphs can not be used to model changes between entities at different time steps. This would be possible with each graph corresponding to one time step only, so one would need as many different graphs as the number of time steps suggests. Otherwise said, graphs (and matrices also) can not present continuous time spatial models, as braids do. Moreover, braids allow us to model and visualize conversions of one spatial element to another that take place repeatedly over the same time interval. For instance, the conversion from forest to agriculture may occur in three different areas of the same landscape and within the same time interval.

iv. In addition, braids present interesting connections with knot theory. A "closure" of a braid is a knot or a "link" that can be created from the unification of the ends of respective strands outside the region of the braid (Fig. 12). The closure of braid R_6 * R_5 represents the higher level of complexity of boundaries than each one of the individual braids R_6 and R_5. This higher level of complications is modelled by a knot or a "link" (the closure of a braid is defined as the link or knot resulting by joining the upper points of the braid's strands to the lower ones).

As shown here, an (arbitrary) knot closure of the product R_5*R_6 may be taken to represent many more interlacings between R_5 and R_6 (Fig. 9), as the level of complexity in the landuse sub-types requires. Whether a closure of two braids produces a knot or a link, is a typical question in topology but to no apparent usefulness for geographical analysis (to answer it, we need to establish a relationship between the braid group and the respective permutation group, so the closure of a braid is a knot iff the permutation associated to the braid generates a cyclic subgroup of order n, Z/nZ, in the permutation group Sn).

v. Braids admit easy algebraic descriptions, therefore making them suitable for quick and easy visualization and analyses (and, most importantly, classifications) of landscape dynamics. This has many hitherto unexplored repercussions, because braids can be studied by means of group theory. For instance, the "word problem" for a braid group is to find an algorithm that determines whether or not any pair of words of the

Fig. 12 The "closure" of braid $R_6 * R_5$ is a set of knots and links closing the ends of the strands emanating from the braid $R_6 * R_5$. (Source: own presentation)

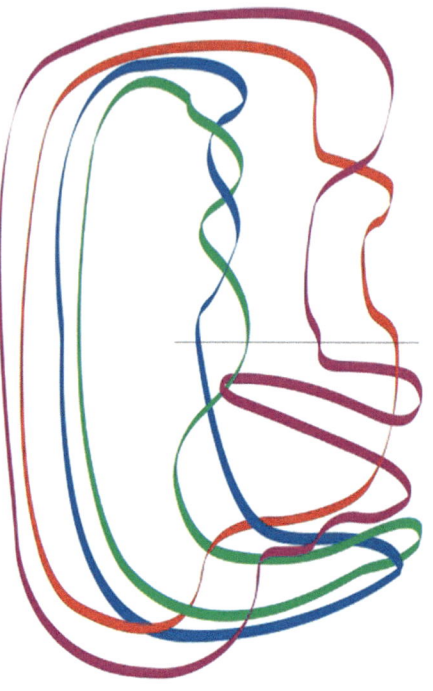

group represents the same element. Concomitant to this problem is the problem of "Artin words" and consists in the following: "For a braid of N strings, find an algorithm which generates an Artin word of minimal length". The length of an Artin word equals the number of crossings in a braid diagram. The word problem in N-braid groups has a complexity $O(\omega > \log n)$ as shown by Hamidi-Tehrani (2000), but no efficient algorithm exists so far for solving the word problem using group-theoretical methods for N > 3. At this point, it is interesting to also notice that Garber et al. (2002) used the term "braids complexity" for the braids word problem.

Solving the word problem (which is solvable for commutative groups) is equivalent to saying that a word in the group can be transformed to another word of the group simply by applying the group's transformations. Expectedly, this extends to the use of other algebraic methods, such as the eigenvalues method. For instance, the complexity of the landscape L_1 of North Attica is measured from the braids product

$$b_1^{-1} b_2^{-1}$$

which, in matrix format, is equivalent to the product:

$$b_1^{-1} b_2^{-1} = \begin{pmatrix} 1 & 0 \\ 1 & 1 \end{pmatrix} \begin{pmatrix} 1 & -1 \\ 0 & 1 \end{pmatrix} = \begin{pmatrix} 1 & -1 \\ 1 & 0 \end{pmatrix}$$

and therefore the eigenvalues are calculated from the equation:

$$\det(b_1^{-1}b_2^{-1}) = \lambda^2 - \lambda + 1$$

and thus they are

$$\lambda_1 = \frac{1}{2} + i\frac{\sqrt{3}}{2}, \lambda_2 = \frac{1}{2} - i\frac{\sqrt{3}}{2}$$

Interestingly, they are exactly the same if the alternative characteristic word for L_1 were used instead:

$$b_2 b_1$$

as in this case, although the matrices are different, the eigenvalues of the product matrix are still the same, since the resulting polynomial is identical to the previous one:

$$\det(b_2 b_1) = \det\left[\begin{pmatrix} 1 & 1 \\ 0 & 1 \end{pmatrix}\begin{pmatrix} 1 & 0 \\ -1 & 1 \end{pmatrix}\right] = \det\begin{pmatrix} 0 & 1 \\ -1 & 1 \end{pmatrix} = \lambda^2 - \lambda + 1$$

vi. Braids can belong to classes, so landscape transformations can be classified in broader categories. It has to be noticed however, that this classification probably extends further than what a landscape ecologist or a geographer would need to analyse, as the classification of braids in classes has been a long-standing problem in braid theory. So, it suffices to say that an isotopy class might correspond to cases of similar entanglements. Hence, the isotopic simplification of a braid for a given class is endowed with the same general algebraic characteristics and belongs to the same algebraic group, thus making it a potentially useful model for a wide class of braids of landscape transformations.

5   Conclusions

Braid theory can serve in visualizing changes in geographical space through time in a qualitative way (regardless of the spatial extent of these changes) and can also serve as a basis for measuring the complexity of landscape transformations. A braid model helps to visualize the complexity of boundary changes between landscape types (or even countries) at different stages in time, either in complex and spatially messy situations, or in resolved and simple (as isotopic simplifications). Also, the product of two braids can serve as a model of qualitative changes in landscapes with time. Among the advantages of braid theory (compared with graph and network approaches) for complexity analysis is the fact that braids can easily be put in algebraic form and their complexity can be calculated algebraically. An isotopic simplification of a braid of intra-landscape relationships and interactions is topologically equivalent to the resolution of conflictual landscape functions between the land units at that level.

References

Aavik, T., & Liira, J. (2010). Quantifying the effect of organic farming, field boundary type and landscape structure on the vegetation of field boundaries. *Agricutlure, Ecosystems and Environment, 135*(3), 178–186.

Aavik, T., Augenstein, I., Bailey, D., Herzog, F., Zobel, M., & Liira, J. (2008). What is the role of local landscape structure in the vegetation composition of field boundaries? *Applied Vegetation Science, 11*(3), 375–386.

Artin, E. (1947). Theory of braids. *Annals of Mathematics, 48,* 101–126.

Atiyah, M. F. (1990). *The geometry and physics of knots.* Cambridge: Cambridge University Press.

Baez, J., & Muniain, J. P. (1994). *Gauge fields, knots and gravity.* Singapore: World Scientific Publishers.

Bangert, P. D., Berger, M. A., & Prandi, R. (2002). In search of minimal random braid configurations. *Journal of Physics A: Mathematical and General, 35*(1), 43–59.

Barenghi, C. F., Ricca, R. L., & Samuels, D. C. (2001). How tangled is a tangle? *Physica D, 157*(3), 197–206.

Bassa, M., Boutin, C., Chamorro, L., & Sans, F. X. (2011). Effects of farming management and landscape heterogeneity on plant species composition of Mediterranean field boundaries. *Agriculture, Ecosystems and Environment, 141*(3–4), 455–460.

Boothby, J. (2004). Lines, boundaries and ontologies in planning: Addressing wildlife and landscape. *Planning Practice and Research, 19*(1), 67–80.

Batty, M. (2005). *Cities and complexity.* Cambridge: MIT Press.

Berger, M. A. (1994). Minimum crossing numbers for 3-Braids. *Journal of Physics A: General Physics, 27*(18), 6205–6213.

Berr, K. (ed.) (2018). *Transdisziplinäre Landschaftsforschung.* [Transdisciplinary Landscape Research].Wiesbaden: Springer.

Berr, K., & Jenal,C. (2019). Grenzüberschreitende Zusammenarbeit auf Interreg-Projektebene: Aushandlungsprozesse und Konflikte um ‚peri-urbane Landschaften‘. [Cross-border cooperation at the Interreg project level: Negotiation processes and conflicts over peri-urban landscapes]. In K. Berr & C. Jenal (Eds.), *Landschaftskonflikte* (pp. 665–685). Wiesbaden: Springer VS.

Blanchet, C., Habegger, N., Masbaum, G., & Vogel, P. (1995). Topological quantum field theories derived from the Kauffman bracket. *Topology, 34,* 883–927.

Bolliger, J., Lischke, H., & Green, D. (2005). Simulating the spatial and temporal dynamics of landscapes using generic and complex models. *Ecological Complexity, 2*(2), 107–116.

Boniolo, G., Faraldo, R., & Saggion, A. (2009). On spatial and temporal ex mensura boundaries. *Foundations of Science, 14*(3), 181–193.

Boyland, P. L., Aref, H., & Stremler, M. A. (2000). Topological fluid mechanics of stirring. *Journal of Fluid Mechanics, 403,* 277–304.

Boyland, P. L., Stremler, M. A., & Aref, H. (2003). Topological fluid mechanics of point vortex motions. *Physica D, 175,* 69–95.

Cadenasso, M. L., Pickett, S. T. A., & Grove, J. M. (2006). Dimensions of ecosystem complexity: Heterogeneity. *Connectivity and History. Ecological Complexity, 3*(1), 1–12.

Chang, Y., Bu, R., Hu, Y., Xu, C., & Wang, Q. (2004). Dynamics of forest landscape boundary at Changbai Mountain. *Chinese Journal of Applied Ecology, 15*(1), 15–20.

Christodoulou, M., & Nakos, G. (1990). An approach to comprehensive land use planning. *Journal of Environmental Management, 31*(1), 39–46.

Croissant, C. (2004). Landscape patterns and parcel boundaries: An analysis of composition and configuration of land use and land cover in south-central Indiana. *Agriculture, Ecosystems and Environment, 101*(2–3), 219–232.

Crowell, R. H., & Fox, R. H. (1963). *Introduction to Knot Theory.* Waltham: Blaisdell.

Diao, Y., & Ernst, C. (1998). The complexity of lattice knots. *Topology and its Applications, 90*(1), 1–9.

Drinfeld, V. G. (1985). Hopf algebras and the quantum Yang – Baxter equation. *Soviet Mathematics Doklady, 32,* 254–258.

Du, S., Qin, Q., Wang, Q., & Ma, H. (2008). Evaluating structural and topological consistency of complex regions with broad boundaries in multi-resolution spatial databases. *Information Sciences, 178*(1), 52–68.

Dynnikov, I., & West, B. (2007). On complexity of Braids. *Journal of the European Mathematical Society, 9*(4), 801–840.

Edler, D., Kühne, O., Jenal, C., Vetter, M., & Dickmann, F. (2018). Potenziale der Raumvisualisierung in Virtual Reality (VR) für die sozialkonstruktivistische Landschaftsforschung [The Potentials of Spatial Visualization in Virtual Reality (VR) for the Social Constructivist Landscape Research]. *Kartographische Nachrichten, 5,* 245–254.

Edler, D., Keil, J., Wiedenlübbert, T., Sossna, M., Kühne, O., & Dickmann, F. (2019). Immersive VR Experience of Redeveloped Post-Industrial Sites: The Example of "Zeche Holland" in Bochum-Wattenscheid. *KN – Journal of Cartography and Geographic Information, 69*(4), 267–284.

Edler, D., Keil, J., & Dickmann, F. (2020). From Na Pali to Earth—An 'Unreal' engine for modern geodata? In D. Edler, C. Jenal, & O. Kühne (Eds.), *Modern approaches to the visualization of landscapes* (pp. 279–291). Wiesbaden: Springer VS.

Fiedler, B., & Rocha, C. (1999). Realization of meander permutations by boundary value problems. *Journal of Differential Equations, 156*(2), 282–308.

Fiege, M. (2005). The weedy west: Mobile nature, boundaries, and common space in the Montana landscape. *Western Historical Quarterly, 36*(1), 22–47.

Fonstad, M. (2006). Cellular automata as analysis and synthesis engines at the geomorphology-ecology interface. *Geomorphology, 7*(7), 217–234.

Fontaine, D. (2020). Landscape in computer games—The examples of GTA V and Watch Dogs 2. In D. Edler, C. Jenal, & O. Kühne (Eds.), *Modern Approaches to the Visualization of Landscapes.* (pp. 293–306). Wiesbaden: Springer VS.

Gabriel, D., Thies, C., & Tscharntke, T. (2005). Local diversity of arable weeds increases with landscape complexity. *Perspectives in Plant Ecology, Evolution and Systematics, 7*(2), 85–93.

Garber, D., Kaplan, S., & Teicher, M. (2002). A new algorithm for solving the word problem in braid groups. *Advances in Mathematics, 167*(1), 142–159.

Gordon, C. M. (1978). Some aspects of classical Knot Theory. *Lecture Notes in Mathematics, 685,* 1–60.

Gruninger, F. (2011). No landscape units without boundaries! Ecotones and their importance in landscape ecology | [Keine landschaftseinheiten ohne grenzen!: Ökotone und ihre bedeutung in der landschaftsökologie].*Geographische Rundschau, 63*(9), 4–11.

Haines-Young, R., Barr, C. J., Firbank, L. G., Furse, M., Howard, D. C., McGowan, G., et al. (2003). Changing landscapes, habitats and vegetation diversity across Great Britain. *Journal of Environmental Management, 67*(3), 267–281.

Hamidi-Tehrani, H. (2000). On complexity of the word problem in braid groups and mapping class groups. *Topology and its Applications, 105*(3), 237–259.

Hemion, G. (1992). *The Classification of Knots and 3–dimensional Spaces.* Oxford: Oxford University Press.

Herzon, I., & O'Hara, R. B. (2006). Effects of landscape complexity on farmland birds in the Baltic states. *Agriculture, Ecosystems and Environment, 118*(1–4), 297–306.

Hochschild, V., Braun, A., Sommer, C., Warth, G., & Omran, A. (2020). Visualizing landscapes by geospatial techniques. In D. Edler, C. Jenal, & O. Kühne (Eds.), *Modern approaches to the visualization of landscapes* (pp. 47–78). Wiesbaden: Springer VS.

Huber, P. R., Greco, S. E., & Thorne, J. H. (2010). Boundaries make a difference: The effects of spatial and temporal parameters on conservation planning. *Professional Geographer, 62*(3), 409–425.

Jaeger, J. A. G., Bertiller, R., Schwick, C., Muller, K., Steinmeier, C., Ewald, K. C., et al. (2008). Implementing landscape fragmentation as an indicator in the Swiss Monitoring System of Sustainable Development (MONET). *Journal of Environmental Management, 88*(4), 737–751.

Jimbo, M. (1989). On Knot invariants related to some statistical mechanics models. *Pacific Journal of Mathematics, 137,* 311–334.

Karstens, S. A. M., Bots, P. W. G., & Slinger, J. H. (2007). Spatial boundary choice and the views of different actors. *Environmental Impact Assessment Review, 27*(5), 386–407.

Kauffman, L. H. (1987). On Knots. *Annals of mathematical studies*, (p. 115). Princeton: Princeton Univ. Press.

Keil, J., Edler, D., & Dickmann, F. (2019). Preparing the hololens for user studies: An augmented reality interface for the spatial adjustment of holographic objects in 3d indoor environments. *KN – Journal of Cartography and Geographic Information, 69*(3), 205–215.

Keil, J., Edler, D., Kuchinke, L., & Dickmann, F. (2020). Effects of visual map complexity on the attentional processing of landmarks. *PLoS ONE, 15*(3), e0229575.

Kent, M., Gill, W. J., Weaver, R. E., & Armitage, R. P. (1997). Landscape and plant community boundaries in biogeography. *Progress in Physical Geography, 21*(3), 315–353.

Kholodenko, A. L., & Rolfsen, D. P. (1996). Knot complexity and related observables from path integrals for semiflexible polymers. *Journal of Physics A: Mathematical and General, 29*(17), 5677–5691.

Kleber, A., Edler, D., & Dickmann, F. (2020). Cartography and the sea: A javascript-based web mapping application for managing maritime shipping. In D. Edler, C. Jenal, & O. Kühne (Eds.), *Modern Approaches to the Visualization of Landscapes* (pp. 173–186). Wiesbaden: Springer VS.

Kolasa, J. (2005). Complexity, system integration and susceptibility to change: Biodiversity connection. *Ecological Complexity, 2*(4), 431–442.

Kühne, O. (2012). Urban Nature between modern and Postmodern Aesthetics: Reflections based on the social constructivist approach. *Questiones Geographicae, 31*(2), 61–70.

Kühne, O., & Schönwald, A. (2018). Hybridisierung und Grenze: das Beispiel San Diego/ Tijuana. [Hybridization and Borders: the example of San Diego/Tijuana]. In M. Heintel, R. Musil, & N. Weixlbaumer (Eds.), *„Grenzen. Theoretische, konzeptionelle und praxisbezogene Fragestellungen zu Grenzen und deren Überschreitungen"* (pp. 401–417). Wiesbaden: Springer VS.

Kühne, O., Weber, F., & Jenal,C. (2018). Der Begriff ‚Landschaft' sowie essentialistisch und positivistisch orientierte Zugänge. [The term 'landscape', essentialist and positivist orientated approaches] In O. Kühne, F. Weber, & C. Jenal (Eds.), *"Neue Landschaftsgeographie. Ein Überblick"* (pp. 5–10). Wiesbaden: Springer VS.

Kühne, O., & Jenal, C. (2020). The threefold landscape dynamics—Basic considerations, conflicts and potentials of virtual landscape research. In D. Edler, C. Jenal, & O. Kühne (Eds.), *Modern approaches to the visualization of landscapes* (pp. 389–402). Wiesbaden: Springer VS.

Le Coeur, D., Baudry, J., Burel, F., & Thenail, C. (2002). Why and how we should study field boundary biodiversity in an agrarian landscape context. *Agriculture, Ecosystems and Environment, 89*(1–2), 23–40.

Li, L., He, X., Li, X., Wen, Q., & He, H. S. (2007). Depth of edge influence of the agricultural-forest landscape boundary, Southwestern China. *Ecological Research, 22*(5), 774–783.

Ligtenberg, A., Wachowicz, M., Bregt, A. K., Beulens, A., & Kettenis, D. L. (2004). Modelling land use change and environmental impact. *Journal of Environmental Management, 72*(1–2), 43–55.

Lütjens, M., Kersten, T., Dorschel, B., & Tschirschwitz, F. (2019). Virtual reality in cartography: Immersive 3D visualization of the Arctic Clyde Inlet (Canada) using digital elevation models and Bathymetric data. *Multimodal Technologies and Interaction, 3,* 1–9.

Makowsky, J. A., & Marino, J. P. (2003). The parametrized complexity of knot polynomials. *Journal of Computer and System Sciences, 67*(4), 742–756.

Markoff, A. (1945). Foundations of the algebraic theory of braids. *Trudy Mathematical Institute Steklov, 16.*

Martin, M. J. R., De Pablo, C. L., & De Agar, P. M. (2006). Landscape changes over time: Comparison of land uses, boundaries and mosaics. *Landscape Ecology, 21*(7), 1075–1088.

Martin-Lopez, B., Gomez-Baggethun, E., Lomas, P. L., & Montes, C. (2009). Effects of spatial and temporal scales on cultural services valuation. *Journal of Environmental Management, 90*(2), 1050–1059.

Maurer, B. (1999). *Untangling ecological complexity.* Chicago: The University of Chicago Press.

Merckx, T., Van Dyck, H., Karlsson, B., & Leimar, O. (2003). The evolution of movements and behaviour at boundaries in different landscapes: A common arena experiment with butterflies. *Proceedings of the Royal Society B: Biological Sciences, 270*(1526), 1815–1821.

Metzger, J. P., & Muller, E. (1996). Characterizing the complexity of landscape boundaries by remote sensing. *Landscape Ecology, 11*(2), 65–77.

Meyer-Heß, F. (2020). Discovering forgotten landscapes. In D. Edler, C. Jenal, & O. Kühne (Eds.), *Modern approaches to the visualization of landscapes* (pp. 33–45). Wiesbaden: Springer VS.

Mielke, E. W. (1977). Knot wormholes in geometrodynamics? *General Relativity and Gravitation, 8,* 175–196.

Murray, B., & Fonstad, M. (2007). Preface: Complexity (and simplicity) in landscapes. *Geomorphology, 91*(3–4), 173–177.

Nan, L., Renyi, L., Guangliang, Z., & Jiong, X. (2006). A spatial-temporal system for dynamic cadastral management. *Journal of Environmental Management, 78*(4), 373–381.

Nechaev, S. K., Grosberg, A. Y., & Vershik, A. M. (1996). Random walks on braid groups: Brownian bridges, complexity and statistics. *Journal of Physics A: Mathematical and General, 29*(10), 2411–2433.

Nechaev, S., & Voituriez, R. (2005). Conformal geoemtry and invariants of 3-Strand Brownian braids. *Nuclear Physics B 714* (FS), 336–356.

Orlandini, E., Tesi, M. C., & Whittington, S. G. (2005). Entanglement complexity of semiflexible lattice polygons. *Journal of Physics A: Mathematical and General, 38*(47), L795–L800.

Papadimitriou, F. (2002). modelling indicators and indices of landscape complexity: An approach using GIS. *Ecological Indicators, 2,* 17–25.

Papadimitriou, F. (2009). Modelling spatial landscape complexity using the levenshtein algorithm. *Ecological Informatics, 4*(1), 51–58.

Papadimitriou, F. (2010a). Mathematical modelling of spatial ecological complex systems: An evaluation. *Geography Environment Sustainability, 1*(3), 67–80.

Papadimitriou, F. (2010b). Conceptual modelling of landscape complexity. *Landscape Research, 35*(5), 563–570.

Papadimitriou, F. (2012a). The algorithmic complexity of landscapes. *Landscape Research, 37*(5), 591–611.

Papadimitriou, F. (2012b). Artificial intelligence in modelling the complexity of mediterranean landscape transformations. *Computers and Electronics in Agriculture, 81,* 87–96.

Papadimitriou, F. (2012c). Modelling landscape complexity for land use management in Rio de Janeiro, Brazil. *Land Use Policy, 29*(4), 855–861.

Papadimitriou, F. (2013). Mathematical modelling of land use/landscape complexity with ultrametric topology. *Journal of Land Use Science, 8*(2), 234–254.

Papadimitriou, F. (2015). Synoriology-A Science for the environment, peace, infrastructures and cross-border management. In M. Culshaw, V. Osipov, S. J. Booth, & A. S. Victorov (Eds.), *Environmental security of the European cross-border energy supply infrastructure* (pp. 187–191). Dordrecht: Springer.

Papadimitriou, F., & Mairota, P. (1996). Spatial scale – Dependent policy planning for land management in Southern Europe. *Environmental Monitoring and Assessment, 39,* 49–60.

Papadimitriou, F. (2020). *Spatial Complexity. Theory, mathematical methods and applications.* Springer.

Petit, S., Stuart, R. C., Gillespie, M. K., & Barr, C. J. (2003). Field boundaries in Great Britain: Stock and change between 1984, 1990 and 1998. *Journal of Environmental Management, 67*(3), 229–238.

Prisille, C., & Ellerbrake, M. (2020). Virtual Reality (VR) and geography education: Potentials of 360° 'Experiences' in secondary schools. In D. Edler, C. Jenal, & O. Kühne (Eds.), *Modern approaches to the visualization of landscapes* (pp. 321–332). Wiesbaden: Springer VS.

Prouix, R., & Parrott, L. (2008). Measures of structural complexity in digital images for monitoring the ecological signature of an old-growth forest ecosystem. *Ecological Indicators, 8*(3), 270–284.

Raymer, D. M., & Smith, D. E. (2007). Spontaneous Knotting of an agitated String. *Proceedings of the National Academy of Sciences of the USA, 104*(42), 16432–16437.

Siepmann, N., Edler, D., & Kühne, O. (2020). Soundscapes in Cartographic Media. In D. Edler, C. Jenal, & O. Kühne (Eds.), *Modern approaches to the visualization of landscapes* (pp. 247–263). Wiesbaden: Springer VS.

Simsek, H., Bayram, M., & Can, I. (2003). Automatic calculation of minimum crossing numbers of 3-Braids. *Applied Mathematics and Computation, 144*(2–3), 507–516.

Stintzing, M., Pietsch, S., & Wardenga, U. (2020). How to Teach "Landscape" through Games? In D. Edler, C. Jenal, & O. Kühne (Eds.), *Modern approaches to the visualization of landscapes* (pp. 333–349). Wiesbaden: Springer VS.

Thiffeault, J.-L. (2005). Measuring topological chaos. *Physics Review Letters, 94,* 084502.

Thiffeault, J.-L., & Finn, M. D. (2006). Topology, braids, and mixing in fluids. *Philosophical Transactions of the Royal Society of London A, 364,* 3251–3266.

Vetter, M. (2020). Technical potentials for the visualization in virtual reality. In D. Edler, C. Jenal, & O. Kühne (Eds.), *Modern approaches to the visualization of landscapes* (pp. 307–317). Wiesbaden: Springer VS.

Vikhansky, A. (2004). Simulation of topological chaos in laminar flows. *Chaos, 14*(1), 14–22.

Werner, B. T. (1999). Complexity in natural landform patterns. *Science, 284*(5411), 102–104.
Wiens, J. A., Crawford, C. S., & Gosz, J. R. (1985). Boundary dynamics: A conceptual framework for studying landscape ecosystems. *Oikos, 45*(3), 421–427.
Yodzis, P. (1988). The indeterminacy of ecological interactions as perceived from perturbation experiments. *Ecology, 69,* 508–515.

Dr. Dr. Fivos Papadimitriou studied geology (B.Sc.), physics (M.Sc.), environmental resources (M.Sc.) and education (M.Ed.), and gained a doctorate in Geography from the University of Budapest (Ph.D.) and another one from the University of Oxford (D.Phi.Oxon.). He has taught at Universities for several years and has accomplished cooperations or field researches in several countries. He is member of the Editorial Boards of ISI-listed journals, and has received numerous prizes, grants, awards, fellowships and distinctions. His main contributions to science consist in the creation of new algebraic, algorithmic and topological mathematical models and formulas for landscape analysis. His papers papers have been cited by scientists from sixty-five countries. Aside of these, he also maintains vivid research interests in geographical education and cyber-geography.

Using the Matrix as a Qualitative Data Display for Landscape Research and a Reflection Based on the Social Constructivist Perspective

Mohammed Al-Khanbashi

Abstract

The qualitative approach is useful in many fields, such as in landscape research, to assist in understanding the complexity of a specific topic. Landscape architects, planners, and researchers seek to find a way to display their data in the initial analysis and results stages. There are many visual displays available to present data. The matrix is the most frequently used, adaptable, and resourceful of all types of displays in qualitative research. To a certain point, it helps to organize, summarize, simplify, and correlate relevant data. However, landscape is subjective, from the social constructivist perspective, being based on the feelings and opinions of people that are outcomes of a very complex interpretation process, where present and past are intermixed. Thus, arises the question regarding a matrix, to what extent can deductions be produced from this complex process regarding this age of globalization, postmodernization, and hybridization. This article explores the use of the matrix as a display tool for qualitative landscape research and projects, presenting matrices for the stages of literature, analysis, and development strategy involving a case study of Riyadh, Saudi Arabia and its Al-Masif neighborhood to achieve sustainable future development as well as concluding with a reflection based on the social constructivist perspective.

Keywords

Matrix · Qualitative data · Data display · Landscape · Social constructivism · Green infrastructure · Riyadh · Sustainable neighborhood

M. Al-Khanbashi (✉)
Forschungsbereich Geographie, Stadt- und Regionalentwicklung (SRE), Eberhard Karls Universität Tübingen, Tübingen, Germany
e-mail: mohammed.al-khanbashi@uni-tuebingen.de

1   Introduction

A qualitative approach assists in exploring, explaining, and understanding the context-specific phenomenon of a particular topic that includes many complex and dynamically interacting effects by evaluating via contrasting, comparing, and classifying the object of the study (Kasim and Dzakaria 2005). In landscape research, the qualitative approach is useful for understanding different complex issues related to a specific topic, and researchers seek to find a way to display their data in order to understand, draw conclusions, and compare literature, analysis, and development. A visual display, in qualitative studies, can be useful and helpful for many purposes at different stages. It can be used to present the initial process such as exploratory and initial datasets, and the analysis process, which can help in displaying detailed and causal explanations, and as a way of creating research hypotheses plus developing theories (Burke et al. 2005; Miles and Huberman 1994; Verdinelli and Scagnoli 2013). Visual displays such as matrices and networks help in organizing, summarizing, simplifying, and transforming data as well as showing connections between different pieces of relevant data (Verdinelli and Scagnoli 2013). Other data could be displayed using text, diagrams, drawings, concept maps, charts, graphs, and photographs. Representing the data graphically, as when using a matrix, gives the reader the opportunity to gain insights, develop a deeper understanding, and appreciate new knowledge (Verdinelli and Scagnoli 2013). However, a visual display was defined by Miles and Huberman (1994, p. 11) as "an organized, compressed assembly of information that permits conclusion drawing and action". Additionally, Lengler and Eppler (2007, p. 1) pointed out that a visualization method is "a graphic representation that depicts information in a way that is conducive to acquiring insights, developing an elaborate understanding, or communicating experiences".

However, a matrix display is a tabular format, which is "essentially the 'intersection' of two lists, set up as rows and columns." (Miles et al. 2013, p. 109), and as mentioned by Lofland et al. (2006, p. 214), it is based on a "cross-classification of two or more dimensions, variables, or concepts of relevance to the topic or topics of interest", which requires thoughtfulness about what data and relevant information could be useful to present and requires organizing a coherent and focused dataset, which should lead to a comprehensive understanding and analysis (Kasim and Dzakaria 2005; Miles and Huberman 1994). A matrix can be filled by text, colors, or numbers as values. Verdinelli and Scagnoli (2013) found, in a comparison between qualitative articles in three valued journals during 2007–2009, that a matrix is the most frequently used visual display, and it is the most adaptable and resourceful of all types of displays. Their study shows that matrices are used in different sections of qualitative articles, found most often in the 'Results/Findings' sections, to enhance categories, themes and phrases as well as compare and contrast experiences, phenomenon, and results. They are second most common in 'Method' sections, including description, instruments, and data analysis, and third most frequently in 'Introduction' sections to simplify and visualize concepts and definitions.

However, there are no fixed principles for creating a matrix, thus, the goal is to build helpful matrices rather than a "correct one" (Miles et al. 2013). In this sense, in order to help the reader to get the intended message, many important criteria must to be considered so as to be, according to Verdinelli and Scagnoli (2013), as uncomplicated as possible, balancing the important information and minimum detail, and also the avoidance of unnecessary content and information. Additionally, criteria must to be considered as to be, according to Scagnoli and Verdinelli (2017) and Suter (2012), at a deeper and more advanced level, using suitable colors and styles possessing logic and meaning can help readers regarding better understanding and guidance. While visuals are part of a larger work, they must fit together and not cause confusion.

In landscape research and projects, matrices have been used often to study interrelations between different issues and principles, to show the compatibility, and for other purposes. For example, Smith et al. (1997) used the matrix, as a conclusion for their study's analysis, to simply illustrate a strong or weak relationship between a number of physical forms and principles of physical and social quality, without placing numerical weightings on each relationship. Another example from an earlier period Ian McHarg, who was one of the most influential landscape architects in the discipline of environmental and regional planning, made use of natural systems and the concept of overlay layers, employing the matrix to visualize the degree of compatibility among diverse land uses and many natural determinants, as well as potential conflicts and their consequences (McHarg and American Museum of Natural History 1969). This matrix involved weighted overlay GIS-Tools to produce "suitability" maps for agriculture, forestry, recreation, and urbanization, for a Potomac River Basin Study in 1965–1966. The matrix was presented using four colors for four classifications, where "Incompatible" refers to bad; "Low Compatibility" as poor; "Medium Compatibility" as fair; and finally, "Full Compatibility" indicates good.

However, the meaning of landscape as territory or scenery has extended to emphasize people, culture, and society—to be conceived as a scene of action and an expression of people's thoughts, beliefs, ideas, and feelings (Antrop 2015; see in this volume: McLean 2020). Landscape is, from the social constructivist perspective, as defined by Kühne (2018, p. 16) "not an objective, univocally definable entity existing within a physical, material world: it is the sociocultural product of a process of mediation". It is, as Cosgrove (1984) explained, not just to see the world visually but it is a construction and a way of seeing it. Thus, the knowledge of its concept itself, its meaning, evaluation, and assessment is the result of "sedimented experiences" (Schütz and Luckmann 2003 [1975]), which are built based on the social interaction with others rather than the direct facing with physical elements (Edler et al. 2018, 2019; see in this volume: Fontaine 2020; Edler et al. 2020a, b). In this perspective, construction becomes a result of a very complex interpretation process, not isolated, but where present and past perceptions are comingled (Schütz 1971 [1962]; Kühne 2018). There are two types of social interaction, 'non-symbolic' and 'symbolic' (Mead and Morris 1967; Blumer 1973; Kühne 2018). Symbolic interaction involves a process of social negotiation or definition using signs, which become symbols when both the recipient and the giver create the same meaning. Therefore, symbolic interactions are linked with 'things' that are perceived by people in their world, which could be, as was mentioned, physical objects, people, institutions, ideals and principles, the actions of others, and whatever situations an individual encounters in daily life. In this sense, symbols and signs are developed differently from person to person and from group to group. The varying constructs of landscape depend on many factors such as social and cultural backgrounds as well as the knowledge, preferences, experiences, emotions, bonds, sense of belongings, norms, values, and patterns of interpretation and evaluation of both groups and individuals (Al-Khanbashi 2020a; see for socialization of landscape also Kühne and Jenal 2020 in this volume; and see also related to landscape theory Kühne 2019; Winchester et al. 2003; Wylie 2007).

This article aims to present the use of a matrix as a display tool for qualitative landscape research and includes a reflection based on the social constructivist perspective to see in what extent this simple tool can be used in our complex world. However, the article begins with an introduction about displaying qualitative data including the matrix and its use in previous landscape research and projects, as well as the meaning of landscape from a social constructivist perspective. Then, it presents, as a case study, the use of matrices in the landscape research project that analyze and implement the principles of green infrastructure (GI) in Riyadh city and Al-Masif neighborhood in Saudi Arabia (Al-Khanbashi 2017). In the case study, firstly, an overview about the research area will be introduced. Followed by an explanation of how these matrices were used in different research efforts to draw deductions and display the data of literature, analysis, and sustainable development strategy. Finally, in this article, a reflection and conclusion regarding the use of matrices, as a display tool, in landscape research will be included.

2   The Case Study

2.1   Overview About the Research Area

In the recent years, green infrastructure has been taking its place as the present and future trend leading to sustainable planning and design for cities around the world. This multifunctional approach—that balances between environmental, social, and economic aspects and issues—is one of the most important paradigms for planners and designers to rescue their cities and the whole world from many recent problems, particularly, mitigation and adaptation of global warming and urban heat islands as well as developing livable cities that improve the quality of life (Austin 2014; Benedict and MacMahon 2002; Rouse and Bunster-Ossa 2013).

Related to this issue, Riyadh, as the capital of Saudi Arabia, is one of the main population centers in the Middle East and Arabian Peninsula. The city had a great experience in green infrastructure previously in its history, but this has been lost to rapid urban growth especially after the oil boom. In the last years, the city, through Arriyadh Development Authority, has been trying to return to this great experience by implementing a variety of projects (ADA-1 2015; ADA-Al-Bujairy 2015; ADA-KAfAPT 2014; Al-Khanbashi 2020b).

So, by renovating previous structures, the research project implements the approach of green infrastructure in Riyadh focusing on the Al-Masif neighborhood, which is one of the typical 2 km × 2 km neighborhoods in Riyadh. The study's methodology involved examining and understanding the concept of green infrastructure and neighborhood development including international and local case studies. Followed by deep analysis of Riyadh and the Al-Masif neighborhood. Then, sustainable neighborhood strategies, concepts, and guidelines were developed for both Riyadh city and the Al-Masif neighborhood, which are to balance the environmental, social, and economic aspects. Each part of these three stages, will present deductions using matrices explaining the interconnected relations of green infrastructure components and benefits.

2.2   Using Matrix Display

The idea was to use many matrices to visually present the conclusion of each research part and to be easier to understand. Then to compare the concept of GI in the three parts, which involve literature, analysis, and sustainable development strategy. These matrices, in each research part, will be explained in detail in the following paragraphs.

Literature matrices are based on studying different resources and analyzing many case studies. Analysis matrices are based on the site observations and analysis as well as analyzing reports, documents, and maps of Riyadh and the Al-Masif neighborhood. Sustainable development matrices are based on proposed ideas and concepts, which are a balance between the typical GI approach and current situation of GI in both Riyadh and the Al-Masif neighborhood.

Literature Matrices
The green infrastructure concept can be used as a multifunctional approach to develop cities in different scales and improve the quality of life by providing multiple environmental, social, and economic benefits (Al-Khanbashi 2017). Matrices, in Figs. 1, 2, and 3, present the outcomes of interconnected relations between different issues of green infrastructure approach. These matrices are based on several literature resources as well as analysis of international case studies in different scales, environments, and cultures.

As a conclusion of the literature part, a list of thirty-four GI components was prepared in a matrix as rows to display their suitability regarding which scales can be used, by which GI principles these can be achieved, which values can contribute, and what the columns of the matrix represent (see Fig. 1). In the columns of the matrix, there are five categories of scales, these represent national, regional, metropolitan and city, district and neighborhood, and site. Also, there are six principles of GI, according to Rouse and

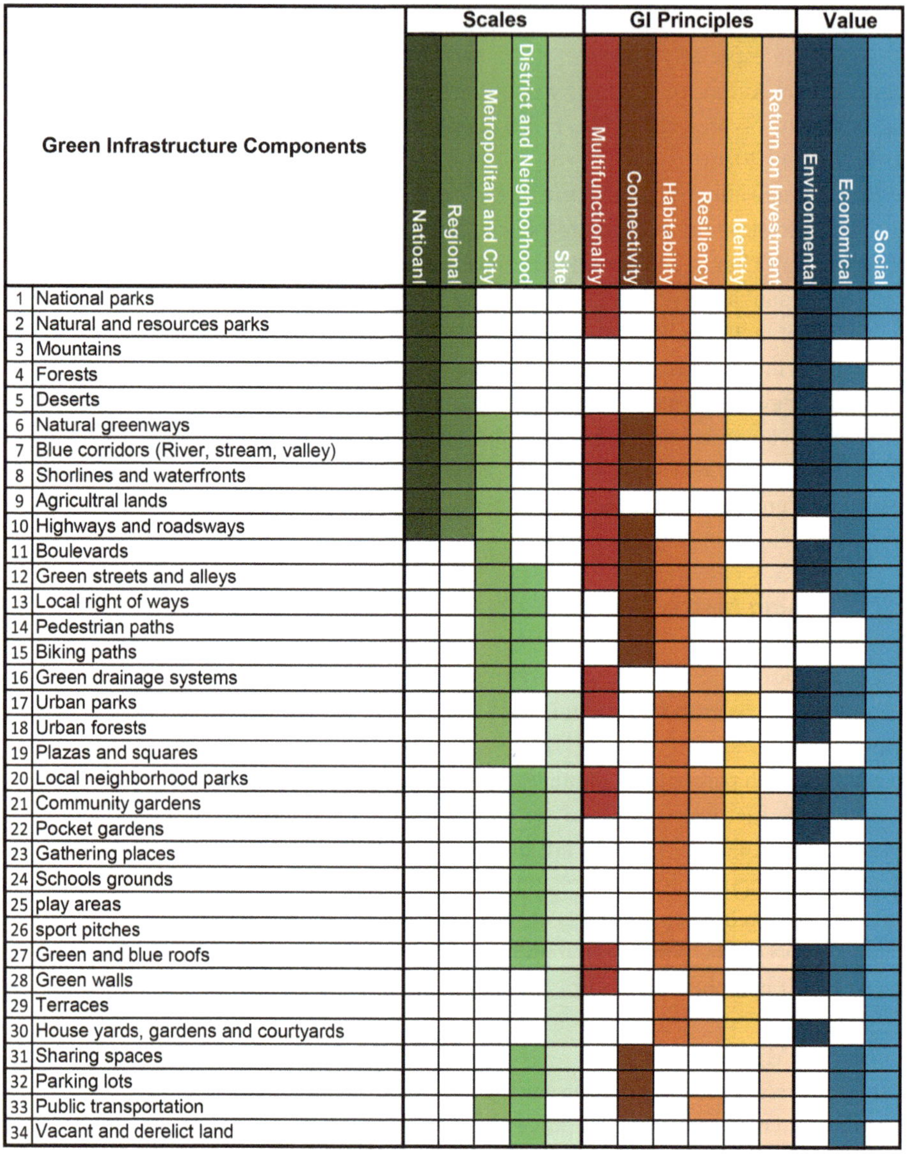

Green Infrastructure Components	Scales					GI Principles						Value		
	Natioanl	Regional	Metropolitan and City	District and Neighborhood	Site	Multifunctionality	Connectivity	Habitability	Resiliency	Identity	Return on Investment	Environmental	Economical	Social
1 National parks														
2 Natural and resources parks														
3 Mountains														
4 Forests														
5 Deserts														
6 Natural greenways														
7 Blue corridors (River, stream, valley)														
8 Shorlines and waterfronts														
9 Agricultral lands														
10 Highways and roadsways														
11 Boulevards														
12 Green streets and alleys														
13 Local right of ways														
14 Pedestrian paths														
15 Biking paths														
16 Green drainage systems														
17 Urban parks														
18 Urban forests														
19 Plazas and squares														
20 Local neighborhood parks														
21 Community gardens														
22 Pocket gardens														
23 Gathering places														
24 Schools grounds														
25 play areas														
26 sport pitches														
27 Green and blue roofs														
28 Green walls														
29 Terraces														
30 House yards, gardens and courtyards														
31 Sharing spaces														
32 Parking lots														
33 Public transportation														
34 Vacant and derelict land														

Fig. 1 Matrix of green infrastructure components. (Source: Al-Khanbashi 2017, p. 70)

Bunster-Ossa (2013), these are multifunctionality, connectivity, habitability, resiliency, identity, and return on investment. Additionally, there are environmental, economic, and social values. Colors, in this matrix, mention if GI components match any scales, principles, and values. For example, local neighborhood parks, as a GI component, can be implemented in two scales, which are 'district and neighborhood' and 'site'. Also, it can

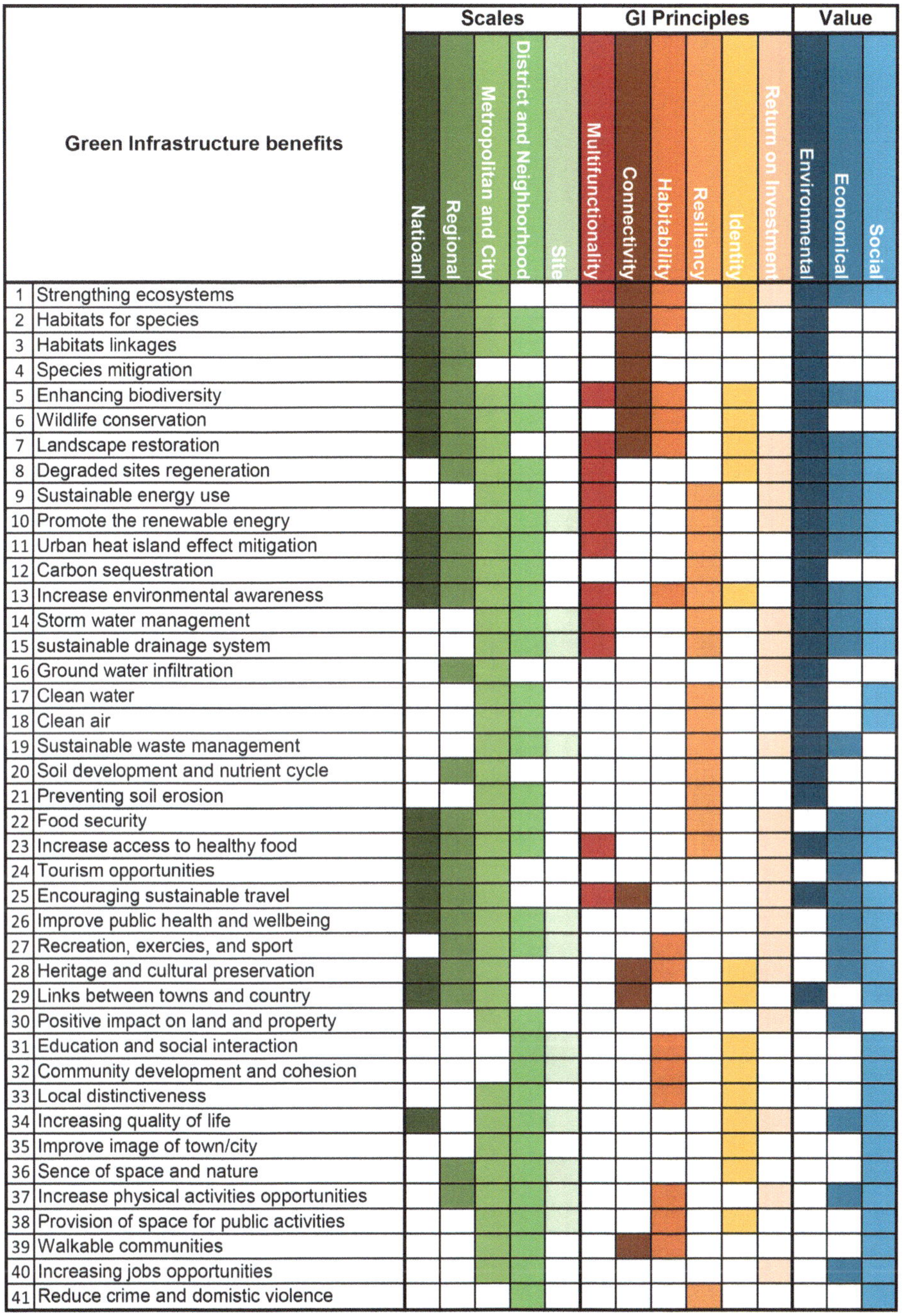

	Green Infrastructure benefits	Scales					GI Principles						Value		
		National	Regional	Metropolitan and City	District and Neighborhood	Site	Multifunctionality	Connectivity	Habitability	Resiliency	Identity	Return on Investment	Environmental	Economical	Social
1	Strengthing ecosystems	■		■	■		■		■		■				
2	Habitats for species			■	■			■	■		■		■		
3	Habitats linkages			■	■			■					■		
4	Species mitigation			■				■							
5	Enhancing biodiversity			■			■			■					
6	Wildlife conservation	■						■					■		
7	Landscape restoration			■			■								
8	Degraded sites regeneration				■		■								
9	Sustainable energy use									■		■			
10	Promote the renewable enegry	■								■		■			
11	Urban heat island effect mitigation			■						■		■			
12	Carbon sequestration			■						■					
13	Increase environmental awareness				■		■		■			■			
14	Storm water management				■		■			■					
15	sustainable drainage system				■		■								
16	Ground water infiltration		■												
17	Clean water				■				■						■
18	Clean air				■					■					
19	Sustainable waste management				■					■		■			
20	Soil development and nutrient cycle			■						■					
21	Preventing soil erosion			■						■			■		
22	Food security	■		■									■	■	
23	Increase access to healthy food			■			■						■		
24	Tourism opportunities				■									■	
25	Encouraging sustainable travel				■		■								
26	Improve public health and wellbeing			■								■			■
27	Recreation, exercies, and sport			■					■						■
28	Heritage and cultural preservation				■			■			■				■
29	Links between towns and country			■				■			■		■		
30	Positive impact on land and property			■							■			■	■
31	Education and social interaction				■				■		■				■
32	Community development and cohesion				■			■			■				■
33	Local distinctiveness			■							■				■
34	Increasing quality of life	■			■						■				■
35	Improve image of town/city				■						■				■
36	Sence of space and nature			■	■						■				■
37	Increase physical activities opportunities			■						■				■	
38	Provision of space for public activities			■						■	■				■
39	Walkable communities				■			■							■
40	Increasing jobs opportunities				■							■		■	
41	Reduce crime and domistic violence				■					■					■

Fig. 2 Matrix of green infrastructure benefits. (Source: Al-Khanbashi 2017, p. 71)

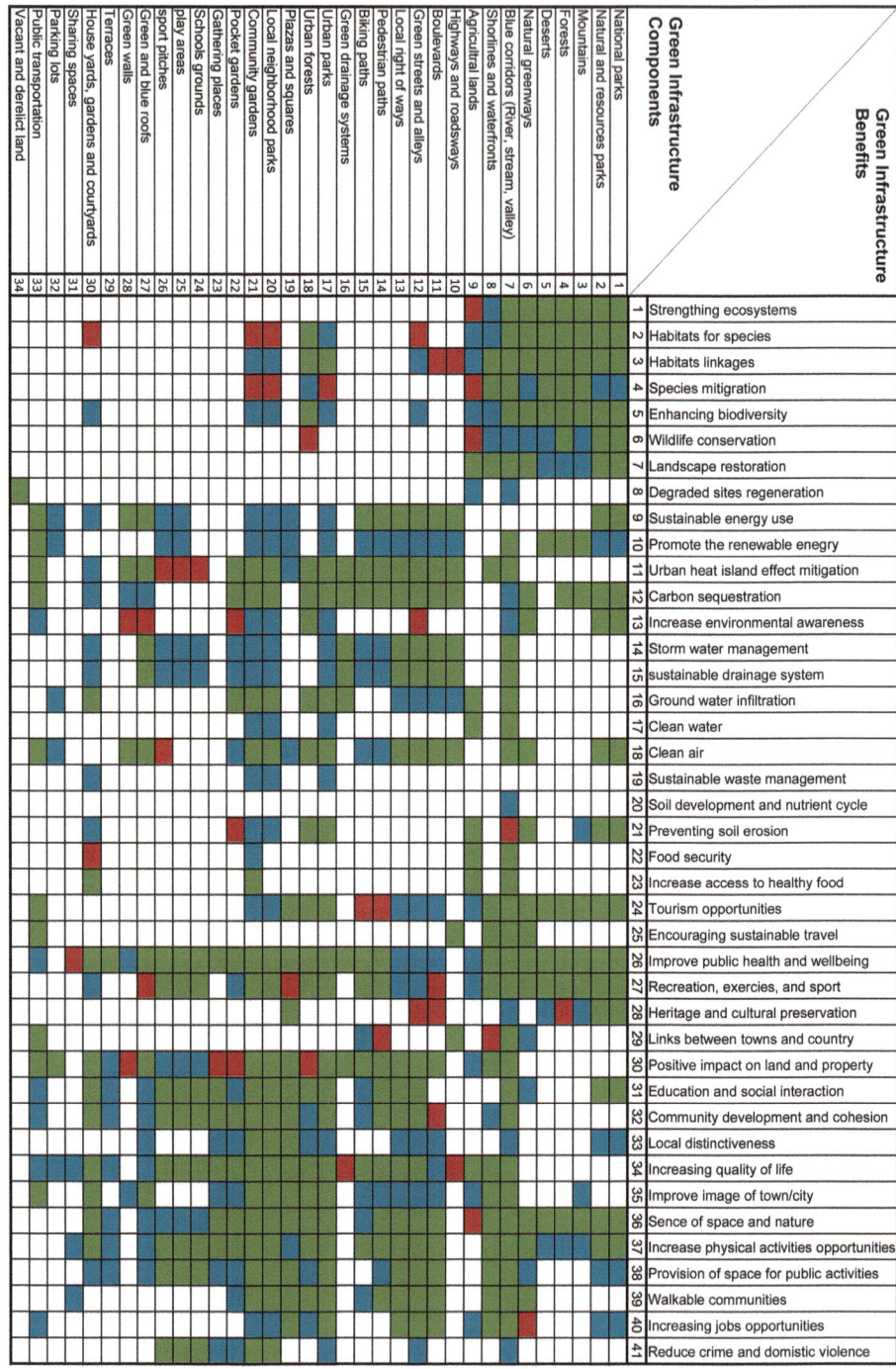

Fig. 3 Matrix of interconnected relations of green infrastructure components and benefits. (Source: Al-Khanbashi 2017, p. 72)

achieve four principles which are multifunctionality, habitability, resiliency, and identity, as well as provide environmental, economic, and social values.

The second matrix presents a list of forty-one GI benefits correlating to the mentioned scales, GI principles, and values (see Fig. 2). Colors, also in this matrix, display if GI benefits match any scales, principles, and values. For example, promoting renewable energy can be a benefit in all scales and can achieve principles such as multifunctionality, resiliency, and return to investment, as well as providing environmental, economic, and social values.

The third matrix presents the interconnected relations between these GI components and benefits (see Fig. 3). In this matrix, there are three colors representing the degree of beneficial effects that can be provided by GI components. So, these GI components can provide a strong beneficial effect as shown by the color of green, medium effect is shown by blue, and weak effect is shown by red. For example, a local neighborhood park can strongly affect urban heat island mitigation, as shown in green. But it has a weak effect in providing habitat for species, as shown in red, while the medium effect in linking habitats is shown in blue color.

These three matrices show the typical situation of GI, which are then used in the analysis part as a base to present the evaluation conclusion of the current situation of both Riyadh and the Al-Masif neighborhood (see Fig. 4). Understanding the typical situation of GI as well as the current situations helped to balance the future development strategies, which are presented visually by matrices for both Riyadh and the Al-Masif neighborhood (see Fig. 5).

Analysis Matrices

In Riyadh, the current condition of green infrastructure concept is still growing via some environmental and urban projects in the city scale, yet are obviously lacking inside the neighborhoods such as the Al- Masif Neighborhood, which is a typical pattern of neighborhoods in the city (Al-Khanbashi 2017). Matrices, in Fig. 4, present a list of green infrastructure components that have been used in different scales in Riyadh or the Al-Masif Neighborhood and which principles have been achieved, as well as defining the value of these components in environmental, social, and economic aspects. For example, green and blue roofs were never used in Riyadh and the Al-Masif Neighborhood. Parking lots are used in neighborhood and site levels, but only with a focus on the investment/return principle and economic value. Local neighborhood parks, as shown in the matrix, only focus on the habitability principle and to provide social benefit.

In addition, there are matrices, in Fig. 4, that explain the interconnected relations of these components and benefits. As shown, components providing benefits with a strong effect are green in color, medium effect are blue, and weak effect are red.

So, these matrices help to provide understand of the current condition of green infrastructure in Riyadh. Additionally, they address the strategic plan and guidelines that could be implemented to improve the Al-Masif Neighborhood in a sustainable manner. This would be accomplished by defining suitable green infrastructure components and

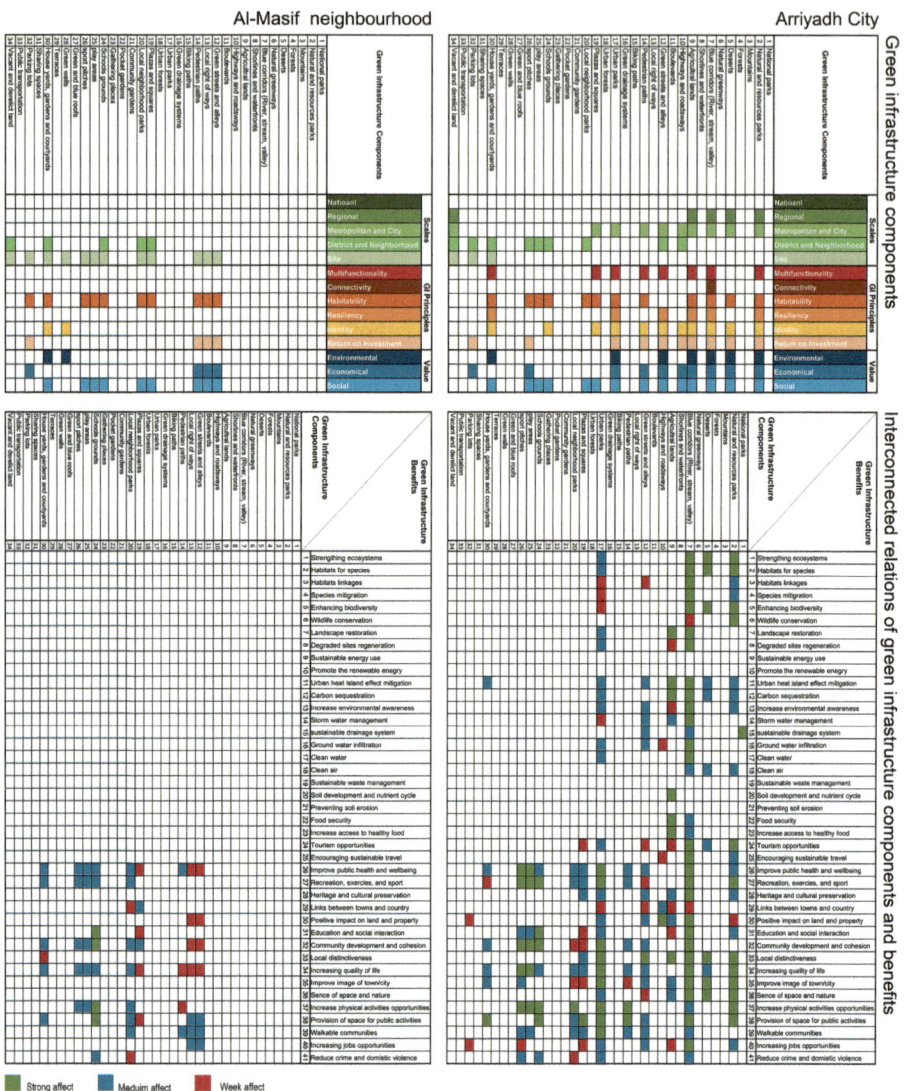

Fig. 4 Matrices analyzing the current conditions of green infrastructure in Riyadh and the Al-Masif neighborhood. (Source: Merged based on Al-Khanbashi 2017, pp. 165–168)

principles that are appropriate for Riyadh and provide multiple environmental, social, and economic benefits, as will be shown in the matrices of the sustainable development strategy.

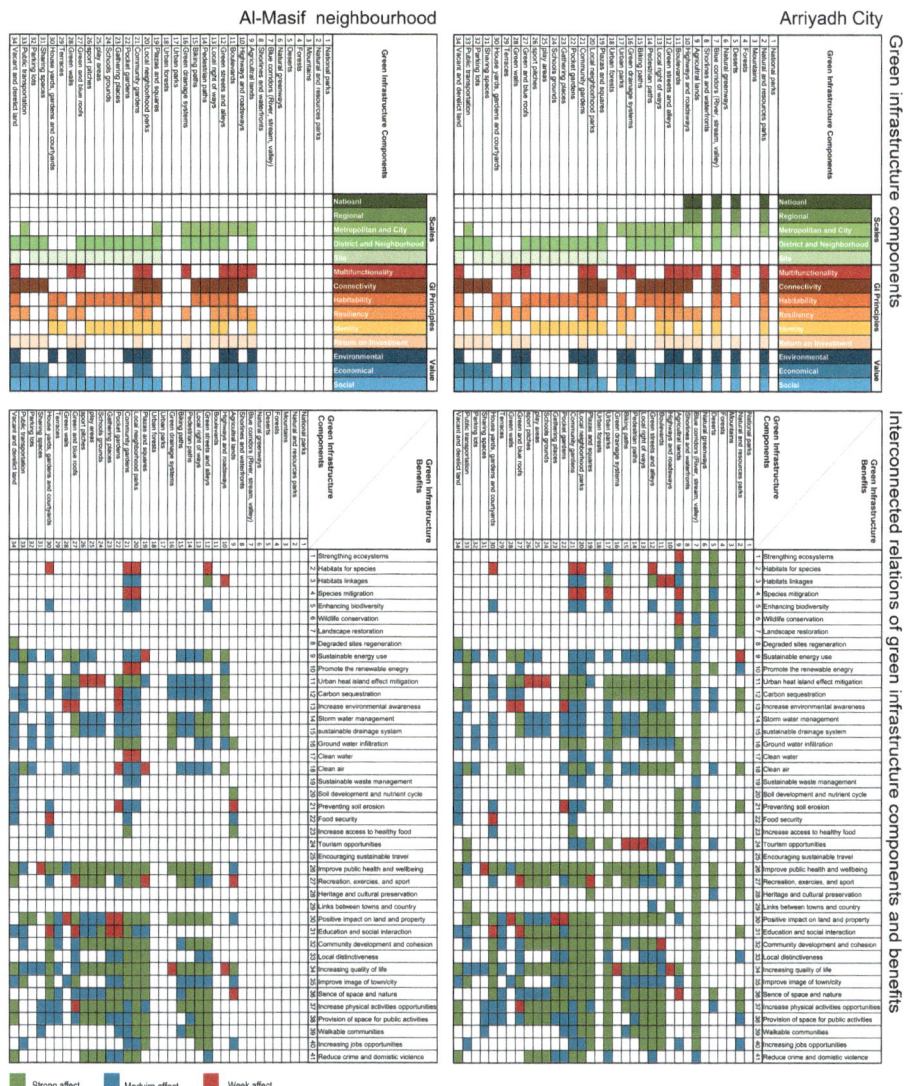

Fig. 5 Matrices of proposed green infrastructure approach in Riyadh and the Al-Masif neighborhood. (Source: Merged based on Al-Khanbashi 2017, pp. 221–224)

Sustainable Development Matrices

Matrices of GI components and benefits found in literature, as well as matrices of Riyadh and the Al-Masif neighborhood GI components and benefits in the current situation, helped to clarify the concept of GI approach in its typical implementation and explain the GI condition in Riyadh as well as the Al-Masif neighborhood. Building on these previous matrices and after explaining the sustainable development approach, the next matrices (see Fig. 5) present the outcomes of interconnected relations between different issues of GI approach through its components and benefits for the future situation of both Riyadh and the Al-Masif neighborhood after implementing the sustainable development concepts and ideas.

As shown, matrices display a list of GI components, where some are recommended to use in different scales in Riyadh and the Al-Masif Neighborhood, and which principles can achieve—as well as defining the value of these components – in terms of environmental, social, and economic aspects. In addition, other matrices, in Fig. 5, explain the interconnected relations of these components and benefits. As shown, recommended components that would provide strong benefits are green in color, medium effect are blue, and weak effect are red.

For example, as shown in the matrices (see Fig. 5), the development strategy proposes to use green and blue roofs at both neighborhood and sites scales in order to achieve many GI principles such as multifunctionality, habitability, resiliency, and return on investment, as well as, to provide environmental, economic, and social benefits. Also, improving the functionality of parking lots was recommended to achieve the connectivity principle as well as to provide social and economic benefits. In addition, improving local neighborhood parks to achieve multiple principles, instead of only habitability, is shown in the analysis matrices, and to expand its benefits for environmental, economic, and social aspects.

In conclusion, these matrices help, to a certain point, to address the strategic plan and guidelines intended to improve the Al-Masif Neighborhood in Riyadh in a sustainable approach by defining suitable GI components and principles that can fit the Riyadh context and provide multiple environmental, social, and economic benefits. In addition, these matrices can help researchers, planners, and designers when using the GI concept to improve their cities and communities.

3   Reflection and Conclusion Based on the Social Constructivist Perspective

Landscape, from the social constructivist perspective being subjective not objective, is based on personal feelings and opinions, influenced by several changeable factors such as the combined knowledge of groups and individuals, their social and cultural backgrounds, preferences, experiences, emotions, bonds, sense of belonging, norms, values, and patterns of interpretation and evaluation. It is constructed differently as the result of a complex interpretation process, where present and past perceptions become interrelated, as mentioned in the introduction by several scholars as well as by the author. In addition to sociocultural interactions, other related issues such as power, political, economic, and environmental aspects should be considered. So, there is no single definition and therefore landscape cannot be defined by one group or individual nor in a very simple way.

In this sense, people construct landscape differently and this matrix can visually force them to see it from one perspective. Many landscape architects, planners, designers, and researchers use the matrix as a visual tool to simplify concepts and study interrelations between landscape components and principles and compatibility of spaces or lands without empirical basis or deep investigations. This leads to losing the meaning of the complex process, while an in-depth examination of the local patterns of interpretation and evaluation of spatial and socioeconomic contexts seems necessary due to the enormous cultural variability of the interpretation and evaluation of spaces. However, the simplicity of the matrix as a visual tool can lead to misunderstanding from different perspectives in how certain landscape components can be perceived from one person to another or from one community to others. So, even though a certain aspect in the matrix is weak to one person that aspect can be perceived as a strong value to another person or a community's perspective, and compatibility can be different as well. In this perspective, there is no fixed, typical, or one model fits all cases, so no generalization, and each matrix for a certain set of relationships or compatibility study can be created differently. The data interpretation in the matrix is based on the person's understanding and can be read differently by the reader. In this sense, it is important to mention that the matrix should not be used singularly but supported with a deep investigation and description in order to communicate the understanding process and meaning as much as possible between the matrix's designer and the reader.

As a conclusion, despite the effectiveness and usefulness, to a certain point, of using the matrix by landscape architects, planners, designers, and researchers in organizing, visualizing, and facilitating the explanation of interrelations and compatibility, there are questions that may arise and need to be investigated.

These include:

- To what extent can a matrix cover the intricacies of the complex processes of a landscape
- is it based on a deep empirical investigation?
- does it really cover related in-depth concepts and conditions?
- how much is the gap and agreement between the understanding of the maker of the matrix and the understanding of the reader?
- does a matrix indicating a good relationship mean it is also good for the reader's concerns considering the continuous changeable factors that could affect the understanding of the matrix's designer himself, can a good relationship itself change?
- is it helpful to use matrix classifications based on dichotomies, where, in this age of globalization, postmodernism, and hybridization (Kühne 2016; Weber and Kühne 2017; Al-Khanbashi 2019; Al-Khanbashi 2020a, b), there is no always fixed good or bad, weak or strong, appropriate or inappropriate, while individuals and groups construct landscapes and appropriate spaces differently according to many factors as mentioned above?

References

ADA-1. (2015). Wadi Hanifah. Official website for Ar-Rriyadh development authority. https://www.ada.gov.sa/ADA_e/DocumentShow_e/?url=/res/ADA/En/Projects/Wadi_Hanifah/index.html. Accessed 28 Nov 2015.

ADA-Al-Bujairy. (2015). AlBujairi, heart of the call. Official website for Ar-Rriyadh development authority. https://www.ada.gov.sa/ADA_A/Prints_ADA_A/index.htm (pdf). Accessed 15 November 2015.

ADA-KAfAPT. (2014). King Abdulziz project for Arriyadh public transportation. Official website for Ar-Rriyadh development authority. https://www.ada.gov.sa/ADA_A/Prints_ADA_A/index.htm (pdf). Accessed 30 Nov 2015.

Al-khanbashi, M. (2017). *Sustainable neighborhood: Improving cities by green infrastructure.* Germany: LAP LAMBERT Academic Publishing.

Al-khanbashi, M. (2019). Urban/rural hybrids and conflicts: New research perspective in Jeddah, Saudi Arabia. In C. Jenal & K. Berr (Eds.), *Landschaftskonflikte* (pp. 617–635). Wiesbaden: Springer VS.

Al-Khanbashi, M. (2020a). *The social construction and use of landscape and public space in the age of migration: Arab immigrants in Berlin.* Wiesbaden: Springer VS.

Al-Khanbashi, M. (2020b). The transformation of Ar-Riyadh's landscape from constructivism perspective. In R. Duttmann, O. Kühne, & F. Weber (Eds.), *Landschaft als Prozess.* Wiesbaden: Springer VS.

Antrop, M. (2015). Interacting cultural, psycological and geographical factors of landscape preference. In D. Bruns, O. Kühne, A. Schönwald, & S. Theile (Eds.), *Landscape culture—Culturing landscapes: The differentiated construction of landscapes* (pp. 53–66). Wiesbaden: Springer VS.

Austin, G. (2014). *Green infrastructure for landscape planning: Integrating human and natural systems.* New York: Routledge.

Benedict, M., & MacMahon, E. (2002). *Green infrastructure: Smart conservation for the 21st century.* Washington: Sprawl Watch Clearinghouse.

Blumer, H. (1973). Der methodologische Standort des symbolischen Interaktionismus. In A. B. Soziologen (Ed.), *Alltagswissen, Interaktion und gesellschaftliche Wirklichkeit* (Vol. 1, pp. 80–146). Reinbek bei Hamburg: Rowohlt.

Burke, J., O'Campo, P., Peak, G., Gielen, A., McDonnell, K., & Trochim, W. (2005). An introduction to concept mapping as a participatory public health research method. *Qualitative Health Research, 15*(10), 1392–1410.

Cosgrove, D. E. (1984). *Social formation and symbolic landscape.* London: University of Wisconsin Press.

Edler, D., Kühne, O., Jenal, C., Vetter, M., & Dickmann, F. (2018). Potenziale der Raumvisualisierung in Virtual Reality (VR) für die sozialkonstruktivistische Landschaftsforschung. *Journal of Cartography and Geographic Information, 5,* 245–254 (Kirschbaum Verlag).

Edler, D., Keil, J., Wiedenlübbert, T., Sossna, M., Kühne, O., & Dickmann, F. (2019). Immersive VR experience of redeveloped post-industrial sites: The example of "Zeche Holland" in Bochum-Wattenscheid. *Journal of Cartography and Geographic Information, 69*(4), 267–284 (Kirschbaum Verlag).

Edler, D., Jenal, C., & Kühne, O. (2020a). Modern approaches to the visualization of landscapes—An introduction. In D. Edler, C. Jenal, & O. Kühne (Eds.), *Modern approaches to the visualization of landscapes* (pp. 103–118). Wiesbaden: Springer VS.

Edler, D., Keil, J., & Dickmann, F. (2020b). From Na Pali to Earth – An 'unreal' engine for modern geodata? In D. Edler, C. Jenal, & O. Kühne (Eds.), *Modern approaches to the visualization of landscapes* (pp. 279–291). Wiesbaden: Springer VS.

Fontaine, D. (2020). Virtuality and landscape. In D. Edler, C. Jenal, & O. Kühne (Eds.), *Modern approaches to the visualization of landscapes* (pp. 267–278). Wiesbaden: Springer VS.

Kasim, A., & Dzakaria, H. (2005). Applying matrix display in qualitative data analysis. In *3rd international qualitative research convention* (Vol. 54). Malaysia: Universiti Utara Malaysia.

Kühne, O. (2016). Urban/rural hybrids: The urbanisation of former suburbs (URFSURBS). *Quaestiones Geographicae, 35*(4), 23–34 (Poznań: Bogucki Wydawnictwo Naukowe).

Kühne, O. (2018). *Landscape and power in geographical space as a social-aesthetic construct.* Cham: Springer International Publishing AG.

Kühne, O. (2019). *Landscape theories. A brief introduction.* Wiesbaden: Springer VS.

Kühne, O., & Jenal, C. (2020). The threefold landscape dynamics—Basic considerations, conflicts and potentials of virtual landscape research. In O. Kühne, D. Edler, & C. Jenal (Eds.), *Modern approaches to the visualization of landscapes* (pp. 389–402). Wiesbaden: Springer VS.

Lengler, R., & Eppler, M. (2007). Towards a periodic table of visualization methods for management. In *IASTED proceedings of the conference on graphics and visualization in engineering*, Clearwater.

Lofland, J., Snow, D., Anderson, L., & Lofland, L. (2006). *Analyzing social settings. A guide to qualitative observation and analysis.* Belmont: Wadsworth/Thomson Learning.

McHarg, I. L., & American Museum of Natural History. (1969). *Design with nature.* Garden City: Published for the American Museum of Natural History [by] the Natural History Press.

McLean, K. (2020). Temporalities of the smellscape: Creative mapping as visual representation. In D. Edler, C. Jenal, & O. Kühne (Eds.), *Modern approaches to the visualization of landscapes* (pp. 217–246). Wiesbaden: Springer VS.

Mead, G., & Morris, C. (1967). *Mind, self & society from the standpoint of a social behaviorist.* Chicago: University of Chicago Press.

Miles, M., & Huberman, A. (1994). *Qualitative data analysis.* Thousand Oaks: SAGE Publications Inc.

Miles, M., Huberman, A. & Saldana, J. (2013). Designing matrix and network displays. In M. Miles, A. Huberman, & J. Saldana (Eds.), Qualitative Data Analysis: A Methods Sourcebook (3rd ed.), (Vol. 5, pp. 107–119). Thousand Oaks: Sage Publications.

Rouse, D., & Bunster-Ossa, I. (2013). *Green infrastructure: A landscape approach.* Chicago: American Planning Association.

Scagnoli, N., & Verdinelli, S. (2017). Editors' perspective on the use of visual displays in qualitative studies. *The Qualitative Report, 22*(7), 1945–1963 (TQR).

Schütz, A. (1971). Gesammelte Aufsätze: Das Problem der Wirklichkeit. Springer, Netherlands (First publication 1962).

Schütz, A., & Luckmann, T. (2003). *Strukturen der Lebenswelt.* Konstanz: UTB (First publication 1975).

Smith, T., Nelischer, M., & Perkins, N. (1997). Quality of an urban community: A framework for understanding the relationship between quality and physical form. *Landscape and Urban Planning, 39,* 229–241 (Elsevier Science Ltd).

Suter, W. (2012). Qualitative data, analysis, and design. In *Introduction to educational research: A critical thinking approach* (Vol. 12, pp. 342–386). Thousand Oaks: SAGE Publications, Inc.

Verdinelli, S., & Scagnoli, N. (2013). Data display in qualitative research. *International Journal of Qualitative Methods, 12,* 359–381 (SAGE Journals).

Weber, F., & Kühne, O. (2017). Hybrid suburbia: New research perspectives in France and Southern California. *Quaestiones Geographicae, 36*(4), 17–28 (Poznań: Bogucki Wydawnictwo Naukowe).

Winchester, H. P. M., Kong, L., & Dunn, K. (2003). *Landscapes. Ways of imagining the world.* London: Routledge.

Wylie, J. (2007). *Landscape*. Abingdon: Routledge.

M.Eng. Mohammed Al-Khanbashi is a landscape architect and currently a doctorate candidate and works as a researcher at the Institute of Geography (Field of Urban and Regional Development), University of Tübingen, Germany. He has an International Master of Landscape Architecture from Weihenstephan-Triesdorf University of Applied Sciences (HSWT) and Nuertingen-Geislingen University of Applied Science (HfWU), Germany as well as a B.Sc. in landscape architecture from King Abdulaziz University, Jeddah, Saudi Arabia. He participated in many academic and professional projects and workshops in Germany, Saudi Arabia, Yemen, Tunisia, Turkey, and Bosnia. His research focuses on social constructivist landscape research, landscape and immigrants, sustainable urban development and green infrastructure.

Part III

Landscape Visualization and the Internet

The Social Construction of Space and Landscape in Internet Videos

Olaf Kühne

Abstract

The social science approach to spatial Internet videos has so far only been rudimentary, although this form of production, transmission, and consumption of moving images and acoustic information is of great social relevance. Film, as well as video, has a considerable significance in the social construction of the world (in this case in relation to spaces, landscapes, places, etc.). The analysis of Internet videos can draw on an elaborate methodology of qualitative and quantitative media content analysis. In the Internet videos examined, three essential aspects could be identified: 1) The contents are dominated by the themes of travel, regional studies, the world of life, music, and sport (supplemented by possibly current events). 2) There is a tendency to propagate spatial stereotypes. 3) There is a tendency to spread spatial conspiracy theories.

Keywords

Internet video · Web video · Social constructivism · Constructivism · Internet · Video · Film · Landscape · Stereotypes

O. Kühne (✉)
Forschungsbereich Geographie, Stadt- und Regionalentwicklung (SRE), Eberhard Karls
Universität Tübingen, Tübingen, Germany
e-mail: olaf.kuehne@uni-tuebingen.de

© Springer Fachmedien Wiesbaden GmbH, part of Springer Nature 2020   121
D. Edler et al. (eds.), *Modern Approaches to the Visualization of Landscapes*,
RaumFragen: Stadt – Region – Landschaft, https://doi.org/10.1007/978-3-658-30956-5_7

1   Introductory Remarks

While the study of films in human geography has become an established subject (Aitken and Dixon 2006; Aitken and Zonn 1994; Davies 2003; Escher and Zimmermann 2001; Forrest et al. 2017; Lukinbeal 2005; Zimmermann 2019) and 'film geography' is often practiced with a regional focus (among many: Bollhöfer 2007; Fine 2004; Ludewig 2011, 2019; Lukinbeal 2012; Strüver 2009) and the Internet has also gained increasing attention from a geographical perspective (e.g.: Graham 2013; Graham et al. 2019), the study of videos uploaded to video portals is not one of the central objects (or methods) of human geography research (Burgess and Green 2018; Garrett 2011; in this volume: Bellini and Leonardi 2020). Since the expansion of the high-speed Internet, even on mobile devices, videos have experienced significant, scientifically relevant developments:

1. It developed into a widespread medium of global construct.
2. With the development of comparatively inexpensive and easy-to-use smartphones, the technical and financial thresholds for video production have also fallen.
3. The videos produced in this way can be disseminated, advertised, and used to gain social and economic capital (in the sense of Bourdieu 1989) on relevant platforms in the field of social media (see detailed Burgess 2008).

With the boom of Internet videos (almost 5 billion videos were viewed on YouTube in 2017; Videonitch 2017) their importance for the social construction of spaces, landscapes, places, etc. has also increased (in general in this volume: Edler et al. 2020a; Kühne and Jenal 2020b). Videos as travel guides or travel reports are just as much a part of this as descriptions of the country, flights of drones, excerpts from television programs, disruptions in regions or cities, artistic disputes, the simple documentation of one's own everyday life, videos introducing 3D environments in immersive Virtual Reality (VR) and much more (Kühne 2012; Kühne and Schönwald 2015; Edler et al. 2018; Kühne and Weber 2019; see in this volume: Edler et al. 2020b; Prisille and Ellerbrake 2020; Siepmann et al. 2020; Vetter 2020). If landscapes are understood not as objectively given materially but as socially and individually constructed (for the social construction of landscape see among many: Cosgrove 1984; Greider and Garkovich 1994; Kühne 2019), the question of how 'landscape' is constructed in films, what meaning is attached to it, whether the medium is used to question or confirm common stereotypes becomes contagious.

A clear conceptual distinction between film and video seems difficult to make. In this respect, I outline the distinction for this contribution as follows—not as a dichotomous separation, but as poles of a wide transition range: videos show a lower formalization than films, in terms of length (they are usually significantly shorter), editing technique, setting, etc. while films are more often produced by professional personnel (directors, actors, cameramen, editors, etc.), videos often have a lower degree of professionalism (with the exception of music videos), films are primarily produced for distribution in return for monetary consideration, which is not or rather indirectly (advertising revenue) found in videos.

Before—in the absence of a more comprehensive approach to the subject—the results of the investigation of Internet videos in relation to 'landscape' are presented, the question of how landscape is negotiated in films and what specifics can be found in Internet videos in contrast to this is first dealt with. The present paper ends with a conclusion in which, in addition to evaluating the results of the paper, open research questions in particular are formulated.

2   Landscape and Film in General

Film, whether feature film or documentary, can be understood as an essential medium for the creation of landscape stereotypes (Beuka 2004; Gold and Gold 2013; Lefebvre 2006; Papadimitriou 2021; in this volume: Papadimitriou 2020). Films do not simply depict physical spaces. Physical spaces are selected, dressed, retouched, or produced according to the desired effect on the viewer, which is particularly evident in computer-generated film landscapes, combined with a higher idealization, which in turn forms the basis for stereotyping (Krebs 2008, p. 137; in this volume also: Loda et al. 2020): "The combination of previously separate fragments of reality and props with the aid of camera and editing desk or computer creates a new landscape condensed to the essential". Sergei Eisenstein (1949, 2014) assigns a similarly central importance to the cinematic landscape as film music, which is the freest element in the film, but also the one that has to bear the heaviest burden of transporting narrative tasks and changing moods, emotional states, and spiritual experiences.

Film, like media content, thus does not represent an image of 'reality', but rather the result of "selection, evaluation, processing and interpretation of social events" (Werlen 1997, p. 383). According to Fröhlich (2007, p. 342), films represent, in relation to cities, "an artistic form of expression in whose collective production processes manifold urban facets are taken up and reflected". A film is therefore an essential component of the simulation of the world, whereby it must always follow on from (presumed) previous experiences of the target audience if it is to have an effect (Plien 2017), whereby these previous experiences in turn are increasingly shaped by the staging of the world (among other things by film or the Internet; see Kühne 2018). The result is that the reference for the evaluation of media stagings of landscape is in turn other media stagings of the world. Luhmann (1996) goes so far as to state that everything we know about the world we know via the mass media. Films become Simulacra as sign contexts (Baudrillard 1978) by referring to other signs and thus creating a reference structure of signs among themselves that cannot be comprehensively grasped, thus turning them into 'realities' themselves (Behrens 2008). Thus postmodern films such as 'Pulp Fiction' or 'Blade Runner' no longer refer to a (stereotypical historical) 'reality', but rather quote other films, ironize connections, and have a cynical relationship to social values and norms (cf. also Moore-Colyer and Scott 2005). In these films, for example, 'City' loses a cohesive context and becomes an incoherent archipelago of living environments that symbolize both (relative) security and crime, social disintegration, but also progress (Kühne 2013). These stereotypical cinematic landscapes condensed into simulacra can, in turn, serve as a basis for the adaptation of material spaces to these cinematic landscapes, just as Hollywood as a physical arrangement is adapted to its glamorous image in order not to make the disappointment of individual confrontation with 'Hollywood' too great in comparison with social landscape ideals (cf. Davis 1998; Vester 1993).

The landscape references of film can be divided into eight different patterns of representation of films (an extension of: Escher and Zimmermann 2001; cf. also Higson 1987):

1. Filmic landscape offers a spatial framework for action and represents an appropriate location of the action in the form of a backdrop: Objects represented in the film are arranged in such a way that they can be interpreted as 'city', 'village', 'mountains', 'ocean' etc. for the viewer. If a slum is portrayed in 'Slum Dog Millionaire', it should not have the character of a white suburb of Baton Rouge. This brings us to the next point:
2. Landscape conveys 'authenticity' and credibility of the action by providing a spatially and temporally consistent frame of reference according to generally available codes. Discussions about metropolitan habitus in the series 'How I Met Your Mother' are more credible in the backdrop of New York than in Vilsbiburg in Lower Bavaria.
3. Landscape is metaphorically as well as symbolically charged. In the process, landscape stereotypes are partly confirmed, such as the kiss in the rose garden (Bollywood), the monotony of everyday life in the 'dreary' industrial city or the carefree life on the beach, partly reinterpreted, such as the dream of living in a suburban home as a place of control and lack of freedom (as in 'American Beauty', for example) or the former industrial district is populated by likeable, bizarre little crooks and life artists, as in 'Bang Boom Bang'.
4. Landscape is mythically charged, as in Westerns with the prairie as a place of struggle for freedom or in the German-language sentimental regional films (Heimatfilme) of the 1950s and 1960s the mountains as a place of the 'ideal' (= pre-modern) world.
5. Landscape can be presented "for its own sake in a feature film" (Escher and Zimmermann 2001, p. 233). It is given a central function in the film and becomes an actor (e.g. in 'Waterworld' or 'The Treasure in Silver Lake').
6. In the production of films, the landscape and the location ('location' or 'set') often do not correspond to each other. Numerous films whose plot is located elsewhere, e.g. in New York or Southeast Asia, were shot in Los Angeles due to the proximity of Hollywood studios.
7. The level of social landscape production on the basis of films also has effects on the spatial actions of people, thus the landscape represented in the film becomes

the destination of tourism. Tourists visit places that are linked to films, such as the film studios in Los Angeles or the locations of the Lord of the Rings trilogy in New Zealand.

8. The representation of place and landscape in film/series can contribute to place making and the identification of people with spaces (Natter 1994). In Saarland, for example, the series 'Familie Heinz Becker', a parody of the bourgeoisie in its regional form, still serves as a frequently used frame of reference today.

9. This in turn is taken up by regional marketing, film locations and series are used to construct local and regional images, as in the case of crime scene episodes acting in Münster (Westphalia) or Wilsberg television films (Bollhöfer and Strüver 2005) or the staging and commercialisation of the Swiss Heidi myth (Leimgruber 2004).

These scenic references can also be found in Internet videos, but there are also other specifics that make it worthwhile to deal with Internet videos scientifically. These will be discussed below.

3   Landscape and Film in Particular: Specifics of Internet Videos

In his essay, 'Videographic geographies: Using digital video for geographic research', Bradley L. Garrett outlines five approaches to scientific engagement with videos/films (Garrett 2011):

a) Writing about movies/videos (analysis)
b) Production for an audience (documentaries with geographical content)
c) Film material as recording (data acquisition)
d) Reflective filmmaking (experiential filmmaking)
e) Participative video (collaborative filmmaking)

The present article focuses on point a). Lastly, the potential of Internet videos for the geographical analysis of virtual landscapes will be presented, not videos as a means of presenting research results or as a method of capturing and reflecting individual contributions to the theme of 'landscape'. The study of Internet videos takes account of the growing importance of user-generated web content (Münker 2009; Nagle 2017; Schmidt 2011).

Internet videos, especially in their diversity, form a differentiated basis for social constructions of landscapes. They have no thematic diversity in which spatial references are, more or less, intensively negotiated. They offer a lower-threshold access to media representations than written texts such as: "Video is ideal for recording the immaterial, even used to create memories of places one has never experienced" (Garrett 2011: 533). Thus, they provide an insight into the internal perspective on spatial arrangements and their attributions of meaning which would not, or only with difficulty, be possible by other methods (including qualitative interviews, e.g. as a result of reservations towards interviewers on the basis of ethnic or gender-specific attributions). They also allow—if in sufficient number—a comparison from a socially or spatially internal and external perspective, from a more lifelike and systemic perspective—enacted with varying intensity—(e.g. tourism advertising; for the approach of system and life-world see: Habermas 1981; on the social construction of tourism landscapes: Aschenbrand 2017). All methods of qualitative and quantitative media content analysis are available to the social sciences in dealing with Internet videos (see e.g.: Altheide and Schneider 2013; Berger 2018; Bonfadelli 2002; Maurer and Reinemann 2006), with which they can fall back on an elaborate set of instruments.

In the following, these potentials of researching Internet videos will be explained using examples.

4   The Representation of Landscape in Internet Videos— Accesses and Examples

The presentations in this paragraph are based on the following empirical studies and theoretical classifications of Internet videos of Los Angeles (Kühne 2012), comparatives between San Diego and Tijuana (Kühne and Schönwald 2015), comparatives between German and English videos on California (Kühne and Weber 2019), the expansion of electricity grids in Germany (Kühne and Weber 2015), and on Baton Rouge (Kühne and Jenal 2020a). The videos are always available via Youtube. The Internet video portal *YouTube* was founded in 2005 and, with over one billion users is the most frequented provider among the video portals worldwide (Tamblé 2012, no. p.; YouTube 2016, no. p.).

The keyword-based search for Internet videos takes place in order to exclude progression-related preselections by the site operator—ideally—with a newly set up or completely new computer. The number of videos examined depends 1) on the availability of analyzable videos, 2) on the type of access (qualitative and/or quantitative) and 3) on the desired sophistication of the search statements. Here, aspect 1) can represent the limiting factor as well as the (co-)determining factor for the further investigation. If only a small number of videos on a topic are available, the possibilities of a quantified approach are limited if not impossible, accordingly either a qualitative evaluation or an extension of the video corpus by a more general search term (e.g. San Diego instead of Cortez Hill) remains, provided that the investigation parameter permits this. If (aspect 2) interpretation patterns are to be recorded in the Internet videos, a detailed examination of the plot, the cutting technique, the dramatic composition, etc. is to be carried out, a qualitative approach is chosen. If, on the other hand, it is necessary to compare aspects of thematic classifications, the degree of 'professionalism' of the videos, the frequency of occurrence of certain objects and object constellations, etc., then a quantitative approach is

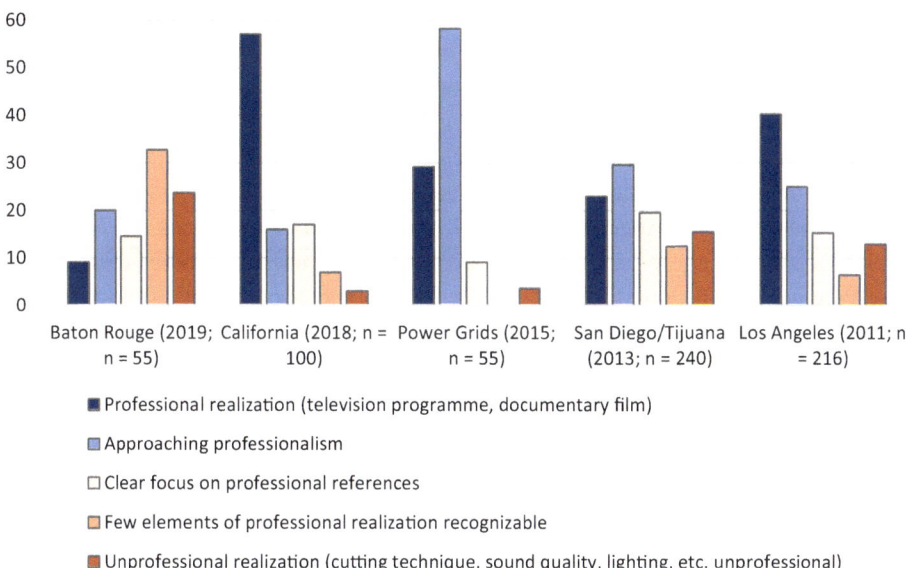

Fig. 1 The degree of professionalism in the implementation of the videos in the above studies (figures in %). (Source: own presentation)

recommended. Both approaches can also be combined if, for example, the quantitatively evaluated categories were inductively generated via qualitative approaches or if, after the quantitative survey, videos that appear to be particularly relevant are subjected to a qualitative approach. Regarding item 3), if (quantitative) statements are to be particularly differentiated categorically, a comprehensive video corpus is indispensable. The evaluation categories can be structured inductively and deductively.

The study of Internet videos can first be divided into the investigation of technical as well as content-related aspects. Technical aspects are, for example, the length of the videos (also as an indicator of the potential detail of the presentation) and the number of hits (in particular as a relative number in relation to the duration of the setting as an indicator of the range). In the above-mentioned study on California in December 2018, the number of views varied between 231 and 1,065,657 (videos detected as German) and 213 and 446,024,343 (videos detected as English; n = 50 each). The length ranged from 53 s to 1 h, 29 min and 53 s. The degree of professionalism of the production can also be estimated. The degree of professionalism of the examined videos differs considerably in the respective study areas (and times) (Fig. 1). As a reference for a 'professional implementation', the presentation of the standard of a supraregional news channel or a documentary film produced in this way serves as a reference in terms of editing, image and sound quality, and balance of presentation, while the other side of the pole is marked by a standard such as that found in an uncut spontaneously generated mobile phone video.

Significant spatial symbols represent an essential element of image-related, some-times also linguistic communication in Internet videos (see Fig. 2, Los Angeles). Their function is at least to create a credible framework for action, in which, for example, the Los Angeles space is marked by the "fading in" of the Hollywood sign, but often objects are staged for their own sake, such as the freeways or, in order to correspond too closely to the stereotype of the 'American vertical metropolis', the (comparatively modest) sky-scraper scenery of Los Angeles is depicted in such a way that it does not appear too incomplete (diagonally to its chessboard layout). At this point it becomes clear that the stereotype-based will to portray Los Angeles as an 'American metropolis with a strong vertical orientation' weighs more heavily than the scientific interpretation of Los Angeles as a 'horizontal city' (see for instance: Fogelson 1993 [1967]; Jacobs 1992 [1961]; Keil 2001).

In terms of content, the videos range considerably in the investigations concerning certain locations. Three inductively obtained categories can be identified as constants:

1. Travel related videos. A distinction can be made between those videos that deal with the representation of sights and those videos that document a journey of one's own.
2. Videos related to regional studies. These deal with the idea of the spatial, social, eco-nomic, ecological, and political specifics of a space. This ranges very selectively from the compilation of material obtained by drone flights to a multi-layered and multi-perspective examination of certain sub regions and social aspects (such as California's economic development and complexity). These form an insight into the life of the people in the locale concerned, provided with a more-or-less significant self-represen-tation component. This ranges from depicting the daily routine of immigrants through subcultural self-portrayals or the (partly unannotated) documentation of everyday sit-uations (such as standing in a traffic jam).
3. Music videos. In these videos a different intensity of engagement with the examined environment takes place. The intensities range from an intensive engagement with life on location, to a mystical-utopian imbuing and a clear yet interchangeable references (if the logic of the song had worked with any other southern city as well), to record-ings of live concerts on smartphones, which happen to take place in the examined location.
4. Sports videos. This includes, in particular, excerpts from sports programs on televi-sion and sequences of sports events filmed with smartphones.

From the last two points in particular, it becomes clear that the frequency of occurrence of the specific locations is sometimes rather low (Fig. 2). It shows a high frequency in videos that explicitly thematize the specific locales (travel videos and geography videos); in the other categories that appear again and again, the specific spatial reference may be low or non-existent. This is particularly evident in the videos on networks (not consid-ered in the categorization above), in which detailed technical questions are frequently negotiated without spatial references being established.

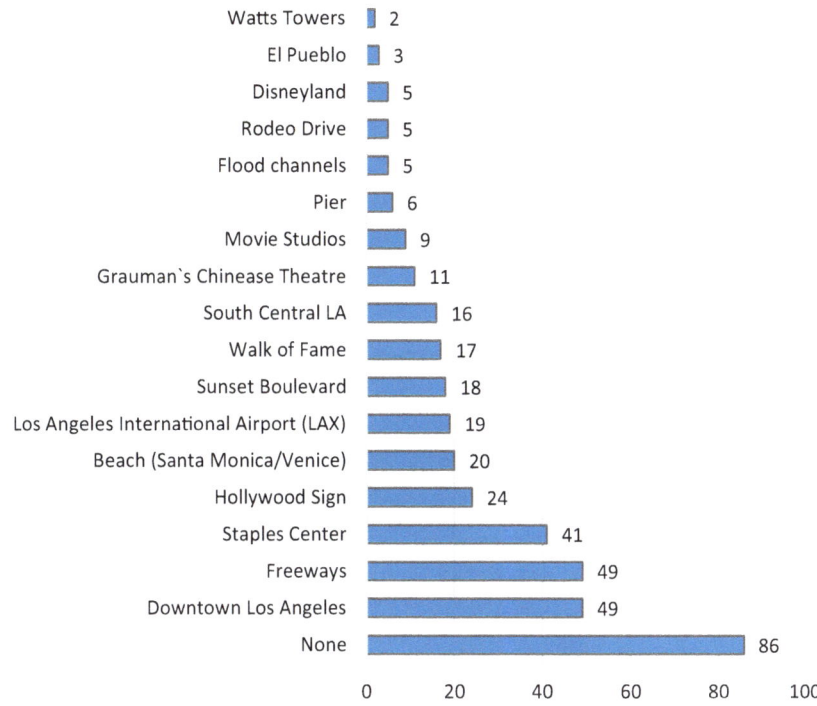

Fig. 2 Significant regional symbols in the Los Angeles videos. (Source: own presentation)

In addition to the above-mentioned categories, which can be found in videos oriented to all specific areas and are usually dealt with in a comparable way, there are certain characteristics to the theming. An example of this can be found in Fig. 3 relating to violence, where the San Diego and Tijuana investigations were evaluated separately as there are significant differences: In three-fifths of the San Diego videos, physical or verbal violence plays no role. In Tijuana, on the other hand, violence is represented in almost seven-tenths. However, the reference to violence in videos about Baton Rouge becomes even more pronounced: in 45.5% of the videos verbal or physical violence is constitutive (Fig. 4). In these videos gang fights, drug offenses, police assaults on blacks, murders, or rapes form the central theme. In contrast, videos in the context of Baton Rouge presenting local history, sights, or sporting events do so without any reference to violence. More than in the other communities treated, Baton Rouge appears in the videos as a dichotomously divided city: on the one hand, black, violent, marked by drug abuse, on the other, white, almost contemplative, emancipative. The only interference between these two worlds is by the police, who are also portrayed as violent.

In addition to spatial specifics in relation to the videos, thematic specifics can also be found that can be traced back to certain current thematic occurrences. One example of this is the forest fires in summer and autumn 2018, which were reflected in the videos

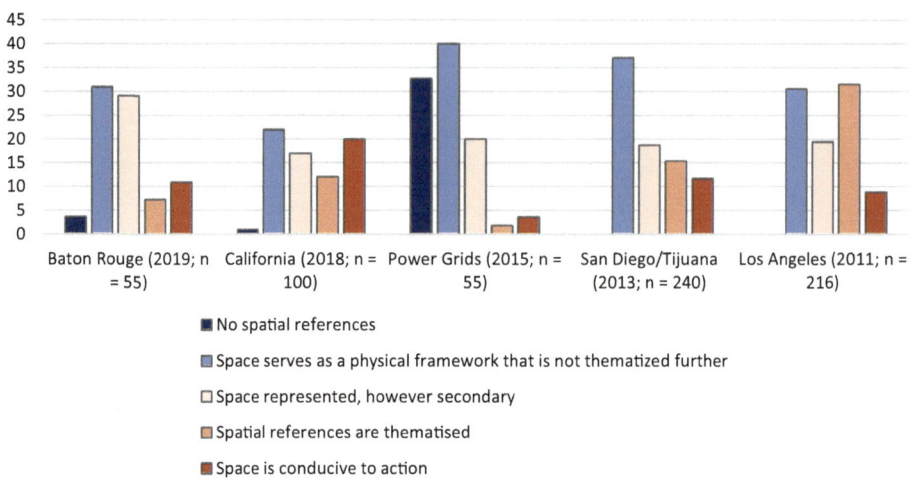

Fig. 3 The intensity of the specific spatial references of the videos in the above studies (in %). (Source: own presentation)

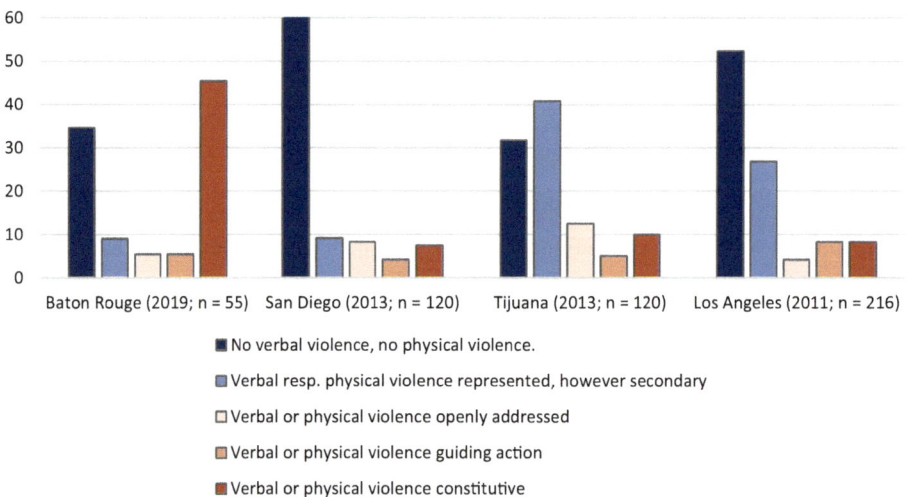

Fig. 4 The intensity of the violent references of the videos in the aforementioned studies (figures in %). (Source: own presentation)

examined. Two different interpretations of the causes and effects of the fires became clear (more detailed in: Kühne and Weber 2019):

1. In 'classic' journalistic reporting, for example from television stations or the web channels of major newspapers, the 'state of affairs' is usually briefly presented. This representation is underlaid with images of burning forests, cars, and houses, as well

as efforts to contain fires (fire engines, fire-fighting helicopters and planes). The presentation of the causes of the fires ranges from brief clues to extensive background reports. The chain of argumentation usually presented here can be (re)constructed as follows: As a result of climate change, many years of drought have occurred so soils and forests are dry. The consequences: a small trigger (e.g. the careless handling of fire) is enough to ignite the fires, which, fanned by strong winds, spread rapidly and despite all the efforts of the (heroic) firefighters now threaten and destroy communities, which—as a result of the rapid spread of the fires—cost many lives. Scientific results regarding fire ecology (fire as a condition for the regeneration of vegetation; see Agee 1996, 2006) are—if at all—mentioned at the margin. Rather, the focus is on questions of how controlled burning of the undergrowth cannot be implemented due to resistance from local residents.

2. Even though the images chosen to deal with the forest fires are, by means of alternative interpretations, comparable to those dealt with in the previous one, their content (especially as regards the interpretations of the causes of the fires), as well as the professionalism of the setting, sound, and editing techniques, are clearly distant from them. The cause (in particular the fires in communities) is not associated with climate change, but with the use of energy weapons. This mission is justified—with recourse to the deep-state theory—by a conspiracy of the military, politics, and science (in short: the elites) who would have conspired against 'the people'. The main indicator is the destruction of houses, while higher trees would have survived the fires (more frequently). The results of fire-ecological research (conducted by 'the elites') are interpreted[1] as part of the conspiracy.

The last example in particular shows how prominent 'alternative' patterns of interpretation are on the Internet and what impact they can have on YouTube regarding certain events, for example 7 out of 50 videos on the keyword 'California' (in German) were in this category (18 on 'classical' reporting) on the fires.

5   Bottom Line

The last example in particular shows that even in spatial contexts, conspiracy theories presented on film are successful in attracting attention to the Internet. Although a difference to 'classical' representations can still be found with regard to the degree of professionalism, this difference can also be seen as 'authenticity' in the eyes of the public that is inclined towards these interpretations of the world. The polarity of information

---

[1]Even beyond the findings of fire ecology research, simple observation could have led to the conclusion that the formation of thicker bark in old trees, for example, reduces the risk of damage caused by forest fires in the corresponding ecosystems.

and disinformation can also be seen in the context of the processing of the world view in the form of images. A responsible handling of this kind of transmission of information and world interpretation requires a high degree of media competence on the part of the user. This is not achieved by the consumption of media, but by the informed and reflected interaction with it (Ferrés and Piscitelli 2012). A basis for a critical examination of the function and effect of new media is also the (critical) examination of stereotypes, in this context spatial stereotypes. The questioning of 'classical' stereotypes is indeed to be found in the videos examined (especially the events around the Californian fires questioning the common stereotypes around the 'Golden State' and likewise the stereotype of Tijuana as a city of cheap pleasure and crime is more often questioned), yet the reproduction of stereotypical points of view dominates (especially Baton Rouge), see Fig. 5.

In Internet videos, the combination of moving images and words provides a particularly immersive opportunity for presentations and self-portrayals. These productions, however, demonstrate quite different degrees. In this respect, the scientific study of Internet videos cannot claim to be 'authentic' (whatever that may mean) to investigate living environments. Rather, they examine the more or less conscious construction of the living world up to a conscious construction of expert world interpretations.

With regard to the categories of the function of landscape in film presented previously, in particular following Escher and Zimmermann (2001), the following connection can be drawn for the Internet videos examined: In the videos examined, the appropriate location of the plot in the form of a backdrop dominates, especially where the specific space is not represented around itself or with the goal of tourist valuation or

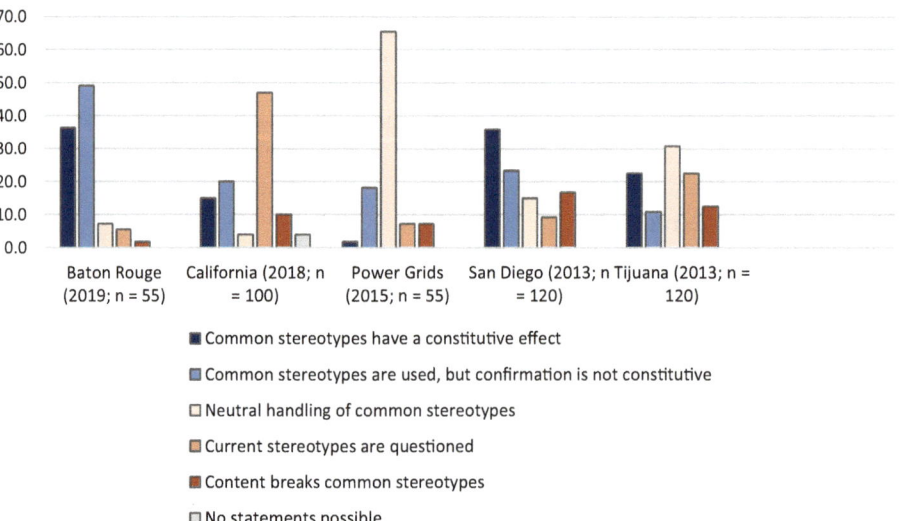

Fig. 5 The degree of stereotypicity of the videos in the above studies (in %). (Source: own presentation)

documentation. A mystical imbuing occurs only in exceptions, a connotation with 'freedom' is rare, rather the struggle for 'respect', especially in underprivileged communities (especially in videos, located in Baton Rouge). Overall, a striving for 'authenticity' on the part of most video makers can be discerned; divergences between spatial connection and location cannot be discerned in the videos examined.

From a spatial science perspective, there is still considerable potential for research in different language areas, including international comparative research. Comparisons accordingly offer the key to the recording and reproduction of different spatial constructions. It is also possible to make comparisons here with 'classic' media approaches, such as newspapers or travel guides. The analysis of Internet videos can use the more comprehensive treatment of regional geography as an element of the social construction of spaces in the sense of a 'new regional geography' (e.g. Gilbert 1988; Holmén 1995; Paasi 2009; Thrift 1991). A greater depth would also be achieved by dealing with Internet videos if both the production side (e.g. by questioning the producers as well as the reception side (beyond the analysis of comments on the videos) were subjected to a systematic investigation. Regarding the presentation of the contents, methods of innovative thematic cartography are also suitable.

References

Agee, J. K. (1996). *Fire ecology of Pacific Northwest forests*. Washington: Island Press.

Agee, J. K. (2006). *Fire in California's ecosystems*. Berkeley: University of California Press.

Aitken, S. C., & Dixon, D. P. (2006). Imagining geographies of film. *Erdkunde, 60*(4), 326–336.

Aitken, S. C., & Zonn, L. (Eds.). (1994). *Place, power, situation, and spectacle. A geography of film*. Lanham: Rowman & Littlefield Publishers.

Altheide, D. L., & Schneider, C. J. (2013). *Qualitative media analysis (qualitative research methods)* (2nd ed., Vol. 38). Los Angeles: SAGE Publications.

Aschenbrand, E. (2017). *Die Landschaft des Tourismus. Wie Landschaft von Reiseveranstaltern inszeniert und von Touristen konsumiert wird*. Wiesbaden: Springer VS.

Baudrillard, J. (1978). *Agonie des Realen* (Merve-Titel, Vol. 81). Berlin: Merve.

Behrens, R. (2008). *Postmoderne (2* (2nd ed.). Hamburg: Europäische Verlaganstalt.

Bellini, A., & Leonardi, L. (2020). Prato: The social construction of an industrial city facing processes of cultural hybridization. In D. Edler, C. Jenal, & O. Kühne (Eds.), *Modern approaches to the visualization of landscapes* (pp. 549–572). Wiesbaden: Springer VS.

Berger, A. A. (2018). *Media analysis techniques* (6th ed.). Los Angeles: SAGE Publications.

Beuka, R. (2004). *SuburbiaNation. Reading suburban landscape in twentieth-century american fiction and film*. New York: Palgrave Macmillan.

Bollhöfer, B. (2007). *Geographien des Fernsehens. Der Kölner Tatort als mediale Verortung kultureller Praktiken* (Kultur- und Medientheorie). Bielefeld: transcript.

Bollhöfer, B., & Strüver, A. (2005). Geographische Ermittlungen in der Münsteraner Filmwelt: Der Fall Wilsberg. *Geographische Revue, 7,*(1/2), 25–42.

Bonfadelli, H. (2002). *Medieninhaltsforschung. Grundlagen, Methoden, Anwendungen* (UTB Medien- und Kommunikationswissenschaft, Vol. 2354). Konstanz: UVK Verlagsgesellschaft mbH.

Bourdieu, P. (1989). Social space and symbolic power. *Sociological theory, 7,*(1), 14–25.

Burgess, J. (2008). 'All your chocolate rain are belong to us'? Viral video, Youtube and the dynamics of participatory culture. In G. Lovink & S. Niederer (Eds.), *Video vortex reader. Responses to Youtube* (INC Reader, Vol. 4, pp. 101–109). Amsterdam: Institute of Network Cultures.

Burgess, J., & Green, J. (2018). *YouTube. Online video and participatory culture (digital media and society series)* (2nd ed.). Cambridge: Polity Press.

Cosgrove, D. E. (1984). *Social formation and symbolic landscape.* London: University of Wisconsin Press.

Davis, M. (1998). *Ecology of fear. Los Angeles and the imagination of disaster.* New York: Metropolitan Books.

Davies, G. (2003). Researching the networks of natural history television. In A. Blunt, P. Gruffudd, J. May, M. Ogborn, & D. Pinder (Eds.), *Cultural geography in practice* (pp. 202–217). London: Arnold.

Edler, D., Kühne, O., Jenal, C., Vetter, M., & Dickmann, F. (2018). Potenziale der Raumvisualisierung in Virtual Reality (VR) für die sozialkonstruktivistische Landschaftsforschung. *Kartographische Nachrichten, 68*(5), 245–254.

Edler, D., Jenal, C., & Kühne, O. (2020a). Modern approaches to the visualization of landscapes—An introduction. In D. Edler, C. Jenal, & O. Kühne (Eds.), *Modern approaches to the visualization of landscapes* (pp. 3–15). Wiesbaden: Springer VS.

Edler, D., Keil, J., & Dickmann, F. (2020b). From Na Pali to Earth—An 'unreal' engine for modern geodata? In D. Edler, C. Jenal, & O. Kühne (Eds.), *Modern approaches to the visualization of landscapes* (pp. 279–291). Wiesbaden: Springer VS.

Eisenstein, S. (1949). *Film form. Essays in film theory (a harvest book)* (Vol. 153). New York: Harcourt, Brace & World (edited and translated by Jay Leyda).

Eisenstein, S. (2014). *Film form. Essays in film theory* . New York: Houghton Mifflin Harcourt (edited and translated by Lay Leyda).

Escher, A., & Zimmermann, S. (2001). Geography meets Hollywood. Die Rolle der Landschaft im Spielfilm. *Geographische Zeitschrift, 89*(4), 227–236.

Ferrés, J., & Piscitelli, A. (2012). Media competence: An articulated proposal of dimensions and indicators. *Comunicar: Revista Científica de Comunicación y Educación, 19*(38), 75–82. https://doi.org/10.3916/C38-2012-02-08.

Fine, D. M. (2004). *Imagining Los Angeles. A city in fiction.* Reno: University of Nevada Press.

Fogelson, R. M. (1993). *The fragmented metropolis. Los Angeles, 1850–1930 (classics in urban history)* (Vol. 3). Berkeley: University of California Press (First publication 1967).

Forrest, D., Harper, G., & Rayner, J. (Eds.). (2017). *Filmurbia. Screening the suburbs.* London: Palgrave Macmillan.

Fröhlich, H. (2007). *Das neue Bild der Stadt. Filmische Stadtbilder und alltägliche Raumvorstellungen im Dialog* (Erdkundliches Wissen, Bd. 142). Stuttgart: F. Steiner.

Garrett, B. L. (2011). Videographic geographies: Using digital video for geographic research. *Progress in Human Geography, 35*(4), 521–541.

Gilbert, A. (1988). The new regional geography in English and French-speaking countries. *Progress in Human Geography, 12*(2), 208–228. https://doi.org/10.1177/030913258801200203.

Gold, J. R., & Gold, M. M. (2013). The field and the frame: Landscape, film and popular culture. In P. Howard, I. H. Thompson, & E. Waterton (Eds.), *The Routledge companion to landscape studies* (pp. 210–219). London: Routledge.

Graham, M. (2013). Geography/internet: Ethereal alternate dimensions of cyberspace or grounded augmented realities? *the Geographical Journal, 179*(2), 177–182.

Graham, M., Ojanperä, S., & Dittus, M. (Eds.). (2019). *Internet geographies. Society and the Internet: How networks of information and communication are changing our lives.* Oxford: Oxford University Press.

Greider, T., & Garkovich, L. (1994). Landscapes: The social construction of nature and the environment. *Rural Sociology, 59*(1), 1–24. https://doi.org/10.1111/j.1549-0831.1994.tb00519.x.

Habermas, J. (1981). *Theorie des kommunikativen Handelns*. Frankfurt (Main): Suhrkamp.

Higson, A. (1987). The landscapes of television. *Landscape Research, 12*(3), 8–13. https://doi.org/10.1080/01426398708706232.

Holmén, H. (1995). What's new and what's regional in the 'new regional geography'? *Geografiska Annaler: Series B, Human Geography, 77*(1), 47–63. https://doi.org/10.1080/04353684.1995.11879680.

Jacobs, J. (1992). *The death and life of great American cities*. New York: Vintage Books (First publication 1961).

Keil, R. (2001). Consolidation and secession in Los Angeles: The dialectics of urban governance reform at the end of the twentieth century. *European Journal of American Culture, 20*(1), 22–35. https://doi.org/10.1386/ejac.20.1.22.

Krebs, S. (2008). Von Landscape One zu Heimat 3—Zur Realität filmischer Landschaften. In H. Küster (Ed.), *Kulturlandschaften. Analyse und Planung (Stadt und Region als Handlungsfeld)* (Vol. 5, pp. 131–142). Frankfurt: Lang.

Kühne, O. (2012). *Stadt—Landschaft—Hybridität. Ästhetische Bezüge im postmodernen Los Angeles mit seinen modernen Persistenzen*. Wiesbaden: Springer VS.

Kühne, O. (2013). *Landschaftstheorie und Landschaftspraxis. Eine Einführung aus sozialkonstruktivistischer Perspektive*. Wiesbaden: Springer VS.

Kühne, O. (2018). *Landscape and power in geographical space as a social-aesthetic construct*. Dordrecht: Springer International Publishing.

Kühne, O. (2019). *Landscape theories. A brief introduction*. Wiesbaden: Springer VS.

Kühne, O., & Jenal, C. (2020a). *Baton rouge—The multivillage metropolis. A neopragmatic landscape biographical approach on spatial pastiches, hybridization, and differentiation*. Wiesbaden: Springer VS.

Kühne, O., & Jenal, C. (2020b). The threefold landscape dynamics—Basic considerations, conflicts and potentials of virtual landscape research. In D. Edler, C. Jenal, & O. Kühne (Eds.), *Modern approaches to the visualization of landscapes* (pp. 389–402). Wiesbaden: Springer VS.

Kühne, O., & Schönwald, A. (2015). *San Diego. Eigenlogiken, Widersprüche und Hybriditäten in und von ,America's finest city'*. Wiesbaden: Springer VS.

Kühne, O., & Weber, F. (2015). Der Energienetzausbau in Internetvideos—Eine quantitativ ausgerichtete diskurstheoretisch orientierte Analyse. In S. Kost & A. Schönwald (Eds.), *Landschaftswandel—Wandel von Machtstrukturen* (pp. 113–126). Wiesbaden: Springer VS.

Kühne, O., & Weber, F. (2019). *Hybrid California. Annäherungen an den Golden State, seine Entwicklungen, Ästhetisierungen und Inszenierungen*. Wiesbaden: Springer VS.

Lefebvre, M. (Ed.). (2006). *Landscape and film*. New York: Routledge.

Leimgruber, W. (2004). Heidi und Tell. Schweizerische Mythen in regionaler, nationaler und globaler Perspektive. In K. Hanika & B. Wagner (Eds.), *Kulturelle Globalisierung und regionale Identität. Beiträge zum kulturpolitischen Diskurs*. Dokumentation des Kulturpolitischen Kongresses vom 5. bis 7. September 2002 in Ludwigsburg (Edition Umbruch, 17: Texte zur Kulturpolitik, pp. 32–44). Essen: Klartext.

Loda, M., Kühne, O., & Puttilli, M. (2020). The social construction of Tuscany in the German and English speaking world—Presented by the analysis of Internet images. In D. Edler, C. Jenal, & O. Kühne (Eds.), *Modern approaches to the visualization of landscapes* (pp. 157–171). Wiesbaden: Springer VS.

Ludewig, A. (2011). *Screening Nostalgia. 100 years of German Heimat film*. Bielefeld: transcript.

Ludewig, A. (2019). Ostalgie im Dresdner *Tatort* (2016/2017). In M. Hülz, O. Kühne, & F. Weber (Eds.), *Heimat. Ein vielfältiges Konstrukt* (pp. 279–297). Wiesbaden: Springer VS.

Luhmann, N. (1996). *Die Realität der Massenmedien*. Opladen: Westdeutscher.

Lukinbeal, C. (2005). Cinematic landscapes. *Journal of Cultural Geography, 23*(1), 3–22.

Lukinbeal, C. (2012). "On location" filming in San Diego county from 1985–2005: How a cinematic landscape is formed through incorporative tasks and represented through mapped inscriptions. *Annals of the Association of American Geographers, 102*(1), 171–190. https://doi.org/10.1080/00045608.2011.583574.

Maurer, M., & Reinemann, C. (2006). *Medieninhalte. Eine Einführung*. Wiesbaden: VS Verlag für Sozialwissenschaften.

Moore-Colyer, R., & Scott, A. (2005). What kind of landscape do we want? Past, present and future perspectives. *Landscape Research, 30*(4), 501–523. https://doi.org/10.1080/01426390500273254.

Münker, S. (2009). *Emergenz digitaler Öffentlichkeiten. Die Sozialen Medien im Web 2.0* (Edition Unseld, Vol. 26). Frankfurt (Main): Suhrkamp.

Nagle, A. (2017). *Kill all normies. The online culture wars from Tumblr and 4chan to the alt-right and Trump*. Winchester, UK: Zero Books.

Natter, W. (1994). The city as cinematic space: Modernism and place in *Berlin, Symphony of a City*. In S. C. Aitken & L. Zonn (Eds.), *Place, power, situation, and spectacle. A geography of film* (pp. 203–228). Lanham: Rowman & Littlefield Publishers.

Paasi, A. (2009). Regional geography I. In R. Kitchin & N. Thrift (Eds.), *International encyclopedia of human geography* (Vol. 9, pp. 214–227). Amsterdam: Elsevier.

Papadimitriou, F. (2020). Visualization of future landscapes, postmodern cinema and geographical education. In D. Edler, C. Jenal, & O. Kühne (Eds.), *Modern approaches to the visualization of landscapes* (pp. 351–369). Wiesbaden: Springer VS.

Papadimitriou, F. (2021). *Spatial Complexity. Theory, mathematical methods and applications*. Cham: Springer.

Plien, M. (2017). Filmisch imaginierte Geographien Jugendlicher. *Erdkundliches Wissen, 160*(1).

Prisille, C., & Ellerbrake, M. (2020). Virtual reality (VR) and geography education: Potentials of 360° 'experiences' in secondary schools. In D. Edler, C. Jenal, & O. Kühne (Eds.), *Modern approaches to the visualization of landscapes* (pp. 321–332). Wiesbaden: Springer VS.

Schmidt, J.-H. (2011). *Das neue Netz. Merkmale, Praktiken und Folgen des Web 2.0*. Konstanz: UVK Verlagsgesellschaft.

Siepmann, N., Edler, D., & Kühne, O. (2020). Soundscapes in cartographic media. In D. Edler, C. Jenal, & O. Kühne (Eds.), *Modern approaches to the visualization of landscapes* (pp. 247–263). Wiesbaden: Springer VS.

Strüver, A. (2009). Tatort Münster: Urbane Inszenierung im/durch Film. *Berichte zur deutschen Landeskunde, 83*(4), 331–348.

Tamblé, M. (2012). Was ist eigentlich: YouTube? http://www.marketing-boerse.de/Fachartikel/details/Was-ist-eigentlich-YouTube/33879. Accessed 25 Feb 2016.

Thrift, N. (1991). For a new regional geography 2. *Progress in Human Geography, 15*(4), 456–466. https://doi.org/10.1177/030913259101500407

Vester, H.-G. (1993). *Soziologie der Postmoderne*. München: Quintessenz.

Vetter, M. (2020). Technical potentials for the visualization in virtual reality. In D. Edler, C. Jenal, & O. Kühne (Eds.), *Modern approaches to the visualization of landscapes* (pp. 307–317). Wiesbaden: Springer VS.

Videonitch. (2017). 36 mind blowing Youtube facts, figuers and statistics—2017 (re-post). http://videonitch.com/2017/12/13/36-mind-blowing-youtube-facts-figures-statistics-2017-re-post//2017/12/13/36-mind-blowing-youtube-facts-figures-statistics-2017-re-post/. Accessed 17 July 2019.

Werlen, B. (1997). *Sozialgeographie alltäglicher Regionalisierungen. Band 2 Globalisierung, Region und Regionalisierung* (Erdkundliches Wissen Schriftenreihe für Forschung und Praxis, Bd. 119). Stuttgart: Steiner.

YouTube. (2016). YouTube-Statistik. https://www.youtube.com/yt/press/de/statistics.html. Accessed 25 Feb 2016.

Zimmermann, S. (2019). Filmlandschaft. In O. Kühne, F. Weber, K. Berr, & C. Jenal (Eds.), *Handbuch Landschaft* (pp. 623–629). Wiesbaden: Springer VS.

Dr. Dr. Olaf Kühne, born in 1973, is Professor of Urban and Regional Development at the University of Tübingen. He received his doctorate in geography (on an urban climatological topic) in Saarbrücken and sociology (on the social construction of landscape) in Hagen. His habilitation in Mainz dealt with ecological and social transformation processes in Poland. Today, he is particularly concerned with social issues of energy system transformation, landscape theory formation and urban transformation processes. These are the focus of his interest in feedback relationships between the material, individual and social world.

Landscape in Internet Pictures

Simone Linke

Abstract

Visualizations of physical spaces known as landscapes are ubiquitous today, but social-scientific access to these pictures has been limited so far. Pictures in social and cultural geography are by no means new. Pictures as social-scientific data or the critical examination of the creation and appropriation of pictures, however, represent a relatively new field of research. This contribution addresses this gap in research and therefore poses the question of how media landscape constructs are aesthetically constructed. This article deals with the editing and selection of these representations, because pictures do not show representations of a so-called reality, but selected, individual constructions that are subject to personal influence. The search for pictures on the Internet serves as an instrument for examining the medial construction of physical spaces described as landscapes. Subsequently, the question is posed as to what everyday constructions these pictures contribute to and what social effects these media constructions can bring with them.

Keywords

Landscape constructs · Landscape aesthetics · Internet pictures · The power of pictures · Picture analysis · Depicted landscapes

S. Linke (✉)
Technische Universität München, Munich, Germany
e-mail: s.linke@tum.de

© Springer Fachmedien Wiesbaden GmbH, part of Springer Nature 2020   139
D. Edler et al. (eds.), *Modern Approaches to the Visualization of Landscapes*,
RaumFragen: Stadt – Region – Landschaft, https://doi.org/10.1007/978-3-658-30956-5_8

1   Introduction: Landscape in Pictures

Landscape can be viewed from many different perspectives: from science and from everyday life. Not infrequently, these considerations are currently related to the change in the physical foundations of landscape. Various competing uses and increasing technical possibilities mean that this change happens quickly and is not always positively accepted by all parts of society. What *landscape* is, how *landscape* should be and how *landscape* is evaluated varies from person to person. Answers to this question are sought in science and everyday life. However, there are no simple answers and the questions must also be differentiated—depending on the scientific perspective. This contribution pursues a social constructivist perspective. That means that *landscape* is not to be understood as an object, but as a social construction (see also in this volume: Al-Khanbashi 2020; Berr 2020; Edler et al. 2020a; Fontaine 2020a, b; Jenal 2020; Kühne 2020; Kühne and Jenal 2020; Loda et al. 2020; Roßmeier 2020; Schenk 2020; Siepmann et al. 2020; Weber 2020). In this understanding, "the construct *landscape* is the result of socially formed patterns of interpretation and evaluation, on the basis of which an internal synthesis of observed material objects and their connection with symbolic meanings take place" (Kühne 2019, p. 17, emphasis in original; for more details see among many Berger and Luckmann 1980; Burckhardt 2008; Cosgrove 1998; DeLue 2008; Kazig 2013; Kühne 2006, 2012, 2013; Leibenath et al. 2013; Schütz 1972; Weber 2015, 2016). This means that the questions must be adjusted: "How are landscapes constructed?" In order to approach this question, different methods can be used, such as surveys. Since in today's digitalized world, pictures and visualizations also play an important role, this article reffers to this visual datas and examines pictures of landscape constructs. In order to limit this object of research and also to refer to the currently pronounced social appropriation of information via the Internet, the question is further specified: "How are landscape constructs visually represented on the Internet?" At present, only a few authors focus on visual objects of investigation in the field of landscape. These few authors are, among others, B. Kühne (2002) with her work "Das Naturbild in der Werbung" (The picture of Nature in Advertising), or the anthology by Schlottmann and Miggelbrink (2015c): "Visuelle Geographien" (Visual Geographies). Some other current authors focus on other themes in relation to *landscape*, but also have studies of pictures in their works, such as Gailing (2013), Kost (2017) or Kühne (2012, 2018). Schlottmann and Miggelbrink also point out that "the turn to pictures and pictorialities and the visual in geography is not new" (2015a, p. 14) but the critical examination of the creation and appropriation of pictures is new (2015a, p. 14). An interesting new field of research in the context of landscape constructs on the Internet, which is not examined in detail here, is the gaming community, which is very much concerned with *landscape* (see in this volume: Edler et al. 2020b; Fontaine 2020a; Prisille and Ellerbrake 2020; Stintzing et al. 2020; Vetter 2020). First investigations show potentials of these visualizations of spaces for social constructivist landscape research (see Edler et al. 2018, 2019).

The investigation of pictures from the Internet represents a contribution to this still young field of research. My doctoral dissertation "The Aesthetics of Medial Landscape Constructs" was published 2019. Within the framework of this work, a study from 2017/2018 was published, which also dealt with these pictures. This article takes up these results, updates and expands the investigation. In order to investigate the question of the media representation of landscape constructs here in this article, I will first briefly discuss the understanding of *landscape* used here in relation to the delimitation of content, to clarify the object of this work. Following this, theoretical aspects of media landscape constructs on the Internet will be mentioned, before landscape constructs depicted on the Internet will be examined by way of example.

2   Of Depicted Landscape Constructs and Pictures on the Internet

Landscape in this contribution is understood as a social construction. In addition to this theoretical perspective, another approach at a different level is also important. It is about the boundaries of the understanding of *landscape* and the breaking down of these boundaries. In the scientific context, either a *narrow* or *extended* concept of landscape is spoken of. If the reference to a *narrow* (thus limited) landscape concept is chosen, this assumes so-called natural spaces or cultivated natural spaces. In this understanding, pre-industrial rural cultural landscapes are regarded as an idealized type of *landscape*. The *extended* landscape concept allows not only undeveloped, so-called near-natural but also built-up areas and so-called non-natural spaces as landscapes (Hokema 2009, p. 239), thus exists without fixed (mental or content-related) boundaries. This contribution is based on the *extended* landscape concept, since the constructivist perspective only permits the extended perspective. The term *landscape* cannot be generally and comprehensively defined in the sense of constructivism and therefore the term cannot make any statements as to which spaces belong to *landscape* and which do not. The extended perspective of *landscape* negates the narrow version of the concept and allows the content to be largely left open.

Since the present article is not only about *landscape*, but also about internet pictures, this object must also be examined more closely. Pictures are not just pictures—the concept of pictures is confusing and blurred, because very different phenomena are often described by them (Klinke 2013, p. 11). Especially in the German language there is only the term *Bild*—here at first nobody knows whether we are talking about an artistic painting, a photograph or an mental image (Klinke 2013, p. 11 f.). The English language distinguishes at least between *image* and *picture* (Klinke 2013, p. 13). While the term *image* refers often to something immaterial, e. g. impression, mental image, imagination—the term *picture* often goes in the direction of something material or physical, e. g. drawing, photography, also description (cf. Belting 2011; Mitchell 1995). Even if it is arguable whether a virtual picture represents a material picture or not (cf. Edler et al. 2018), the term picture is used in the further course due to the differentiation of meaning. In this context it is necessary to briefly discuss which picutres are considered here and how they are to be understood. Internet pictures in this article are more or less edited photographs that can be seen on certain websites on the Internet. There is no mention of unedited photographs, since according to Harper pictures are always "subjectively perceived representations" (2008) and cannot be regarded as objective material (Harper 2008, p. 406).[1] In the following analysis, those internet pictures will be examined that show landscape constructs in a broader understanding. Limiting oneself to the narrow concept of landscape would simplify the investigation but does not reflect the current development and negates the constructivist perspective, as has already been mentioned.

Nevertheless, it is not enough just to ask questions about the pictorial representation of landscape constructs on the Internet. The far-reaching significance of this question is only apparent from the realization that pictures on the Internet are an object of communication for a broad mass of people. On the one hand this is the production, but on the other hand also the consumption of this content on the Internet, which is available to a large public. This dissemination and public effectiveness of various contents on the Internet has far-reaching consequences: "Like few global changes, the Internet has changed individual, social and political forms of communication and action" (Thimm 2017, p. 433). Not only global changes can be observed, but also the everyday life changes by the medium Internet, so Thimm (2017). In this context, media are not only to be understood as neutral messengers, in the sense of the generativity of media they always add something to the message and shape it. The media are therefore also embossing instances (Wagner 2014, p. 19 f.).

Especially pictures in the internet have to be understood as a user-generated content (cf. Thimm 2017). This means that many users actively produce content and become producers (Thimm 2017, p. 435). The pictures examined therefore show how these average users or producers construct the term *landscape*. According to Popitz, this active generation of data turns users into data setters who can exercise object-mediated power. Through the production of pictures, they "add a new fact to the reality of the world" (Popitz 1992, p. 30). This "power of production and producers" (Popitz 1992, p. 31) has consequences, for pictures are "in many respects both a product and a central construction element of social and spatial realities" (Felgenhauer 2015, p. 67). Gailing speaks of the fact that "spatial pictures have great suggestive power and, as informal institutions, have a decisive influence on collective action" (2013, p. 238). The production and effect of pictures are thus in constant interaction with each other and influence each other.

The pictures on the Internet, such as on Google, communicate with a more or less broad public. They are not private in the sense that they stand for their selves or only for the producers. In order to be able to use pictures as social science data, it plays an

---

[1]The subjectively selected section of the depiction, for example, is an illustration of the objectivity of depicted landscape constructs.

important role, depending on the question posed, to be aware of the meaning, construction or influence of the picture producers (Harper 2008, p. 406). Many of the pictures examined have obvious economic advertising intentions, as shown in the previous dissertation (Linke 2019). If an overview of the producers or broadcasters of the pictorial material is compiled, the following groups are to be found in the dissertation just mentioned as well as in the current study: Photographers, providers of travel or accommodation (including marketing for cities or regions), providers of entertainment platforms, bloggers, picture databases, online journals, expert portals and associations that provide general information (health, nature conservation), educational centres or universities. Advertising intentions are usually, but not only economically motivated, and can also, for example, be influenced by health policy or voluntary work or, in the case of social marketing, be an advertisement for regions, cities, political parties or cultural institutions (B. Kühne 2002, p. 21). Behrens describes advertising in a wider context: "In the broadest sense, *advertising* is a certain form of influence in the interpersonal sphere; it is therefore not a specific economic phenomenon, but a general social phenomenon" (1976, p. 11, emphasis in original). According to this statement, all pictures of this search can be unterstood as advertising, with material consumption character or not, because they all communicate, add something to society and thus influence it. After these explanations, a small qualitative analysis will now answer the question of how landscape constructs are represented in pictures on the Internet.

3   The Representations of Landscape Constructs on the Internet

The question about landscapes in internet picutres is very general, as the internet simply stands as an umbrella term for a seemingly infinite number of different digitized representations (see in this volume: Kleber et al. 2020; Kühne 2020; Loda et al. 2020). Therefore, the scope of the study on the topic of landcape in internet picutres must be limited. First, the Internet is currently a very important source of information for large population groups around the world. In order to better understand what users are doing on the Internet, the results of various studies from Germany are briefly presented here. According to these studies, around 64 million people in Germany used the Internet in 2018 (Brandt 2019). Of these, around 24.89 million people aged 14 (and older) used the Internet every day to search for information (Pawlik 2019). It is interesting to note that the consumption of entertainment-related content (music, video and pictures) on the Internet has risen sharply, especially among young people: in 2008, the usage time for entertainment was 18%, in 2018 already 31% (Weidenbach 2019). In order to find answers to the question about the visual representation of landscape constructs on the Internet, picture searches on the topic of landscape in Google are presented below.

Fig. 1 Pictures of the Google picture search. (Source: https://pixabay.com/de/photos/fr%C3%BChling-landschaft-blumen-4022427/; labeled for reuse)

The terms *Landschaft* and *landscape*[2] were entered into the Google picture search on 12.08.2019. The first 100 pictures of the Google picture search were subjected to a content analysis (after Früh 2015) as well as a supplementary picture interpretation and an aesthetic fine analysis. Afterwards, spatial pictures of landscape are also briefly taken up in other areas of the Internet.

3.1 Landscape in Google Picture Search

In the Google picture search, both the German term *Landschaft* and the English term *landscape* were searched. The results were very similar, even if other pictures were displayed. For the analysis, the results of the picture search with the German term *Landschaft* were used, because the location of the search was in Germany and the German term for *Landschaft* is more meaningful here.

Three sample pictures[3] of a google picture search can be seen in Figs. 1, 2 and 3.

---

[2]It should be noted here that the meanings of *Landschaft* and *landscape* are not identical; see Drexler (2013). Among other things, the terms differ in the material dimension. The German term also contains a materiality, the English term has its own words for the material dimension of landscape, such as *land* and *country*; Kühne (2019, p. 67). Furthermore the term *Landschaft* shows according to Kühne "a special emotional charge in the context of *Heimat*" (2019, p. 67, emphasis in original).

[3]For licensing reasons not all results of the search can be shown. Here you will find exemplary pictures which can be used freely and which are very similar to the majority of the remaining pictures.

Fig. 2 Pictures of the Google picture search. (Source: https://pixabay.com/de/photos/landschaft-deutschland-natur-3378774/; labeled for reuse)

Fig. 3 Pictures of the Google picture search. (Source: https://www.flickr.com/photos/93243867@N00/29223632383; labeled for reuse)

First, there is the question of which spaces are constructed in the pictures. In general, it can be said that almost without exception the first 100 pictures show motifs that can be described as romanticised and idyllic landscapes. This can first be concluded from the fact that, with very few exceptions, there are no physical elements depicted in the pictures that can be described as disturbing in relation to the theme of landscape—only two physical elements can clearly be described as disturbing: an asphalted road (barely visible) in picture No. 20 and a nuclear power plant in picture No. 92 (the classification of disturbing and stereotypical physical elements in relation to the theme of landscape is based on previous research done by Hokema 2013; Kook 2009; Kühne 2006, 2017; Kühne et al. 2013; Linke 2018; Micheel 2012). However, not only are there elements missing that are described as disturbing, there are also at least two, usually more physical elements in each individual picture that are described as positive, as can be seen from the quantitative evaluation, which is only presented here in summary form. Examples of these positive elements in connection with the theme of landscape are meadows, blue skies, trees or forests, bushes, mountains, country lanes, sun, water or farms (based on previous research done by Hokema 2013; Kook 2009; Kühne 2006, 2017; Kühne et al. 2013; Linke 2018; Micheel 2012).

In addition to the content of the motifs, the question of how the landscape constructs are aesthetically constructed will also be discussed here. Various aspects such as depicted physical elements, colour effects and picture editing contribute to answering this question. Almost without exception, the physical elements shown, as well as the overall view of the elements, can be attributed to the aesthetically positive categories of beauty and sublimity, but also to kitsch: For example, colorful flower meadows, individual farms, trees, sunsets, streams can be attributed to the category of beauty, while mountains, steep cliffs, the sea, seemingly endless fields can stand for a sublime aesthetic. Assigned to the category of kitsch or even romantic, there are elements to be mentioned which, although they also occur in the category of beauty and sublimity, are then overdrawn and strongly idealized (cf. Gelfert 2000). These can be strongly overdrawn sunsets or a rainbow in combination with other stereotyped elements (cf. Linke 2019).

Almost all pictures show particularly intense, strong and saturated colours, which, according to Camgöz et al. (2002), Felser (2015) and Gorn et al. (1997), are initially generally described as positive. Usually it is the colours blue and green that characterise the pictures. Green stands next to vegetation and the terms *nature* and *landscape*, furthermore in general for life, origin, new beginning, growth and hope (B. Kühne 2002; Welsch and Liebmann 2012), green and blue as colour of sky and water for distance, longing, purity and eternity (B. Kühne 2002; Welsch and Liebmann 2012). Both colours are mostly positively connotated (Welsch and Liebmann 2012). In addition, red/orange/yellow tones can be seen again and again, mostly in warm shades. The colours stand for power, fire, energy, conviviality, cheerfulness, wealth, optimism and joy (B. Kühne 2002; Welsch and Liebmann 2012). Irrespective of the choice of colour, more than two thirds of the pictures appear revised. It is not possible to determine exactly when a picture was revised (e.g. in terms of colours, textures or light source). A comparison of the pictures shows, however, that either the colour saturation varies greatly, in some cases a second incidence of light, or even strongly altered surface textures can be seen.

The aestheticized representation of the pictures pursues various purposes, some of which are mentioned below as examples: The bright and strong colours should be regarded as beautiful or generally as aesthetically positive, physical elements or motifs described as sublime should impress, the calm composition of the picture should have a harmonious or calming effect, and the absence of sources of noise should suggest calm (cf.. B. Kühne 2002). *Landscape* should therefore reflect various feelings, such as relaxation, detachment, happiness, joy, security, but also reverence and overwhelming (cf. Felser 2015; Linke 2019).

A comparable picture search was already carried out in November 2017 as part of my dissertation. Even two years later, it can be stated that the results have not changed fundamentally. The motifs, the color representations, and the absence of physical elements, which have been described as stereotypical, in relation to the theme of *landscape* remain the same. Despite worldwide events in the climate change debate, such as the worldwide protest action "Fridays for Future" and considerable gains in votes by green parties not only in Germany, the pictures are similar. Neither in *Landschaft* nor in *landscape* are there new energy landscapes in the first 100 pictures, as in the picture search of 2017. However, pictures of wind turbines can be seen further ahead in the picture search: Whereas in 2017 it was only the 238th picture on which a wind turbine could be seen, today, in 2019, it is the 127th and the 131st picture. But these are the only pictures of wind turbines in the first 300 pictures. Whether this is a trend and whether pictures of renewable energies can be seen more frequently and "further ahead" in the Google picture search in the future can only be said after further investigations in the next few years. Two studies within two years are not meaningful enough to make a statement here. Despite these two pictures, the term *landscape* in Google picture search remains a stereotype.

3.2   The Term Landscape in Other Online Sections

This article focuses on the pictures in Google picture search, and there are good reasons for this, as explained at the beginning of Sect. 3. At this point, however, other areas will also be considered here: How are pictures of landscape constructions displayed in other digital spaces on the Internet? In this section already existing investigations will be used and a short look at YouTube will be taken. In my doctoral dissertation I have looked at the detailed analysis of results of the Google picture search as well as pictures of landscape constructs on the start pages of Lower Bavarian communities. Here, it could be established that these pictures did not*neutrally* represent the entire municipality, including the physical areas called landscape. Rather, here, too, a strong selection was made: On almost all pictures motifs are represented, which are particularly characteristic, like e.g. churches, village squares or also as attractively designated undeveloped areas, like e.g. the Bavarian Forest. These pictures, however, are much less frequently heavily edited, as was the case with Google picture search. Nevertheless, it is noticeable that hardly any picture shows physical elements that have been described as disturbing (only one picture shows a wind turbine in the background, photovoltaic systems are not visible at all). The editing of the pictures or the selection of the room sections is therefore less pronounced than the spatially independent[4] results of the Google picture search, but still clearly available. The study by Erik Aschenbrand in "Die Landschaft des Tourismus" (The Landscape of Tourism, 2017) showed that stereotypical constructions of landscape are also dominant in the tourism industry on the Internet. He examined, among other things, catalogues of tour operators, who "publish the same content, but with additional information on their websites" (2017, p. 91). Similar to the results of the Google picture search, the tourism industry also refers to the narrow concept of landscape and associates the understanding of landscape with *beautiful nature* (2017, p. 248). This so-called beautiful nature is intended to evoke emotions in readers (Aschenbrand 2017, p. 249). Finally, a look at the video platform YouTube shows that stereotypical landscape constructs are also shown here when the term "landscape" is entered in the search bar. Even if YouTube does not provide pictures but videos as results, these also show visual representations of landscape constructs and can thus also contribute to answering the question about landcape in internet pictures. Of the first ten videos,[5] five have *relaxation* as a theme,[6] two a *spectacular, sublime landscape*, two a *beautiful landscape* and one a drawing course on *landscape*. The pictures in the videos are comparable with the pictures of the Google picture search. They are characterized by a strong stereotyping, an equally strict editing and a fading out of disturbing physical elements.

These various other areas show that landscape constructs in the various virtual spaces of the Internet are predominantly represented as stereotypical landscapes. In addition, a narrow concept of landscape dominates. If the explanations on the power of pictures (Sect. 2) and the results of the representations of landscape constructs on the Internet are thought together here, the question of the effects of these pictures arises.

4   Is this Landscape? The Question of the Impact of These Pictures

Already in Sect. 2, the cycle was presented that the media content of the Internet arises on the one hand from social processes and on the other hand affects social processes and changes social practices (cf. Felgenhauer 2015; Gailing 2013; Popitz 1992; Wagner 2014). According to Kühne, these media contents thus do not represent reality (2012, p. 376), but are always the result of "selection, evaluation, editing and interpretation of

---

[4]Spatially independent, because the entered term *landscape* did not have to be assigned to a spatial location as in the search on the start pages of Lower Bavarian municipalities. Without exception, rooms of the respective municipality were shown here.

[5]The search was carried out like the Google picture search on 12.08.2019. The term *Landschaft* was entered into the search bar.

[6]The topic of the videos can be determined on the one hand by the title and on the other hand by the contents of the videos.

social events" (Werlen 1997, p. 383). The depicted pictures do not show *landscape*, but only a selected space, which the picture producer calls landscape. And these landscape constructs depicted in the media are not particularly versatile, as we know landscapes from found spaces. Only highly selected sections of physical spaces, particularly stereotyped as landscapes, are shown. The motifs show a very one-sided picture of so-called landscapes. For clarification here a comparison of pictures of the direct surroundings of my hometown in Lower Bavaria with pictures of the Google picture search:

vs.

Fig. 4 Pictures of the Google picture search. (Source: https://pixabay.com/de/photos/italien-landschaft-landschaftlich-1758193/; labeled for reuse)

Fig. 5 Möding, Lower Bavaria. (Source: own photo 2018)

vs.

Fig. 6 Pictures of the Google picture search. (Source:www.commons.wikimedia.org/wiki/File:Landschaft_bei_Kürten.JPG, labeled for reuse)

Fig. 7 Hebertsfelden, Lower Bavaria. (Source: own photo 2014)

vs.

Fig. 8 Pictures of the Google picture
search. (Source:https://pixnio.com/de/
landschaften/sonnenaufgang/himmel-
landschaft-meer-strand-stein-sand-wasser,
labeled for reuse)

Fig. 9 Isar bei Zulling, Lower Bavaria.
(Source: own photo 2018)

This picture comparison (Figs. 4, 5, 6, 7, 8 and 9) shows exemplarily the discrepancy of the representations of medially represented and found landscape constructs. Also for the amateur eye the partly strong editing of the medial pictures becomes apparent. Here the question arises whether Google's picture search shows constructs that correspond to the extended concept of landscape? Probably not, the pictures correspond more to the narrow concept of landscape and are often even a strongly stereotyped idea of landscape. Characteristics for it are among other things: The selection of the sections of space shown, characterised by the absence of physical elements described as disturbing, ugly or anaesthetic; an intense, sometimes strongly artificially heightened colour saturation; colour changes and artificial illumination.

On the one hand, by hiding or ignoring existing landscape constructs as well as new developments (such as the new so-called energy landscapes or a mechanized form of agriculture), on the other hand, the picture of the idealized and stereotyped construction of landscapes in the media is strengthened.

As already mentioned in Sect. 2, these media pictures not only represent something, they also produce something, they shape social constructions (cf. Gailing 2013; Schlottmann and Miggelbrink 2015b; Schneider 2015; Wintzer 2015). Of course, a single picture has no far-reaching social consequences, but the totality of the results does. After all, it is not only stereotyped representations that are dealt with in the Google picture search, but, as Sect. 3.2 has shown, also in other areas on the Internet. These pictures contribute to the fact that the construct landscape is often understood as a stereotyped landscape. Thus, these media pictures can contribute to an acceptance problem, e.g. of new energy landscapes. Stereotyped pictures therefore not only explain existing acceptance problems—they also contribute to them to a certain extent. Because they also have a socializing function.

5   Conclusion

Landscape in Internet pictures—as this contribution has shown, these are mostly stereotypical landscape constructs that arise from a narrow understanding of *landscape*. The main focus is on "well-known subjects that formulate nature as an intact nature, large and sublime, small and idyllic, transparent and manageable in both small and large" (B. Kühne 2002, p. 218). B. Kühne had already stated this in 2002 (B. Kühne 2002). This conclusion is still extremely relevant today and can also be applied to other media areas. Only a few pictures in the media examined show motifs that do not originate in stereotypical or narrow notions of landscape: the so-called new energy landscapes, mechanized agriculture, the numerous sprawling agglomerations with supermarkets outside the outskirts of town (Linke 2017) and much more are missing. In percentage terms, these spaces should also be visible in the pictures, because these are the landscape constructs that surround us as well and and that are generally referred to as *landscapes* (although in a narrow sense). On view are depictions of spaces that want to convey the impression that they are untouched by human hands. Nothing has changed about that to this day. All these stereotypical pictures also have an impact on our idea of *landscape*. For these pictures—to repeat it once more—not only represent the concept of *landscape*, they also create the idea of *landscape*. They shape the associations with the term *landscape* (cf. Gailing 2013; Schlottmann and Miggelbrink 2015b; Schneider 2015; Wintzer 2015). So the more we see these stereotypical landscape constructs, the more they remain a content filling to the umbrella term *landscape* in our minds. And if landscape is then understood as a stereotypical landscape or defined as a social landscape, the found landscapes have a problem—because in the conclusion these are flawed, insufficiently enriched with stereotypical physical elements in relation to landscape. A change in the positive acceptance of new energy landscapes, for example, also requires a change in the media representations of *landscapes*. This does not necessarily only involve the subsequent editing of these pictures, but also the motifs and interpretation of the concept in terms of content. As long as the expanded concept of *landscape* with all its developments, such as so-called new energy landscapes, does not find its way into the media as a landscape construct, it will also have difficulty with broad social acceptance. The same is true of the reverse—without a broader social acceptance, little will change in the pictures. The question that arises is, where should we begin?

References

Al-Khanbashi, M. (2020). Using matrix as a qualitative data display for landscape research and a reflection based on the social constructivist perspective. In D. Edler, C. Jenal, & O. Kühne (Eds.), *Modern approaches to the visualization of landscapes* (pp. 103–118). Wiesbaden: Springer VS.

Aschenbrand, E. (2017). *Die Landschaft des Tourismus*. Dissertation, Eberhard Karls Universität Tübingen. https://doi.org/10.1007/978-3-658-18429-2.

Behrens, K. C. (1976). *Absatzwerbung*. Wiesbaden: Gabler.

Belting, H. (2011). *Bild-Anthropologie. Entwürfe für eine Bildwissenschaft* (Bild und Text, 4th ed.). Paderborn: Fink.

Berger, P. L., & Luckmann, T. (1980). *Die gesellschaftliche Konstruktion der Wirklichkeit. Eine Theorie der Wissenssoziologie* (Conditio humana, 5th ed.). Frankfurt/M.: Fischer.

Berr, K. (2020). Visuality, Aesthetics and landscape. For the enlightenment and self-enlightenment of constructivist landscape research. In D. Edler, C. Jenal, & O. Kühne (Eds.), *Modern approaches to the visualization of landscapes* (pp. 189–215). Wiesbaden: Springer VS.

Brandt, M. (statista, Ed.). (2019). Das machen die Deutschen im Netz. Retrieved July 30, 2019, from https://de.statista.com/infografik/3307/internetaktivitaeten-in-deutschland/.

Burckhardt, L. (2008). *Warum ist Landschaft schön? Die Spaziergangswissenschaft* (2nd ed.). Kassel: Schmitz.

Camgöz, N., Yener, C., & Güvenç, D. (2002). Effects of hue, saturation, and brightness on preference. *Color Research & Application, 27*(3), 199–207. https://doi.org/10.1002/col.10051.

Cosgrove, D. E. (1998). *Social formation and symbolic landscape (originally Croom Helm historical geography series)*. Madison: University of Wisconsin Press.

DeLue, R. Z. (2008). Elusive landscapes and shifting grounds. In R. Z. DeLue & J. Elkins (Eds.), *Landscape theory (the art seminar)* (pp. 3–14). New York: Routledge.

Drexler, D. (2013). Landscape, Paysage, Landschaft, Táj. The cultural background of landscape perceptions in England, France, Germany, and Hungary. *Journal of Ecological Anthropology, 16*(1). https://doi.org/10.5038/2162-4593.16.1.7.

Edler, D., Kühne, O., Jenal, C., Vetter, M., & Dickmann, F. (2018). Potenziale der Raumvisualisierung in Vitual Reality (VR) für die sozialkonstruktivistische Landschaftsforschung. *Kartographische Nachrichten, 68*(5), 245–254.

Edler, D., Keil, J., Wiedenlübbert, T., Sossna, M., Kühne, O., & Dickmann, F. (2019). Immersive VR experience of redeveloped post-industrial sites. The example of "Zeche Holland" in Bochum-Wattenscheid. *KN—Journal of Cartography and Geographic Information, 69*(4), 267–284. https://doi.org/10.1007/s42489-019-00030-2.

Edler, D., Jenal, C., & Kühne, O. (2020a). Modern approaches to the visualization of landscapes—An introduction. In D. Edler, C. Jenal, & O. Kühne (Eds.), *Modern approaches to the visualization of landscapes* (pp. 3–15). Wiesbaden: Springer VS.

Edler, D., Keil, J., & Dickmann, F. (2020b). From Na Pali to Earth—An 'unreal' engine for modern geodata? In D. Edler, C. Jenal, & O. Kühne (Eds.), *Modern approaches to the visualization of landscapes* (pp. 279–291). Wiesbaden: Springer VS.

Felgenhauer, T. (2015). Die visuelle Konstruktion gesellschaftlicher Räumlichkeit. In A. Schlottmann & J. Miggelbrink (Eds.), *Visuelle Geographien. Zur Produktion, Aneignung und Vermittlung von RaumBildern* (Sozial- und Kulturgeographie, Vol. 2, pp. 67–83). Bielefeld: transcript.

Felser, G. (2015). *Werbe- und Konsumentenpsychologie* (4th ed.). Berlin: Springer.

Fontaine, D. (2020a). Landscape in computer games—The examples of GTA V and watch dogs 2. In D. Edler, C. Jenal, & O. Kühne (Eds.), *Modern approaches to the visualization of landscapes* (pp. 293–306). Wiesbaden: Springer VS.

Fontaine, D. (2020b). Virtuality and landscape. In D. Edler, C. Jenal, & O. Kühne (Eds.), *Modern approaches to the visualization of landscapes* (pp. 267–278). Wiesbaden: Springer VS.

Früh, W. (2015). *Inhaltsanalyse. Theorie und Praxis* (UTB,, 8th ed., Vol. 2501). Konstanz: UVK Verlagsgesellschaft mbH; UVK/Lucius.

Gailing, L. (2013). *Kulturlandschaftspolitik. Die gesellschaftliche Konstituierung von Kulturlandschaft durch Institutionen und Governance*. Dissertation, Technischen Universität Dortmund, Dortmund.

Gelfert, H.-D. (2000). *Was ist Kitsch?* (Kleine Reihe V & R, Vol. 4024). Göttingen: Vandenhoeck & Ruprecht.

Gorn, G. J., Chattopadhyay, A., Yi, T., & Dahl, D. W. (1997). Effects of color as an executional cue in advertising. They're in the shade. *Management Science, 43*(10), 1387–1400. https://doi.org/10.1287/mnsc.43.10.1387.

Harper, D. (2008). Fotografien als sozialwissenschaftliche Daten. In Flick, U.., Kardorff, E. V., & Steinke, I. (Eds.), *Qualitative Forschung. Ein Handbuch* (Rororo, 55628: Rowohlts Enzyklopädie, 6th ed., pp. 402–416). Reinbek bei Hamburg: Rowohlt Taschenbuch.

Hokema, D. (2009). Die Landschaft der Regionalentwicklung: Wie flexibel ist der Landschaftsbegriff? *Raumforschung und Raumordnung, 67*(3), 239–249. https://doi.org/10.1007/BF03183009.

Hokema, D. (2013). *Landschaft im Wandel? Zeitgenössische Landschaftsbegriffe in Wissenschaft, Planung und Alltag* (RaumFragen—Stadt—Region—Landschaft, Vol. 7). Wiesbaden: Springer (Techn. Univ., Diss.—Berlin, 2012).

Jenal, C. (2020). Visualizations of 'landscape' in protest movements. On exclusive and inclusive patterns of vision and interpretation using the example of resistance to the expansion of the electricity grid in Germany. In D. Edler, C. Jenal, & O. Kühne (Eds.), *Modern approaches to the visualization of landscapes* (pp. 427–445). Wiesbaden: Springer VS.

Kazig, R. (2013). Landschaft mit allen Sinnen. Zum Wert des Atmosphärenbegriffs für die Landschaftsforschung. In D. Bruns & O. Kühne (Eds.), *Landschaften: Theorie, Praxis und internationale Bezüge* (Institut Norddeutsche Kulturlandschaft, Lübeck, H. 5, pp. 221–232). Schwerin: Oceano.

Kleber, A., Edler, D., & Dickmann, F. (2020). Cartography and the sea: A Javascript-based web mapping application for managing maritime shipping. In D. Edler, C. Jenal, & O. Kühne (Eds.), *Modern approaches to the visualization of landscapes* (pp. 173–186). Wiesbaden: Springer VS.

Klinke, H. (2013). Bildwissenschaften ohne Bildbegriff. In H. Klinke & L. Stamm (Eds.), *Bilder der Gegenwart. Aspekte und Perspektiven des digitalen Wandels* (1st ed., pp. 11–28). Göttingen: Graphentis.

Kook, K. (2009). *Landschaft als soziale Konstruktion. Raumwahrnehmung und Imagination am Kaiserstuhl.* Dissertation, Albert-Ludwigs-Universität Freiburg im Breisgau. Freiburg im Breisgau. Retrieved August 23, 2016, from https://www.freidok.uni-freiburg.de/fedora/objects/freidok:7117/datastreams/FILE1/content.

Kost, S. (2017). Raumbilder und Raumwahrnehmung von Jugendlichen. In O. Kühne, H. Megerle, & F. Weber (Eds.), *Landschaftsästhetik und Landschaftswandel* (RaumFragen, pp. 69–85). Wiesbaden: Springer Fachmedien Wiesbaden.

Kühne, B. (2002). *Das Naturbild in der Werbung. Über die Emotionalisierung eines kulturellen Musters.* Frankfurt am Main: Anabas (Teilw. zugl.: Hannover, Univ., Diss., 2000).

Kühne, O. (2006). *Landschaft in der Postmoderne. Das Beispiel des Saarlandes.* Wiesbaden: Deutscher Universitäts.

Kühne, O. (2012). *Stadt—Landschaft—Hybridität. Ästhetische Bezüge im postmodernen Los Angeles mit seinen modernen Persistenzen* (RaumFragen Stadt—Region—Landschaft). Wiesbaden: VS Verlag für Sozialwissenschaften.

Kühne, O. (2013). *Landschaftstheorie und Landschaftspraxis. Eine Einführung aus sozialkonstruktivistischer Perspektive* (RaumFragen—Stadt—Region—Landschaft). Wiesbaden: Springer Fachmedien.

Kühne, O. (2017). *Landschaft und Wandel. Zur Veränderlichkeit von Wahrnehmungen.* Wiesbaden: Springer VS.

Kühne, O. (2018). *Landscape and power in geographical space as a social-aesthetic construct.* Cham: Springer International Publishing.

Kühne, O. (2019). *Landscape theories* (RaumFragen: Stadt—Region—Landschaft). Wiesbaden: Springer Fachmedien Wiesbaden.

Kühne, O. (2020). The social construction of space and landscape in internet videos. In D. Edler, C. Jenal, & O. Kühne (Eds.), *Modern approaches to the visualization of landscapes* (pp. 121–137). Wiesbaden: Springer VS.

Kühne, O., & Jenal, C. (2020). The threefold landscape dynamics—Basic considerations, conflicts and potentials of virtual landscape research. In D. Edler, C. Jenal, & O. Kühne (Eds.), *Modern approaches to the visualization of landscapes* (pp. 389–402). Wiesbaden: Springer VS.

Kühne, O., Weber, F., & Weber, F. (2013). Wiesen, Berge, blauer Himmel. Aktuelle Landschaftskonstruktionen am Beispiel des Tourismusmarketings des Salzburger Landes aus diskurstheoretischer Perspektive. *Geographische Zeitrschrift, 101*(1), 36–54.

Leibenath, M., Heiland, S., Kilper, H., & Tzschaschel, S. (Eds.). (2013). *Wie werden Landschaften gemacht? Sozialwissenschaftliche Perspektiven auf die Konstituierung von Kulturlandschaften* (Kultur- und Medientheorie). Bielefeld: transcript.

Linke, S. (2017). Räumliche Wandlungsprozesse in ländlich bezeichneten Regionen im Kontext des gesellschaftlichen Wertewandels. In P. Droege & J. Knieling (Eds.), *Regenerative Räume. Leitbilder und Praktiken nachhaltiger Raumentwicklung* (pp. 281–296). München: oekom.

Linke, S. (2018). Ästhetik der neuen Energielandschaften—Oder: „Was Schönheit ist, das weiß ich nicht". In O. Kühne & F. Weber (Eds.), *Bausteine der Energiewende* (pp. 409–429). Wiesbaden: Springer VS.

Linke, S. (2019). *Die Ästhetik medialer Landschaftskonstrukte. Theoretische Reflexionen und empirische Befunde.* Wiesbaden: Springer VS.

Loda, M., Kühne, O., & Puttilli, M. (2020). The social construction of Tuscany in the German and English speaking world—Presented by the analysis of internet images. In D. Edler, C. Jenal, & O. Kühne (Eds.), *Modern approaches to the visualization of landscapes* (pp. 157–171). Wiesbaden: Springer VS.

Micheel, M. (2012). Alltagsweltliche Konstruktionen von Kulturlandschaft. *Raumforschung und Raumordnung, 70*(2), 107–117. https://doi.org/10.1007/s13147-011-0143-x.

Mitchell, W. J. T. (1995). *Picture theory. Essays on verbal and visual representation.* Chicago: University of Chicago Press.

Pawlik, V. (statista, Ed.). (2019). Anzahl der Personen in Deutschland, die das Internet zur Informationssuche (Suchmaschinen) nutzen, nach Häufigkeit von 2015 bis 2018 (in Millionen). Retrieved July 30, 2019, from https://de.statista.com/statistik/daten/studie/183133/umfrage/nachrichten-und-informationen---internetnutzung/.

Popitz, H. (1992). *Phänomene der Macht* (2nd ed.). Tübingen: Mohr.

Prisille, C., & Ellerbrake, M. (2020). Virtual reality (VR) and geography education: potentials of 360° 'experiences' in secondary schools. In D. Edler, C. Jenal, & O. Kühne (Eds.), *Modern approaches to the visualization of landscapes* (pp. 321–332). Wiesbaden: Springer VS.

Roßmeier, A. (2020). Urban/rural hybridity in pictures—The creation of neighborhood images using the example of San Diego's urbanizing inner-ring suburbs east village and Barrio Logan. In D. Edler, C. Jenal, & O. Kühne (Eds.), *Modern approaches to the visualization of landscapes* (pp. 479–498). Wiesbaden: Springer VS.

Schenk, W. (2020). Visualization of the fundamental dimensions of "landscape" in landscape paintings around 1500 A.D. In D. Edler, C. Jenal, & O. Kühne (Eds.), *Modern approaches to the visualization of landscapes* (pp. 19–32). Wiesbaden: Springer VS.

Schlottmann, A., & Miggelbrink, J. (2015a). Ausgangspunkte. Das Visuelle in der Geographie und ihrer Vermittlung. In A. Schlottmann & J. Miggelbrink (Eds.), *Visuelle Geographien. Zur Produktion, Aneignung und Vermittlung von RaumBildern* (Sozial- und Kulturgeographie, Vol. 2, pp. 13–25). Bielefeld: transcript.

Schlottmann, A., & Miggelbrink, J. (2015b). Teil ll: Praktiken visueller Geopgraphien. Einleitende Bemerkungen. In A. Schlottmann & J. Miggelbrink (Eds.), *Visuelle Geographien. Zur Produktion, Aneignung und Vermittlung von RaumBildern* (Sozial- und Kulturgeographie, Vol. 2, pp. 86–89). Bielefeld: transcript.

Schlottmann, A., & Miggelbrink, J. (Eds.). (2015c). *Visuelle Geographien. Zur Produktion, Aneignung und Vermittlung von RaumBildern* (Sozial- und Kulturgeographie, Vol. 2). Bielefeld: transcript.

Schneider, A. (2015). RaumBilder und Bildung. In A. Schlottmann & J. Miggelbrink (Eds.), *Visuelle Geographien. Zur Produktion, Aneignung und Vermittlung von RaumBildern* (Sozial- und Kulturgeographie, Vol. 2, pp. 91–102). Bielefeld: transcript.

Schütz, A. (1972). *Gesammelte Aufsätze. Band 1: Das Problem der sozialen Wirklichkeit.* Dordrecht: Springer Netherlands.

Siepmann, N., Edler, D., & Kühne, O. (2020). Soundscapes in cartographic media. In D. Edler, C. Jenal, & O. Kühne (Eds.), *Modern approaches to the visualization of landscapes* (pp. 247–263). Wiesbaden: Springer VS.

Stintzing, M., Pietsch, S., & Wardenga, U. (2020). How to teach "landscape" through games? In D. Edler, C. Jenal, & O. Kühne (Eds.), *Modern approaches to the visualization of landscapes* (pp. 333–349). Wiesbaden: Springer VS.

Thimm, C. (2017). Internet. In L. Kühnhardt, & T. Mayer (Eds.), *Bonner Enzyklopädie der Globalität* (1st ed., pp. 433–442). Wiesbaden: Springer VS

Vetter, M. (2020). Technical potentials for the visualization in virtual reality. In D. Edler, C. Jenal, & O. Kühne (Eds.), *Modern approaches to the visualization of landscapes* (pp. 307–317). Wiesbaden: Springer VS.

Wagner, E. (2014). *Mediensoziologie* (UTB Soziologie, Vol. 4224). Konstanz: UVK Verlagsgesellschaft.

Weber, F. (2015). Landschaft aus diskurstheoretischer Perspektive. Eine Einordnung und Perspektiven. *MORPHĒ, 1*(1). Retrieved February 24, 2016, from http://www.hswt.de/fileadmin/Dateien/Hochschule/Fakultaeten/LA/Dokumente/MORPHE/MORPHE-Band-01-Juni-2015.pdf.

Weber, F. (2016). The potential of discourse theory for landscape research. Dissertations of cultural landscape commission 31. Retrieved March 7, 2017, from http://www.krajobraz.kulturowy.us.edu.pl/publikacje.artykuly/31/6.weber.pdf.

Weber, F. (2020). Blurring the boundaries of landscape visualization: Welcome to fabulous Las Vegas. In D. Edler, C. Jenal, & O. Kühne (Eds.), *Modern approaches to the visualization of landscapes* (pp. 461–478). Wiesbaden: Springer VS.

Weidenbach, B. (statista, Ed.). (2019). Anteil von Kommunikation, Spielen, Informationssuche und Unterhaltung an der Internetnutzungszeit Jugendlicher in den Jahren 2008 bis 2018. Retrieved July 31, 2019, from https://de.statista.com/statistik/daten/studie/29694/umfrage/inhaltliche-verteilung-der-internetnutzung-durch-jugendliche-ab-2008/.

Welsch, N., & Liebmann, C. C. (2012). *Farben. Natur Technik Kunst* (3rd ed.). Heidelberg: Spektrum Akademischer.

Werlen, B. (1997). *Gesellschaft, Handlung und Raum. Grundlagen handlungstheoretischer Sozialgeographie* (3rd ed.). Stuttgart: Steiner.

Wintzer, J. (2015). „…wie in der folgenden Abbildung zu sehen ist…". Nachvollsehbarkeit und Bevölkerung. In A. Schlottmann & J. Miggelbrink (Eds.), *Visuelle Geographien. Zur Produktion, Aneignung und Vermittlung von RaumBildern* (Sozial- und Kulturgeographie, Vol. 2, pp. 103–119). Bielefeld: transcript.

Dr. rer. nat. Simone Linke studied landscape architecture at the University of Applied Sciences in Weihenstephan and then completed a master's degree in Urban Design at the Technical University in Berlin. Since 2013 she has been working on the change and aesthetics of landscape constructs and received her doctorate in geography at the Eberhard Karls University in Tübingen. She is currently working as a research assistant at the Technical University of Munich, where she is also investigating the current potentials and challenges of landscapes.

The Social Construction of Tuscany in the German- and English-Speaking World— Presented by the Analysis of Internet Images

Mirella Loda, Olaf Kühne and Matteo Puttilli

Abstract

The essay examines aspects of the social construction of Tuscany, with a focus on the German-speaking regions. This is contextualized in relation to the social construction of Tuscany in English and Italian. The basis of the investigation is a picture search with Google using the word Toscana in German, English, and Italian. The results of the German and English language search clearly point to the considerably different stereotyping of Tuscany in terms of landscape, while the images determined using the Italian version of the search word are clearly more commonplace and have a greater variation in content. The results refer to the complexity of the social construction of landscape in particular and space in general, which carries a great significance for the image, the touristic potential, the self-description of regions, and yet have only been scientifically processed in rudimentary form.

M. Loda (✉) · M. Puttilli
Dipartimento di Storia, Archeologia, Geografia, Arte e Spettacolo (SAGAS),
 Università degli studi Firenze, Florence, Italy
e-mail: mirella.loda@unifi.it

M. Puttilli
e-mail: matteogirolamo.puttilli@unifi.it

O. Kühne
Forschungsbereich Geographie, Stadt- und Regionalentwicklung (SRE),
Eberhard Karls Universität Tübingen, Tübingen, Germany
e-mail: olaf.kuehne@uni-tuebingen.de

© Springer Fachmedien Wiesbaden GmbH, part of Springer Nature 2020   157
D. Edler et al. (eds.), *Modern Approaches to the Visualization of Landscapes*,
RaumFragen: Stadt – Region – Landschaft, https://doi.org/10.1007/978-3-658-30956-5_9

Keywords

Tuscany · Landscape · Social constructivism · Internet pictures · Internet photos · Stereotypes · German · English · Italian

1   Introductory Remarks

More than nearly any other region, Tuscany, in the German-speaking world and beyond, is subject to an idealization (Lehmann 1976). This idealization is particularly promoted via poetry, travelogues, illustrations, earlier paintings, later photos, and today, frequently published on the Internet. Besides the historical genesis of the landscape stereotype of Tuscany in the German-speaking world, this article focuses on the representation of Tuscany in Internet images. The images generated with the German search word 'Toskana' are examined alongside those found with the Italian search word 'Toscana' and the English 'Tuscany' (with the descriptions of the images kept in their respective language).

This article follows the social constructivist landscape theory that has established itself in social and cultural landscape research in recent decades. According to this approach, the constitutive level of landscape is not to be found in physical space, but in social and individual conceptions of landscape. This aspect will be discussed briefly in this essay, followed by a description of the genesis of Tuscan perceptions in the German-speaking and English-speaking world. The investigation of the Internet images is followed by a conclusion that correlates the results relative to the theory and, subsequently, presents future research potentials.

2   Notes on the Social Construction of Landscape

In the recent decades, human geography, as well as associated cultural and social science disciplines, has been able to gain the acceptance of a perspective of perceiving landscape as not only a physical object, but as a social and individual construction (among many: Cosgrove 1984; Greider and Garkovich 1994; Kühne 2009, 2018a; in this volume: Kühne and Jenal 2020). Central to this perspective is the shift in the constitutive level of landscape: it is not the material world but social conventions that form the constitutive level. Landscape is a physical space(s) envisioned by the individual in accordance with social patterns of generation, interpretation, and evaluation. However, these social patterns are greatly differentiated in terms of cultural and social origins (for example: Bruns et al. 2015; Bruns and Kühne 2013; Hunziker et al. 2008; Stotten 2015).

In the German-speaking world, the history of the term 'landscape' goes back to the Middle Ages. Landscape was initially understood to mean a larger settlement area, which was later defined by certain common customs and norms, although not precisely defined. Prior to the High Middle Ages, the understanding of landscape was condensed into a synopsis of laws and institutions defining an area located on 'this side of the forest' (Berr and Schenk 2019; Gruenter 1975 [1953]; Müller 1977; Schenk 2001, 2013). With the Renaissance, the concept of landscape in the German-speaking world was given the aesthetic connotation that it had acquired in Western Europe in the context of the spread of landscape painting, and the distinctive ability to see landscapes in physical spaces (in this volume: Berr 2020). The German concept of landscape developed a special path in Romanticism: Landscape was seen as beyond the city and became a symbol of a normatively preserved synthesis of culture and nature, of homeland, which was to be defended against the tendencies of industrialization, rationalization, and ultimately enlightenment (Berr and Kühne 2020; Berr and Schenk 2019; Körner 2005, 2006; Kühne 2018c). After the First World War, National Socialists, who saw a normative ideal of spatial design in the 'German cultural landscape', had appropriated this motif. Still later, the concept was then ecologized in the 1960s. In the course of de-industrialization, the 'landscape' conceived now as 'old industrial landscape' was once again charged with a symbol of loss, fed by the experience of the loss of a secure, predictable 'modern' life in an industrial society. This gave way to an individualized postmodern society characterized equally by opportunities and uncertainties (Hauser 2001; Höfer and Vicenzotti 2013; Jenal 2019; Kühne 2007; Schönwald 2015; see in this volume: Papadimitriou 2020).

The stereotypical notions of landscape developed in this way are accompanied by notions made through individual experiences of space in childhood and adolescence, through their living environment, which form an essential basis for the interpretation and evaluation of spaces, thusly perceiving such landscapes as a 'normal native landscape' (Kook 2008; Kost 2017; Kühne 2008; Stotten 2013). A third level of assessment is selectively added when people acquire 'special knowledge' related to landscape, especially via scientific studies. These special knowledge stocks are highly selective in terms of their subject matter and characterize a specific reductive view. When spaces, interpreted as landscapes, are compared with one's own professional norms, from whose ideals the 'real landscapes' deviated to a large extent, this deviation in turn is marked as an inadequacy (Kühne 2013, 2019a, b, 2020; Stemmer et al. 2019; Wojtkiewicz and Heiland 2012).

Presently, for the physical spaces in which landscape concepts are formed in the German-speaking context, they are subject to varying patterns of interpretation and evaluation. If these ideas are brought to spaces that are neither everyday nor where social landscape norms were formed, a notable difference arises. In the following we will examine these in more detail in the context of a 'sacralized European landscape'—Tuscany.

3 The Social Construction of Tuscany in the German-Speaking World

For the composite German tradition of traveling to Italy, through which the aristocracy and then the upper bourgeois classes completed their education (*Grand Tour*, then called *Bildungsreise—educational tour*), Tuscany was for a long time simply a transit station on the way to the legacies of the Greek-Roman classic culture in Rome, Naples, and Sicily, as demonstrated by the scarce attention dedicated to it by Goethe, Moritz, Schinkel, Seume, and Winckelmann (see in general: Borsig 1938, Günter 1985, Heitmann and Scamardi 1993 and Loda 1997).

It was only during the middle of the nineteenth century that Tuscany began to occupy a more precise position in the German collective imaginary. In that period the tradition of the trip to Italy underwent a deep transformation. One aspect was Ferdinand Gregorovius's decision to move from Rome to Germany in 1874, which symbolically concluded the classic tradition of the *Bildungsreise*. Another influence being the interest in Italy—in part arising from the recent political change—found new and diversified forms and expressions, such as memory books, epistolaries, and the initiation of travel guides (Tresoldi 1975).

To this stream belong the numerous trips to Italy that Jacob Burckhardt started to make between 1838–1839. Burkhardt radically revaluated the art of the "Renaissance" as both renewal and opposed to the classical Greek-Roman model. He considered it as the model of the pre-capitalistic liberal, an enlightened and thus a-moral society that put an end to the stratified medieval one. His book *Die Cultur der Renaissance in Italien*, published in 1860, greatly contributed to shaping the concept of "Renaissance" as the domain of individual freedom where the "man of Renaissance" (the "uomo universal") can fully display his potential and creativity.

A few decades later, this image was reinforced by the Viennese art historian Heinrich Wölfflin, who represents the second milestone in the creation of the German concept of Renaissance, and who greatly contributed to circulating it among the well-educated German bourgeoisie. In his book *Renaissance und Barock* (1888) he developed the theory of the continuous alternation of the classic and the baroque styles in history.

Opposing the classical antiquity and the Renaissance in contrast to the Baroque, Wölfflin reinforced the idea of the Renaissance as the realm of individual freedom, harmony, and "humanism". Because of its almost didactic straightforwardness, Wölfflin's aesthetic criteria rooted deeply within the well-educated German bourgeois, and were often used outside of the artistic sphere, as well, to describe the regions with the most important legacy of the Renaissance art period.

Along this line, Tuscany—having hosted some of the most important monuments of Renaissance art—together with the "Republican" history of its cities established themselves in the German collective imagination as a sort of ideal and mythical place, as the empirical realization of the synthesis between man and nature. It was therefore Burckhardt's view of the Renaissance, seen as the paradigm of the enlightened society, that set the interpretative and aesthetic key concepts for the reception of Tuscany in German speaking countries.

The region became the object of a new and specific cultural interest itself, changing from a culturally marginal place to a major destination Loda (1997).

A very effective vehicle and at the same time the climactic manifestation of such a reception is the German concept of "Tuscan Landscape", that developed at the turn of the twentieth century as a projection of aesthetic criteria over the physical space of the region. "Tuscan Landscape" became an equivalent to 'harmonious place', to place "on a human scale".

Merging aesthetic appreciation (of Renaissance art) and perception of the region itself, the concept of "Tuscan Landscape" (*Toskanische Landschaft*) represents a paradigmatic example of the semantic migration of the German concept '*Landschaft*' (Landscape) from the very original meaning of "region", through that of "pictorial landscape" (*gemalte Landschaft*) acquired from the seventeenth century onwards in the artistic domain, towards the designation of a physical space ("real landscape"). The aesthetic meaning of the concept 'Landschaft', however, did not disappear while still referring it to the description of a physical space and yet forging the way to how a real place is perceived.

In the common language, 'Landscape' is therefore equivalent to 'beautiful landscape' (Hard 1985). What went lost in shifting from the aesthetic level to the designation of a real space is the awareness concerning the role of the observing subject who perceives the landscape. The subjective perception guided by aesthetic criteria becomes an 'objective' feature of a real space. In the case of Tuscany, at the turn of the twentieth century subjective aesthetic criteria consistent with Burkhardt's appreciation of Renaissance art began to be transferred through the concept of "Tuscan Landscape" on to the "real" world, being perceived as the "objective" features of the region: Tuscany becomes synonymous with "harmony" and its landscape a symbol of absolute "beauty".

In the first half of the twentieth century, this process of semantic migration can even be observed in the scientific and geographic literature. The description of Tuscany provided by Herman Lehman after the Second World War and published in the *Geographische Zeitschrift* (Lehmann 1976) clearly reflects this aesthetic approach:

> "One of the main features of the Tuscan landscape [...] is the harmonic stream of all its lines to create a generous and balanced rhythm [...]; [the Tuscan cities; note by the authors] are perfectly integrated into the landscape, they sum up its formal style, at the same time enriching it with a specific sign, that of the individual, conscious leap within the flow of lines all around" (p. 2).

The same gaze on the region is to be found in the travel guides, where adjectives like "harmonious", "balanced", and "ideal" abound. Iconic is the famous guidebook of Klaus Zimmermanns (1980), who describes as following the background of his work on the hill settlements:

> "The great emotion of a trip to Tuscany (consists in) meeting in their purest expression the forms that coined our culture, of which general value we have an 'idea', in seeing them realized in the landscape, in the stones and in the light of their origin" (p. 13).

This image of Tuscany as a sort of man-made Arcadia and of its landscape as its visual projection has been reproduced and even strengthened in the following years through contemporary tourism, conceivable as the continuation—even if not as exclusive—of the nineteenth-century *Bildungsreise*.

According to empirical research conducted on German speaking tourists in Tuscany (Krüger and Loda 1993), the image of the region as a harmonious place, a place "on a human scale" is still crucial in the perception of Tuscany. What attracts most is the idea of Tuscany as a sort of Arcadia directly reflected in the landscape, that comes up as the main factor of attraction for German speaking tourists in Tuscany, well above mention of the cities of art or the beach.

The concept of Tuscan landscape that emerged through the interviews with tourists is in most cases a very abstract and idealistic one, even after their direct experience of travel to Tuscany. The idea of Tuscany landscape remains very near to the eighteenth century *veduta* model, and evokes a "fantastic", "wonderful" postcard landscape. That a Burckhardtian conception still determines the perception of Tuscan landscape, in German speaking tourists, is confirmed by their automatic association of the concept with the hill-landscape of the Chianti area.

The great variety of physical space and of the anthropic structures in other parts of the region remain neglected, not to mention the highly impacting contemporary phenomena like industrialization or urban sprawl.

4   The Social Construction of Tuscany in the English-Speaking World

As it is widely known, Tuscany has a privileged place in the English-speaking imagination. This stems from a long-lasting relationship between the English-speaking world and Tuscan society, nurtured by a never-ending cultural production that encompasses all of the performing and fine arts, from painting to photography, fiction, and cinema. It can even be argued that Tuscany played a role in the shaping of the Anglo-Saxon idea of landscape itself. At the same time, though, the cultural construction and circulation of the idea of Tuscany in the English-speaking world and beyond is at the base of the sacralization of the Tuscan landscape as one of the most archetypal landscapes worldwide.

Indeed, the construction of the Tuscan landscape in the English-speaking world results from a double movement, first inward, and then outward.

On the one hand, the principles of proportional and harmonic composition elaborated and codified by the Italian Renaissance, once they had been imported into England, were a primary source for the establishment of the aesthetic conception of landscape during the 17th Century. As it was masterfully shown by Denis Cosgrove (1984), upon these principles was shaped the central iconography of the English idea of landscape up to the mid-eighteenth century: the view of the country-house and its prospect, the predominance of picturesque rural scenes with all its related components of landscape gardening. It is not a case that some of the most representative examples of landscape gardens in England at the time (such as William Kent's Rousham House in Oxfordshire) also made use of Tuscan references (in this specific case, the Villa di Pratolino in Vaglia, close to

Florence). An expression of landscape that reflected the emerging social role of the class of gentry landowners, their economic power, and their political aspirations.

On the other hand, the British travelers who visited Tuscany at different times could easily recognize these same components in the shape of the land and find confirmation of their cultural imaginings. In their view, in fact, Tuscany recalled a garden produced by history, the result of a harmonious and balanced relationship between man and nature through the centuries, where the latter takes the form of a "work of art". As it was reported in 1819 by the Irish novelist and traveler Lady Morgan (quoted in Brilli 2004, p. 478, translated by the authors)—"the scene of Florence from the Incisa is that of an English garden; many farmhouses seen from a distance resemble the ancient, aristocratic country residences of Queen Anne's time". Likewise, according to the English essayist William Hazlitt, the city of Florence appeared like a city *transplanted* into a garden (Brilli 2004).

Tuscany was not just seen as a garden, but it was also shaped as such. Indeed, the aesthetic use of some of the most iconic, and pre-existing, elements of what were to become the symbols of the Tuscan landscape around the world—like the rows of cypresses or the white Serpentines—are products of the creations and interventions initially promoted by representatives of the English colony in Tuscany, like John Temple Leader in the midst of the 19th Century and Cecil Pinsent at the beginning of the 20th (Zoppi 2009).

In the process of the cultural construction of Tuscany, as counterpart of the countryside, stands the city. From the eighteenth century onwards, Tuscany, and Florence in particular, are home of a populous and influential English—and later on American—community (estimated to count more than 50,000 at the beginning of twentieth century), attracted by the artistic heritage as well as the lively and cosmopolitan cultural activity. In this milieu, the city views and its atmospheres become the setting of stories—"Portrait of a Lady", "A Room with a View"—targeted to an international audience, positioning Florence, and the Tuscan region, as an archetype that merges the appetites for classicism and exoticism.

During the course of the twentieth century, the idea of Tuscany as a "harmonic and complete universe" (Brilli 2004) made of an invaluable cultural heritage and a countryside looked after and well-tended by man continued to enrich the English-speaking view of the territory. New artistic expressions such as cinema helped to consolidate the place of Tuscan landscape among the most famous and recognizable iconographies worldwide.

5   The Visual Construction of Tuscany in Internet Images: Comparing the Results of the Search Words 'Tuscany', 'Toskana' and 'Toscana'

The Internet has developed as an essential medium for the social construction of landscape. As a result of the great importance of combined comparative viewing—images, in the form of photographs or paintings, have a special significance for the social communication of landscape and in particular for the updating of landscape stereotypes. In this respect, the investigation of Internet communication, specifically of Internet images, becomes a worthwhile object of research into the social construction of landscape (for more details, see: Kühne 2012, 2018b; Linke 2019; in this volume: Jenal 2020; Linke 2020).

With the aim of capturing visual stereotypes representing Tuscany, the first 100 hits of the Google image search, sought with the German spelling "Toskana", were examined. To facilitate contextualizing these hits, the first 100 hits with the English ("Tuscany") and the Italian ("Toscana") spellings were also evaluated. The image search was performed with computer settings that suppressed referencing of previous searches and was tested with four other computers to ensure that the same results were displayed in each. Only pictures with inscriptions in the respective language and spatial arrangements that can generally be understood as 'landscape' (, i.e. no sofas named 'Toscana',) were evaluated. The classification of the image contents was inductively based on the preliminary examination of the contents of the German-language search results and was then (in the case of 'Mensch/Menschen') later supplemented and post-coded (for a more detailed methodology, see Kühne 2012; Kühne and Schönwald 2015).

While most of the categories are highly selective, the categorization of 'agricultural land' sometimes proved difficult. While the term 'pasture' can be categorized quite clearly in the presence of livestock and/or pasture fences, meadows can only be clearly distinguished from fields in certain vegetation periods, it is difficult to distinguish between them in the early vegetation period. Misclassifications may have occurred here. The category 'rural buildings' also needs to be explained. Generally, these are buildings of agricultural origin either standing individually or in small groups, but in individual cases also sacred buildings. Fog was classified as fog if the visibility in the image or part thereof was (estimated) less than one kilometer (WMO 2019). The degree of image processing was estimated, primarily, according to the color saturation, but also according to the use of soft focus or the use of contour sharpening (see more details: Linke 2019). In addition to current photos, for which little processing is available, historical photos with traces of ageing were also assigned to the category of 'low processing' as well as maps, since these apparently were not subsequently processed. The thematic classification was made according to the main motif, which means that even when classified in the category 'small town'—rural buildings (of agricultural origin) could not be found.

The thematic orientation of the examined pictures (Fig. 1) shows an accumulation of the motif of the individually standing rural buildings for the German-language, and even more so for the English-language hits, in each case to be about two-thirds. Small towns, on the other hand, are much less represented, as are (almost) uninhabited areas. Metropolitan representations are hardly to be found in the German-speaking context and almost not at all in the English-speaking context (only one photo). Cartographic representations can be found in the results based on the German and Italian search terms. Overall, the Italian-language selection of images is much less thematically focused than the German and, especially, the English ones. Tuscany is not primarily perceived here as

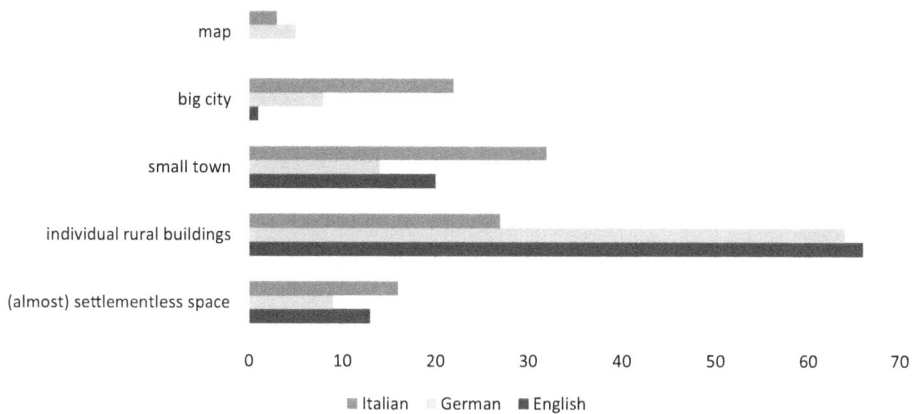

Fig. 1 Thematic classification of the images examined, broken down by hits in German, English, and Italian (own survey and evaluation)

possessing a separate agricultural building in a space dominated by agricultural activities (as in the German and English contexts), which will be explained in more detail below, but primarily as having a small town context. The motif of the single rural building is followed by the depiction of large cities.

A clear differentiation between the hits resulting from language-specific searches can also be found with regard to the degree of image processing (Fig. 2). Especially those images that emerged from the English search (followed by the German one) show a very high degree of processing. This is particularly evident in the fact that preferred sunrise or sunset situations were selected that were represented with a distinct over-saturation of the color. Images that were acquired via 'Toscana' usually have a much lower post-processing rate. More rarely are 'sacralizing' or idealized representations found here, but rather commonplace, less stereotypical motifs, such as meadows (without cypresses) or new development areas. In addition, historical photos are often found here, which were not to be found in the English and German searches.

In consideration of what has been said so far about motifs and image processing, a dominant landscape stereotype of Tuscany can be reconstructed for the images with a German, and even more so with an English, search term yielding small variations with regard to the representation of the object constellations (Fig. 3): In the middle ground, centrally located on a hill, there is a single, usually two-story, building surrounded by trees or groups of trees that cannot be specified more precisely. The areas with fruit tree stands and grapevines are adjacent to this. Between the observer and the building runs a curved, one-lane street, lined with cypress trees. On the left and right side of the road there are meadows and fields. The background is formed by sparsely wooded hills, in rarer cases also wooded mountains. The cloudiness is—within representations with higher sun position—of fair-weather cumulus (Cumulus Humilis), within representations

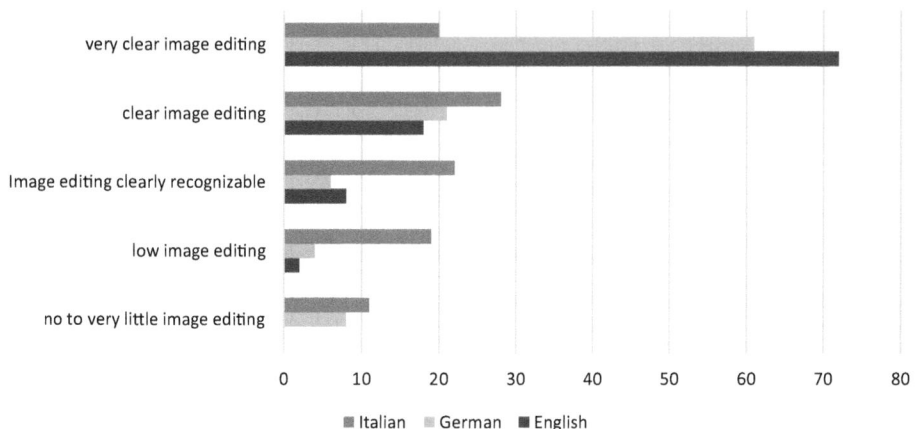

Fig. 2 The degree of image processing of the examined images, broken down by hits in German, English and Italian (own survey and evaluation)

with low standing sun (rising or setting) of intensely colored staged Cirrus, more frequently also Cirrocumulus. The scenery is deserted of mankind, present only through his physical manifestations. While in the German context small and metropolitan scenes and coastal representations are added, which mostly do not contain any people. In the English context a far-reaching limitation to the presented main motif can be found. The results of the Italian-language image search are not only differentiated regarding the motifs, but also as regards the depicted objects. The depiction of metropolitan scenes is also accompanied by the depiction of metropolitan buildings, of large streets and squares both inhabited with people. Tuscany is presented here not only as a rural idyll, but also as a multifunctional and multi-structural space, whereby 'sacralized' contents also occur, albeit to a lesser extent and frequently in a tourist context.

6   Conclusion: Summary, Conclusions and Further Research Needs

The representation of Tuscany in Internet pictures shows—depending on the language of the selected search term—interesting differences. A rural idyll with stereotypical furnishings dominates in German and even more so in English. The images generated using the Italian search term show, on the contrary, a much more differentiated and often more lifelike image of Tuscany. Even if traces of a regressive interpretation of Tuscany as a "golden mean" between the industrialized Northern and the underdeveloped Southern Italy (Becattini 1986) still survive, here Tuscany appears less a 'sacralized' rural scenery than a lived-in everyday space.

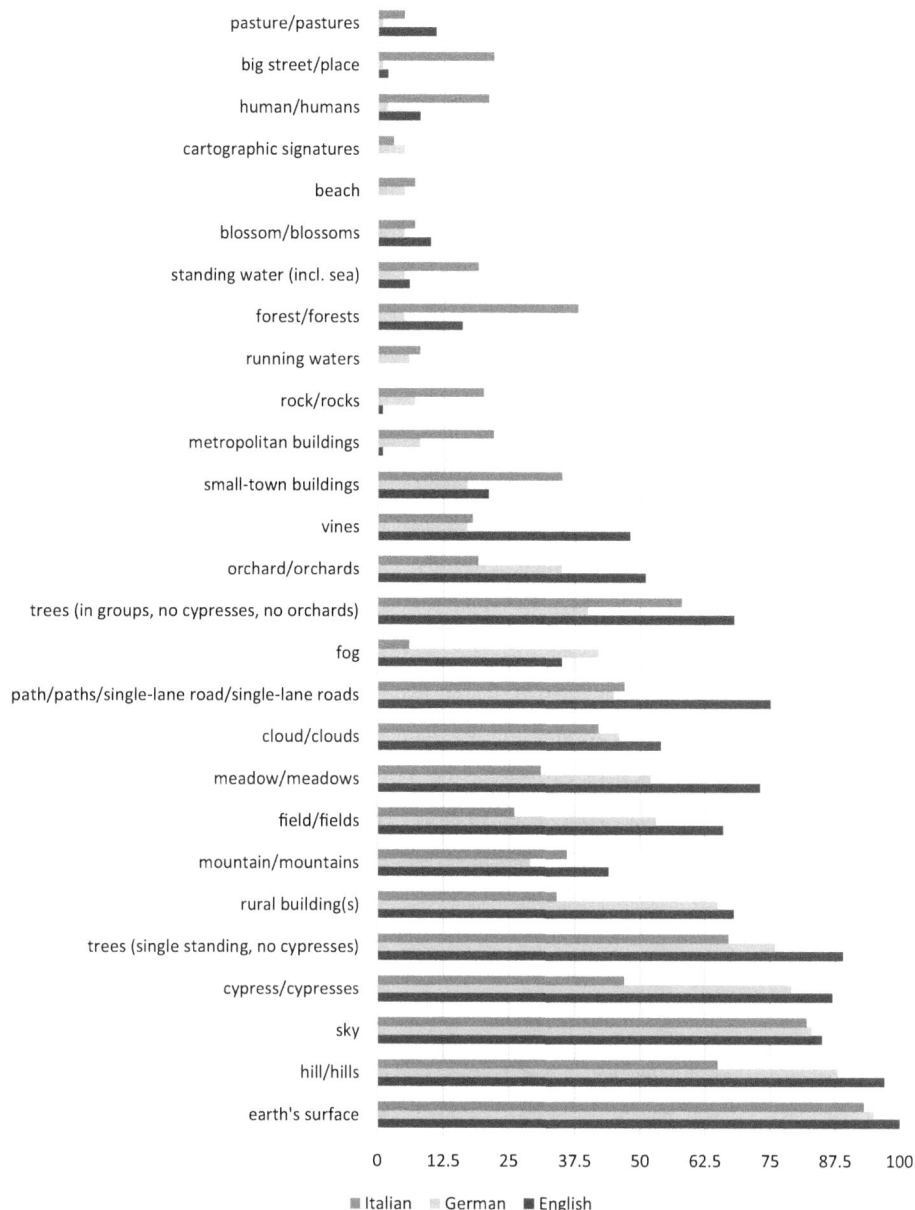

Fig. 3 Objects depicted in the images examined, broken down by hits in German, English, and Italian (own survey and evaluation)

One approach to interpreting these different constructs lies in contrasts of the living-environment connectedness of the local population and, conversely, in the different stereotypes of the Tuscan landscape. These stereotypes, in turn, are shaped by fundamentally different understandings of landscape in the different language areas examined: In German, the understanding of landscape is strongly representational, both morally and normatively charged, while in the English an aesthetic component dominates. The connection between the generated images can thus be assigned to two different levels: The images generated with the German and English terms refer to specific landscape stereotypes, the images generated with the Italian term refer more to an everyday construction of Tuscany—landscape stereotypes, which also refer to the Italian concept of landscape, are clearly less present.

Overall, the variations in the images evaluated according to linguistic differentiation show how differently the same physical space is socially constructed. This happens, on the one hand, in the difference between the everyday world (Italian) and the stereotype (German and English), while on the other hand, also according to different stereotypical patterns (German and English) which refer to different perceptions of landscape. However, the stereotyping of certain landscapes is also subject to a specific, culturally sedimented, linguistically defined history. This is clearly shown by the historical development and stereotyping of the "Tuscan Landscape" in German: it was not until after its function as merely a transit area that it became the symbol of a "harmonious landscape".

A more precise and methodically differentiated investigation could further refine these rough results. Especially in a comparison between the approaches to Tuscany and other 'sacralized landscapes', such as Provence, the Middle Rhine Valley, and Transylvania, different stereotypes and everyday approaches could be identified, plus the patterns of stereotypical and everyday approaches to landscape could be investigated more closely.

References

Becattini, G. (1986). Riflessioni sullo sviluppo socio-economico della Toscana. In G. Mori (a cura di). La Toscana (pp. 899–924). Torino: Einaudi.

Berr, K. (2020). Visuality, aesthetics and landscape. For the enlightenment and self-enlightenment of constructivist landscape research. In D. Edler, C. Jenal, & O. Kühne (Eds.), *Modern approaches to the visualization of landscapes* (pp. 189–215). Wiesbaden: Springer VS.

Berr, K., & Kühne, O. (2020). *„Und das ungeheure Bild der Landschaft...". The genesis of landscape understanding inthe German-speaking regions* (RaumFragen: Stadt—Region—Landschaft). Wiesbaden: Springer VS.

Berr, K., & Schenk, W. (2019). Begriffsgeschichte. In O. Kühne, F. Weber, K. Berr, & C. Jenal (Eds.), *Handbuch Landschaft* (pp. 23–38). Wiesbaden: Springer VS.

Brilli, A. (2004). Il paesaggio toscano e lo sguardo del viaggiatore. In C. L. Bonelli, A. Brilli, & G. Cantelli (Eds.), *Il paesaggio toscano. Storia e rappresentazione* (pp. 463–525). Milano: Silvana.

Bruns, D., & Kühne, O. (Eds.). (2013). *Landschaften: Theorie, Praxis und internationale Bezüge. Impulse zum Landschaftsbegriff mit seinen ästhetischen, ökonomischen, sozialen und philosophischen Bezügen mit dem Ziel, die Verbindung von Theorie und Planungspraxis zu stärken.* Schwerin: Oceano.

Borsig, von A. (1938). *Die Toskana. Landschaft, Kunst und Leben im Bild*. München: Anton Schroll.

Bruns, D., Kühne, O., Schönwald, A., & Theile, S. (Eds.). (2015). *Landscape culture—Culturing landscapes. The differentiated construction of landscapes*. Wiesbaden: Springer VS.

Cosgrove, D. E. (1984). *Social formation and symbolic landscape*. London: University of Wisconsin Press.

Greider, T., & Garkovich, L. (1994). Landscapes: The social construction of nature and the environment. *Rural Sociology, 59*(1), 1–24. https://doi.org/10.1111/j.1549-0831.1994.tb00519.x.

Gruenter, R. (1975). Zum Problem der Landschaftsdarstellung im höfischen Versroman. In A. Ritter (Ed.), *Landschaft und Raum in der Erzählkunst* (Wege der Forschung, Vol. 418, pp. 193–335). Darmstadt: WBG (First publication 1953).

Günter, R. (1985). *Toskana. Ein Reisebuch*. Gießen: anabas

Hard G. (1985). Die Landshaft der Künstler und die der Geographen. In Evangelische Akademie Loccum. Landshaftsbilder, Landshaftswahrnehmung, Landschaft. Die Rolle der Kunst in der Geschichte der Wahrnehmung unserer Landschaft (pp. 122–139). Rehburg-Loccum.

Hauser, S. (2001). *Metamorphosen des Abfalls. Konzepte für alte Industrieareale*. Frankfurt (Main): Campus.

Heitmann, K., & Scamardi, T. (Eds.). (1993). *Deutsches Italienbild und italienisches Deutschlandbild im 18. Jahrhundert*. Tübingen: Max Niemeyer.

Höfer, W., & Vicenzotti, V. (2013). Post-industrial landscapes: Evolving concepts. In P. Howard, I. Thompson, & E. Waterton (Eds.), *The routledge companion to landscape studies* (pp. 405–416). London: Routledge.

Hunziker, M., Felber, P., Gehring, K., Buchecker, M., Bauer, N., & Kienast, F. (2008). Evaluation of landscape change by different social groups. Results of two empirical studies in Switzerland. *Mountain Research and Development, 28*(2), 140–147. https://doi.org/10.1659/mrd.0952.

Jenal, C. (2019). (Alt)Industrielandschaften. In O. Kühne, F. Weber, K. Berr, & C. Jenal (Eds.), *Handbuch Landschaft* (pp. 831–841). Wiesbaden: Springer VS.

Jenal, C. (2020). Visualizations of 'landscape' in protest movements. On exclusive and inclusive patterns of vision and interpretation using the example of resistance to the expansion of the electricity grid in Germany. In D. Edler, C. Jenal, & O. Kühne (Eds.), *Modern approaches to the visualization of landscapes* (pp. 427–445). Wiesbaden: Springer VS.

Kook, K. (2008). Zum Landschaftsverständnis von Kindern: Aussichten—Ansichten—Einsichten. In R. Schindler, J. Stadelbauer, & W. Konold (Eds.), *Points of view. Landschaft verstehen—Geographie und Ästhetik, Energie und Technik* (pp. 107–124). Freiburg: modo.

Körner, S. (2005). Landschaft und Raum im Heimat- und Naturschutz. In M. Weingarten (Ed.), *Strukturierung von Raum und Landschaft. Konzepte in Ökologie und der Theorie gesellschaftlicher Naturverhältnisse* (pp. 107–117). Münster: Westfälisches Dampfboot.

Körner, S. (2006). Heimatschutz, Naturschutz und Landschaftsplanung. In Institut für Landschaftsarchitektur und Umweltplanung—Technische Universität Berlin (Ed.), *Perspektive Landschaft* (pp. 131–142). Berlin: wvb Wissenschaftlicher.

Kost, S. (2017). Raumbilder und Raumwahrnehmung von Jugendlichen. In O. Kühne, H. Megerle, & F. Weber (Eds.), *Landschaftsästhetik und Landschaftswandel* (pp. 69–85). Wiesbaden: Springer VS.

Krüger, R., & Loda, M. (1993). *Quale turismo per la Toscana minore? Indagine sulla struttura motivazionale die turisti tedeschi nell'area delle colline pisane*. Milano: Irpet/F. Angeli.

Kühne, O. (2007). Soziale Akzeptanz und Perspektiven der Altindustrielandschaft. Ergebnisse einer empirischen Untersuchung im Saarland. *RaumPlanung, 132/133*, 156–160.

Kühne, O. (2008). Die Sozialisation von Landschaft—sozialkonstruktivistische Überlegungen, empirische Befunde und Konsequenzen für den Umgang mit dem Thema Landschaft in Geographie und räumlicher Planung. *Geographische Zeitschrift, 96*(4), 189–206.

Kühne, O. (2009). Grundzüge einer konstruktivistischen Landschaftstheorie und ihre Konsequenzen für die räumliche Planung. *Raumforschung und Raumordnung, 67*(5/6), 395–404. https://doi.org/10.1007/BF03185714.

Kühne, O. (2012). *Stadt—Landschaft—Hybridität. Ästhetische Bezüge im postmodernen Los Angeles mit seinen modernen Persistenzen.* Wiesbaden: Springer VS.

Kühne, O. (2013). Macht und Landschaft: Annäherungen an die Konstruktion von Experten und Laien. In M. Leibenath, S. Heiland, H. Kilper, & S. Tzschaschel (Eds.), *Wie werden Landschaften gemacht? Sozialwissenschaftliche Perspektiven auf die Konstituierung von Kulturlandschaften* (pp. 237–271). Bielefeld: transcript.

Kühne, O. (2018a). *Landscape and power in geographical space as a social-aesthetic construct.* Dordrecht: Springer International Publishing.

Kühne, O. (2018b). *Landschaft und Wandel. Zur Veränderlichkeit von Wahrnehmungen.* Wiesbaden: Springer VS.

Kühne, O. (2018c). *Landschaftstheorie und Landschaftspraxis. Eine Einführung aus sozialkonstruktivistischer Perspektive* (2., aktualisierte und überarbeitete ed.). Wiesbaden: Springer VS.

Kühne, O. (2019a). Der dreifache Landschaftswandel. *Forum Raumentwicklung, 1,* 18–19.

Kühne, O. (2019b). *Landscape theories. A brief introduction.* Wiesbaden: Springer VS.

Kühne, O. (2020). Landscape conflicts—a theoretical approach based on the three worlds theory of Karl Popper and the conflict theory of Ralf Dahrendorf, illustrated by the example of the energy system transformation in Germany. *Sustainability, 12*(17), 6772.

Kühne, O., & Jenal, C. (2020). The threefold landscape dynamics—Basic considerations, conflicts and potentials of virtual landscape research. In D. Edler, C. Jenal, & O. Kühne (Eds.), *Modern approaches to the visualization of landscapes* (pp. 389–402). Wiesbaden: Springer VS.

Kühne, O., & Schönwald, A. (2015). *San Diego. Eigenlogiken, Widersprüche und Hybriditäten in und von ‚America's finest city'.* Wiesbaden: Springer VS.

Lehmann, H. (1976). Toskanische Landschaft. In *Villa und Villegiatura in der Toskana. Eine italienische Institution und ihre gesellschaftsgeographische Bedeutung. Mit einer einleitenden Schilderung "Toskanische Landschaft" von Herbert Lehmann* (Erdkundliches Wissen, Vol. 44, pp. 1–10). Wiesbaden: Steiner.

Linke, S. I. (2019). *Die Ästhetik medialer Landschaftskonstrukte. Theoretische Reflexionen und empirische Befunde.* Wiesbaden: Springer VS.

Linke, S. (2020). Landscape in Internet pictures. In D. Edler, C. Jenal, & O. Kühne (Eds.), *Modern approaches to the visualization of landscapes* (pp. 139–156). Wiesbaden: Springer VS.

Loda, M. (1997), Alcune considerazioni sul cencetto di „paesaggio toscano" nella cultura e nella geografia tedesca. In *Toscana, Paesaggio, Ambiente. Scritti dedicati a Giuseppe Barbieri, Atti dell'Istituto di Geografia, Quaderno* (Vol. 18, pp. 146–157). Firenze: Università di Firenze.

Müller, G. (1977). Zur Geschichte des Wortes Landschaft. In A. Hartlieb von Wallthor & H. Quirin (Eds.), *„Landschaft" als interdisziplinäres Forschungsproblem. Vorträge und Diskussionen des Kolloquiums am 7./8. November 1975 in Münster* (pp. 3–13). Münster: Aschendorff.

Papadimitriou, F. (2020). Visualization of future landscapes, postmodern cinema and geographical education. In D. Edler, C. Jenal, & O. Kühne (Eds.), *Modern approaches to the visualization of landscapes* (pp. 351–369). Wiesbaden: Springer VS.

Schenk, W. (2001). Landschaft. In H. Beck, D. Geuenich, & H. Steuer (Eds.), *Reallexikon der Germanischen Altertumskunde* (Vol. 17, pp. 617–630). Berlin: de Gruyter.

Schenk, W. (2013). Landschaft als zweifache sekundäre Bildung—Historische Aspekte im aktuellen Gebrauch von Landschaft im deutschsprachigen Raum, namentlich in der Geographie. In D. Bruns & O. Kühne (Eds.), *Landschaften: Theorie, Praxis und internationale Bezüge. Impulse zum Landschaftsbegriff mit seinen ästhetischen, ökonomischen, sozialen und philosophischen Bezügen mit dem Ziel, die Verbindung von Theorie und Planungspraxis zu stärken* (pp. 23–36). Schwerin: Oceano.

Schönwald, A. (2015). Die Transformation von Altindustrielandschaften. In O. Kühne, K. Gawroński, & J. Hernik (Eds.), *Transformation und Landschaft. Die Folgen sozialer Wandlungsprozesse auf Landschaft* (pp. 63–73). Wiesbaden: Springer VS.

Stemmer, B., Philipper, S., Moczek, N., & Röttger, J. (2019). Die Sicht von Landschaftsexperten und Laien auf ausgewählte Kulturlandschaften in Deutschland—Entwicklung eines Antizipativ-Iterativen Geo-Indikatoren-Landschaftspräferenzmodells (AIGILaP). In K. Berr & C. Jenal (Eds.), *Landschaftskonflikte* (pp. 507–534). Wiesbaden: Springer VS.

Stotten, R. (2013). Kulturlandschaft gemeinsam verstehen—Praktische Beispiele der Landschaftssozialisation aus dem Schweizer Alpenraum. *Geographica Helvetica, 68*(2), 117–127. https://doi.org/10.5194/gh-68-117-2013.

Stotten, R. (2015). *Das Konstrukt der bäuerlichen Kulturlandschaft. Perspektiven von Landwirten im Schweizerischen Alpenraum* (alpine space—man & environment, Vol. 15). Innsbruck: Innsbruck University Press.

Tresoldi, L. (1975). *Viaggiatori tedeschi in Italia: 1452–1870. Saggio bibliografico*. Roma: Bulzoni.

Wojtkiewicz, W., & Heiland, S. (2012). Landschaftsverständnisse in der Landschaftsplanung. Eine semantische Analyse der Verwendung des Wortes „Landschaft" in kommunalen Landschaftsplänen. *Raumforschung und Raumordnung, 70*(2), 133–145. https://doi.org/10.1007/s13147-011-0138-7.

Zimmermanns, K. (1980). *Toscana: Das Hügelland und historischen Stadtzentren*. Köln: Dumont.

Zoppi, M. (2009). Il paesaggio toscano: mito, icone, realtà. In *TRIA—Rivista Internazionale Semestrale di Cultura Urbanistica* (Vol. 3, pp. 67–74). Napoli: Federico II University Press.

Dr. Mirella Loda is Professor of Geography at the University of Florence. She received her doctorate in geography (on long term consequences of earthquake and reconstruction processes) at the Technical University in Munich. Her habilitation in Munich dealt with the impact of the tightening in the environmental policy upon the leather manufacturing in Italy. Today, she is particularly concerned with social and economic change in contemporary urban spaces (with a special focus on management and fruition of urban public space), tourism development and cultural heritage, both in Western countries and in the global South.

Dr. Dr. Olaf Kühne, born in 1973, is Professor of Urban and Regional Development at the University of Tübingen. He received his doctorate in geography (on an urban climatological topic) in Saarbrücken and sociology (on the social construction of landscape) in Hagen. His habilitation in Mainz dealt with ecological and social transformation processes in Poland. Today, he is particularly concerned with social issues of energy system transformation, landscape theory formation and urban transformation processes. These are the focus of his interest in feedback relationships between the material, individual and social world.

Matteo Puttilli, Ph.D. in Regional Planning and Local Development at the Polytechnic of Turin, is associate professor in the Department of History, Archaeology, Geography, Fine and Performing Arts at the University of Florence. His research interests involve social, political and environmental geography, and include: environmental policies, with a specific focus on energy transition and agri-food networks; urban marginality and socio-spatial justice; policies and practices of local and urban development; and geography education and didactics.

Cartography and the Sea: A JavaScript-Based Web Mapping Application for Managing Maritime Shipping

Alexander Kleber, Dennis Edler and Frank Dickmann

Abstract

The Sea has attracted researchers from various disciplines over centuries. To administer and manage maritime shipping, people have traditionally relied on paper maps. Thanks to modern web technologies, cartographic expert libraries programmed in JavaScript (leaflet.js) and freely available spatial data sets, modern sea cartography can lead to interactive and animated information systems. These systems can be adjusted to specific usage scenarios and, in contrast to printed maps, irrelevant information can be temporarily omitted. This article focuses on a web cartographic application that aims to support maritime shipping in the Western Scheldt area in the Netherlands. The article introduces the main functionalities of the application and discusses their advantages over traditional practical approaches. The application was developed in consultation with practitioners, i.e. professionals working on ships or in the management department. In this article, the term landscape is rather understood in the positivistic sense. Landscape is represented through graphical symbols which are used to deliver information and quantities for efficient management purposes.

A. Kleber (✉) · D. Edler · F. Dickmann
Geographisches Institut, Geb. IA, Ruhr-Universität Bochum, Bochum, Germany
e-mail: alexander.kleber@ruhr-uni-bochum.de

D. Edler
e-mail: dennis.edler@ruhr-uni-bochum.de
F. Dickmann
e-mail: frank.dickmann@ruhr-uni-bochum.de

© Springer Fachmedien Wiesbaden GmbH, part of Springer Nature 2020   173
D. Edler et al. (eds.), *Modern Approaches to the Visualization of Landscapes*, RaumFragen: Stadt – Region – Landschaft,
https://doi.org/10.1007/978-3-658-30956-5_10

Keywords

Maritime landscape · Web cartography · Nautical chart · OpenSeaMap · Leaflet.js

1   Introduction

The sea has fascinated geographers and cartographers for centuries. The sea as well as its coasts, as cross-border regions between water and land, have attracted people from many perspectives. They can include signifiers of multiple meanings which have changed in societies over the centuries (Kühne and Weber 2019; Ratter and Walsh 2019; Schönwald 2017; Steinberg 2001). For example, interpretations of maritime and coastal landscapes were provided by landscape painters, such as the German Romantic painter Caspar David Friedrich (1774–1840), who is famous for his works on the contemplation of nature, and the meaningful combination of landscape and atmosphere (Aschenbrand 2017; Kühne 2018a, b).

Long before the individual and social meaning of maritime and coastal landscapes received a new attention through arts and literature in Europe at the end of the eighteenth century, the economic potential of the sea has been discovered and expanded. An important component of maritime trade is a successful navigation, which implies knowledge of naval routes and hydrography. Useful tools which contain and graphically represent the required spatial information are maps. The cartographic representations used for these administrative and economic purposes rather take a very factual view on maritime and coastal landscapes. As coordinates, going along with accurate object locations, angles and directions, are the core elements of interest in nautical charts, the landscape concept in these charts is commonly based on a positivistic understanding.

In (nautical) cartography, the 1569 world map (*Nova et Aucta Orbis Terrae Descriptio ad Usum Navigantium Emendate Accommodata*) published by Gerardus Mercator (1512–1594) is an important milestone for maritime navigation and trade. The underlying projection has become a standard component for many nautical charts. In the same century, hydrography became an established field of research. In Europe that time, maritime trade and map-making was significantly shaped by Swedish and Dutch seafarers (Glete 2010).

Today, nautical charts still belong to the relevant set of cartographic media maintained and published by state agencies, such as the Federal Maritime and Hydrographic Agency of Germany (BSH). In addition, the volunteering community of openseamap.org—a thematic group contributing to the *OpenStreetMap* (OSM) project (openstreetmap.org)— acquires and shares open nautical information and geodata. The overarching aim of *OpenSeaMap* is to create and continue a freely available and usable nautical chart. The data and charts are offered worldwide and can be used by everyone, incl. private households, state agencies and private companies.

To administer and manage maritime shipping today, cartographic media are still considered as important tools providing spatial information of coasts and the Sea. Traditionally, nautical charts are (paper) maps with a high density of information and experience is required to read and extract information properly. Earlier approaches of multimedia cartography have already addressed sea-related (map) themes (e.g. Pulsifer et al. 2005, 2007; Retchless 2014; Monmonier 2008). Thanks to modern web technologies, nautical (geo-)data and information can be managed and visualized based on new approaches of interactive and animated cartography and WebGIS (other approaches to the GIS-based visualization in this volume: Poplin et al. 2020; Stratmann et al. 2020). This article presents an application developed on the interest of the "Loodswezen Regio Scheldemonden" (Western Scheldt, Netherlands). This application includes open data sets offered as Volunteered Geographic Information (VGI), provided by the OpenSeaMap community. Methodologically, the application is based on the possibilities of the open source JavaScript library leaflet.js, a widespread API used for building interactive web mapping applications for common desktop and mobile browsers (see also Donohue et al. 2014; Edler and Vetter 2019; Horbinski and Lorek 2020).

2   Context of the Application

Nautical charts contain complex nautical information about coordinates, angles, directions, waterways (e.g. buoys, water depths etc.), object locations (e.g. wrecks, sandbanks etc.) or traffic separation schemes. The nautical charts serve as a cartographic reference and as a resource for spatial orientation in sea areas (Wrenger 2015, p. 12). To point out the importance of nautical charts for commercial shipping, the International Maritime Organisation (IMO) emphasizes that two independent but redundant systems of nautical charts are compulsory on any seagoing vessel, either in a digital or paper version. The IMO is the 'United Nations specialized agency with responsibility for the safety and security of shipping and for the prevention of marine and atmospheric pollution by ships' (IMO 2020a).

Even though most ships still have paper charts onboard, commercial shipping nowadays relies on the Electronic Chart Display and Information System (ECDIS). These systems digitally display all crucial navigational information for nautical officers on Electronic Navigational Charts (ENC) (BSH 2020). They combine information from different sources, such as ENCs, the Global Positioning System (GPS), the internal systems of a ship and the Automatic Identification System (AIS). ECDIS cannot only display real-time information on the nautical charts but also interact with them, as it compares the position of a ship, speed and course with other ship positions, charted objects, navigation aids or hazards in order to acoustically alert the nautical personnel in case of unforeseen circumstances or dangerous situations (Weintrit 2009, p. 128).

The complexity of navigational information systems as ECDIS can, however, lead to several issues in the usage of nautical charts. Processing all information simultaneously is challenging and may result in 'cognitive overload' (for other examples of 'overloaded' cartographic media, see Edler 2020). The processing of the information (depth) on ENCs combined with the additional information provided by ECDIS can be highly challenging, even for experts of the seafaring personnel. Situations occur in which the captains and officers have difficulties to extract crucial information about a specific sea area.

To reduce the amount of information that is simultaneously displayed in a map, modern opportunities of interactive web cartography can help. The maps can be adjusted to the individual demands in specific situations and not required (visual) information can be deactivated, which results in more user-oriented media and a lower visual complexity to be processed at the same time.

An area where these issues occur is the Western Scheldt estuary between Belgium and the Netherlands. The Belgian seaports of Antwerp, Gent and Zeebrugge as well as the Dutch seaports Terneuzen and Vlissingen—last-mentioned named Flushing in nautical English—lie in this area. It comprises 150 km from east to west and 75 km from north to south. In addition, it contains five seaports, two pilot stations, seven radio areas, three main and over ten secondary fairways. Combined with the busy shipping traffic of the English Channel (which is crossed by the Western Scheldt area), shallow waters, wind parks and local restrictions, the Western Scheldt is known as a difficult navigational area. To guarantee safe and efficient navigation in those waters, the Dutch *Loodswezen Regio Scheldemonden* and the Belgian *DAB Loodswezen* are responsible to provide pilots that advice the nautical personnel to choose and sail on the most efficient and safest route. This is set into law by the so called *Scheldereglement*, whose roots can be traced back to the *Scheldetractaat* between Belgium and the Netherlands in 1843 (Weber 1994).

In the past, incidents have shown that captains who approached the area had several issues in extracting information from nautical charts. For example, it seemed difficult to locate the northerly positioned *Steenbank Pilot Station* on the charts. The *Steenbank Pilot Station* is 'squeezed in' between Dutch and Belgian wind parks, the port entrance and anchorage areas of the port of Rotterdam, the Dutch peninsula of Walcheren and shallow waters. The pilot station was rather concealed on charts and covered by represented objects of these surrounding areas at larger scales (Fig. 1). Ships sometimes accidentally sailed passed the *Steenbank Pilot Station* to approach the southerly positioned and more clearly marked *Wandelaar Pilot Station* (Fig. 1). By going there, ships had to sail an extra distance of 73–98 nautical miles, depending on the specific route they took. Since *Loodswezen* is obliged to plan in pilots' multiple hours in advance, they sometimes had the pilot for a specific ship already going to the *Steenbank Pilot Station*, whereas the ship went to the *Wandelaar Pilot Station*. Either the pilot or the ship had then to be shifted back to the other pilot station, which caused delays in the chain logistics of the ports.

Even if the nautical personnel found and approached the *Steenbank Pilot Station* correctly, they had further issues to determine which was the best route to take afterwards. A third issue was to correctly follow the rules of the Vessel Traffic Service (VTS). Radio communication between ships and shore-based radio centrals takes place in different

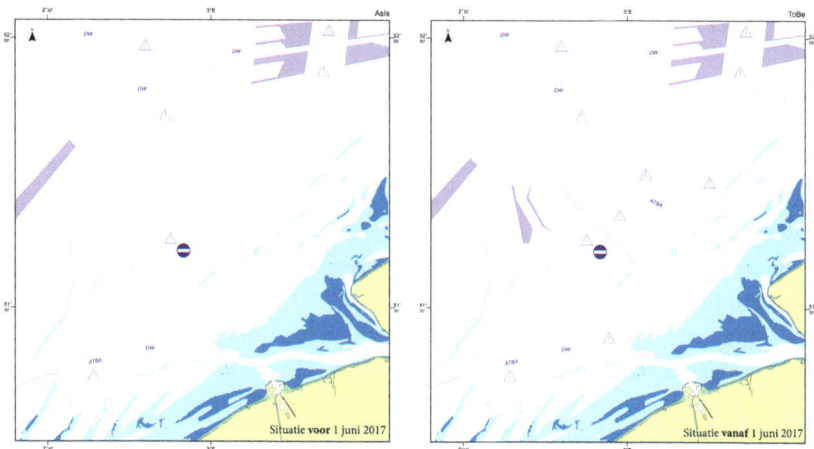

Fig. 1 Change of traffic schemes at the Western Scheldt area at the 1st of July 2017 to clarify the position of the Steenbank Pilot Station and the connected fairways from to Flushing. (Source: Mobility and Transport and Rijkswaterstaat 2017)

VTS-areas on Very High Frequency (VHF) radio channels. Inside a VTS-area, the VHF-channels a ship has to use differ depending on the subsection where the ships are positioned. If the nautical personnel misses out to adjust the correct VHF-channel, it is very hard to get in contact with the ship, since the traffic centres have to figure out the channel which they can use to get in contact with the vessel in the first place.

Due to the mentioned problems in finding the right pilot station or being aware of the different fairways and VHF-channels, delays in the chain logistics occurred frequently. Therefore, the authorities decided to entirely rebuilt the general route layout scheme at the *Steenbank Pilot Station*, to clarify the traffic situation, visualize the positions of the new wind parks and assure a quicker and safer access to the North Sea (Fig. 1) (Mobility and Transport and Rijkswaterstaat 2017). The route layout change took place on 1st of July 2017.

In order to support the pilots of Loodswezen Regio Scheldemonden in better understanding the route layout change, an interacive cartographic web application was developed, as a part of an M.Sc. thesis at the Ruhr-University Bochum (RUB), Germany (Geomatics/Cartography Group). This was developed in spring 2019. The purpose of this application was to demonstrate the potential of the JavaScript-Library leaflet.js for supporting sea navigation. It could improve the digital administration and use of required spatial data. It may also help to support map reading in navigational contexts, as interactive maps can be individually adjusted to specific situations, such as a reduction of information. The application was tested in usability study with 15 experts. Their comments and feedbacks were used to create an application meeting the demands of the practice.

3   Functionalities

The application has become an interactive and multifunctional tool, which combines the conventional information and advantages of ENCs with added value information, real-time data and a flexible database, while keeping it tuneable for the purpose of any user. It is hosted on a webserver, publicly accessible on www.cargo-ships.de. It is suited for the use of mobile devices as well. It was developed with the (open source) JavaScript library (leaflet.js) and the included geodata are almost exclusively open data and Volunteered Geographic Information (VGI).

A base map of the application is an ENC offered by the Italian sea chart-provider Navionics, which can be accessed as open data. Furthermore, the less information packed nautical chart from the OpenStreetMap (OSM) project—*OpenSeaMap*—was included as a second layer. It includes less information than the regular ENC but gives a first overview about a chosen sea area. As two non-nautical base maps, the OpenStreetMap itself and a satellite image from the Environmental Systems Research Institute (ESRI) are implemented as well. The key idea in including multiple different base maps (which all have a different level of detail) is the opportunity for any user, to choose exactly the level of chart detail which is considered necessary while using the programmed chart.

The application contains two major improvements compared to the already existing regular nautical charts and ECDIS. Firstly, grouped thematic layers which provide additional information are included. Secondly, an improved interactivity (going beyond activating and deactivating layers or layer groups) offers new options for the users to adjust the chart to their individual purposes. Depending on the sea area which a user is interested in, area of interest can be preselected. This causes that displayed information is spatially reduced to a specific part of the available map. The thematic layers themselves visually include data on, for example, main and secondary fairways, anchorage areas, pilot areas, VHF-areas, VTS-areas, locks, wind parks, wrecks, and lighthouses. They include data which provide important information and reference objects which had not been given a high relevance in the past. To get such information and objects, the nautical personnel had to extract them manually from the conventional nautical charts. By highlighting these elements in the web application (in the first place), a faster access is provided.

In addition, by hovering over the visualized thematic layers, additional information is joint in a database and displayed in a summarized, compact form (Fig. 2). This may not only support orientation but also the extraction of relevant information, without manual processing steps. The sources of the database are information from official documents maintained by *Loodswezen Regio Scheldemonden* as well as regular ENCs, VGI or laws—if restrictive patterns for a certain section must be kept in mind.

Furthermore, the application provides more interactive tools, such as the inclusion of real-time weather and tidal information. Physical aspects such as wind force, wind

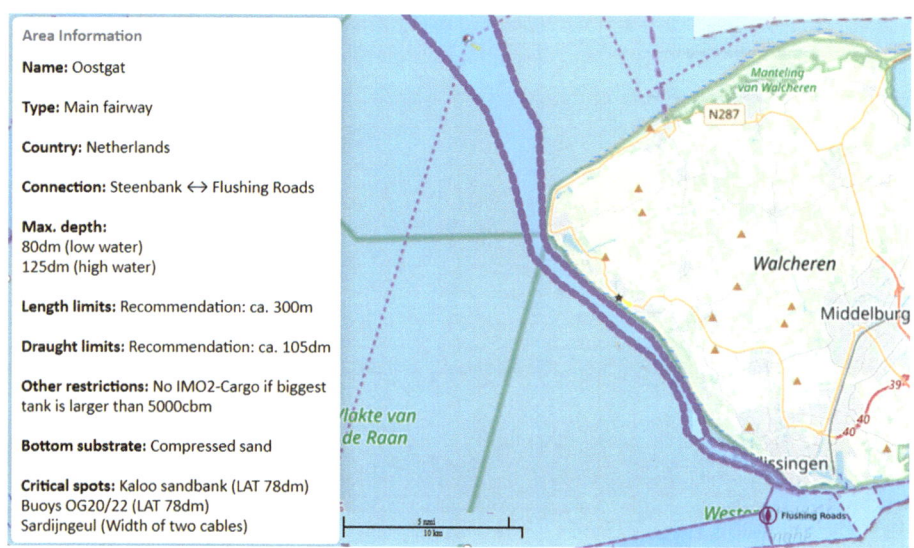

Fig. 2 Highlight of a chosen shipping fairway with compact nautical information generated from a background database, while only having toggled on the OpenSeaMap base layer. (Screenshot of application, base map: © OpenSeaMap-Contributors)

direction, visibility or water levels at certain geographical locations are crucial in the nautical navigation but are neither included on ENCs nor in ECDIS. The sources are *OpenWeatherMaps* (which also belong to the *OpenStreetMap* project) as well as official data provided by the governmental Dutch organization *Rijkswaterstaat*.

As another interactive function, the user has the option to use a tool which can be individualized for any ship. An actual value for the draught of a ship can be entered. The application will then compare the given figures with the current tidal situation at the shallowest spots in the area (Fig. 3). Afterwards, it will display the current space between the keel of a ship and the bottom of the fairway, which is the so called *Under Keel Clearance* (UKC) and which is as a safety margin in maritime shipping (IMO 2007, p. 12).

As shipping companies include regulations for a minimum UKC for their ships in their company policy, the user will immediately get the information about the predicted clearance of the ship when it reaches the shallowest spots in the fairways. Even though this safety margin applies to all ships and shipping companies, the policies for a 'safe' UKC differ from company to company and from the type of fairway a ship is inside. For the Western Scheldt area, a safe UKC is considered 15% of the actual draught of a ship in the fairways to and from the pilot stations and 12.5% of the draught on the river Scheldt itself (Eloot et al. 2008).

The critical spots themselves and their key peculiarities are highlighted separately in the application by hovering over them with the cursor. In this way, the user gets

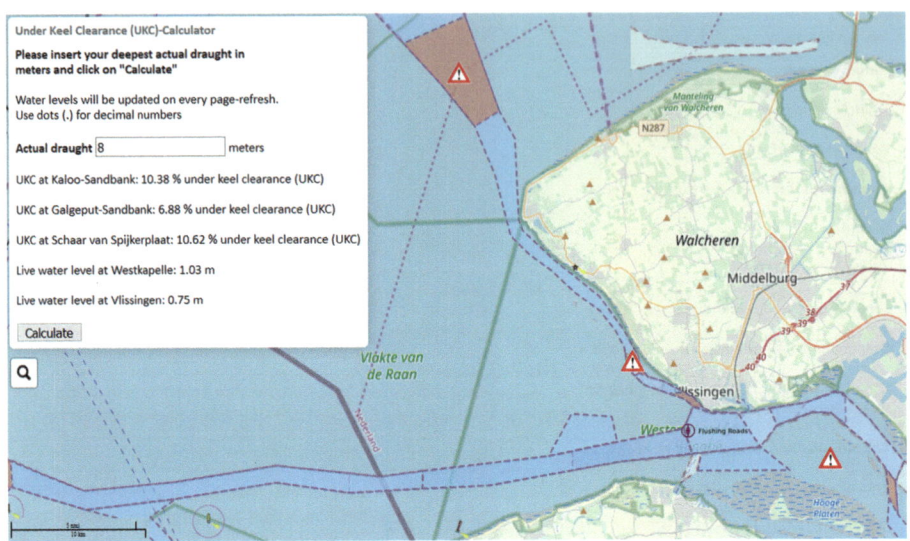

Fig. 3 UKC-calculation tool with the current draught of a ship compared with the tide while highlighting all relevant critical spots in the area. (Screenshot of application, base map: © OpenSeaMap-Contributors)

summarized information why the spots are critical, which bottom substrate prevails, whether there are any restrictions and which advices exist to safely navigate in these areas. The sailing advices can also be visualized in an animation, where a virtual vessel passes through a selected fairway and displays sailing instructions on several spots via popups. The animations of different fairways can also be activated all at once, so that the user gets a visualized comparison of the ad- and disadvantages of the fairways.

The third added function which has not yet been implemented in ENCs nor ECDIS is the option to directly search for any nautical aspect (e.g. pilot stations, fairways, buoys, berths, locks etc.). This query is linked to another database in the background of the application (Fig. 4). After typing in the term of interest, the application makes suggestions and offers possible fitting terms in a drop-down menu list. By choosing and clicking on the desired aspect, the user will directly be guided to the position of the required object and will get summarized information about it.

As a trial function, the application is also technically able to obtain and display information about all ships in the area by being connected to the Automatic Identification System (AIS). It provides information, such as a ships position, its dimension, draught, destination, course or speed (IMO 2020b). In the map, the ships are represented as animated objects and their positions are updated in real-time. Additional information can be compiled in a popup window (Fig. 5) In contrast to the many open data used in this project, AIS data must be purchased.

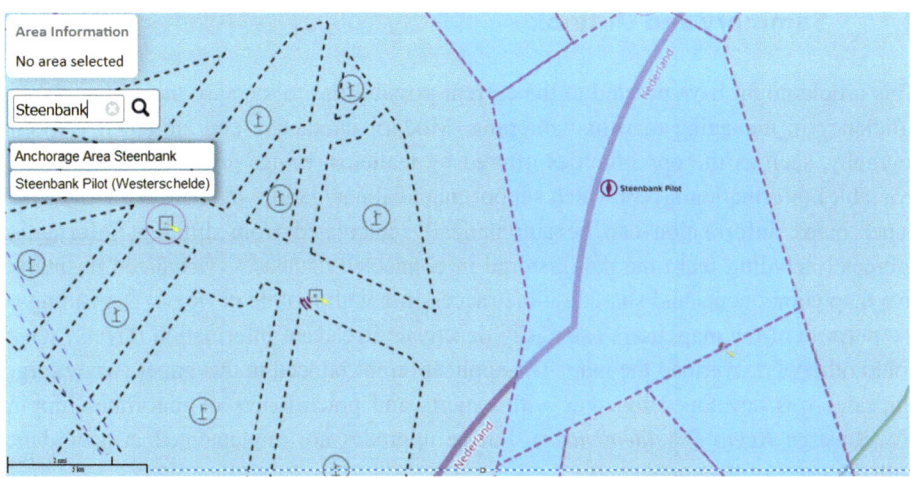

Fig. 4 Search tool with autocomplete function which automatically navigates the user to the desired spot and providing backup-information by hovering over the area. (Screenshot of application, base map: © OpenSeaMap-Contributors)

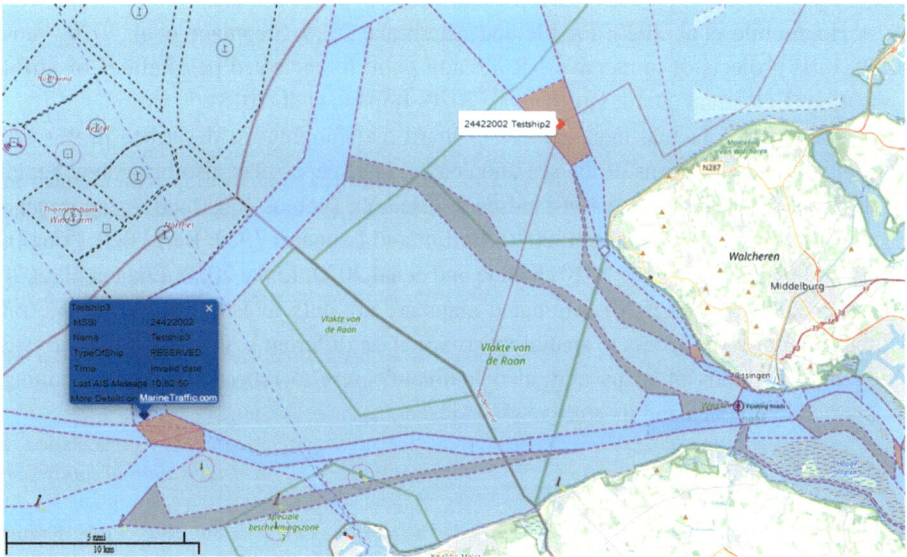

Fig. 5 AIS data and compiled information (popup) visualized in the animated map. The red and blue symbols represent current positions of two ships. (Screenshot of application, base map: © OpenSeaMap-Contributors)

4   Summary and Outlook

This article might have pointed to the current possibilities to increase the effectivity and efficiency in managing maritime shipping. Modern interactive and animated web cartography, such as the opportunities offered by leaflet.js, brings together several (freely available) information layers which support nautical navigation. In contrast to traditional paper maps, information can be automatically calculated from different information sources (including real-time data), stored in connected databases, visualized in interactive map components and shared online (accessible with mobile devices). Depending on the purpose of the map, users can easily deactivate irrelevant information (layers) which could otherwise overload the map. The application introduced in this paper (www.cargo-ships.de) was developed together with experts and practitioners of maritime shipping (*Loodswezen Regio Scheldemonden*). Future upgrades are implemented in accordance with the users of the application.

This project also indicates the potentials for a use of modern cartographic media and digital spatial information in maritime shipping. Future enhancements could also consider other modern visualization approaches to enhance the map use experience. Methods and techniques of Augmented Reality and Virtual Reality are currently under study (e.g. Çöltekin et al. 2019; Kersten and Edler 2020; see in this volume: Edler et al. 2020; Hochschild et al. 2020; Prisille and Ellerbrake 2020; Siepmann et al. 2020; Vetter 2020). First projects of immersive VR already point to enhanced possibilities of virtual navigation (Edler et al. 2018; Hruby et al. 2020; Kersten et al. 2018).

In addition to a factual management of maritime shipping (as described in this article), modern visualizations of the sea and coastal landscapes may also bring new benefit for different scientific accesses of landscape research, such as constructivist approaches (see in this volume: Al-Khanbashi 2020; Bellini and Leonardi 2020; Berr 2020; Fontaine 2020a, b; Jenal 2020; Kühne 2020; Kühne and Jenal 2020; Linke 2020; Loda et al. 2020; Roßmeier 2020; Weber 2020). Animated maps are not only tools to administer sea traffic; they also provide access to spatial information on different scales and in a vivid way. These modern kinds of maps may also 'animate' users to reflect individual meanings transported by the Sea and its coasts.

References

Al-Khanbashi, M. (2020). Using matrix as a qualitative data display for landscape research and a reflection based on the social constructivist perspective. In D. Edler, C. Jenal, & O. Kühne (Eds.), *Modern approaches to the visualization of landscapes* (pp. 103–118). Wiesbaden: Springer VS.

Aschenbrand, E. (2017). *Die Landschaft des Tourismus. Wie Landschaft von Reiseveranstaltern inszeniert und von Touristen konsumiert wird*. Wiesbaden: Springer VS.

Bellini, A., & Leonardi, L. (2020). Prato: The social construction of an industrial city facing processes of cultural hybridization. In D. Edler, C. Jenal, & O. Kühne (Eds.), *Modern approaches to the visualization of landscapes* (pp. 549–572). Wiesbaden: Springer VS.

Berr, K. (2020). Visuality, aesthetics and landscape. For the enlightenment and self-enlightenment of constructivist landscape research. In D. Edler, C. Jenal, & O. Kühne (Eds.), *Modern approaches to the visualization of landscapes* (pp. 189–215). Wiesbaden: Springer VS.

Bundesamt für Seeschifffahrt und Hydrographie (BSH). (Ed.). (2020). Elektronische Seekarten. https://www.bsh.de/DE/THEMEN/Vermessung_und_Kartographie/Seekartographie/Elektronische_Seekarten/elektronische-seekarten_node.html. Accessed 6 Mar 2020.

Çöltekin, A., Oprean, D., Wallgrün, J. O., & Klippel, A. (2019). Where are we now? Re-visiting the digital earth through human-centered virtual and augmented reality geovisualization environments. *International Journal of Digital Earth, 12*(2), 119–122. https://doi.org/10.1080/17538947.2018.1560986

Donohue, R. G., Sack, C. M., & Roth, R. E. (2014). Time series proportional symbol maps with Leaflet and jQuery. *Cartographic Perspectives, 76*, 43–66. https://doi.org/10.14714/CP76.1248.

Edler, D. (2020). Where spatial visualization meets landscape research and "Pinballology": Examples of landscape construction in pinball games. *KN—Journal of Cartography and Geographic Information.* https://doi.org/10.1007/s42489-020-00044-1.

Edler, D., & Vetter, M. (2019). The simplicity of modern audiovisual web cartography: An example with the open-source javascript library leaflet.js. *KN—Journal of Cartography and Geographic Information, 69*(1), 51–62. https://doi.org/10.1007/s42489-019-00006-2.

Edler, D., Husar, A., Keil, J., Vetter, M., & Dickmann, F. (2018). Virtual reality (VR) and open source software: A workflow for constructing an interactive cartographic VR environment to explore urban landscapes. *KN—Journal of Cartography and Geographic Information, 68*(1), 3–11. https://doi.org/10.1007/BF03545339.

Edler, D., Keil, J., & Dickmann, F. (2020). From Na Pali to Earth—An 'unreal' engine for modern geodata? In D. Edler, C. Jenal, & O. Kühne (Eds.), *Modern approaches to the visualization of landscapes* (pp. 279–291). Wiesbaden: Springer VS.

Eloot, K., Verwilligen, J., & Vantorre, M. (2008). *An overview of squat measurements for container ships in restricted waters.* Glasgow: SOCW.

Fontaine, D. (2020a). Landscape in computer games—The examples of GTA V and Watch Dogs 2. In D. Edler, C. Jenal, & O. Kühne (Eds.), *Modern approaches to the visualization of landscapes* (pp. 293–306). Wiesbaden: Springer VS.

Fontaine, D. (2020b). Virtuality and landscape. In D. Edler, C. Jenal, & O. Kühne (Eds.), *Modern approaches to the visualization of landscapes* (pp. 267–278). Wiesbaden: Springer VS.

FPS Mobility and Transport, Van Craeyvelt, E., & Rijkswaterstaat. (Eds.) (2017). New shipping routes along the Dutch and Belgian coast. https://www.varendoejesamen.nl/storage/app/media/downloads/EN_new-shipping-routes-north-sea-coast.pdf.

Glete, J. (2010). *Swedish Naval Administration, 1521–1721. Resource flows and organisational capabilities.* Leiden and Boston: BRILL.

Hochschild, V., Braun, A., Sommer, C., Warth, G., & Omran, A. (2020). Visualizing landscapes by geospatial techniques. In D. Edler, C. Jenal, & O. Kühne (Eds.), *Modern approaches to the visualization of landscapes* (pp. 47–78). Wiesbaden: Springer VS.

Horbinski, T., & Lorek, D. (2020). The use of Leaflet and GeoJSON files for creating the interactive web map of the preindustrial state of the natural environment. *Journal of Spatial Science.* https://doi.org/10.1080/14498596.2020.1713237.

Hruby, F., Sánchez, L. F. Á., Ressl, R., & Escobar-Briones, E. G. (2020). An empirical study on spatial presence in immersive geo-environments. *PFG—Journal of Photogrammetry, Remote Sensing and Geoinformation Science.* https://doi.org/10.1007/s41064-020-00107-y.

International Maritime Organisation (IMO). (Ed.). (2007). Annex 30. Resolution MSC.252(83) (adopted on 8th October 2007)—Adoption of the revised performance standards for integrated navigation systems (INS). https://www.imo.org/en/KnowledgeCentre/IndexofIMOResolutions/Maritime-Safety-Committee-%28MSC%29/Documents/MSC.252%2883%29.pdf.

International Maritime Organisation (IMO). (Ed.). (2020a). About IMO. https://www.imo.org/en/About/Pages/Default.aspx. Accessed 6 Mar 2020.

International Maritime Organisation (IMO). (Ed.). (2020b). AIS transponders. https://www.imo.org/en/OurWork/Safety/Navigation/Pages/AIS.aspx. Accessed 6 Mar 2020.

Jenal, C. (2020). Visualizations of 'landscape' in protest movements. On exclusive and inclusive patterns of vision and interpretation using the example of resistance to the expansion of the electricity grid in Germany. In D. Edler, C. Jenal, & O. Kühne (Eds.), *Modern Approaches to the Visualization of Landscapes* (pp. 427–445). Wiesbaden: Springer VS.

Kersten, T. P., & Edler, D. (2020). Special issue "methods and applications of virtual and augmented reality in geo-information sciences". *PFG—Journal of Photogrammetry, Remote Sensing and Geoinformation Science.* https://doi.org/10.1007/s41064-020-00109-w.

Kersten, T. P., Deggim, S., Tschirschwitz, F., Lindstaedt, M., & Hinrichsen, N. (2018). Segeberg 1600—Eine Stadtrekonstruktion in virtual reality. *KN—Journal of Cartography and Geographic Information, 68*(4), 183–191. https://doi.org/10.1007/BF03545360.

Kühne, O. (2018a). *Landscape and power in geographical space as a social-aesthetic construct.* Dordrecht: Springer.

Kühne, O. (2018b). *Landschaftstheorie und Landschaftspraxis. Eine Einführung aus sozialkonstruktivistischer Perspektive* (2., aktualisierte und überarbeitete ed.). Wiesbaden: Springer VS.

Kühne, O. (2020). The social construction of space and landscape in internet videos. In D. Edler, C. Jenal, & O. Kühne (Eds.), *Modern approaches to the visualization of landscapes* (pp. 121–137). Wiesbaden: Springer VS.

Kühne, O., & Weber, F. (2019). *Hybrid California: Annäherungen an den Golden State, seine Entwicklungen, Ästhetisierungen und Inszenierungen.* Wiesbaden: Springer VS.

Kühne, O., & Jenal, C. (2020). The threefold landscape dynamics—Basic considerations, conflicts and potentials of virtual landscape research. In D. Edler, C. Jenal, & O. Kühne (Eds.), *Modern approaches to the visualization of landscapes* (pp. 389–402). Wiesbaden: Springer VS.

Linke, S. (2020). Landscape in internet pictures. In D. Edler, C. Jenal, & O. Kühne (Eds.), *Modern approaches to the visualization of landscapes* (pp. 139–156). Wiesbaden: Springer VS.

Loda, M., Kühne, O., & Puttilli, M. (2020). The social construction of Tuscany in the German and English speaking world—Presented by the analysis of internet images. In D. Edler, C. Jenal, & O. Kühne (Eds.), *Modern approaches to the visualization of landscapes* (pp. 157–171). Wiesbaden: Springer VS.

Monmonier, M. (2008). Web cartography and the dissemination of cartographic information about coastal inundation and sea level rise. In M. P. Peterson (Ed.), *International perspectives on maps and the internet* (pp. 49–71). Berlin: Springer.

Poplin, A., de Andrade, B., & Mahmud, S. (2020). Exploring tangible and intangible landscapes of evocative places: Case study of the city of Vitória in Brazil. In D. Edler, C. Jenal, & O. Kühne (Eds.), *Modern approaches to the visualization of landscapes* (pp. 519–547). Wiesbaden: Springer VS.

Prisille, C., & Ellerbrake, M. (2020). Virtual reality (VR) and geography education: Potentials of 360° 'experiences' in secondary schools. In D. Edler, C. Jenal, & O. Kühne (Eds.), *Modern approaches to the visualization of landscapes* (pp. 321–332). Wiesbaden: Springer VS.

Pulsifer, P. L., Parush, A., Lindgaard, G., & Taylor, D. R. F. (2005). The development of the cyber-cartographic atlas of Antarctica. In D. R. F. Taylor (Ed.), *Cybercartography: Theory and practice* (pp. 461–490). Amsterdam: Elsevier.

Pulsifer, P. L., Caquard, S., & Taylor, D. R. F. (2007). Toward a new generation of community atlases—The cybercartographic atlas of Antarctica. In W. Cartwright, M. P. Peterson, & G. Gartner (Eds.), *Multimedia cartography* (2nd ed., pp. 195–216). Berlin: Springer.

Ratter, B. M. W., & Walsh, C. (2019). Küstenlandschaften. In O. Kühne, F. Weber, K. Berr, & C. Jenal (Eds.), *Handbuch Landschaft* (pp. 699–710). Wiesbaden: Springer VS.

Retchless, D. P. (2014). Sea level rise maps: How individual differences complicate the cartographic communication of an uncertain climate change hazard. *Cartographic Perspectives, 77*, 17–32. https://doi.org/10.14714/CP77.1235.

Roßmeier, A. (2020). Urban/rural hybridity in pictures—The creation of neighborhood images using the example of San Diego's urbanizing inner-ring suburbs east village and Barrio Logan. In D. Edler, C. Jenal, & O. Kühne (Eds.), *Modern approaches to the visualization of landscapes* (pp. 479–498). Wiesbaden: Springer VS.

Schönwald, A. (2017). Mehr Bedeutungsoffenheit für Landschaften durch Hybridisierungen. In O. Kühne, H. Megerle, & F. Weber (Eds.), *Landschaftsästhetik und Landschaftswandel* (pp. 161–176). Wiesbaden: Springer VS.

Siepmann, N., Edler, D., & Kühne, O. (2020). Soundscapes in cartographic media. In D. Edler, C. Jenal, & O. Kühne (Eds.), *Modern approaches to the visualization of landscapes* (pp. 247–263). Wiesbaden: Springer VS.

Steinberg, P. E. (2001). *The social construction of the ocean*. Cambridge: CUP.

Stratmann, J., Ristea, A., Leitner, M., & Paulus, G. (2020). Exploring urban "blightscapes" applying spatial video technology and geographic information system: A case study from Baton Rouge, USA. In D. Edler, C. Jenal, & O. Kühne (Eds.), *Modern approaches to the visualization of landscapes* (pp. 499–517). Wiesbaden: Springer VS.

Vetter, M. (2020). Technical potentials for the visualization in virtual reality. In D. Edler, C. Jenal, & O. Kühne (Eds.), *Modern approaches to the visualization of landscapes* (pp. 307–317). Wiesbaden: Springer VS.

Weber, W. (1994). Schipperen op de Schelde: 150 jaar Scheldereglement. *Den Spiegel, 1994*(2), 2–7.

Weber, F. (2020). Blurring the boundaries of landscape visualization: Welcome to fabulous Las Vegas. In D. Edler, C. Jenal, & O. Kühne (Eds.), *Modern approaches to the visualization of landscapes* (pp. 461–478). Wiesbaden: Springer VS.

Weintrit, A. (2009). *The electronical chart display information system (ECDIS). An operational handbook*. Boca Raton: CRC Press.

Wrenger, K. (2015). *Kartengestützte Orientierung im Realraum unter besonderer Berücksichtigung der Einflussgröße Raum*. Münster: hgd.

Alexander Kleber studied English philology and Geography (B.A., 2013–2017) at the Ruhr-University Bochum, after being at nautical college in Cuxhaven from 2012 to 2013. Since 2013 he is working as a freelancing editorial sport journalist. As a master student (M.Sc., 2017–2019), he specialized in Cartography and transport geography, due to a simultaneous collaboration with the Dutch pilot organisation *Loodswezen Regio Scheldemonden* in Vlissingen. In cooperation with the pilot organisation and the Ruhr-University in Bochum, he is currently working in Vlissingen on this doctoral thesis about the economical, ecological and sociological value of the *Oostgat*-fairway for the Western Scheldt area.

Dennis Edler studied English philology and Geography (B.A., 2006–2009) at the Ruhr-University Bochum (RUB) and University College Cork (UCC). He also spent several study periods at other European universities and research institutes: Regent's Park College (University of Oxford), Riga Technical University (RTU), Mediterranean Agronomic Institute of Chania

(MAICh) and German Aerospace Center (DLR). As a master student (M.Sc., 2009–2011), he specialized in Cartography, GIS and Remote Sensing. In his doctoral thesis (2012–2015), he conducted experimental studies on the effects of artificial map elements, such as grids, on the accuracy of spatial memory. Since 2015, he holds a permanent staff position as a senior lecturer at the Ruhr-University Bochum (Geography Department). His current research and teaching activities involve modern analytical and visualization methods based on open source software and open geodata.

Frank Dickmann is Professor of Cartography and Geo-Information Science at the Ruhr-University Bochum. His main interests are 3D cartography, cognitive cartography, and exploring map efficiency using modern integrated media.

Part IV
Multisensory Landscape Presentations

Visuality, Aesthetics, and Landscape: For the Clarification and Self-Awareness of Constructivist Landscape Research

Karsten Berr

Abstract

The essay traces the connections between seeing, aesthetics, and landscape. The connection between seeing and aesthetics refers to the historical connection between 'pre-scientific perception' and aesthetics as a philosophical discipline. Aesthetics and landscape are connected through the question of natural beauty and the literary topos *Arcadia*. The connection between seeing and landscape results from the assignment of the composition of landscape on the visibility and visualising of what is constituted as landscape. The 'landscape eye' necessary to accomplish this had to be trained in a historical socio-cultural process. Seeing is reconstructed as 'practice' and as 'cognitive looking'. It is therefore necessary to further investigate and clarify the epistemological processes of landscape components as well as the socio-cultural 'conditions' of landscape research in the context of a critical self-awareness. The potentials of such clarification and self-awareness are each sounded out using two examples from discourses concerned with 'nature' and 'landscape'.

Keywords

Seeing · Aesthetics · Landscape · Virtuality · Constitution · Constructivism · Composition · Epistemology · Clarification · Self-awareness

K. Berr (✉)
Forschungsbereich Geographie, Stadt- und Regionalentwicklung (SRE), Eberhard Karls
Universität Tübingen, Tübingen, Germany
e-mail: karsten.berr@uni-tuebingen.de

© Springer Fachmedien Wiesbaden GmbH, part of Springer Nature 2020   189
D. Edler et al. (eds.), *Modern Approaches to the Visualization
of Landscapes*, RaumFragen: Stadt – Region – Landschaft,
https://doi.org/10.1007/978-3-658-30956-5_11

1   Introduction

It took centuries of social and cultural history (among many: Berr and Kühne 2020; Berr and Schenk 2019; Hard 1977; Kirchhoff 2017; Kühne 2018c; Müller 1977; Schenk 2017) in order to learn to interpret nature or a "viewed slice of nature" (Schenk 2017, p. 676) as landscape (see Hammerschmidt and Wilke 1990; Hard 2002, p. 177). And it required immense scientific preparations and efforts to first abstract these historically conveyed preconditions into the formula of the 'landscape eye' (Riehl 1996) and, following on from this, not only to grasp their 'landscape' constituent function, but also to explicate and operationalize it theoretically for landscape theory and landscape practice (for the English-speaking world, especially Cosgrove 1984; Greider and Garkovich 1994; for the German-speaking world especially Kühne 2005, 2006a, b, c, 2008, 2018b, c, 2019b, c, 2020; Kühne et al. 2019; Weber et al. 2018).

Notwithstanding this insight into the 'world' and the landscape-constituting function of seeing, and in view of new phenomena such as the Internet and the possibilities of 3D visualisation of 'landscape', it is necessary to further investigate the epistemological processes of landscape composition. And in view of some of the follow-up problems connected with the constructiveness of seeing, which have so far led to ideological trench warfare, it is also necessary to further investigate the socio-cultural 'conditions' of landscape research in the course of a critical self-awareness.

First, the connection between aesthetics and landscape aesthetics is more closely examined. For this purpose, connections between the three concepts of seeing, aesthetics, and landscape are established. The connection between seeing and aesthetics refers to the historical connection between 'pre-scientific perception' and aesthetics as a philosophical discipline (Sect. 2.1). Aesthetics and landscape are connected through the question of natural beauty and the literary topos Arcadia (Sect. 2.2). The connection between seeing and landscape results from the assignment of the composition of landscape on the visibility and visualisation of what constitutes landscape (Sect. 2.3). On this basis the connection between 'virtuality' and 'landscape' is examined (Sect. 3) and the fundamental act of landscape composition is historically located and described (Sect. 4). In an interim result, it is explained why the investigations of 'seeing' lead to the demand for a clarification and self-awareness of landscape research (Sect. 5). Then the potentials of such clarification (Sect. 6) and self-awareness (Sect. 7) are each sounded out using two examples from discourses concerned with 'nature' or 'landscape'. In conclusion, research needs are expressed (Sect. 8).

2   Aesthetics and Landscape Aesthetics

2.1   Visuality and Aesthetics

The connection between the terms 'visuality' and 'aesthetics' may be easy to see. 'Aesthetics' refers both colloquially and from the ancient Greek word origin (*aisthesis*) to 'sensory perception' (cf. e.g. Liessmann 2009a, p. 13; Majetschak 2016, p. 10; Reicher 2015, p. 9 and many more), and thus, in addition to hearing, smelling, tasting, feeling, and especially to the sense of sight, i.e. the human eye and vision. The sense of sight often stands as pars pro toto for the visual and sensory perception par excellence. Of course, the original colloquial meaning of '*aisthesis*' also includes the other senses and the sum of all sensations creating their overall effect.

In scientific language, this etymological background is found in the, likewise ancient, Greek term '*Aistetike Episteme*' for the 'science of sensory knowledge' (Henckmann 1992a, p. 20), which was established in the middle of the eighteenth century. In the middle of the eighteenth century, Alexander Gottlieb Baumgarten named it for the first time as a science in § 1 of the 'Prolegomena' in his 'Aesthetica' (Baumgarten 2009 [1750–1758]) and introduced it as an independent philosophical sub-discipline into philosophy: "Aesthetica [...] est scientia cognitionis sensitivae" ("Aesthetics [...] is the science of knowledge of the senses"). Although the history of philosophical aesthetics as an independent sub-discipline only begins in the eighteenth century (cf. exemplary Ritter ([1971] 2019), Baumgarten is part of a long tradition of philosophical inquiries that, although not in terms of the concept ('Aesthetica'), have been asked since the early beginnings of philosophy—admittedly "mostly not in terms of the main theme, but in the context of superordinate metaphysical questions, and also not under the explicit, modern title of an 'aesthetic'" (Majetschak 2016, p. 10). This context includes, in particular, the question of the ontological, epistemological, and practical status of 'beauty', art, and perceptions of the senses. Accordingly, research has reconstructed three traditional basic meanings that currently serve as definitions of 'aesthetics' in philosophical discourses: Aesthetics is 1) theory of beauty, 2) theory of art, and 3) theory of sensory perception or cognition (Gilbert and Kuhn 1953; Majetschak 2016; Peres 2013; Pöltner 2008; Reicher 2015; Scheer 2015 [1997]; Nida-Rümelin 1998 and many more).

2.2   Aesthetics and Landscape

The connection between the terms 'Aesthetics' and 'Landscape' is perhaps less obvious. This combination of terms refers to landscape aesthetics as a kind of 'areal aesthetics'. As with the term 'aesthetics', the question of an aesthetic approach to landscape can be placed as part of a long-standing tradition, which is essentially embedded in two contexts. One context being the metaphysical and post-metaphysical question of 'natural beauty', while the other pertains to the tradition of the literary topos 'Arcadia'.

Before the 'subjectivation of aesthetics' or the 'beautiful' (Gadamer 2010, pp. 48–86) by Kant (1959b [1790]) and the modern "view that beauty is the expression of a subjective taste" (Liessmann 2009b, p. 13), in antiquity and the Middle Ages the 'beautiful' was regarded as an ideal transcendent entity, which was regarded as the objective basis and measure for all earthly beauty—in antiquity as an 'idea', in the Middle Ages as God's 'primal splendour' (see e.g. Assunto 1963; Büttner 2006; Liessmann 2009b; Perpeet 1977, 1988; Pochat 1986; Scheer 2015 [1997]; Plumpe 1993; Tatarkiewicz 2003;

Zimmermann 1991). In the Renaissance, the individual was increasingly liberated from religious convictions and constraints, art from sacral purposes, man discovered himself and the *secular* world or *secular* nature (see e.g. Hauskeller 1995, 2005; Kristeller 1980; Perpeet 1987; Scheer 2015 [1997]). From then on, phenomena can be understood as independent entities, since they are no longer measured by the standard of metaphysical or religious ideas, but by the "regularity that is shown by the phenomena themselves" (Scheer 2015 [1997], p. 29). In the seventeenth and eighteenth century, the 'beauty of *nature' is* also increasingly being addressed; the term 'natural beauty' "was developed from two different points of view: 1. from a metaphysical and scientific intellectual view of the perfection of the structure of the world or of some elements of it, [...] 2. from a direct, concrete and (directly or indirectly) visual view of nature, as sky, as landscape or special objects in nature (plants, animals, etc.). Sometimes both points of view are connected" (Tonelli 2019, column 623). Physicotheologists like Shaftesbury praised the "majestick [sic] Beautys of this Earth" as "Supreme Creator" (quoted from Tonelli 2019, column 624) and thus concluded from the harmony and beauty of nature to the world builder God: Experience of nature as experience of God. In the course of the eighteenth century, in this tradition, nature poets such as Barthold Hinrich Brockes (1680–1747), Albrecht von Haller (1708–1777), Salomon Geßner (1730–1788), Johann Heinrich Voss (1751–1826) and Gottlieb Klopstock (1724–1803) conveyed a sensitive ideal of nature in the context of idyll poetry (cf. Böschenstein-Schäfer 1977; Garber 2017). The 'natural beauty' of landscapes increasingly came into focus, for example, the Alps and thus the "beauty of the mountains" were thematized (Tonelli 2019, column 624). After a thousand years of socio-cultural development of the term 'landscape' (among many: Berr and Kühne 2020; Berr and Schenk 2019; Kirchhoff 2017; Kühne 2018c), the period around 1800 is the time when the discovery and enthusiasm for European landscapes, conveyed by artists, writers, and travel writers, is reaching a climax (cf. exemplary: Dinnebier 2001, Küster 2002; 2012b, 2013a [1995], b [1998]; Tümmers 1999; Bätzing 2005; Beck 1996; Börsch-Supan 2002).

The second tradition, which is linked to landscape aesthetics, places landscape in the literary context of an ideal that is connected with the landscape '*Arcadia*' known in antiquity and transfigured by writers (see Curtius 1954 [1948]; Frizell 2009; Gruenter 1975 [1953]; Küster 2012a; Panofsky 1975 [1936]; Roters 1995; Snell 1975 [1945]; Highet 1964; Maisak 1981; Garber 1974, 1976) and at the same time with ideas of *paradise*. In the final analysis, these are *two* models of the utopian 'Golden Age' (see Schneider 2009, p. 8), which "came from very different traditions, but which have often been mixed up in the history of reception" (Schneider 2009, p. 9). Paradise is therefore often depicted as a 'primeval landscape' (as for example in '*Das Paradies*' by Lucas Cranach the Elder from 1530), which (Schneider 2009, p. 8) "clearly shows Arcadian features" (Schneider 2009, p. 9). Transposed into our time, it is a 'landscape' that promises the "happiness of a harmonious symbiosis of nature and man, free of guilt and suffering. It can be felt in the moment of a moving L[andscape] experience or be visualised in artistic representation, but it can also express itself in the lasting basic feeling of a homeland connection that preserves life and the community" (Henckmann 1992b, p. 146). In more recent scientific discussions about the concept, theories, and concepts of 'landscape', the term 'Arcadian associations' (see e.g. Berr 2019; Berr and Kühne 2020; Eisel and Körner 2009; Hard 1991; Hokema 2009, 2013; Kühne 2018a; Linke 2019; Prominski 2004) plays a decisive role. For the term 'landscape' is still associated with traditional notions of a 'beautiful' or 'lovely', as well as, 'natural' and 'harmonious' scenery (Jackson 1984).

2.3 Visuality and Landscape

The connection between the terms 'visuality' and 'landscape' is even less obvious. However, this connection has already been addressed in various scientific traditions, especially with regard to the implemented relationality of what is seen in general, i.e. the general dependence of perception on a specific 'aesthetic' act of composing—and this general dependence then also concerns that of a section of the physical environment as '*landscape*'. In art philosophy and art theory, the function and effect of this act of composing has been investigated since the Renaissance, particularly in the analysis of central perspective and the picture frame in painting (see also Schenk 2020 in this volume). As since the Renaissance, phenomena can be viewed in their inherent properties, it is also possible "to represent the objects and their context in the picture surface in exactly the same way as they are perceived in the eye of the beholder during the process of seeing. This attempt leads to the discovery of the central perspective (it is attributed to Brunelleschi) and to the development of a uniform, constructible, infinite space of optics" (Scheer 2015 [1997], p. 30). Against this thesis, "that the rules of pictorial representation imitate the laws of the human visual process" (Krämer 1998, p. 28), Sybille Krämer, following Erwin Panofsky (1980 [1927]), argues that central perspective is a 'symbolic form' in the sense of Ernst Cassirer. This means that the central perspective "pretends to imitate the order of the natural process of seeing, but that this claim is illusory: the space of central perspective […] does not at all coincide with the psychophysical space of human corporeality […]" (Krämer 1998, p. 28). The central perspective develops a "topos-forming function" in that people "generally perceive something as natural and right" (ibid.), *if* it is presented in central perspective.

The central perspective captures what is seen in a geometric grid which, from the eye of the observer, spreads out a net-like structure of spatial points over everything seen: "The act of seeing is made calculable and culturally stereotyped" (Krämer 1998, p. 28). In this way, the three-dimensional illusion of space is created on a two-dimensional surface. Everything that is grasped by the central perspective is classified according to mathematical laws into this spatial structure and, in the course of this localisation, gains its existence, as it were, only through the fact that it is dependent on the observer and on the relationships within the spatial structure. In this way, the observer or—in epistemological terms—the perceiving subject "becomes, so to speak, the transformer or medium of reality, which itself has become a fixed relational structure of spatial locations. Thus, a context of the real is created" (Scheer 2015 [1997], p. 31). Ultimately, the thematisation of the perspective of seeing and the development of the central perspective already led to the solution of an epistemological problem that Immanuel Kant described in the eighteenth century as the 'Copernican Turn' (Kant 1959a [1781])—the dependence of human knowledge on subjective conditions or prerequisites (see Scheer 2015 [1997], pp. 31–32). Even before Kant, the philosophy of the French thinker René Descartes prepared this epistemological turn, when the eye as the focus of the central perspective corresponds to the unquestionable 'ego cogito' of a subject methodically destroying itself and the world as 'fundamentum inconcussum' of a continuous chain of deduction of humanly possible knowledge and a scientific reconstruction of the 'world' (Descartes 2008).

Further analyses of the painting process have shown that the artist, in the process of painting, must *set* the depicted section of the world by *himself*; the picture must be able to limit or *frame* the field of vision. The means for this is accordingly the *picture frame* (cf. Hegel 2003 [1823], p. 44), with which an "accentuation of the execution character" (Gethmann-Siefert 2005, p. 281) of perception, for example that of landscape, is connected. In the context of his analysis of Dutch landscape painting of the seventeenth century, the art historian Kurt Bauch also sees the framed view of 'nature' as the central constituting act of 'landscape': "[…] a *part* of it, a detail is brought", namely "where we turn, what we consider" (Bauch 1957 [1937], p. 127). 'Nature' is now "present as a detail, as the visible, the seen" (Bauch 1957 [1937], p. 127). In his essay on the 'Philosophy of Landscape', the sociologist Georg Simmel detached this componential achievement from the observation of nature as landscape in art and referred to landscape observation as such. 'Landscape' is the product of a synthesis in the sense of a synopsis of "individual natural objects" (Simmel 1957 [1913], p. 147). What is also required is a specific "mood" that makes it possible to see "a piece of ground with what is on it as a landscape" (Simmel 1957 [1913], p. 142)—and not as the "sum of individual natural objects" (Simmel 1957 [1913], p. 147), i.e. as part of a physical space. With a view to the famous ascent of Mont Ventoux on April 26, 1335 by Francesco Petrarca (Petrarca 1995), who stands on the summit "like a stunned man", "moved" by the "completely free panoramic view" (Petrarca 1995, p. 17) of the mountain landscape lying at his feet, Jacob Burckhardt had also emphasized "the importance of landscape for the excitable soul" (Burckhardt 1976 [1859], p. 277).

In the sense of a synthesis of individual things ('objects of nature' or portions of physical space) into a unity ('landscape'), the observation of nature as landscape is an *image*. It owes its existence to the 'constructive' gaze of an observer, whose preconditions for composing were described by Wilhelm Heinrich Riehl with the well-known formulation of a 'landscape eye' (Riehl 1996). Simmel clearly states that this perceptual-aesthetic synthesis is an act of formulating the perception, which is related to the act of composing in art: "Where we really see landscape and no longer a sum of individual natural objects, we have a work of art in statu nascendi" (Simmel 1957 [1913], p. 147). With Hegel, this fact can also be expressed in such a way that landscape is a "reflex of the spirit" (Hegel 2004 [1826], p. 2), and again in the words of Simmel in such a way that it is "itself already a spiritual entity" (Simmel 1957 [1913], p. 150). And Hegel considered this 'spirituality' in landscape painting long before Burckhardt, Riehl, and Simmel as an *atmosphere-induced* execution of nature: "Landscape painting views nature with soul and spirit and arranges its structures according to the purpose of expressing a mood" (Hegel 2003 [1823], pp. 255–256). Rainer Piepmeier has described this way of 'constructing' 'nature' or physical space to landscape, which is induced by mood in landscape painting as well as in landscape perception and is thus bound to perspective, briefly and precisely: "To see nature as L[andscape], a subject belongs to the seeing of nature as L[andscape] in such a correlative way, which makes nature into L[andscape] in a special act of seeing" (Piepmeier 2019, column 17).

The compositional gaze using a 'landscape eye' (Riehl 1996) admittedly requires a variety of cultural prerequisites, initiated especially via landscape painting, which made a decisive contribution to the prehistory of landscape perception, before the aesthetic relationship to nature *initiated* by Petrarca was first individually and later socio-culturally communicated in such a way that nature could be constructed into landscape on a broad social basis. According to Burckhardt, some artists of the Italian Renaissance already sensed and anticipated that "besides research and knowledge, there was another way of approaching nature" (Burckhardt 1976 [1859], p. 274). In the Renaissance, the ancient and medieval Arcadian ideal was taken up and developed further in the landscape paintings of Claude Lorrain and Nicolas Poussin, for example. The natural backdrops depicted in these paintings do not represent a landscape that is still 'realistic' but are the product of aesthetic idealisation and thus 'virtual' reality. It was only in the context of a long social and cultural history (among many others: Berr and Kühne 2020; Berr and Schenk 2019; Hard 1977; Kirchhoff 2017; Kühne 2018c; Müller 1977; Schenk 2017) that it was possible to learn to interpret nature or a "viewed slice of nature" (Schenk 2017, p. 676) as a landscape (cf. Hammerschmidt and Wilke 1990; Hard 2002, p. 177). 'Landscape' is therefore not a 'naturally given landscape', but as an image of a 'visual figure' (Hard 1991, p. 14) in the sense of a visual whole (see Schneider 2009; Steingräber 1985; Tesdorpf 1984), later also a symbolic '*sensory-image*' (Kirchhoff 2017; Kortländer 1977; Kühne 2018a; Schneider 2009). In the 'walk-in pictures' of English landscape gardens (Buttlar 1989, p. 14), the idealized two-dimensional images of nature (Hard 1991) in the eighteenth and nineteenth centuries gain a three-dimensional 'virtual' reality. In the literature and painting of the Romantic period, such images were also specifically linked to the perception and emotionality of an individual (Gruenter 1975 [1953]; Langen 1975 [1953]; Ritter 1996; Spanier 2006), who at the end of this long history of mediation saw what was initially 'virtual' reality as now being 'real'. This leads to a hypostasis of media induced phenomenality in the form of the 'reification' of an image or viewing space into a spatial section of reality (Hard 1991; Körner 2006; Kühne 2018c). This means that the construction and representation scheme developed in art, which synthesises physical objects into an artificial ('virtual') viewing space as 'landscape' (cf. Burckhardt 1976 [1859]; Lehmann 1968; Ritter 1996; Simmel

1990), is 'imitated' by art and transferred to 'reality' outside art (Hard 1991; cf. Hauck 2014; Kühne 2018c). This "aesthetically induced mechanism of reification" (Berr and Schenk 2019, p. 30) led to the fact that "world, nature and earth (surface) can potentially be regarded as landscape" (Hard 1991, p. 14) "and—with the exception of aesthetic mediation—as an objective fact independent of the viewer" (Berr and Schenk 2019, p. 30).

In the *English*-speaking world, the basic idea of the aesthetic constructing of landscape was taken up in the second half of the twentieth century by Denis Edmund Cosgrove, in particular, and translated into the catchy formula of 'landscape' as "a way of seeing" (Cosgrove 1984, p. 13). T. Greider and Lorraine Garkovich have made the same basic idea fruitful for a science-theoretical approach to 'landscape', which no longer has to define 'landscape' in an 'essentialist' or 'positivist' way via an objective 'being' or objectively quantifiable properties as a 'real object' independent of the observer (cf. Kühne et al. 2018), but instead in a constructivist perspective as an observer-dependent social construct (see on this subject, among others, Aschenbrand 2017; Cosgrove 1984; Greider and Garkovich 1994; Ipsen 2006; Kühne 2006a, 2012, 2017, 2018b, 2019f; Kühne and Schönwald 2015; Leibenath et al. 2013; Weber 2015, 2018; on the social constructivist perspective see also Bellini and Leonardi 2020; Fontaine 2020; Jenal 2020; Kühne 2020; Kühne and Jenal 2020; Linke 2020; Loda et al. 2020; Roßmeier 2020; Weber 2020 in this volume) as "a way of seeing" (Cosgrove 1984, p. 13). In the *German*-speaking world, Olaf Kühne has introduced this social constructivist approach to landscape constitution into the discussion in numerous theoretical works, developed it in a scientifically elaborated way and tested and proven it in many empirical studies (among others Kühne 2005, 2006a, b, c, 2008, 2018b, c, 2019b, c, 2020; Kühne et al. 2019; Weber et al. 2018).

3   Virtuality and Landscape

Among the results already achieved is the insight that 'landscape' is to be understood as an 'image' or even as a 'frame' through which 'subjects' or people look at the physical spatial environment and perceive it as identifying 'landscape'. It is thus ultimately written as *virtual* and as a special form of 'space' which can be reconstructed as a specific form of virtual reality. In the following, the concept of virtuality will therefore be examined more closely and its connection with 'landscape' will be discussed and developed further.

'Virtuality' is derived from the Latin 'virtus' ('vir' man), which means a) 'male efficiency, bravery, steadfastness' (Regenbogen and Meyer 1998, p. 708), b), 'effective force', 'according to force', and thus refers to 'the relationship between effect and cause' (Knebel 2019, p. 1062), c) to be understood in the sense of neo-Latin 'virtualiter' "according to the system, according to the possibility, in contrast to real, really existing" (Regenbogen and Meyer 1998, p. 708). In today's language 'virtuality' means 'inherent power or possibility', 'virtually' "according to the power or possibility available, apparently" (Duden Editorship 2000, p. 1043). Since the late 1980s, the term 'virtual reality' has been used to describe "reality simulated by the computer" (Duden Editorship 2000, p. 1043) or "audiovisual and tactile simulation technologies" (Grötker 2019, p. 1066). In this way, 'virtuality' is connected ontologically, or rather 'reality theorising', with the terms 'possibility' and 'reality', as well as cultural-philosophically and aesthetically (Grötker 2019, p. 1066) with the thesis of 'dissolution' (Vattimo 1998) or "virtualisation of reality" (Welsch 1998, p. 239). In terms of media theory, Luhmann, for example, still understands *every* medium as "pure virtuality" (Luhmann 1993, p. 356) or "potentiality", whereas the later technical development of digital media led to exclusively associating these *digital* media with "virtual reality", and thus also with "simulation" and "appearance".

Sybille Krämer places the 'technology of virtual realities' in the line of development from central perspective and calculation as a formal language to the algorithms of the simulation technologies of virtual reality, in which an 'ontologising' of calculation becomes apparent: "As soon as the techniques of artificial intelligence and virtual realities take the rank of explanatory models for what 'spirit' or 'reality' means in each case, formality no longer remains a mere property of a description system, but is itself hypostasised as a property of the processes to be described" (Krämer 1998, p. 34). Wolfgang Welsch does not deny this thesis of the hypostasis of media-induced phenomenality in the course of a "virtualisation of our consciousness of reality", but opposes a frequently "blanket negative evaluation" of this development with a positive thesis of progress: "In my opinion, the experience of the media and artificial worlds also has an almost enlightening effect, provided that it makes us aware of the fundamentally constructivist character of reality, the interpretativity of all our views of reality. (…) Thanks to the way we deal with the realities of the media we understand that reality has always been (at least to a large extent) a construction, that people just did not want to admit this to themselves earlier" (Welsch 1998, p. 241).

Thus Thomas Metzinger shows in the context of a 'self-model theory of subjectivity' (Metzinger 2005) that the 'I' or 'self' is already a specific construct, that is, "that there is no such thing as self in the world: Selves and subjects are not among the irreducible basic components of reality. What exists is the experienced ego and the different, constantly changing contents of our self-consciousness—what philosophers call the 'phenomenal self'" (2005, o. S.). Other philosophers have pointed out, for example, that people and their possessions are not only inexorably situational and individually 'entangled' in stories that construct their 'identity' (Schapp 1953), but "also involved in scenes" through which alone "they can gain their identity" (Waldenfels 2005, p. 197). Further empirical studies, on which Metzinger relies, allow the interpretation that consciously experienced 'time' and 'space' are also 'constructions': "In this sense, the phenomena of 'Now' is also itself a representational construct, it is a virtual presence, and at this point one can for the first time make clear what it means to say that the space of phenomena is a virtual space: its content is a possible reality. The realism of the experience of phenomena arises from the fact that in it a possibility - the best hypothesis that exists at the moment - is presented as a reality - a topicality - in an inescapable way" (2005, o. S.). The phenomena's space in which a person moves or stays as a 'subject' or 'I' can therefore be understood as virtual space. This diagnosis not only leads to an emphasis on the insight into the constructional character of our 'realities', but also enables research into the formational and intermediary procedures linked to media: "Contemporary enthusiasm for human penetration into *artificial* virtual worlds overlooks the fact that we have always been [in] a *biologically* generated 'phenospace': Within a virtual reality created by mental simulation. Nevertheless, the technological metaphor of *cyberspace* is an important pointer because it can give us interesting intuitions about our own phenomenon states. Artificial systems that generate interactive virtual realities in real time give us a first feeling of how complete worlds of experience can be created from pure information processing" (Metzinger 1994, p. 66).

These research results on 'virtuality' in general and on the virtuality of 'space' in particular are connectable to a philosophical tradition of approaches to constitutional theory regarding landscape and to a social science understanding that perceives 'landscape' as a special form of spatial experience, whereby spatial experience is preceded by landscape experience (Kühne 2006a, 2008, 2018c; Lehmann 1968; Piaget and Inhalder 1975 [1947]). Following Kant (1959a [1781]), who was able to show that space and time are not objective properties of entities, but rather object-constituting forms of perception as conditions of the possibility of spatial and temporal experience in general, according to the "pragmatic tendency" of "consciousness to action" (Gethmann 1987), individual and social experiences of space can be understood as achievements of abstraction from original body-bound (Husserl 2007 [1936]) life-worldly ways of dealing with things and other people (Heidegger 1993 [1927]; Bollnow 1994; Läpple 1992). Whether as a container or container-space (for the first time: Heidegger 1993 [1927], § 14) or other spatial constructions (cf. Kühne 2018c, pp. 15–22)—the "space" is individually and socially, as it were, pragmatically "granted", i.e. constituted or constructed by people.

4   The Construction of Landscape

In order to better understand the historical background and the conditions of validity of this fundamental compositional act, some lines of argumentation from the "classical" (Vietta 1995, p. 217) landscape essay by Joachim Ritter (Ritter 1974 [1963], pp. 141–190) are reproduced below. Ritter seeks an explanation for the "aesthetic constitution of landscape" (Ritter 1974 [1963], p. 183) in modern times, for which there has been no equivalent in the pre-modern world. This irritating "absence of nature as landscape" is "objectively justified" (Ritter 1974 [1963], p. 149). Ritter now argues that landscape can convey the "natural whole (…)" that was removed from the concept of metaphysical theory (theoría tou kósmou) in the course of the modern scientific "reification and objectification of nature" (Ritter 1974 [1963], p. 155) and keep it aesthetically current for man" (Ritter 1974 [1963], p. 153). For the pre-modern natural relationship, visible nature remained "to a certain extent *without virulence*" (Ritter 1974 [1963], p. 149; accentuation by the author). The "whole" of nature, which was formerly accessible in the medium of the "reasonable concept", was transformed by the analytical and empirical methods of modern natural sciences into an abstract, mathematically composed "object world" (Ritter 1974 [1963], p. 153). In the course of this development, the previously known order *of* 'nature' and 'world' as a 'whole' is being pushed "away from socially irrelevant remnants from pre-modern times" (Schweda 2013, p. 185) and can no longer be conveyed through the reality of human experience. The aesthetic view of nature as a landscape therefore has the function of conveying a "new shape and form" (Ritter 1974 [1963], p. 148) of the view of nature, i.e. a different and new relationship to nature. In this aesthetic attention to nature, it is neither theoretically analysed or researched in a new way nor used in practice. Instead, nature becomes "in 'freely' enjoying contemplation […] the great, sublime and beautiful" (Ritter 1974 [1963], p. 151). Ritter summarizes this process in a precise formulation: "Landscape is nature that is aesthetically present in the view of a feeling and sentient observer" (Ritter 1974 [1963], p. 150).

Following Burckhardt (Burckhardt 1976 [1859], pp. 292–303), Francesco Petrarca and his ascent of Mont Ventoux on April 26, 1336 (cf. Petrarca 1995) are the chief witnesses of the first time this *aesthetic* reference to nature was made *possible*. In research, this document is regarded as a classic topos for the new relationship with nature (cf. exemplarily: Jauß 1982; Steinmann 1995, pp. 39–49; Stierle 1979; critically: Groh and Groh 1992, 1996; Perpeet 1987, pp. 102–114; Seel 1996, pp. 220–230; Sieferle 1986; intermediary: Gil 2000; Kirchhoff 2017; Schweda 2013, pp. 180–194; Vietta 1995; cf. also Berr and Schenk 2019). A decisive characteristic is the "panoramic view" (Vietta 1995, p. 217) of a *feeling* and *sentient* observer (Ritter 1974 [1963]) "into the *distance*" (Burckhardt 1976 [1859], p. 274), *free of* utilitarian and religious considerations (Ritter 1974 [1963]). This 'view' from the mountain summit, which is "decidedly no longer medieval", but announces "a modern view of and over the landscape" (Vietta 1995, p. 217), Petrarca could *enjoy* "as aesthetic value" (Wimmer 2004, p. 32)—in contrast to the then prevailing theological prohibition of 'curiositas' (curiosity or 'lusting with the eyes') (see Flasch 1980; Groh and Groh 1992, 1996).

Ritter makes it clear that landscape requires a constituent observer in an aesthetic perspective. Accordingly, the *constituent of* 'landscape' is the subject (individual). The question now is which structural conditions lead to an aesthetic subject succeeding in making *visible* a phenomenon that is undetectable without aesthetic mediation. The question being: What are the *prerequisites of* seeing so that landscape is perceivable? And does Petrarca actually see landscape already? In the report he speaks merely of the irritation of the new visual experience: "like an anaesthetised person" he stands on the summit, "moved" by "the completely free panoramic view" (Petrarca 1995, p. 17) of the space surrounding this mountain. No word of 'landscape'. In reference to Gebser (1966), however, the achievement of Petrarca's "panoramic view" (Vietta 1995, p. 217) consists in "separating landscape from the whole of the visible world (pan-orama) by means of sectoral viewing" (Steinmann 1995, pp. 61), i.e. for the first time a *section of* the physical environment in the absence of practical goals and theoretical interests, "inspired solely by the urge to see this extraordinarily high place" (Petrarca 1995, p. 5) and to *enjoy it*. The aesthetic relationship to nature unintentionally *initiated* by Petrarca, however, remains "a descendant of philosophical theory in the exact sense that it is the presence of the whole of nature (…) as nature itself" (Ritter 1974 [1963], p. 151). On the other hand, Petrarca fails when he tries to tie his aesthetic experience back into the tradition of theoretical contemplation. For he is angry with himself because he admires earthly things for their own sake, instead of making God present in the sight of nature (Petrarca 1995, pp. 23–29). Petrarca's report thus remains "one of the great moments oscillating undecidedly between the epochs" (Blumenberg 1984, p. 142). Petrarca stands on the epochal threshold between the medieval metaphysical and the modern natural relationship. His discovery is, however, the *impetus* for a new aesthetic relationship with nature.

5   Interim Result

The 'landscape eye' has had to develop over the course of centuries of cultural history. This means two things with regard to the seeing of 'landscape' and seeing in general: *Firstly,* seeing is not a passive act of accepting visual data, but a productive act of conveying content in the process of seeing as an *aesthetic* "active form of dealing with the ambiguity of the visible world" (Schürmann 2008, p. 67). In other words: Seeing is a 'practice' in the sense of a 'performative activity', "with whose help we open up the world epistemically, ethically and aesthetically" (Schürmann 2008, p. 2). Insofar as cultural preconditions are constitutively influential as aspects of seeing, "understanding seeing as a practice […] means *making clear to* oneself the social and historical conditionality that *enables* and *disables* the individual view" (Schürmann 2008, p. 64; author's note).

Secondly, these statements mean that the seeing, the gaze, and the contemplation are, as it were, 'formed' or 'shaped'. With Hegel, one can speak in this sense of an "educated mind" (Hegel 1994, p. 188) and subsequently of an "educated gaze" (Halbig 2002, p. 98) or an "educated view" (Berr 2009, pp. 75–78). This means that the seeing or educated perception is not part of a 'machinery' of human consciousness, which processes incoming visual information and passes it on to other parts of this machinery for further processing, but already "*cognitive* perception" (Hegel 1830 [1992], § 445; emphasis in the original). This means that seeing in general, such as seeing landscape, is based on preconditions, such as socio-cultural habits of seeing, memories of what has been seen before, moral principles, aesthetic judgements of taste, religious convictions, scientific knowledge, philosophical insights, and more, which *form* something seen in a creative way and thereby give it *meaning*. An example of this is the psychology of form (Gestalt Psychology).

These two interim results point to two further research objectives. The understanding of seeing as a practice and the associated demand to "make clear" the preconditions of this practice, also with regard to the composition of landscape, requiring a *critical self-awareness of* landscape research on these historically variable socio-cultural "conditionalities". The understanding of seeing as 'cognitive perception' requires a further *epistemological clarification of* the epistemological processes involved in the composition of landscape. With regard to the first mentioned aspect, it would be advisable for landscape research, in the course of critical self-awareness, to repeatedly 'make clear' the 'conditions' of its own research and of landscape understanding, in order to avoid ideological trench warfare, for example. In the following (Sect. 7), two lines of research are presented as examples of this theme. As far as the second aspect is concerned, it would be advisable for landscape research to recognise that the knowledge gained in modern societies through the division of labour and correspondingly scattered knowledge with Popper (1963) and Dahrendorf (1968) can only apply provisionally. For theory(s) *of multimedia approaches to 'landscape'*, this could and should mean using the 'enlightening effect' of experiences with new media to advance the epistemological clarification of cognitive processes of knowledge regarding the composing of 'landscape'. In the following (Sect. 6), two lines of research on this topic are presented as examples.

6 Education About Virtualisation and Visual Media

The previous remarks should have made it clear that seeing as a 'practice' is a constructive act that is ultimately written *virtually* and does not produce exact images of any conceived 'reality', instead yielding constructed *images*. With the investigations of art theorists and philosophers concerning central perspective, this compositional mechanism was not only reflected upon and analysed for the first time in the Renaissance, but also "culturally stereotyped" (Krämer 1998, p. 28). Decisive for further considerations is the fact that in this way seeing, itself, was examined more closely and its mode of action better understood. And not only this, 'Reality' itself was also better understood in this way—namely as a compositional product of human seeing and recognizing. This can be explained in more detail with some reflections by Martin Heidegger and Kurt Bauch.

Jacob Burckhardt had spoken of the 'discovery of the world and man' by modern art (Burckhardt 1976 [1859], Sect. 4). Heidegger speaks of the "world as image", of "world view", i.e. "the world understood as an image" (Heidegger 1977 [1859], p. 89). By 'world view', Heidegger understands that the modern subject *understands* the world—and with this, the existing as 'a totality' is meant—*in the imagination as an image*. In this way, the subject becomes the reference centre of its 'world' and, as this subject, the reference centre of all that exists. When the 'world' is understood as the existing as 'a totality', but the existing is sought and found in the imaginings of the existing as image, then the world becomes the *construction of* the subject. However, this also means: Modern man attains knowledge of and orientation in the 'world' through the *image of* the 'world'. In his analysis of Dutch landscape painting of the seventeenth century, Bauch links the term 'modern world view' with the question of the function of *seeing*. He points out that in the paintings of the Dutch "seeing itself became a problem" (Bauch 1957 [1937], p. 130). For the Dutch artists "first immediately recognized that the actual task of painting was to open the eyes of mankind, to show it what is to be seen, and above all how it is seen" (Bauch 1957 [1937], p. 137). It is thus about the meaning of *seeing itself*. Seeing, as it were, searches for its own possibilities. These 'research efforts' can be observed—as already shown—not only with regard to central perspective and modern painting, but also with regard to the 'new media' such as the Internet and the new possibilities of 'virtual reality' associated with it.

6.1   Research on 3D Visualization of Landscape

If one adds that electronic media also access 'objects' 'specifically', i.e. *"only in their own way"* (Welsch 1998, p. 242), then it is not only not astonishing, but even to be welcomed, if the mechanisms of vision and construction of these media are explored in terms of how they make the 'world' visible and construct 'reality'. In this way, the use of *virtual* reality can be used to research mechanisms of visualisation or virtualisation not only in their mode of operation in general, but above all in a *more realistic way*. This means that one goes back to the fundamental *virtual* ground of the 'constituent level', i.e. the effective, media-induced configuration mechanisms of social reality constructions, in order to investigate the 'virtual' construction of 'reality' 'close to reality' and thus better understand it. This is now also happening in landscape research when it attempts to use the potentials of virtual reality for the investigation of the structural and intermediary mechanisms that constitute 'landscape' (fundamental: Edler et al. 2018; Vetter 2019). The example of the '3D visualisation of landscape' (Edler et al. 2018, 2019)—for example in digital maps—shows that it is precisely through the 'laboratory conditions' made possible in 'virtual reality' that 'real' disturbing factors such as certain irritating noises (e.g. car noise) can be eliminated and thus the comparability of collected data significantly increased. Conversely, non-visual stimuli (e.g. acoustic) can be integrated for practical purposes, which is not possible with two-dimensional images or maps (see also in this volume: Edler et al. 2020; Siepmann et al. 2020; Vetter 2020). The 'realism' of simulated objects and landscapes can then be increased.

6.2   Synaesthesia of Landscape

At present, efforts can be observed in research to break up the traditional 'dominance of the visual' and also to consider 'non-visual landscapes' in terms of the compositional performance of non-visual sensory impressions (Edler and Kühne 2019). The composition of a physical space also derives from the senses of hearing, smell, touch, and taste (see Kazig 2013, 2019; Kühne 2019d). The philosopher Wolfgang Welsch has also pointed out that without media there would be no meanings and thus no realities, that "meaning always owes itself to the inscription in media and that the medium itself does not add to meaning only just afterwards and externally, but is formative for meaning from the very beginning, that it has *productive* meaning for the processes of meaning" (Welsch 1998, p. 236). Media in general are also 'specific', since they "access *all objects*, but - like any other medium - *only in their own way*" (Welsch 1998, p. 242). This then also applies to the perceptual-aesthetic synthesis to landscape, in the execution of which each sense organ productively synthesises specific aspects of the perceived physical space into 'landscape', either on its own or in synaesthesia with other senses (Kühne 2006a, 2018c, 2019a, f; Wojtkiewicz and Heiland 2012; see on the visualization of synaesthetic approaches also McLean 2020; Poplin et al. 2020; Siepmann et al. 2020 in this volume).

7   Self-Education About the Presence and Materiality of Landscape

In recent decades, research has explicitly pointed out two connection problems associated with the structuring of seeing 'landscape'. On the one hand, there is the question of what role 'landscape' actually plays or should play in the present with regard to the conveyance of 'nature'. The second question is what role 'physical materiality' or the 'nature' or 'object' aspect plays in the construction of a physical space to landscape.

7.1   The 'Presence' of Landscape

'Nature' is in full swing. In times of climate change, it is becoming the focus of global attention, since not only the 'climate' and, depending on climate, the 'weather' are affected, but also landscapes. In the general consciousness, it is the forest, the fields, the meadows, the sea level, areas devastated by tornadoes or forest fires, and many other parts of physical environments that have been assembled to form 'landscapes', whose damage and endangerment are being observed. The thesis could therefore be: "Without 'Landscape', No 'Nature'"—or "Without 'Landscape Experiences' No 'Nature Experiences'". It was precisely this thesis that Ritter had advocated in his well-known landscape essay; and this thesis was attacked, for example by Jürgen Habermas and Martin Seel (cf. Schweda 2013, pp. 174–188).

Seel asserts that Ritter's position understands "the modern experience of nature as the restoration of a pre-modern experience" (Seel 1996, p. 230), that it becomes "a nostalgic ritual", which admittedly abandons metaphysics and the belief in a pre-modern "wholeness" of nature as a philosophical or scientific theory, but only in order to "rehabilitate it aesthetically" or to ennoble it as an "aesthetic attitude" (Seel 1996, p. 227). Schweda objects that the criticism of Seel moves in the "footsteps of Jürgen Habermas" (2013, p. 186) and his neo-conservatism reproach to Ritter, according to which within the framework of a 'functionalist traditionalism' "everything cultural that does not directly get into the maelstrom of the dynamics of modernisation is caught up in the perspective of remembering preservation" (Habermas 1996, p. 93). Ritter himself, on the other hand, had made it unmistakably clear that the aesthetic attention to nature as landscape has nothing whatsoever "to do with illusionary flight or the (deadly) dream of returning to the origin as to a still intact world"—on the contrary, it is precisely "the present" (Ritter 1974 [1963], p. 162). According to Ritter, the perspective of 'preserving' criticised by Habermas therefore does not at all consist in a backward-looking "holding on to what sinks back into time", but in "holding on to what is revealed and not seen in the present, as that which belongs to the present and is not considered in it" (Ritter 1974 [1963], p. 95). 'Nature' is 'present' in the aesthetic mediation as landscape, and not as a nostalgic reminiscence of a historically outdated metaphysical image of nature, but as the realization of something that is *scientifically* abstractly grasped by technology and natural sciences, which *can no longer be vividly conveyed* with the 'nature' and 'landscape' observations within the framework of the everyday reality of experience, and which would therefore remain "unsaid" (Ritter 1974 [1963], p. 161) or *invisible without* this aesthetic mediation.

The self-awareness of landscape research should also include not just breaking up ideological disputes about the meaning of 'landscape', but also sounding out the potentials of this concept for understanding and accessing 'nature'. Of course, this can only succeed if landscape researchers can commit themselves to the 'good' goals and purposes of this research by following the guideline of what their research and specific concepts are 'good' for.

7.2   The 'Materiality' of Landscape

A question that has been repeatedly raised in discourses on the constitution of landscape, but also on the aesthetics of nature and in the discussion about so-called 'atmospheres' in landscape, is: What role does 'physical materiality' or the 'nature' or 'object' aspect play in the construction of a physical space to landscape? Can 'physical materiality' be faded out, indeed does it not have to be considered somehow? These new 'tendencies to re-thematise the material' are currently emerging, especially within the framework of the 'actor-network theory' and the 'assemblage theory' (see Kühne 2019e).

A discussion that has been on-going for several decades involving a 'subjectivism' or 'objectivism' regarding the experiencing of landscape or nature induced by the sensitive interpretation of landscape or nature is heading in a similar direction. This can be seen in the current debate on 'atmospheres of the landscape', in which the concept of 'ambience' appears as a counterpart to the concept of 'mood'. At its core, the debate revolves around the question of whether 'atmospheres' are projections or whether they belong to the landscape itself. The question is: "Does the landscape itself have the property of being cheerful or melancholy, or do we attribute these properties to it through projection?" (Großheim 1999, p. 325). According to Michael Großheim, representatives of such an anthropocentric 'constructivism' or 'projectionism' are, for example, Georg Simmel, Ernst Bloch, Ruth and Dieter Groh, as well as August Wilhelm Schlegel, Theodor Lipps, Martin Seel, and Rolf Peter Sieferle. The main accusation levelled at the so-called 'constructivists' and 'projectionists' is that they ignored the danger of a "complete loss of objects"—meaning an object of cultural constructs or projections "over which cultural forms can only be built" (Großheim 1999, p. 359). The "historical research on the aesthetics of nature", which investigates the "constitutional conditions of images, symbols, forms of appropriation of nature", would indeed recognise the "subjective and objective spirit of images of nature", but "at the same time any level of reference and content would be erased" (Böhme 1995, p. 139). Großheim calls for a "phenomenology of atmospheres" (Großheim 1999, p. 361), which understands atmospheres as "entities" (Großheim 1999, p. 343) or as "objective feelings" (Großheim 1999, p. 344). Conversely, more recent debates on the concept of atmosphere attempt to convey the subjectivist and objectivist associations of this concept (for example Hahn 2012; Kazig 2007, 2013, 2019).

A further question in this context, which is also still virulent as a 'perennial issue' in the discussions about 'natural aesthetics', is whether the elaborated dependence of the components of nature to landscape on preceding historical *art experiences* represents "an ideology not only of art dependence, but of the explicit or implicit art-relatedness of all aesthetic perception of nature" (Seel 1996, p. 175). Without wanting to go into this question in more detail in the end, it should at least be recalled that a proposal by Heinz Paetzold, made some time ago, wherein Paetzold demanded that "natural aesthetics should enter into a complementary relationship to art philosophy" (Paetzold 1996, p. 52), that "aesthetics in the sense of a philosophy of art and in the sense of natural aesthetics are mutually related" (Paetzold 1996, p. 57) and that "today it must be about the regulated togetherness and opposition with each other in art philosophy and natural aesthetics" (Paetzold 1996, p. 58).

The three related problems point to a need for further research within the framework of the self-awareness of constructivist landscape research. In each case, the same is true as already said in the dispute between Ritter and Habermas/Seel on the danger of possible ideological trench warfare.

8   Conclusion

It has been shown that seeing is a practice that opens-up the 'world' for the human being. Seeing is a complex process that integrates different pre-scientific and scientific dimensions of 'cognitive viewing' and represents a process of composition that constructs realities that are always conceived virtually. In this sense, 'landscape' could be reconstructed as a special form of virtual reality. The investigation of these compositional achievements of seeing and of various visual media clarifies the fundamental constructional character of 'reality'—for example of 'landscape'. This clarification is at the same time connected with the recognition of the responsibility for one's own and collective seeing, insofar as people do not blindly accept the preconditions of seeing, but want to explore them and process them for practice in an action-oriented way. Within the process of self-awareness, landscape research must take responsibility for the aims and purposes of scientific research, publishing, and action. It does not want to blindly claim the 'conditions' of its own research and understanding of landscape, but to further explore these so to be able to turn against possible dogmatisms, immunisation strategies, and presumptuous interpretative sovereignty.

There is a need for research both in terms of epistemological clarification and critical self-awareness of landscape research. The elucidation of the epistemological processes of seeing, which are relevant in the context of 'landscape', is only just beginning and promises—regarding, for example, 'virtuality' and 3D visualisations—new theoretical and practical insights. The self-awareness of landscape research faces a specific difficulty: Here a common 'regulative idea' would be desirable, which could create a common research horizon with jointly shared general goals and purposes (cf. tentative Berr 2020). As concerns the requirement of assuming responsibility for both clarification and self-awareness, the area of an ethics of science is already being entered which has scarcely found its way into landscape research so far. This appears to be an urgent research desideratum.

References

Aschenbrand, E. (2017). *Die Landschaft des Tourismus. Wie Landschaft von Reiseveranstaltern inszeniert und von Touristen konsumiert wird.* Wiesbaden: Springer VS.
Assunto, R. (1963). *Die Theorie des Schönen im Mittelalter.* Köln: DuMont.
Bätzing, W. (2005). *Die Alpen. Geschichte und Zukunft einer europäischen Kulturlandschaft.* München: Beck.
Bauch, K. (1957). Anfänge der neuzeitlichen Kunst. In Joachim Jungius-Gesellschaft der Wissenschaften (Ed.), *Die Entfaltung der Wissenschaft. Zum Gedenken an Joachim Jungius (1587–1657).* Vorträge gehalten auf der Tagung der Joachim Jungius-Gesellschaft der Wissenschaften, Hamburg, am 31. Okt./1. Nov. 1957 aus Anlaß der 300. Wiederkehr des Todestages von Joachim Jungius (pp. 118–139). Hamburg: Augustin (First publication 1937).
Baumgarten, A. G. (2009). *Ästhetik* (Philosophische Bibliothek, 572a/b, Vol. 2). Hamburg: Meine (First publication 1750–1758).
Beck, R. (1996). Die Abschaffung der Wildnis. Landschaftsästhetik, bäuerliche Wirtschaft und Ökologie zu Beginn der Moderne. In W. Konold (Ed.), *Naturlandschaft—Kulturlandschaft. Die Veränderung der Landschaften nach der Nutzbarmachung durch den Menschen* (pp. 27–44). Landsberg: Ecomed.
Bellini, A., & Leonardi, L. (2020). Prato: The social construction of an industrial city facing processes of cultural hybridization. In D. Edler, C. Jenal, & O. Kühne (Eds.), *Modern approaches to the visualization of landscapes* (pp. 549–572). Wiesbaden: Springer VS.

Berr, K. (2009). *Hegels Bestimmung des Naturschönen. Zur Betrachtung und Darstellung schöner Natur und Landschaft*. Saarbrücken: Südwestdeutscher Verlag für Hochschulschriften.

Berr, K. (2019). Klassiker der Landschaftsforschung und ihre gegenwärtige Wirkung. In O. Kühne, F. Weber, K. Berr, & C. Jenal (Eds.), *Handbuch Landschaft* (pp. 39–53). Wiesbaden: Springer VS.

Berr, K. (2020). Vom Wahren, Schönen und Guten. Philosophische Zugänge zu Landschaftsprozessen. In R. Duttmann, D. Knitter, O. Kühne, & F. Weber (Hrsg.), *Landschaft als Prozess*. Wiesbaden: Springer VS (In press).

Berr, K., & Kühne, O. (2020). *„Und das ungeheure Bild der Landschaft...".* The Genesis of Landscape Understanding in the German-speaking Regions (RaumFragen: Stadt—Region—Landschaft). Wiesbaden: Springer VS.

Berr, K., & Schenk, W. (2019). Begriffsgeschichte. In O. Kühne, F. Weber, K. Berr, & C. Jenal (Eds.), *Handbuch Landschaft* (pp. 23–38). Wiesbaden: Springer VS.

Blumenberg, H. (1984). *Der Prozeß der theoretischen Neugierde*. Frankfurt (Main): Suhrkamp.

Böhme, H. (1995). Materialismus oder Konstruktivismus. Eine falsche Alternative—aus der Sicht der Goethezeit. In M. Großheim (Ed.), *Leib und Gefühl. Beiträge zur Anthropologie* (Lynkeus, Vol. 1, pp. 129–140). Berlin: Akademie.

Bollnow, O. F. (1994). *Mensch und Raum* (7th ed.). Stuttgart: Kohlhammer.

Börsch-Supan, H. (2002). Die künstlerische Entdeckung der Landschaften Europas in der Epoche der Aufklärung und der Romantik. In K. Weschenfelder & U. Roeber (Eds.), *Wasser, Wolken, Licht und Steine. Die Entdeckung der Landschaft in der europäischen Malerei um 1800* (pp. 11–26) [Ausstellung Mittelrhein-Museum Koblenz, 25. August bis 3. November 2002]. Heidelberg: Edition Braus.

Böschenstein-Schäfer, R. (1977). *Idylle*. Stuttgart: Metzler.

Burckhardt, J. (1976). *Die Kultur der Renaissance in Italien. Ein Versuch*. Stuttgart: Kröner (First publication 1859).

Buttlar, A. V. (1989). *Der Landschaftsgarten. Gartenkunst des Klassizismus und der Romantik*. Köln: DuMont.

Büttner, S. (2006). *Antike Ästhetik. Eine Einführung in die Prinzipien des Schönen*. München: Beck.

Cosgrove, D. E. (1984). *Social formation and symbolic landscape*. London: University of Wisconsin Press.

Curtius, E. R. (1954). *Europäische Literatur und lateinisches Mittelalter (Zweite durchgesehene)*. Bern: Francke (First publication 1948).

Dahrendorf, R. (1968). *Pfade aus Utopia. Arbeiten zur Theorie und Methode der Soziologie*. München: Piper.

Descartes, R. (2008). *Meditationes de prima philosophia. Lateinisch—Deutsch* (Philosophische Bibliothek, Vol. 597). Hamburg: Meiner.

Dinnebier, A. (2001). Zur Zukunft der ästhetischen Landschaft. In H. Friesen & E. Führ (Eds.), *Neue Kulturlandschaften* (pp. 55–69). Cottbus: Libri Digital Services.

Dudenredaktion. (2000). *Duden. Die deutsche Rechtschreibung*. Mannheim u. a.: Dudenverlag.

Edler, D., & Kühne, O. (2019). Nicht-visuelle Landschaften. In O. Kühne, F. Weber, K. Berr, & C. Jenal (Eds.), *Handbuch Landschaft* (pp. 599–612). Wiesbaden: Springer VS.

Edler, D., Kühne, O., Jenal, C., Vetter, M., & Dickmann, F. (2018). Potenziale der Raumvisualisierung in Virtual Reality (VR) für die sozialkonstruktivistische Landschaftsforschung. *Kartographische Nachrichten, 68*(5), 245–254.

Edler, D., Keil, J., Wiedenlübbert, T., Sossna, M., Kühne, O., & Dickmann, F. (2019). Immersive VR experience of redeveloped post-industrial sites: The example of "Zeche Holland" in

Bochum-Wattenscheid. *KN—Journal of Cartography and Geographic Information, 69*(4), 267–284. https://doi.org/10.1007/s42489-019-00030-2.

Edler, D., Keil, J., & Dickmann, F. (2020). From Na Pali to Earth – An 'unreal' engine for modern geodata? In D. Edler, C. Jenal, & O. Kühne (Eds.), *Modern approaches to the visualization of landscapes* (pp. 279–291). Wiesbaden: Springer VS.

Eisel, U., & Körner, S. (Eds.). (2009). *Befreite Landschaft. Moderne Landschaftsarchitektur ohne arkadischen Ballast?* (Beiträge zur Kulturgeschichte der Natur, Vol. 18). Freising: Technische Universität München.

Flasch, K. (1980). *Augustin. Einführung in sein Denken* (Reclams Universalbibliothek, Vol. 9962). Stuttgart: Reclam.

Fontaine, D. (2020). Virtuality and landscape. In D. Edler, C. Jenal, & O. Kühne (Eds.), *Modern approaches to the visualization of landscapes* (pp. 267–278). Wiesbaden: Springer VS.

Frizell, B. S. (2009). *Arkadien. Mythos und Wirklichkeit.* Köln: Böhlau (Aus dem Schwedischen übersetzt von Ylva Eriksson-Kuchenbuch).

Gadamer, H.-G. (2010). *Band 1: Hermeneutik. 1: Wahrheit und Methode: Grundzüge einer philosophischen Hermeneutik.* Tübingen: Mohr Siebeck.

Garber, K. (1974). *Der locus amoenus und der locus terribilis. Bild und Funktion der Natur in der deutschen Schäfer- und Landlebendichtung des 17. Jahrhunderts.* Köln, Wien.

Garber, K. (Ed.). (1976). *Europäische Bukolik und Georgik.* WBG: Darmstadt.

Garber, K. (2017). *Literatur und Kultur im Deutschland der Frühen Neuzeit.* Paderborn: Fink.

Gebser, J. (1966). *Ursprung und Gegenwart, Fundamente und Manifestationen der aperspektivischen Welt.* Stuttgart: Deutsche Verlags-Anstalt.

Gethmann, C. F. (1987). Vom Bewusstsein zum Handeln. Pragmatische Tendenzen in der deutschen Philosophie der ersten Jahrzehnte des 20. Jahrhunderts. In H. Stachowiak (Ed.), *Handbuch Pragmatik. Bd. 2: Der Aufstieg des pragmatischen Denkens im 19. und 20. Jahrhundert* (pp. 202–232). Hamburg: Meiner.

Gethmann-Siefert, A. (2005). *Einführung in Hegels Ästhetik.* München: Fink.

Gil, T. (2000). *Der Begriff der ästhetischen Erfahrung.* Berlin: Berlin Verlag.

Gilbert, K. E., & Kuhn, H. (1953). *A history of esthetics.* Bloomington: Indiana University Press.

Greider, T., & Garkovich, L. (1994). Landscapes: The social construction of nature and the environment. *Rural Sociology, 59*(1), 1–24. https://doi.org/10.1111/j.1549-0831.1994.tb00519.x.

Groh, R., & Groh, D. (1992). *Petrarca Und Der Mont Ventoux. Merkur, 46*(4), 290–307.

Groh, R., & Groh, D. (1996). *Die Außenwelt der Innenwelt. Zur Kulturgeschichte der Natur 2.* Frankfurt (Main): Suhrkamp.

Großheim, M. (1999). Atmosphären in der Natur: Phänomene oder Konstrukte? In R. P. Sieferle & H. Breuninger (Eds.), *Natur-Bilder. Wahrnehmungen von Natur und Umwelt in der Geschichte* (pp. 325–365). Frankfurt (Main): Campus.

Grötker, R. (2019). [Artikel] Virtualität. II. Virtuelle Realität. In J. Ritter (Ed.), *Historisches Wörterbuch der Philosophie. Band 11: U—V [Sonderausgabe 2019]* (pp. 1066–1068). Darmstadt: WBG.

Gruenter, R. (1975). Landschaft. Bemerkungen zu Wort und Bedeutungsgeschichte. In A. Ritter (Ed.), *Landschaft und Raum in der Erzählkunst* (Wege der Forschung, Vol. 418, pp. 192–207). Darmstadt: WBG (First publication 1953).

Habermas, J. (1996). *Der philosophische Diskurs der Moderne. Zwölf Vorlesungen.* Frankfurt a. M.: Suhrkamp.

Halbig, C. (2002). *Objektives Denken. Erkenntnistheorie und Philosophy of Mind in Hegels System.* Stuttgart-Bad Cannstatt: Frommann-Holzboog.

Hahn, A. (Ed.) (2012). *Erlebnislandschaft – Erlebnis Landschaft? Atmosphären im architektonischen Entwurf.* Bielefeld: transcript.

Hammerschmidt, V., & Wilke, J. (1990). *Die Entdeckung der Landschaft. Englische Gärten des 18. Jahrhunderts*. Stuttgart: Deutsche Verlags-Anstalt.

Hard, G. (1977). Zu den Landschaftsbegriffen der Geographie. In A. Hartlieb von Wallthor & H. Quirin (Eds.), *„Landschaft" als interdisziplinäres Forschungsproblem. Vorträge und Diskussionen des Kolloquiums am 7./8. November 1975 in Münster* (pp. 13–24). Münster: Aschendorff.

Hard, G. (1991). Landschaft als professionelles Idol. *Garten + Landschaft, 3,* 13–18.

Hard, G. (2002). Zu Begriff und Geschichte von „Natur" und „Landschaft" in der Geographie des 19. und 20. Jahrhunderts [1983 erstveröffentlicht]. In G. Hard (Ed.), *Landschaft und Raum. Aufsätze zur Theorie der Geographie* (Osnabrücker Studien zur Geographie, Vol. 22, pp. 171–210). Osnabrück: Universitätsverlag Rasch.

Hauck, T. E. (2014). *Landschaft und Gestaltung. Die Vergegenständlichung ästhetischer Ideen am Beispiel von „Landschaft"*. Bielefeld: transcript.

Hauskeller, M. (Ed.). (1995). *Was das Schöne sei. Klassische Texte von Platon bis Adorno.* München: Deutscher Taschenbuch.

Hauskeller, M. (2005). *Was ist Kunst? Positionen der Ästhetik von Platon bis Danto (Beck'sche Reihe* (Beck'sche Reihe, 8th ed.). München: Beck.

Hegel, G. W. F. (1992). *Enzyklopädie der philosophischen Wissenschaften im Grundrisse (1830).* In von W. Bonsiepen & H.-C. Lucas (Eds.), *Unter Mitarbeit von Udo Rameil* (Gesammelte Werke. Vol. 20). Hamburg: Meiner (First publication 1830).

Hegel, G. W. F. (1994). *Vorlesungen über die Philosophie des Geistes. Berlin 1827/1828. Nachgeschrieben von Johann Eduard Erdmann und Ferdinand Walter.* In Hrsg. von Franz Hespe und Burkhard Tuschling unter Mitarbeit von Markus Eichel, Werner Euler, Dieter Hüning, Torsten Poths und Uli Vogel (Vorlesungen. Ausgewählte Nachschriften und Manuskripte, Vol. 13). Hamburg: Meiner (First publication 1827/1828).

Hegel, G. W. F. (2003). *Vorlesungen über die Philosophie der Kunst. Berlin 1823.* Hamburg: Meiner (First publication 1823).

Hegel, G. W. F. (2004). *Philosophie der Kunst oder Ästhetik. Berlin 1826.* Nachgeschrieben von Friedrich Carl Hermann Victor von Kehler. Hrsg. von A. Gethmann-Siefert und B. Collenberg-Plotnikov unter Mitarbeit von F. Iannelli und K. Berr. München: Fink (First publication 1826).

Heidegger, M. (1977). Die Zeit des Weltbildes. In M. Heidegger (Ed.), *Gesamtausgabe* (Vol. 5, pp. 75–96). Frankfurt a. M.: Klostermann.

Heidegger, M. (1993). *Sein und Zeit.* Tübingen: Max Niemeyer (First publication 1927).

Henckmann, W. (1992a). [Artikel] Ästhetik. In W. Henckmann & K. Lotter (Eds.), *Lexikon der Ästhetik* (pp. 20–24). München: Beck.

Henckmann, W. (1992b). [Artikel] Landschaft. In W. Henckmann & K. Lotter (Eds.), *Lexikon der Ästhetik* (pp. 146–147). München: Beck.

Highet, G. (1964). *Römisches Arkadien. Dichter und ihre Landschaft: Catull, Vergil, Properz, Horaz, Tibull, Ovid, Juvenal.* München: Goldmann.

Hokema, D. (2009). Die Landschaft der Regionalentwicklung: Wie flexibel ist der Landschaftsbegriff? *Raumforschung Und Raumordnung, 67*(3), 239–249.

Hokema, D. (2013). *Landschaft im Wandel? Zeitgenössische Landschaftsbegriffe in Wissenschaft, Planung und Alltag.* Wiesbaden: Springer VS.

Husserl, E. (2007). *Die Krisis der europäischen Wissenschaften und die transzendentale Phänomenologie. Eine Einleitung in die phänomenologische Philosophie* (Philosophische Bibliothek, 3rd ed., Vol. 292). Hamburg: Meiner (First publication 1936).

Ipsen, D. (2006). *Ort und Landschaft.* Wiesbaden: VS Verlag für Sozialwissenschaften.

Jackson, J. B. (1984). *Discovering the Vernacular Landscape.* New Haven: Yale University Press.

Jauß, H. R. (1982). *Ästhetische Erfahrung und literarische Hermeneutik*. Frankfurt (Main): Suhrkamp.

Jenal, C. (2020). Visualizations of 'landscape' in protest movements. On exclusive and inclusive patterns of vision and interpretation using the example of resistance to the expansion of the electricity grid in Germany. In D. Edler, C. Jenal, & O. Kühne (Eds.), *Modern approaches to the visualization of landscapes* (pp. 427–445). Wiesbaden: Springer VS.

Kant, I. (1959a). *Kritik der reinen Vernunft*. Hamburg: Felix Meiner (First publication 1781).

Kant, I. (1959b). *Kritik der Urteilskraft* (Philosophische Bibliothek, Unveränd. Neudr. der Ausg. von 1924). Hamburg: Meiner (First publication 1790).

Kazig, R. (2007). Atmosphären—Konzept für einen nicht repräsentationellen Zugang zum Raum. In C. Berndt & R. Pütz (Eds.), *Kulturelle Geographien. Zur Beschäftigung mit Raum und Ort nach dem Cultural Turn* (pp. 167–187). Bielefeld: transcript.

Kazig, R. (2013). Landschaft mit allen Sinnen—Zum Wert des Atmosphärenbegriffs für die Landschaftsforschung. In D. Bruns & O. Kühne (Eds.), *Landschaften: Theorie, Praxis und internationale Bezüge. Impulse zum Landschaftsbegriff mit seinen ästhetischen, ökonomischen, sozialen und philosophischen Bezügen mit dem Ziel, die Verbindung von Theorie und Planungspraxis zu stärken* (pp. 221–232). Schwerin: Oceano.

Kazig, R. (2019). Atmosphären und Landschaft. In O. Kühne, F. Weber, K. Berr, & C. Jenal (Eds.), *Handbuch Landschaft* (pp. 453–460). Wiesbaden: Springer VS.

Kirchhoff, T. (2017). Landschaft. In T. Kirchhoff, N. C. Karafyllis, D. Evers, B. Falkenburg, M. Gerhard, G. Hartung, et al. (Eds.), *Naturphilosophie. Ein Lehr- und Studienbuch* (pp. 152–158). Tübingen: Mohr Siebeck; UTB GmbH.

Knebel, S. K. (2019). [Artikel] Virtualität. I. Virtuell. In J. Ritter (Ed.), *Historisches Wörterbuch der Philosophie. Band 11: U—V [Sonderausgabe 2019]* (pp. 1062–1066). Darmstadt: WBG.

Körner, S. (2006). Eine neue Landschaftstheorie? Eine Kritik am Begriff „Landschaft Drei". *Stadt + Grün, 10*, 18–25.

Kortländer, B. (1977). Die Landschaft in der Literatur des ausgehenden 18. und beginnenden 19. Jahrhunderts. In A. Hartlieb von Wallthor & H. Quirin (Eds.), *„Landschaft" als interdisziplinäres Forschungsproblem. Vorträge und Diskussionen des Kolloquiums am 7./8. November 1975 in Münster*. Münster: Aschendorff.

Krämer, S. (1998). Zentralperspektive, Kalkül, Virtuelle Realität: Sieben Thesen über die Weltbildimplikationen symbolischer Formen. In G. Vattimo & W. Welsch (Eds.), *Medien-Welten. Wirklichkeiten* (pp. 27–37). München: Fink.

Kristeller, P. O. (1980). *Humanismus und Renaissance II. Philosophie, Bildung und Kunst*. München: Fink.

Kühne, O. (2005). *Landschaft als Konstrukt und die Fragwürdigkeit der Grundlagen der konservierenden Landschaftserhaltung—eine konstruktivistisch-systemtheoretische Betrachtung. 2005* (Beiträge zur Kritischen Geographie, Vol. 4). Wien: Selbstverlag.

Kühne, O. (2006a). *Landschaft in der Postmoderne. Das Beispiel des Saarlandes*. Wiesbaden: DUV.

Kühne, O. (2006b). Landschaft und ihre Konstruktion. Theoretische Überlegungen und empirische Befunde. *Naturschutz und Landschaftsplanung, 38*(5), 146–152.

Kühne, O. (2006c). Soziale Distinktion und Landschaft. Eine landschaftssoziologische Betrachtung. *Stadt + Grün, 12*, 42–45.

Kühne, O. (2008). *Distinktion – Macht – Landschaft. Zur sozialen Definition von Landschaft*. Wiesbaden: VS Verlag für Sozialwissenschaften.

Kühne, O. (2012). *Stadt—Landschaft—Hybridität. Ästhetische Bezüge im postmodernen Los Angeles mit seinen modernen Persistenzen*. Wiesbaden: Springer VS.

Kühne, O. (2017). Der intergenerationelle Wandel landschaftsästhetischer Vorstellungen—Eine Betrachtung aus sozialkonstruktivistischer Perspektive. In O. Kühne, H. Megerle, & F. Weber (Eds.), *Landschaftsästhetik und Landschaftswandel* (pp. 53–67). Wiesbaden: Springer VS.

Kühne, O. (2018a). *Landscape and power in geographical space as a social-aesthetic construct.* Dordrecht: Springer International Publishing.

Kühne, O. (2018b). *Landschaft und Wandel. Zur Veränderlichkeit von Wahrnehmungen.* Wiesbaden: Springer VS.

Kühne, O. (2018c). *Landschaftstheorie und Landschaftspraxis. Eine Einführung aus sozialkonstruktivistischer Perspektive* (2nd ed.). Wiesbaden: Springer VS.

Kühne, O. (2019a). Die Sozialisation von Landschaft. In O. Kühne, F. Weber, K. Berr, & C. Jenal (Eds.), *Handbuch Landschaft* (pp. 301–312). Wiesbaden: Springer VS.

Kühne, O. (2019b). Heimat Saarland—Deutungen und Zuschreiben. In M. Hülz, O. Kühne, & F. Weber (Eds.), *Heimat. Ein vielfältiges Konstrukt* (pp. 231–243). Wiesbaden: Springer VS.

Kühne, O. (2019c). *Landscape theories. A brief introduction.* Wiesbaden: Springer VS.

Kühne, O. (2019d). Phänomenologische Landschaftsforschung. In O. Kühne, F. Weber, K. Berr, & C. Jenal (Eds.), *Handbuch Landschaft* (pp. 135–144). Wiesbaden: Springer VS.

Kühne, O. (2019e). Sich abzeichnende theoretische Perspektiven für die Landschaftsforschung: Neopragmatismus, Akteur-Netzwerk-Theorie und Assemblage-Theorie. In O. Kühne, F. Weber, K. Berr, & C. Jenal (Eds.), *Handbuch Landschaft* (pp. 153–162). Wiesbaden: Springer VS.

Kühne, O. (2019f). Sozialkonstruktivistische Landschaftstheorie. In O. Kühne, F. Weber, K. Berr, & C. Jenal (Eds.), *Handbuch Landschaft* (pp. 69–79). Wiesbaden: Springer VS.

Kühne, O. (2020). The social construction of space and landscape in internet videos. In D. Edler, C. Jenal, & O. Kühne (Eds.), *Modern approaches to the visualization of landscapes* (pp. 121–137). Wiesbaden: Springer VS.

Kühne, O., & Jenal, C. (2020). The threefold landscape dynamics—Basic considerations, conflicts and potentials of virtual landscape research. In D. Edler, C. Jenal, & O. Kühne (Eds.), *Modern approaches to the visualization of landscapes* (pp. 389–402). Wiesbaden: Springer VS.

Kühne, O., & Schönwald, A. (2015). *San Diego. Eigenlogiken, Widersprüche und Hybriditäten in und von ‚America's finest city'.* Wiesbaden: Springer VS.

Kühne, O., Weber, F., & Jenal, C. (2018). *Neue Landschaftsgeographie. Ein Überblick* (Essentials). Wiesbaden: Springer VS.

Kühne, O., Weber, F., Berr, K., & Jenal, C. (Eds.). (2019). *Handbuch Landschaft.* Wiesbaden: Springer VS.

Küster, H. (2002). *Die Ostsee. Eine Natur- und Kulturgeschichte.* München: Beck.

Küster, H. (2012a). *Arkadien Als Halboffene Weidelandschaft. Merkur, 66*(758), 651–656.

Küster, H. (2012b). *Die Entdeckung der Landschaft. Einführung in eine neue Wissenschaft.* München: Beck.

Küster, H. (2013a). *Geschichte der Landschaft in Mitteleuropa. Von der Eiszeit bis zur Gegenwart.* München: Beck (First publication 1995).

Küster, H. (2013b). *Geschichte des Waldes. Von der Urzeit bis zur Gegenwart* (3rd ed.). München: Beck (First publication 1998).

Langen, A. (1975). Verbale Dynamik in der dichterischen Landschaftsschilderung des 18. Jahrhunderts (1948/49). In A. Ritter (Ed.), *Landschaft und Raum in der Erzählkunst* (Wege der Forschung, Vol. 418, pp. 112–191). Darmstadt: WBG (First publication 1953).

Läpple, D. (1992). Essay über den Raum. Für ein gesellschaftswissenschaftliches Raumkonzept. In H. Häußermann, D. Ipsen, R. Krämer-Badoni, D. Läpple, M. Rodenstein, & W. Siebel (Eds.), *Stadt und Raum. Soziologische Analysen* (2nd ed., pp. 157–207). Pfaffenweiler: Centaurus.

Lehmann, H. (1968). *Formen landschaftlicher Raumerfahrung im Spiegel der bildenden Kunst* (Erlanger Geographische Arbeiten, Vol. 22). Erlangen: Selbstverlag der Fränkischen Geographischen Gesellschaft in Kommission bei Palm & Enke.

Leibenath, M., Heiland, S., Kilper, H., & Tzschaschel, S. (Eds.). (2013). *Wie werden Landschaften gemacht? Sozialwissenschaftliche Perspektiven auf die Konstituierung von Kulturlandschaften.* Bielefeld: transcript.

Liessmann, K. P. (2009a). *Ästhetische Empfindungen. Eine Einführung* (UTB, Vol. 3133). Stuttgart: UTB.

Liessmann, K. P. (2009b). *Schönheit.* Wien: facultas.

Linke, S. (2019). Landschaftsästhetik. In O. Kühne, F. Weber, K. Berr, & C. Jenal (Eds.), *Handbuch Landschaft* (pp. 441–452). Wiesbaden: Springer VS.

Linke, S. (2020). Landscape in internet pictures. In D. Edler, C. Jenal, & O. Kühne (Eds.), *Modern approaches to the visualization of landscapes* (pp. 139–156). Wiesbaden: Springer VS.

Loda, M., Kühne, O., & Puttilli, M. (2020). The social construction of Tuscany in the German and English speaking world—Presented by the analysis of internet images. In D. Edler, C. Jenal, & O. Kühne (Eds.), *Modern approaches to the visualization of landscapes* (pp. 157–171). Wiesbaden: Springer VS.

Luhmann, N. (1993). Die Form der Schrift. In H. U. Gumbrecht & K. L. Pfeiffer (Eds.), *Schrift* (pp. 349–366). München: Fink.

Maisak, P. (1981). *Arkadien. Genese und Typologie einer idyllischen Wunschwelt.* Frankfurt a. M.: Lang.

Majetschak, S. (2016). *Ästhetik zur Einführung* (4th ed.). Hamburg: Junius.

McLean, K. (2020). Temporalities of the smellscape: Creative mapping as visual representation. In D. Edler, C. Jenal, & O. Kühne (Eds.), *Modern approaches to the visualization of landscapes* (pp. 217–246). Wiesbaden: Springer VS.

Metzinger, T. (1994). Schimpansen, Spiegelbilder, Selbstmodelle und Subjekte. In S. Krämer (Ed.), *Geist—Gehirn—künstliche Intelligenz. Zeitgenössische Modelle des Denkens. Ringvorlesung an der freien Universität Berlin* (pp. 41–70). Berlin: De Gruyter.

Metzinger, T. (2005). Die Selbstmodell-Theorie der Subjektivität. Eine Kurzdarstellung in sechs Schritten. https://www.researchgate.net/publication/241328786_Die_Selbstmodell-Theorie_der_Subjektivitat_Eine_Kurzdarstellung_in_sechs_Schritten. Accessed 23 Jan 2020.

Müller, G. (1977). Zur Geschichte des Wortes Landschaft. In A. Hartlieb von Wallthor & H. Quirin (Eds.), *„Landschaft" als interdisziplinäres Forschungsproblem. Vorträge und Diskussionen des Kolloquiums am 7./8. November 1975 in Münster* (pp. 3–13). Münster: Aschendorff.

Nida-Rümelin, J. (1998). Vorwort. In J. Nida-Rümelin & M. Betzler (Eds.), *Ästhetik und Kunstphilosophie. Von der Antike bis zur Gegenwart in Einzeldarstellungen* (pp. X–XIII). Stuttgart: Kröner.

Paetzold, H. (1996). Das neue Interesse an einer Ästhetik der Natur. In J. Zimmermann in Verbindung mit U. Saenger & G.-L. Darsow (Eds.), *Ästhetik und Naturerfahrung* (pp. 43–58). Stuttgart-Bad Canstatt: frommann-holzboog.

Panofsky, E. (1975). *Sinn und Deutung in der bildenden Kunst.* Köln: DuMont Schauberg (First publication 1936).

Panofsky, E. (1980). Die Perspektive als ‚symbolische Form'. In H. Oberer & E. Verheyen (Eds.), *Aufsätze zu Grundfragen der Kunstwissenschaft* (pp. 99–167). Berlin: Spiess (First publication 1927).

Peres, C. (2013). Philosophische Ästhetik. Eine Standortbestimmung. In H. Friesen & M. Wolf (Eds.), *Kunst, Ästhetik, Philosophie. Im Spannungsfeld der Disziplinen* (pp. 13–69). Münster: Mentis.

Perpeet, W. (1977). *Ästhetik im Mittelalter.* Freiburg: Alber.

Perpeet, W. (1987). *Das Kunstschöne. Sein Ursprung in der italienischen Renaissance.* Freiburg (Breisgau): Alber.

Perpeet, W. (1988). *Antike Ästhetik.* Freiburg: Alber.

Petrarca, F. (Ed.). (1995). *Die Besteigung des Mont Ventoux*. Stuttgart: Reclam.

Piaget, J., & Inhalder, B. (1975). *Die Entwicklung des räumlichen Denkens beim Kinde*. Stuttgart: Klett-Cotta (First publication 1947).

Piepmeier, R. (2019). [Artikel] Landschaft. III. Der ästhetisch-philosophische Begriff. In J. Ritter (Ed.), *Historisches Wörterbuch der Philosophie. Band 6: L – Mn [Sonderausgabe 2019]* (pp. 15–28). Darmstadt: WBG.

Plumpe, G. (1993). *Ästhetische Kommunikation der Moderne. Band 1: Von Kant bis Hegel*. Opladen: Westdeutscher.

Pochat, G. (1986). *Geschichte der Ästhetik und Kunsttheorie. Von der Antike bis zum 19. Jahrhundert*. Köln: DuMont.

Pöltner, G. (2008). *Philosophische Ästhetik* (Kohlhammer-Urban-Taschenbücher, Vol. 400). Stuttgart: Kohlhammer.

Poplin, A., de Andrade, B., & Mahmud, S. (2020). Exploring tangible and intangible landscapes of evocative places: Case study of the city of Vitória in Brazil. In D. Edler, C. Jenal, & O. Kühne (Eds.), *Modern approaches to the visualization of landscapes* (pp. 519–547). Wiesbaden: Springer VS.

Popper, K. R. (1963). *Conjectures and refutations. The growth of scientific knowledge*. London: Routledge & Kegan Paul.

Prominski, M. (2004). *Landschaft entwerfen. Zur Theorie aktueller Landschaftsarchitektur*. Berlin: Reimer.

Regenbogen, A., & Meyer, U. (Eds.). (1998). *Wörterbuch der philosophischen Begriffe*. Meiner: Hamburg.

Reicher, M. E. (2015). *Einführung in die philosophische Ästhetik* (Einführungen Philosophie, 3rd ed.). Darmstadt: WBG.

Riehl, W. H. (1996). Das landschaftliche Auge. In G. Gröning & U. Herlyn (Eds.), *Landschaftswahrnehmung und Landschaftserfahrung* (Arbeiten zur sozialwissenschaftlich orientierten Freiraumplanung, pp. 144–162). Münster: LIT.

Ritter, J. (1974). *Subjektivität. Sechs Aufsätze*. Frankfurt (Main): Suhrkamp (First publication 1963).

Ritter, J. (1996). Landschaft. Zur Funktion des Ästhetischen in der modernen Gesellschaft. In G. Gröning & U. Herlyn (Eds.), *Landschaftswahrnehmung und Landschaftserfahrung* (Arbeiten zur sozialwissenschaftlich orientierten Freiraumplanung, pp. 28–68). Münster: LIT.

Ritter, J. (2019). [Artikel] Ästhetik, ästhetisch. In J. Ritter (Ed.), *Historisches Wörterbuch der Philosophie. Band 1: A–C [Sonderausgabe 2019]* (pp. 555–580). Darmstadt: WBG (First publication 1971).

Roßmeier, A. (2020). Urban/rural hybridity in pictures—The creation of neighborhood images using the example of San Diego's urbanizing inner-ring suburbs east village and Barrio Logan. In D. Edler, C. Jenal, & O. Kühne (Eds.), *Modern approaches to the visualization of landscapes* (pp. 479–498). Wiesbaden: Springer VS.

Roters, E. (1995). *Jenseits von Arkadien. Die romantische Landschaft*. Köln: DuMont.

Schapp, W. (1953). *In Geschichten verstrickt. Zum Sein von Mensch und Ding*. Hamburg: Meiner.

Scheer, B. (2015). *Einführung in die philosophische Ästhetik*. Darmstadt: WBG (First publication 1997).

Schenk, W. (2017). Landschaft. und Band 2. In L. Kühnhardt & T. Mayer (Eds.), *Bonner Enzyklopädie der Globalität* (Vols. 1 and 2, pp. 671–684). Wiesbaden: Springer VS.

Schenk, W. (2020). Visualization of the fundamental dimensions of "landscape" in landscape paintings around 1500 A.D. In D. Edler, C. Jenal, & O. Kühne (Eds.), *Modern approaches to the visualization of landscapes* (pp. 19–32). Wiesbaden: Springer VS.

Schneider, N. (2009). *Geschichte der Landschaftsmalerei. Vom Spätmittelalter bis zur Romantik*. Darmstadt: WBG.

Schürmann, E. (2008). *Sehen als Praxis. Ethisch-ästhetische Studien zum Verhältnis von Sicht und Einsicht.* Frankfurt a. M.: Suhrkamp.

Schweda, M. (2013). *Entzweiung und Kompensation. Joachim Ritters philosophische Theorie der modernen Welt.* Freiburg (Breisgau): Karl Alber.

Seel, M. (1996). *Eine Ästhetik der Natur* (Vol. 1231). Frankfurt (Main): Suhrkamp.

Sieferle, R. P. (1986). Entstehung und Zerstörung der Landschaft. In M. Smuda (Ed.), *Landschaft* (pp. 238–265). Frankfurt (Main): Suhrkamp.

Siepmann, N., Edler, D., & Kühne, O. (2020). Soundscapes in cartographic media. In D. Edler, C. Jenal, & O. Kühne (Eds.), *Modern approaches to the visualization of landscapes* (pp. 247–263). Wiesbaden: Springer VS.

Simmel, G. (1957). *Brücke und Tür. Essays des Philosophen zur Geschichte, Religion, Kunst und Gesellschaft.* Stuttgart: K. F. Kohler (First publication 1913).

Simmel, G. (1990). Philosophie der Landschaft. In G. Gröning & U. Herlyn (Eds.), *Landschaftswahrnehmung und Landschaftserfahrung. Texte zur Konstitution und Rezeption von Natur als Landschaft* (pp. 67–79). München: Minerva.

Snell, B. (1975). *Die Entdeckung des Geistes. Studien zur Entstehung des europäischen Denkens bei den Griechen.* Göttingen: Vandenhoeck & Ruprecht (First publication 1945).

Spanier, H. (2006). Pathos der Nachhaltigkeit. Von der Schwierigkeit, „Nachhaltigkeit" zu kommunizieren. *Stadt + Grün, 12,* 26–33.

Steingräber, E. (1985). *Zweitausend Jahre europäische Landschaftsmalerei.* München: Hirmer.

Steinmann, K. (1995). Nachwort. Grenzscheide zweier Welten—Petrarcas Besteigung des Mont Ventoux. In F. Petrarca (Ed.), *Die Besteigung des Mont Ventoux* (pp. 39–67). Stuttgart: Reclam.

Stierle, K. (1979). *Petrarcas Landschaften. Zur Geschichte ästhetischer Landschaftserfahrung.* Krefeld: Scherpe.

Tatarkiewicz, W. (2003). *Geschichte der sechs Begriffe. Kunst, Schönheit, Form, Kreativität, Mimesis, Ästhetisches Erlebnis. Aus dem Polnischen von Friedrich Griese.* Frankfurt a. M.: Suhrkamp.

Tesdorpf, J. C. (1984). *Landschaftsverbrauch. Begriffsbestimmung, Ursachenanalyse und Vorschläge zur Eindämmung. Dargestellt an Beispielen Baden-Württembergs.* Berlin: Tesdorpf.

Tonelli, G. (2019). Naturschönheit/Kunstschönheit. I. In J. Ritter (Ed.), *Historisches Wörterbuch der Philosophie. Band 6: Mo—O [Sonderausgabe 2019]* (pp. 623–626). Darmstadt: WBG.

Tümmers, H. J. (1999). *Der Rhein. Ein europäischer Fluß und seine Geschichte.* München: Beck.

Vattimo, G. (1998). Die Grenzen der Wirklichkeitsauflösung. G. Vattimo & W. Welsch (Eds.), *Medien-Welten. Wirklichkeiten* (pp. 15–26). München: Fink.

Vetter, M. (2019). 3D-Visualisierung von Landschaft—Ein Ausblick auf zukünftige Entwicklungen. In O. Kühne, F. Weber, K. Berr, & C. Jenal (Eds.), *Handbuch Landschaft* (pp. 559–573). Wiesbaden: Springer VS.

Vetter, M. (2020). Technical potentials for the visualization in virtual reality. In D. Edler, C. Jenal, & O. Kühne (Eds.), *Modern approaches to the visualization of landscapes* (pp. 307–317). Wiesbaden: Springer VS.

Vietta, S. (1995). *Die vollendete Speculation führt zur Natur zurück. Natur und Ästhetik.* Leipzig: Reclam.

Waldenfels, B. (2005). *In den Netzen der Lebenswelt* (Suhrkamp-Taschenbuch Wissenschaft, 3rd ed., Vol. 545). Frankfurt (Main): Suhrkamp.

Weber, F. (2015). Landschaft aus diskurstheoretischer Perspektive. Eine Einordnung und Perspektiven. *morphé. rural—suburban—urban, 1,* 39–49. http://www.hswt.de/fileadmin/Dateien/Hochschule/Fakultaeten/LA/Dokumente/MORPHE/MORPHE-Band-01-Juni-2015.pdf. Accessed 30 Aug 2017.

Weber, F. (2018). *Konflikte um die Energiewende. Vom Diskurs zur Praxis*. Wiesbaden: Springer VS.

Weber, F. (2020). Blurring the boundaries of landscape visualization: Welcome to fabulous Las Vegas. In D. Edler, C. Jenal, & O. Kühne (Eds.), *Modern approaches to the visualization of landscapes* (pp. 461–478). Wiesbaden: Springer VS.

Weber, F., Kühne, O., Jenal, C., Aschenbrand, E., & Artuković, A. (2018). *Sand im Getriebe. Aushandlungsprozesse um die Gewinnung mineralischer Rohstoffe aus konflikttheoretischer Perspektive nach Ralf Dahrendorf*. Wiesbaden: Springer VS.

Welsch, W. (1998). Eine Doppelfigur der Gegenwart. Virtualisierung und Revalidierung. In G. Vattimo & W. Welsch (Eds.), *Medien-Welten. Wirklichkeiten* (pp. 229–248). München: Fink.

Wimmer, C. A. (2004). Zur schönen Aussicht. Typologie und Genese einer ästhetischen Erfahrung. In G. Horn (Ed.), *Wege zum Garten. Gewidmet Michael Seiler zum 65. Geburtstag* (pp. 30–36). Leipzig: Koehler & Amelang.

Wojtkiewicz, W., & Heiland, S. (2012). Landschaftsverständnisse in der Landschaftsplanung. Eine semantische Analyse der Verwendung des Wortes „Landschaft" in kommunalen Landschaftsplänen. *Raumforschung und Raumordnung, 70*(2), 133–145. https://doi.org/10.1007/s13147-011-0138-7.

Zimmermann, J. (1991). Das Schöne. In E. Martens & H. Schnädelbach (Eds.), *Philosophie. Ein Grundkurs* (Vol. 1, pp. 348–394). Reinbek: Rowohlt.

Karsten Berr studied landscape management at the Osnabrück University of Applied Sciences and philosophy and sociology at the FernUniversität in Hagen. In 2008 he received his doctorate on the conception of natural beauty at Hegel. 2010–2012 he worked as a teacher for special tasks in philosophy at the University of Vechta. From 2012–2017, Mr. Berr conducted research in a DFG-project on the theory of landscape and landscape architecture as well as on the ethics of architecture and planning at the TU Dresden, BTU Cottbus and the University of Vechta. Since May 2018, he is working at the Eberhard Karls University of Tübingen. His research interests include the theory of landscape, landscape architecture theory, architectural theory; ethics of architecture and planning as well as landscape conflicts; inter- and transdisciplinary architecture and landscape research; philosophy of art and aesthetics, natural and landscape aesthetics; cultural theory and anthropology.

Temporalities of the Smellscape: Creative Mapping as Visual Representation

Kate McLean

Abstract

This chapter responds to debates in olfactory art and urbanism that highlight the challenges inherent in obtaining and sharing a vast, ephemeral and eye-invisible olfactory dataset. Concerned with representation and communication of the smellscape as theorised by J. Douglas Porteous and activated by Victoria Henshaw, the case study, situated in Kyiv, explores how mapping might record temporal qualities of smell and contribute to communication of non-visual olfactory information. Drawing on interdisciplinary methods; sensewalking, agentic mapping, rhythmanalysis and creative practice, I develop and apply original approaches to practices of smellscape mapping as a means of analysing, interpreting and communicating a theorised fragmentary and episodic olfactory landscape. The findings include visualisations to indicate multiscalar temporalities of a city, polyrhythmic relationships between the situated human body and a range of smells, and a series of projective mappings that render visible olfactory-sensed information. Together they examine relationships between smells and space, smells and time, and smells and people.

Keywords

Cartography · Communication · Design · Mapping · Olfactory · Smellscape · Spatialisation · Temporality · Visualisation

K. McLean (✉)
Canterbury Christ Church University, Canterbury, UK
e-mail: mcleankate@mac.com

© Springer Fachmedien Wiesbaden GmbH, part of Springer Nature 2020   217
D. Edler et al. (eds.), *Modern Approaches to the Visualization of Landscapes*, RaumFragen: Stadt – Region – Landschaft,
https://doi.org/10.1007/978-3-658-30956-5_12

1   Introduction

Scholars, designers and smell artists have visualised the smellscape through mapping. Generally, as befits the mapping process and understandings of maps as artefact, the imperative is to mark which smells are located where; a spatial depiction and description of geo-located odours, their range and their combinations. Such cartographic visual spatializations of smellscape qualities render it as eye-visible and modified cartographic symbols serve to establish relationships between smell and place, and smell and space, at any given moment in time. However, the smellscape (Porteous 1985) is dissimilar to a landscape in one major regard—it fluctuates in composition at different times of day, on different days, seasonally, over the years and decades. The smellscape is highly temporal, and it is the visual representation of the fragmented, mutating nature of smells that this chapter will address.

The city of Kyiv, Ukraine was the location for the case study in which I examined the temporal dimension of qualitative smellscapes. I combined two complementary methodologies, research-through-design (Frayling 1993) and rhythmanalysis (Lefebvre 2013), to investigate how a smellwalk in Kyiv, undertaken by a group of seven local inhabitants, might be temporally analysed through visual practice. The resulting five smellscape mapping strategies afford insights into new relationships; individual instances of smell-time perception, the smellwalk as a somatic exercise and a collective exercise, and individual smell-environment rhythms.

Conversations, notes, performance, landscape, photographs and the interplay of theory, practice and analysis of map creation come together to form the basis for five ontogenetic mapping visualisations (Kitchin and Dodge 2007). In combining diagram, text and photography to present the research I take a lead from Wunderlich's development of a conceptual framework for place-rhythm through analysis of the aesthetics of London's Fitzroy Square (Wunderlich 2013) and from Simpson's investigations of the space-times of street performance both of which are communicated and argued through photomontage, photo sequences, spectral diagrams and time-lapse photography (Simpson 2012).

The practice of creatively mapping the smellwalks instigated new questions. As I engaged in the design of one mapping, so my actions prompted the conceptualisation and creation of the next mapping, which in turn informed the original.

2   Logistics and Rationale: Kyiv

The smellwalk featured in this case study was conducted during the winter of 2016 in Kyiv, Ukraine, which as an urban environment, is full of potential encounters with a variety of smells. Tim Edensor theorises the rhythmic environment of the city as, '[…] a series of different paced and orchestrated mobile rhythms produces a collectively constituted choreography that gives temporal shape to place' (2013, p. 167) and it is this collective choreography of human-smell encounters that provides the material for this serendipitous case study. This particular case study is representative of other smellwalks in its structure and has a small enough sample size (seven participants) to enable a comparative analysis of individual walks. The small number of participants also meant that I could recall and geolocate individual trajectories and specific smell encounters after the smellwalk.

At 1 pm on Sunday December 25, 2016 a huddle of warmly dressed Ukrainian residents clustered outside the main Post Office on Independence Square, Kyiv.[1] It was my ninety-fourth smellwalk, and I had seized the opportunity to witness Kyiv through the noses of its residents. Local people with city familiarity are central to my research practice; their tacit knowledge of smell sources, anomalies and expectations contribute to both cultural and place-specific perspectives. In a city other than Kyiv the smells would have been different, but the ethos and structure of the method of their detection, remain consistent. The convener of the smellwalk collaborated with other participants to suggest a route with stopping places. While I was prepared to conduct the smellwalk in translation through the convener all the participants spoke English to a high level, which ensured discussion.

Smells play an important role in Lefebvre's rhythmanalysis, as indicators and markers for other urban rhythms (Lefebvre 2013, p. 31). This study models and synthesises the repeated, yet changing, rhythms and beats of the smellscape (Moore 2013). To borrow from the human geographer, Edensor, rhythmanalysis is particularly useful 'for investigating the patterning of a range of multi-scalar temporalities—calendrical, diurnal and lunar, lifecycle, somatic and mechanical' (2012, p. 55). The smellscape will be shown to operate at many temporal levels and visual mapping strategies deployed to reveal such patterns.

Smell is possibly the ultimate repetitive sense; it is through a series of sniffs that we ingest olfactory information in discrete, repeated units to form an overall experience. When Elden describes Lefebvre's definition of rhythm as being 'the placement of notes and their relative lengths' (2013, p. 5) he could just as easily be talking about the physiological experience of a person smellwalking.

3   Rhythms of the Smellwalk

I divided the smellwalk into three sections punctuated by planned stops during which a facilitated conversation elicited each individual's smell findings and comments. Each section varied considerably from my recommended 15-min increments. The introduction and Smell Catching lasted 49 min, the first stop and Smell Hunting lasted 69 min, the second stop and Smell Research lasted 10 min whereupon we went to a local restaurant

---

[1]Ukraine respects Orthodox Christian holidays, celebrating Christmas on January 6th.

to sketch and discuss the final section. The extended walk length can be attributed to the collective negotiation of a route (by its participants) in contrast to designer-led smellwalks in which I stipulate more precise routes and timings. This particular walk was more akin to a Debordian psychogeographic wandering, as mentioned by Drobnick, enabling participants to experience the 'complex sensory matrix that includes the presence of smell' (2002, p. 35). The route was negotiated between the smellwalkers who were more familiar with their city than me, the researcher.

Following the walk, and prior to creating the mappings, I generated a detailed diary itemising the stages of the walk and bodily movement. The diary forms the basis for understanding how repeated, physical actions enacted during the smellwalk relate to smells. A smellwalker moves both through, and in, space. Comparing the nose (breathing), nose (sniffing), ears, legs, arms, eyes and walking pace reveals inverse correlations between walking speeds and numbers of smells noted. It also suggests how stops, imposed by the walk's tripartite structure, afford more opportunities for smell catching. My overall orchestration of the walk encourages each participant to walk at their own pace while matching that of the group as a whole; in this case the pace varied with accelerations and lingering moments when stopping to sniff. In summary, there was an interweaving of multi-pace trajectories whereby individuals convened in small groups to share sniffing smells before separating to follow odour trails of their own. The physical environments traversed during the route—underground, descents, riverbanks and market interiors combined with the imposed structure of the smellwalk (catching, hunting, free smelling) to generate unique smell detection body positioning. Smellwalking breaks the monotonous repetition of the commute, shopping and other routine walking practices. The environment has a direct impact on smell encounters since 'walking generates a range of possibilities for putting oneself in an experiential flow while simultaneously maintaining a flow of thoughts' (Edensor 2010, p. 72). The following photo series documents the rhythmicity of the smellwalk showing affordances of the environment and human action alongside the smell mentions indicating an interrelation of smell and place.

The smellwalkers gathered, dressed for winter, in the expanse of Independence Square where temperatures hovered around freezing. For the Smell Catching activity we walked through Hreshchtyk, Independence Square's subterranean metro access and low-ceilinged underpass filled with vendors and micro-shops selling flowers, clothing and snacks (Fig. 1) finding a miscellany of smells in confined spaces. Cold rushing winds, at the metro entrances, and warm rushing people mixed, creating an indeterminate season. Smells noted verbatim by smellwalkers included: 'Confectioner sugar hot dog water, Rotten egg and mayo, Oil frying (Balfy metro), Cigarette smoke, Scent of women perfumes, Smooth sweet-like doughnuts, Cheap plastic souvenirs which not interesting to touch, Cheap food and coffee, Humidity fresh dampness'. The smell combinations noted were of people, food and transport.

While Smell Hunting along the deserted banks of the river Dneiper the smellwalkers wove paths in and around each other, seeking smell sources. The communal walking pace aligned to a single rhythm until individuals interrupted; we felt secure keeping

Fig. 1 Smell catching underground (2016). Image courtesy of Tom Bassett

each other in sight. The air was cold. No-one lingered unless it was to explore a potential smell opportunity. Smells recorded verbatim during Smell Hunting included 'Inside light pole—cold steel concrete, wire insulation, Fungus (near river), Fuel and old car, wheels near bus, Rust, Metal'. With the absence of people, the smells were of inanimate structures and machinery.

The fluidity of our walking movement was interrupted. One smellwalker looked at a corrugated metal structure ... he stopped and sniffed, seeking out possible odour sources (Fig. 2) ... his walking pace decreased with an increased sniffing rate as he deliberately sampled odours (Barkham 2016). Finding nothing of note in this venue he moved on.

Another smellwalker stopped, her attention attracted by an anomaly while the group continued onwards. She bent to sniff an item on the ground by the river railings (Fig. 3). Some smells can only be ascertained through deliberate action and proximity. Her body adapted and moved to detect the smell, to verify and, in this case, to physically collect the source of the odour. She engaged in a swift series of actions that did not conform with the group's, creating an arrhythmia 'produced by the distractions and diversions offered by heterogeneous activities and sights' (Edensor 2010, p. 73). The two examples above lead me to suggest how both real-time olfactory stimuli, and the imagination of a possible future smell, are diversions that interrupt the rhythm of the group walk and individual pace and movement.

A spontaneous behaviour noted during smellwalks is how participants share their smell experiences; at times they seek verification, at others simply desire to pass on a particular odour experience to someone else in the group and occasionally it happens prior to sniffing upon sight of a possible odour source. In the instance shown (Fig. 4), the group's walking flow was interrupted by the smelly potential of a mailbox, and yet no smell was noted. The bodies and actions of the smellwalkers reinforced an understanding that the smellscape is in a constant state of flux, 'always a relationship, chemical or mechanical between that which gives off (smells) and the individual who smells (or sniffs)' (Rodaway 1994, p. 71).

Fig. 2 Smellwalker breaks from group to investigate a source (2016). (Image courtesy of Tom Bassett)

Fig. 3 Smellwalker breaks from group and bends to collect a smell source (2016). (Image courtesy of Tom Bassett)

Fig. 4 Sharing potential smell at the mailbox (2016). (Image courtesy of Tom Bassett)

Zhytnyi Market Place, a two-floor mezzanine interior space was half-full of independent vendors. There was an upward shift in temperature by a bare couple of degrees. While a far cry from the potency of a hot, summer city market, the venue afforded plenty of odiferous opportunities; foodstuffs, produce, clothes, drinks, and plants. One pervasive smell mentioned by several smellwalkers was described as 'damp fur of animals' detected on the ground floor. During the 'Free Smelling' activity inside the market the smellwalking group divided and meandered at their own pace following their own interests. Recorded smells included: 'Borved coffee with plastic + raw meat, Rosemary, Smth unclean and damp like floor, Fresh strawberries, Dried summer herbs, Tasty smell of cabbage, pickles with adding of vinaigre, Ukrainian salo, Market smells'. Similar to the underground at the start of the walk it was everyday working activities of people in the space that produced the smells; whiffs of edible produce and those resulting from its preparation, waste by-products, preservation and containment.

One remark during the post-walk conversation was how Zhytnyi market presented a complete seasonal cycle under one roof. Smellwalkers listed the following in evidence: herbs of spring, strawberries of summer, pomegranates of autumn and glintwein of winter.

4   Scales Within the Smellscape: Processes of Visualisation

- Kitchin et al. suggest that a processual understanding of maps as 'mappings', is best understood through the lens of practice examining how maps are produced and consumed in diverse ways 'technically, socially, bodily, aesthetically and politically'

(2013, p. 480). To this end they propose a series of methods through which this might happen. As a mapping practitioner and researcher, I narrate my own unfolding production and present a series of visual mappings of the Kyiv smellwalk to explore how temporal knowledge embedded in the smellscape might be communicated through:

- individual instances of smell perception in the form of participant smell sketch graphic elicitations (Edwards and Holland 2013) using watercolour paints;
- a visualization of the multiple rhythms involved in aspects of the walk;
- an animation tracing the collective smell detection points and smell intensity indicating temporal rhythm of smell occurrences during the smellwalk;
- animations of individual smellwalk trajectories;
- an investigation of the temporo-spatial experiences of seven smellwalkers including reference to local wind conditions affecting the directional flow of smells (available online)[2];
- an animation of the collective smellwalk experience, in which a polyrhythmia of superimposed smellwalk sequences create new temporo-spatial forms (available online).[3]

There are multiple somatic rhythms in operation during a smellwalk; breathing, sniffing, walking and stopping contribute to the overall experience. Rhythmanalysis requires bilateral transition between the lived and the analytical. Lefebvre suggests that 'to grasp a rhythm, it is necessary to have been grasped by it; one must let oneself go, abandon oneself to its duration' (Lefebvre 2013, p. 37). When applied to smell it is beneficial to appreciate and understand the physical rhythms of the body involved in the act of sniffing. Lefebvre continues to state, 'In order to grasp and analyse rhythms it is necessary to get outside them, but not completely' (Lefebvre 2013, p. 37). To analyse smellscapes a level of exteriority facilitates the functioning of intellect. My capacity to grasp the rhythms of the smellwalk is drawn from prior experience of ninety-three smellwalks; I have certainly been grasped. In this case study I undertook practice-led and practice-based analysis after the event, reframing the performances of the smellwalk participants into visual and animated mappings.

The body is key to both rhythmanalysis and to smellwalking. Prior to departure from Independence Square our breathing and hearts were in eurythmia; a rhythmic resonance

---

[2]I: https://vimeo.com/213708419
II: https://vimeo.com/213708796
III: https://vimeo.com/213709054
IV: https://vimeo.com/213709305
V: https://vimeo.com/213709679
VI: https://vimeo.com/213709760
VII: https://vimeo.com/213710044

[3]https://vimeo.com/211929457

with each other. As the walk commenced, so we drew deeper breaths, detecting smells, re-detecting smells, stretching upwards into a breeze, or downwards to ground-level in order to finesse a description through a series of confirmatory sniffs. With these actions breathing and heart rate rhythms changed but eurythmia was maintained as intervals of sniffing superimposed onto regularities of breathing which were themselves modified by the terrain, walking speed, obstacles and temperature (Fig. 5).

The smellscape, as perceived by humans, encompasses multiple rhythms. Our inner cyclical rhythmic activity of approximately twenty-four thousand breaths every day not only keeps us alive but is also a mechanism that enables us to ingest smells. However, to smell we still need to sniff deliberately. This conscious action serves to re-attune normal occular-centric understanding and instead foreground information gleaned from the nose. Breathing is not sniffing; one is a reflex action, the other a conscious act (Horowitz 2016,

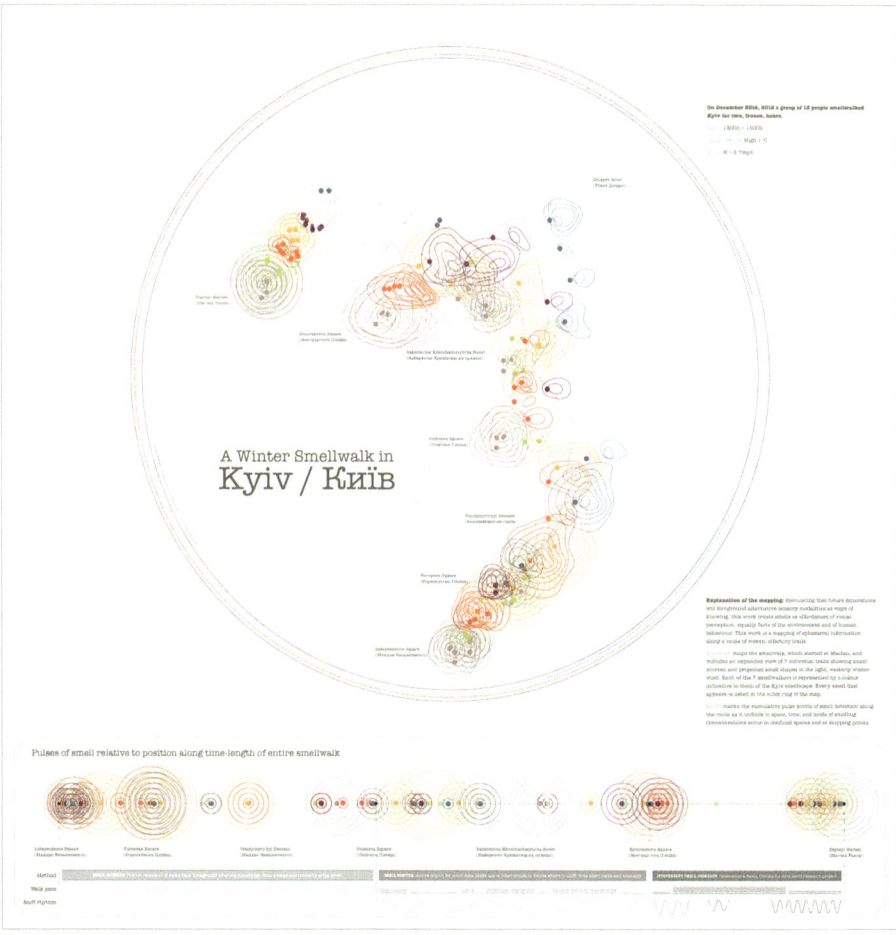

Fig. 5 A Winter Smellwalk in Kyiv (2017). (Image created by the author)

p. 38). So, the sniff rhythm changes according to the stage of the walk (Smell Catching, Smell Hunting, Free Smelling), the terrain, levels of physical exertion and the temperature. Smell Catching invokes a measured, deep and regular breathing cycle, whereas Smell Hunting invokes a more regular breathing pattern with irregular deeper sniffs for short periods of time and Free Smelling might incorporate both as the walker chooses. The respiratory rate for humans varies with age but averages out at between twelve and eighteen breaths per minute, or one breath cycle every three to five seconds, a value I use for the duration of smells in the animated sequences (Barrett et al. 2012, p. 619). The following section analyses relationships; the patterns, rhythms and temporalities that emerge from the smellwalk experience in Kyiv.

5   Analysing and Constructing the Smellscape: Sketches, Symbols, Animations

At the end of the smellwalk, as we were sitting in a restaurant, I asked the participants to create a summary sketch of their smellwalk experience. This served as a communication tool to help describe the highlight of their individual experience to others in the group. Each of the smell sketches alludes to a temporality:

Smellwalker # 5 (Fig. 6) talked about the olfactory impact of Independence Square's underground space. She termed it an 'asylum' alluding to a multi-sensory environment of chaos and painted this experience as a tornado swirl releasing trails in its activity. The

Fig. 6 Smellsketch (2016). A subterranean environment. (Image courtesy of Smellwalker #5)

smell visualisation indicates a powerful mass, a combination of smells constrained in a form by its own velocity, with a seeming capacity to strike anywhere. The colours are dirty, even grimy, with a dark value and mid to weak chroma, reflecting the comments in her smellnotes for this section; 'Cheap food and coffee • CO/CO$_2$ in the air • Parfum, especially near entrance'. I interpret the black/grey to be read as coffee and CO, and the adulterated pink hue as cheap perfume based on patterns of ascribing colour to smell and previous personal experience whereby smell colour equates to the colour of its source; in this case coffee is dark and perfume packaging is frequently pink. The visual environmental context of the walk of a very grey day, as seen in photographs Figs. 5.2–5.8 may equally influence chroma and hue of smell representation. The powerful and dimensional vortex indicates a spiralling cacophony of vaguely unpleasant smells which have collided in a funnelling mass. The representation references coffee; a repeated and a diurnal activity which prompted further questions as to the annual and seasonal patterns of drink sales. The 'parfum' may have emanated from a vendor, a single passerby, or many people walking through the space, and so be considered as either episodic or ephemeral in terms of its tenacity (Henshaw 2014).

Smellwalker #2 (Fig. 7) selected her memorable smell of 'Fungus (near river)' in the smellnote and 'fungus on the street' in her smellsketch, writing 'I would call that a smell of islands of summer in winter'. She visualised this as a seasonal exchange in which a vibrant colour scheme is at odds with the colours of the landscape in which the smell itself was noted. This is a cyclical and seasonal smell reference as summer momentarily manifested itself through a smell sniffed on the paved walkway. This composite smell encompasses a range of 'greenery during winter' sources.

Fig. 7 Smellsketch (2016). '[…] greenery during winter […]'. Image courtesy of Smellwalker #2

usual mix
of winter smells
(with blue
and green
semi-colours)

Fig. 8 Smellsketch (2016). 'Usual mix of winter smells […]'. (Image courtesy of Smellwalker #7)

Smellwalker #7 (Fig. 8) consolidated the smellwalk experience into a single visual creating a smellsketch of an entire season. The writing clearly suggests the idea of
a repetitive and seasonal winter smellscape; the term 'usual mix' contributes a layer of
expectation and a shared understanding to her definition. The colour appears to reflect
the colour of winter skies with 'blue and green semi-colours'. Her smellnotes of 'Fast-
food • European Square • Freshness snow • McDonalds + • Rust • Metal • Leaves • Meat
• Croissant • Straw • Fish' indicate a broad mix of environmental, food, and traffic-
related urban smells.

Smellwalker #6 (Fig. 9) selected his memorable smell as that of 'pine tree'.
Subsequent conversation with him and other smellwalkers indicated the ubiquity of the
species that covers 34% of Ukraine's forest landscape and is also common within the
city, notably along Volodymyrs'kyi Descent from European Square to the Dnieper River
(Internet Encyclopedia of Ukraine 2001). Pine is a resinous evergreen whose fresh smell
is a permanent, background feature of the Kyiv smellscape.

Smellwalker #3 (Fig. 10) noted far more smells overall than any other participant, most of which were low intensity, but the overriding smell was encapsulated as a
momentary counterpoint of 'clear scent of strawberries' against a background of 'like
wet animal fur', visualised as a mist or fog that is punctuated with something fresh and
vibrant. The smellscape is shown to have punctuation points in antithesis to the constancy of the background.

Smellwalker #4 (Fig. 11) deviated from the smellwalk experience to delineate a
progressive, worsening and invasive smell that is affecting some of the city's population. Ukraine incinerates garbage at two sites, in Dnepropetrovsk and Kyiv. These sites

Fig. 9 Smellsketch (2016). '[...] pine tree'. (Image courtesy of Smellwalker #6)

Fig. 10 Smellsketch (2016). A moment in the indoor market. (Image courtesy of Smellwalker #3)

were built in the 1980s and reportedly, in 2015, 'their resources are almost worked out' (Vorotnikov 2015). This smellwalker asked friends and co-workers in advance of the smellwalk about a typical Kyiv smell. Based on their responses he chose to call attention

Fig. 11 Smellsketch (2016). Indicating a progressive and invasive negative smell. (Image courtesy of Smellwalker #4)

Fig. 12 Smellsketch (2016). Indicating a complex, background, location-specific smell. (Image courtesy of Smellwalker #1)

to the garbage treatment plants in his smellsketch. The increasingly pungent, negatively-perceived smell has been building for over two decades as he writes, 'the smell of burning garbage intensifies through the time and getting worse'. The sketch is read from left to right as the initial smell becomes both darker in value and larger in size.

Smellwalker #1 (Fig. 12) depicted a location-specific smell, the abandoned building's window casing was a repository for the combination of stale air and rusty metal. Visualised with a rusty brown hue overlaid by grey, black and deep maroon this smell representation is an abstract shape with a sharp, hooked peak (possibly a top note) and a widening base. The colours appear to refer to the context and the environment; a brick building on a particularly grey winter's day.

The visuals are testament to the multi-scalar temporalities conjured through smell knowledge gleaned on a single walk through the city, demonstrating both a knowledge that derives from smell and the possibilities inherent in relating smell to place through the human body input via the nose and output as olfactory experience artworks.

Many maps of smell can be problematised in that they neglect to take account of the temporal nature of smell experience. In addition, they neglect consideration of alternative angles of view often using a mimetic, scientific cartographic paradigm whereby the map's angle of view is typically bird's eye; a top-down, empirical, all-seeing, recording of visualised objects as theorised in Haraway's 'god-trick' (Haraway 1988, p. 581) and Certeau's 'totalising view' (Certeau 2011, p. 92).

In representing an ephemeral and transient olfactory world as a smellmap the resulting artefact summarises a temporary state only milliseconds in length (Dodt et al. 2017). In this it is similar to an Impressionist painter's rendering of light whereby the subject remains static but the visual sensations change constantly. Since my research understands walking and smelling as down-to-earth—embodied physical activities undertaken at street level—I now turn to my experiments exploring potential for alternative mappings to address these concerns (McBride and Nolan 2017).

In *A Winter Smellwalk in Kyiv* (Fig. 5) I compare rhythms of the smells detected during the smellwalk and expose relational connections between the seven participants, the timing and intensity of their smell observations and the somatic components of the smellwalk experience. I employ olfactory symbolisation and glyphs developed within my own practice whereby each smell detection is symbolised with a location dot and a multi-contour symbol to represent the qualitative smell intensity as assessed by the smellwalker. All smellwalkers started at the same location and followed a similar route, although minor individual deviations occurred. Without geolocation technology, I used my own discretion and memory of the walk to position smell sources on a base-map according to the participant's written description e.g. 'Confectioner sugar hot dog water' (Smellwalker #1) is directly attributable to the subway under Independence Square, just

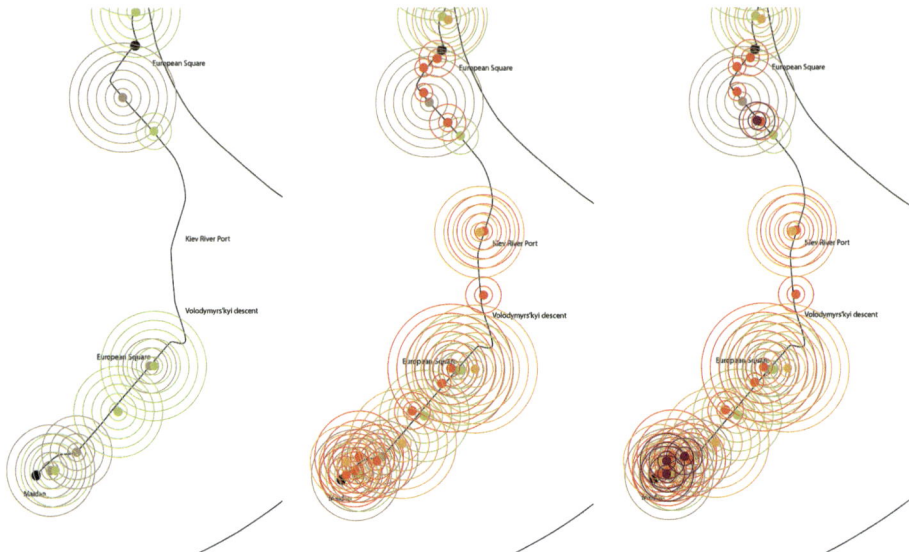

Fig. 13 A Winter Smellwalk in Kyiv (2017). Layering smell narratives on top of one another. (Image created by the author)

as 'Ukrainian salo' (Smellwalker #7) was experienced in Zhytniy Market at the end of the walk. I chose a colour for the smell notes of each smellwalker based on their smell sketches (Figs. 6, 7, 8, 9, 10, 11 and 12).

Patterns emerged from the transcription of individual smell narratives into visual symbols (Fig. 13). Pulses of collective smell detection can be seen as hot spots interspersed with olfactory voids. The surge of smell detection at the start of the walk might be attributed to a novel method of engaging with the world of smell as well as a rich olfactory environment and the warmth of the busy underpass.

The visual vibration at the start of the trail of slightly offset concentric rings indicates a mass of olfactory source stimuli. This visual cacophony of smell detection in the Independence Square underpass (Fig. 14) exhibits both contradictions (arrhythmias) and corroborations (eurythmias) which are echoed by the text descriptions from each smellwalker. In turn this prompts questions as to the similarity and diversity of smell detection and its semantic labelling; does 'cheap food' equate to 'fast food'? Is there a relationship between 'sweet doughnuts' and 'rotten egg'? Is the scent of 'cheap plastic souvenir smells' as far removed from that of 'perfume' as the words might indicate? Smells such as coffee, hot dogs and flowers are episodic in time and can connect us to a diurnal cycle of social human movement. The aromas of coffee will vary according to the numbers of passing consumers, the day of the week and the time of day. The hot dog and flower stalls will experience peaks in activity affecting their volatile release of odours into airspace. When visualised thus, localised olfactory potential is a vibrant pulsation of smell

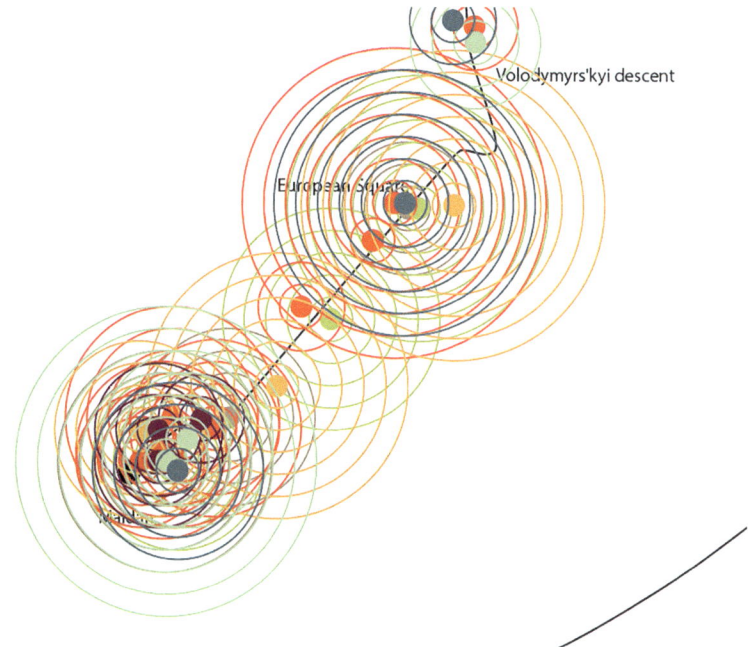

Fig. 14 A Winter Smellwalk in Kyiv (2017). Polyrhythms: olfactory contradictions and corroborations. (Image created by the author)

instances, which are not permanent inscriptions; their concentration dissipates according to air currents, temperature and the odour source itself (Kephart and Mikesell 2000). This temporal environment of smells might be visually conceived either as odour plumes or through time-based media as explained in the following sections.

In considering the fluidity of smell movement in airspace this thesis looks to meteorology at a basic level. Airborne smell molecules are subject to airflows, themselves affected by a combination of the natural landscape, the built environment and changes in atmospheric pressure. Wind patterns are continually subject to change on a massive scale and airflows experienced on the ground during a smellwalk are similarly subject to global fluctuations (Lamb 1975). At an intimate scale, acts of walking, turning, bending and stopping modify localised airflows.

Initial smell detections in Kyiv occurred in the subterranean underpass where air movement at the entry and exit points was stronger (a temperature change generated convection currents as air moved away from a warm space into a colder one). At the centre of the low-ceilinged underpass, eddies formed as turbulence was generated by the heat and movement of vendors, mobile human traffic and food outlets. The bulk of the walk, from Independence Square to Zhytniy Market, took place outdoors, along the banks of a semi-frozen river and through streets largely bereft of any traffic, in a light westerly wind

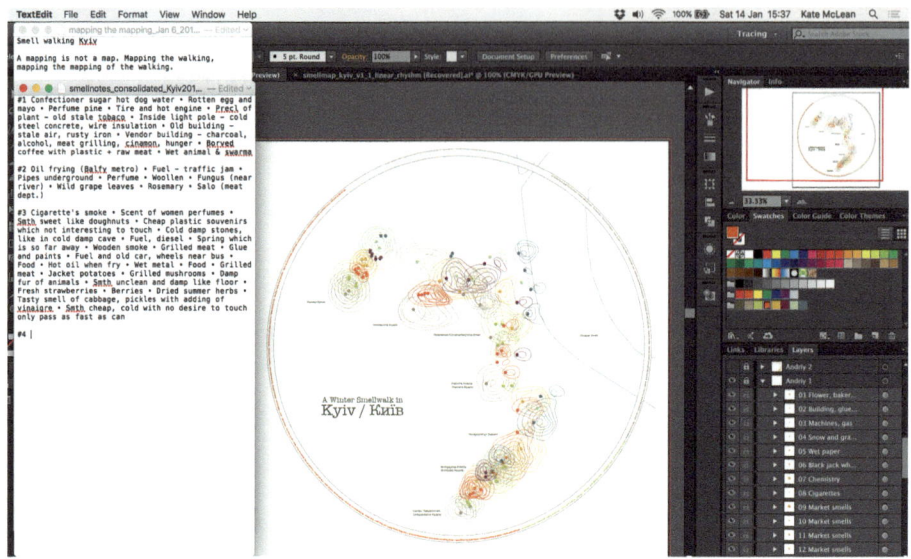

Fig. 15 A Winter Smellwalk in Kyiv (2017). Screen capture showing combination of digital media used to enable practice-theory relationship through reflection-in-practice while mapping Kyiv smellscape. Image created by the author

which carried volatile smell molecules with it. The final section of the walk took place at a farmers' market in a cold, dank, two-storey mezzanine building. The building enabled most smells to move freely but still the smellwalkers noticed a damp, unpleasant, pervasive background aroma into which other, more pleasant, smells pierced fleetingly. The challenge was how to map this movement alluding to the meteorological while maintaining subjectivity of individual experience.

The composite design of smell symbols suggests both the sources of smells and the range of their likely encounter. To account for the contributory air movement in Kyiv, as described above, I manipulated the smell symbols according to the places in which they were noted; smells encountered along the river Dneiper thus drift eastwards whereas the subterranean Independence Square and Zhytniy Market smells circulate (Fig. 5). While air currents affect smell trajectories, the activity at the source of each smell can also be instrumental to their propulsion; molecules from hot glintwine and freshly-made coffee volatilise as visible steam carries odours as it rises. In future smellscape mappings it may be possible to combine modelling algorithms of meteorology with those of fluid dynamics to generate the visuals for an imagined 4D smellscape.

To create the map (Fig. 15), I transitioned between the smellnotes and design software. I typeset smell descriptions in the circular border of the map to draw attention to the everydayness of free, publicly accessible information of smells and in so doing demonstrate how non-specialists might contribute to future olfactory mappings. As contemporary trends for advertising and marketing urban scenting practices develop it

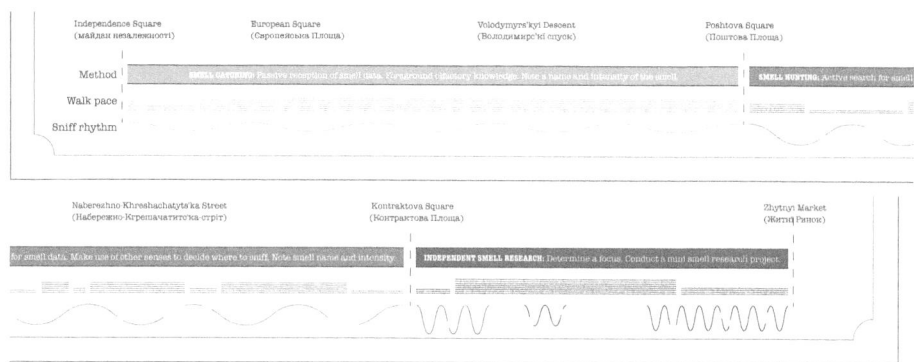

Fig. 16 A Winter Smellwalk in Kyiv (2017). Visual comparison of smellwalk method, walk pace and sniff rhythm. (Image created by the author)

is important to demonstrate how smell detection is definitively not an exclusive or specialist activity, despite the secrecy surrounding its commercial synthesis and fabrication (Davies et al. 2003; Henshaw et al. 2016; Hodson 2013; Lupton and Lipps 2018).

I divided the Kyiv smellwalk into three sections as experienced during the smellwalk—smellcatching, smell hunting and free smelling—each of which affected the somatic rhythms of walking and sniffing. To highlight the reciprocal rhythms, I visualised walking pace with a histogram indicating relative speed, sniff rhythm with an undulating, broken line, and smellwalking method as solid bars with increasing colour value fills. Smellwalker bodily movements involved in perceiving the olfactory city can be compared as shown in Fig. 16.

As can be seen in Fig. 16 sniff rhythm in the initial 'Smell Catching' stage is a gentle, constant wave in line with regular breathing. Walking pace was slowed by staircases and the volume of pedestrians until reaching European Square when walking pace increased. From Poshtova Square, and a change to the smellwalking method to 'Smell Hunting', arrhythmias occurred in sniff rhythms. Subsequent arrhythmias were created by interruptions to 'stop and sniff' which necessitated a break in walking pace and an increase in sniffing rate. Arrythmias are created by interruptions, and interruptions cause arrhythmias. The pace of the walk picked up towards the end when, having spent nearly two hours in the cold, smellwalkers wanted to move quickly resulting in a complete break from sniffing. Finally the walking rhythm slowed due to the populated environment of the indoor market and the sniff rhythm increased with 'Smell Research'.

When the smelling method, walk pace and sniff rhythm are coordinated with smells detected relational patterns emerge (Fig. 17). While the volumes and sizes of smells noted varied according to the individual smellwalker, a vibration of similar intensity is apparent at the start and end (when indoors) of this smellwalk, with greater variation in the central (outdoor) sections. Polyrhythmia occurs as 'bodies modify themselves' so as to detect the smells of the city (Lefebvre 2013, p. 49).

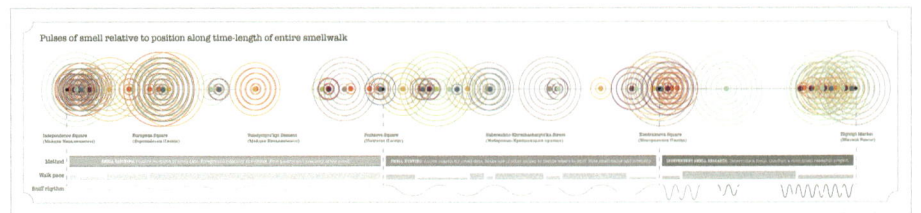

Fig. 17 A Winter Smellwalk in Kyiv (2017). Pulses of smell relative to position along the time-length of the entire walk. (Image created by the author)

Where *A Winter Smellwalk in Kyiv* mapping included Lefebvrian punctuations that mark out a rhythm, *Taking a Line from its Walk* examines their vitality. The former provided an overview, whereas the latter investigated smells' dynamic qualities. Subsequent mappings in this section emphasise when people detected smells over what they smelled. The size of the smell directly references perceived intensity as noted by individual smell-walkers, to whom I allocated one colour per person. Based on the average length of a human breath, each smell duration is four seconds. The title of the work is taken from Klee's description of the drawing of a line free of all restriction (Klee 1961). However, in this instance the unrestricted line of the walk is marshalled into an organised linearity to depict rhythms of smelling.

This animated mapping deliberately despatialised the smellwalk, and in removing place specificity I foregrounded the temporal aspects (Fig. 18). When recast as a straight line, the number and intensity of smell detections early and late in the walk are notable in comparison to the olfactory voids in sections between stopping points. I suggest this despatialised mapping approach reflects the imposed rhythms of the walk's structure more clearly and delineates a polyrhythmic co-existence of seven individual sets of smell encounters as it highlights the human subjectivity of smell perception in Kyiv's smellscape.

In order to detect a smell, we must first inhale odour molecules; each sniff including many smells from a range of sources. In a single sniff we might take in odour molecules from an item and its surroundings; the combination of a strawberry, the interior of the market and animal fur for example comprises hundreds of individual volatile molecules. In a research paper written while undertaking MA Philosophy, The New York for Social Research, Andreas Keller explained how humans perceive lighter smell molecules soonest after inhaling, and heavier smell molecules slightly later (2008). The time taken for each molecule to travel up the nasal cavity is dependent on weight; lighter molecules travel faster. Therefore, from a physiological perspective, the smellscape is in constant motion, invoking multiple temporalities in the physical experience of sniffing and smelling (Fig. 19). My approach to smell visualisation in this case study was based on my understanding of smell detection as parcels of related sniffs with memory acting as a linking mechanism. This frame-by-frame approach enables smell to be conceived as an animated sense and represented using animation.

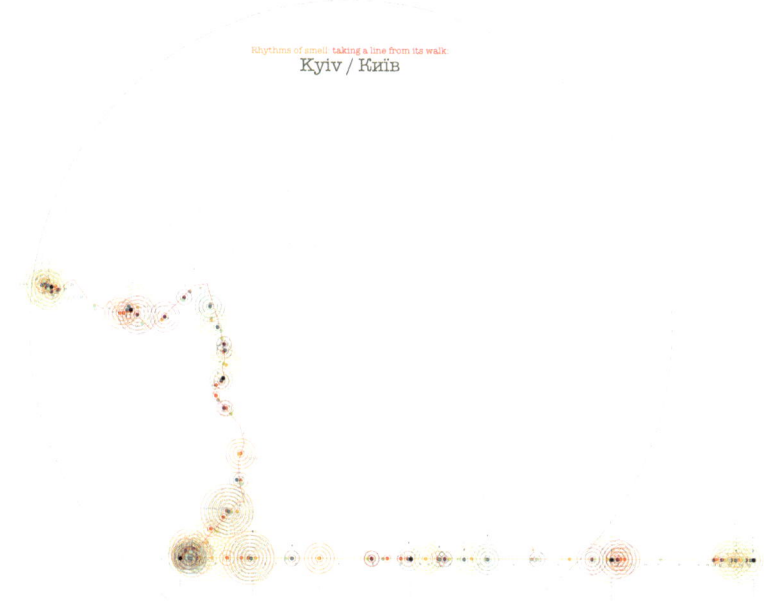

Fig. 18 Taking a Line from its Walk (2017). Creative process: deliberate de-spatialisation. (Image created by the author)

In the final thesis Keller revises his original phrasing to explain the physiology and perception of the temporal nature of the smellscape in which human perception of smell is temporally discontinuous, parcelled into 'odor successions' that are then affected by environmental wind conditions (Keller 2009, p. 37). An understanding of smell that accepts this temporal parcelling, along with the fluctuation of olfactory detection from light to heavy odour molecules within a single sniff, helps to explain just why no two smells are perceived as identical (even when detected by the same person). These are significant details behind my animated smellscape mappings.

To create the animated mapping, *Taking a Line from its Walk* (Fig. 20), I used the Kyiv dataset. The resulting animation accentuates the idea of repetition with difference and highlights rhythms of similarity and diversity as can be seen from screen grabs and online.[4] Moreover, the animation symbols fade, referencing decreasing smell intensity as a result of molecular volatilisation and human adaptation (Engen 1991). Animation

---

[4]The animation mapping entitled *Taking a Line from its Walk* is available at https://vimeo. com/213713916

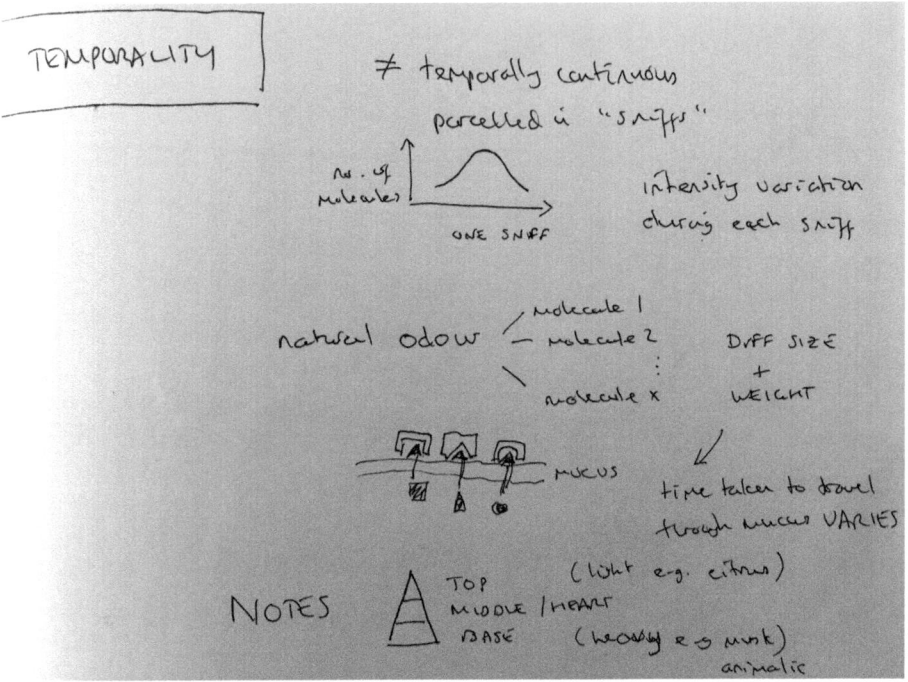

Fig. 19 Sketchbook visualisation (2014). Multiple temporalities of sniffing and smelling experience. (Image created by the author)

highlights the temporal nature of smell experience as sequential where previous mappings represented spatial, sensorial experiences of the city.

I set the animation on a black background to increase visual contrast and each smell instance lasts for the average length of a human sniff, four seconds, reducing the two hour smellwalk to twenty-five seconds (Barrett et al. 2012). The resulting superimposed wave patterns pulse and evince how smells might be subject to interference in the air. *Taking a Line from its Walk* (Fig. 20) provides a model to communicate layering of everyday smellscape and the polyrhythmic contradictions of those detecting it. This may inform future modelling systems for current, historical and future smellscapes.

The series of animations *Walking and Sniffing in the Wind: Smellwalk Rhythms I–VII* consider the spatiality of smells encountered, drawing attention to the temporality of encounter and dynamics of the airspace in which each smell was perceived. In this series of practical works I address participant smellwalks individually. The spatially-located smell symbols morph according to wind speed and direction (Fig. 5) and fade within a timeframe. Early in his critique Lefebvre remarks that:

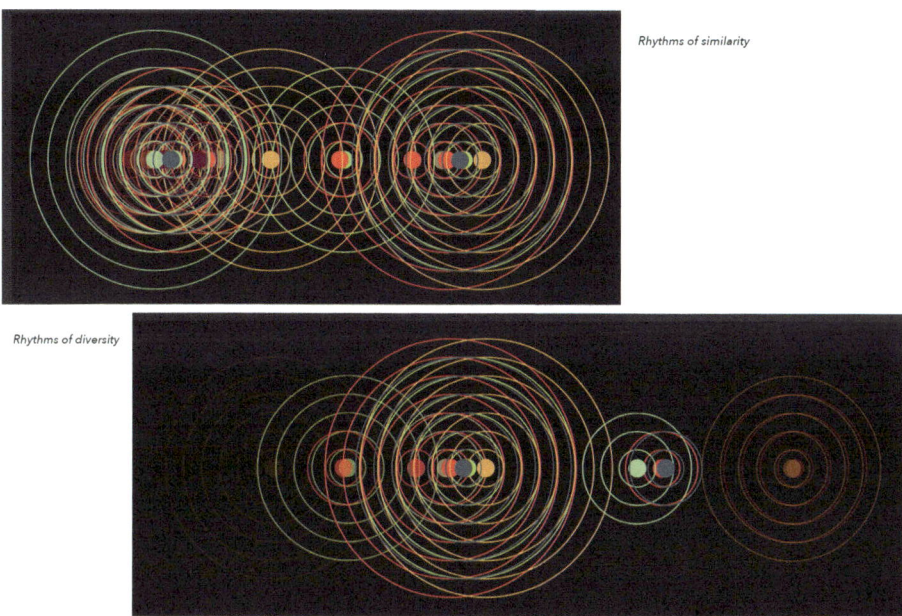

Rhythms of similarity

Rhythms of diversity

Fig. 20 Taking a Line from its Walk (2017). Screen grabs of smells in suspended animated sequence indicating concordance and difference. Image created by the author

We know a rhythm is slow or lively only in relation to other rhythms (often our own: those of our walking, our breathing, our heart). This is the case even though each rhythm has its own and specific measure: speed, frequency, consistency. (Lefebvre 2013, p. 20).

Comparative rhythms within the smellwalk include the smellwalker's linear smell detection sequence, somatic stoops and sniffs, linear duration of the walk, global wind patterns of a winter in Kyiv. Each rhythm is played out individually, and then superposed as theory and practice combine to create new knowledge of multiple smellwalk temporalities.

The following sections should be read alongside respective motion graphic visualisations accessible online with a password as indicated in the footnotes:

- Smell detections from smellwalker #1 are slow and stately. They occur at regular intervals with strong intensity and thus have strong, visible presence with a conflation of two smells at end of the walk[5];
- Smell detections from smellwalker #2 appear in staccato bursts between peaceful interludes of olfactory silence[6];

---

[5]https://vimeo.com/213708419

[6]https://vimeo.com/213708796

- Smell detections from smellwalker #3 are light and lively, occurring with a regular rhythm. The sheer quantity of smell detections results in a more constant beat along the route than other participants. Smell detection ceases towards the end of the walk before a finale that is in keeping with other participants[7];
- Smell detections from smellwalker #4 are steady, sequential and sustained with limited smell interaction until the final sniffing exercise in the market during which four smells mingle[8];
- The sequence of smell detections from smellwalker #5 is marked by subtlety and delicacy; the smells are small in stature, reticent in their apparition and subsequently diminish[9];
- The series of smell detections from smellwalker #6 commence with two intense smells, includes a long smell void in the central section hinting at barely perceptible odours before a greater intensity of smell experiences at the end[10];
- Smell detections from smellwalker #7 appear calm and considered, verging on tentative. The sequence fades to a series of smell sources with zero intensity.[11]

These separate sequences illustrate individual smell detection experience and inform the final stage of visually imagining a collective smellwalk experience.

The final animated mapping, *Polyrhythmias of the Smellwalk*, comprises superposition of individual smellwalk sequences to create a composite animation and an alternative format for communication of an aggregated smellscape.[12] As each of the encountered smells dissipates and disappears, its source remains as a reminder of the original source, an indicator of the path taken and the potential for a re-encounter at another time (Fig. 21). My vision for subsequent work in this field, considered in Chap. "Using the Matrix as a Qualitative Data Display for Landscape Research and a Reflection Based on the Social Constructivist Perspective", will go beyond such a single trail to examine how multiple, simultaneous smell detections in a variety of locations across the city might be recorded and communicated.

---

[7]https://vimeo.com/213709054

[8]https://vimeo.com/213709305

[9]https://vimeo.com/213709679

[10]https://vimeo.com/213709760

[11]https://vimeo.com/213710044

[12]https://vimeo.com/211929457

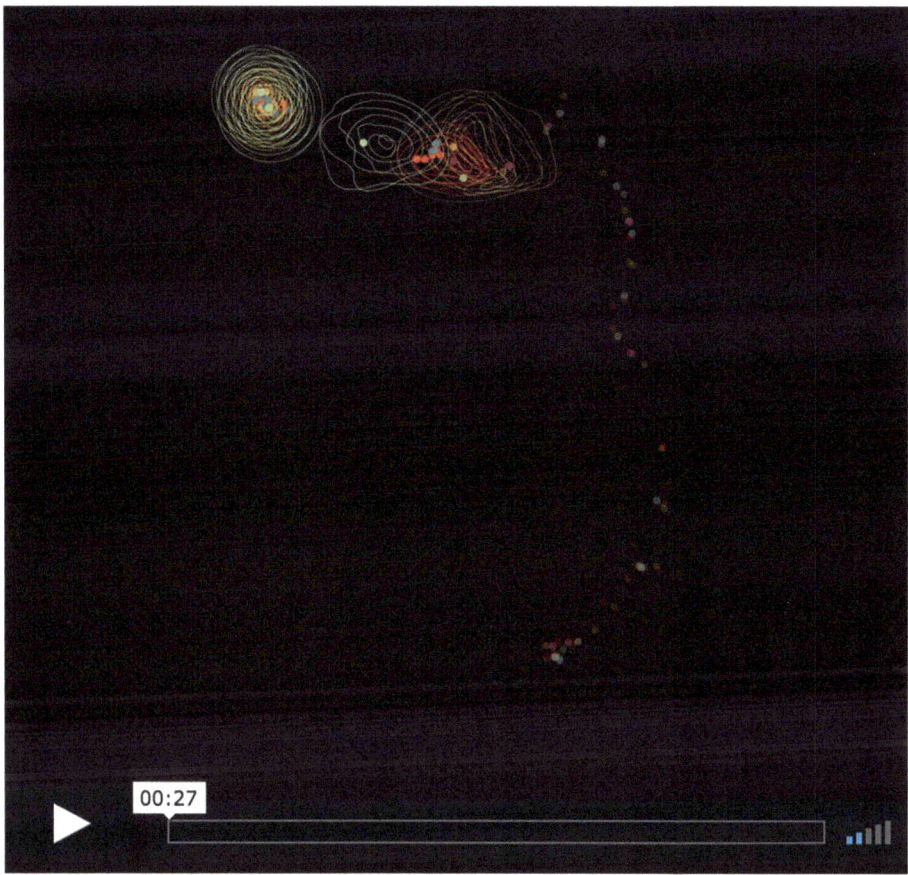

Fig. 21 Polyrhythmias of the Smellwalk (2017). Screen grab from animation. Composite layered series of seven smellwalks in Kyiv. (Image created by the author)

6   Discussion: Motion, Rhythm and Spatiality of Smell

As a practice the smellwalk fulfils a dual role; simultaneously a data collection tool and a designed artistic experience facilitating individual mappings of olfactory airspace. The relationship between smelling and walking is reciprocal, and in eternal flux. When focussed on rhythm, smellscape mapping combines somatic and environmental experiences from multiple viewpoints—top down (the route), lived (walking and sniffing pace), in-motion (the smells in drift mode) and static (materially-embedded smell sources) (McLean 2018). This approach, as Edensor points out, 'emphasises the dynamic and processual and thus circumvents reifications of place and culture which purvey geographical notions of place as static' contributing to arguments for the significance of the temporality of place (Edensor 2012, p. 57). While smellscape space might be conceptualised as constantly in motion, it is only brought into being through bodily encounters and detections. By focussing on activity and methods of engaging with the city 'momentary enactments and rhythms of difference undermine and contradict essentialist thought' enabling new conceived spaces of smells to emerge (Wilson 2017, p. 464).

Despite assertions that the human body does not detect as many smells in cold weather, the seven smellwalkers who completed the walk produced a collective list of nearly ninety identified odours while under guidance to restrict their smellnotes to twelve apiece (DNews 2014). Thus, human experience of smell is shown to be heterogeneous; smells exist, they are both present and identifiable in the environment when we play an active role in seeking them out. Smells elicited through smell-sketching point to multi-scalar temporalities; millennia, seasonal, a single sniff. Diurnal temporalities were witnessed in the 'animal fur' of the market and the 'asylum' both of which are busy places operating regular cycles of repetition over daily and weekly schedules (Shcherbak 2017). Seasonal smells of 'fungus' and 'winter' might be seen as markers of cyclical constancy and reliability, but they also hint of constant change. The smellsketches testify to an understanding of smell that is far from inert or passive, from a momentary sniff of the hundreds of volatile molecules present in a strawberry to the subtle, permanent presence of pine that contributes to Kyiv's background smell. As such smell is neither abstract nor invisible, but rather physically witnessed and recalled over time, especially the worsening, negative smell of burning garbage.

A Winter Smellwalk in Kyiv points to diversity and contingency in human experience. While there are endless opportunities for a repeat smellwalk, following the same route at the same time on a new day, anticipating the spiralling wisps of evanescent smells, it is unlikely that a repeat smellwalk would ever be a direct replication of the first (Lefebvre 2013).

Taking a Line from its Walk reframes space as time to reveal how the small-scale temporalities of breathing and sniffing combine to create repetitive patterns of smell detection. A pulsing timeline of interference amongst detected smells is similar in format to that generated as several stones are thrown into a pond simultaneously or the movements of waves generated as currents oppose the wind direction (Lefebvre 2013). And so the smellscape, as an experienced phenomenon, might be understood to exist in a constant state of potential interruption. Through the representation of smells as ephemeral repetitive patterns of detection unfold, a 'repetition in movement' (Lefebvre 2013, p. 86) captures the rhythms of both the body and the city's odours. Animated mapping of the smellwalk shows how human perceptions of smells have their own beats, pulses and fades whose rhythms interact. Each of the smellwalkers generated singular smell detection rhythms as their unique patterns of sniffing unfolded.

The *Smellwalk Rhythms I–VII* series highlight smell movement and transition; a phenomenon specifically recited to me by one smellwalker smellwalking Volodymyrs'kyi Descent from a city square to the river. At the top of the hill, a distinctive smell of the river was apparent to him. Halfway down the slope the river smell vanished, only to recur at the bottom of the hill. I suggest how the temporal and volumetric space of the smellscape can be imagined through layering individual smell observations on top of each other. A walk through any environment will afford sequential smell detection, noted via a sniff facilitated by an intake of breath. For an individual, the overall smellscape composition results from a piecing together of parcels of sniffs—indeed, individual smellnotes take account of one perspective, precluding those smells at different heights to the sniffer's own nose. The entirety of the smellscape contains many smells at any given moment all of which overlap and interact as they drift in airflows and sink to ground level. The final mapping in this series, *Polyrhythmias of the Smellwalk*, visually examines how such a combined smellscape might be communicated visually.

Smellwalk Rhythms I–VII were layered to create *Polyrhythmias of the Smellwalk* resulting in an organic mass of smell data in motion. The dense smell zone at the beginning of the walk is convoluted and complex. Rhythmic ribbons of smelly interactions occur in the central section of the walk. The finale takes place in the enclosed space of Zhytniy Market where the enclosed space generated a bounded swirl of smells. Visualised in the animation the wake of the walk, a trail of semi-transparent static dots indicating smell sources, call attention to original source locations, and the possibility for detection by another smellwalker. Through agglomeration of multiple sequences a relativity emerges since, as proposed by Lefebvre (2013), rhythm is best measured by comparison to other rhythms, their coincidences marking plurality of interactions and reciprocations. In creating the maps I came to understand how smell experiences might be qualitatively compared, and to discern collective rhythmic patterns created through visual interference.

A potential shape and form of a collectively apprehended smellscape is expressed in *Polyrhythmias of the Smellwalk*. The animation mapping possesses eurythmic qualities demonstrating diversity of intensity and a combination of slow and faster paces. A syncopation, and sense of the offbeat, arise as smell isopleths overlay themselves and each other. Henshaw and Porteous (1990, p. 25) both imagine smellscape from the purview of a single being, 'limited by the height of our noses from the ground, where smells tend to linger'. The Kyiv smellwalk mappings build from such individual perspectives, aggregating isopleth narratives, reframing the smellscape to exist simultaneously from multiple perspectives. We may never experience or fully understand the smellscape from the point-of-smell of another; but mapping it in a collective manner 'has a creative potential to reveal the unseen, ephemeral and imagined' (McLean 2017, p. 76). We may come to greater appreciation of the smellscape's complexity through combinatorial mappings of subjective experience. The resulting work is an abstraction that, as yet, only hints to the dimensional possibilities of representing the smellscape as a never-ending series of intertwining, morphing, volatilising, tumbling smell molecules in physical space.

This case study commenced with a static map depicting a recorded smellwalk. Through analytical practice I have shown how linking theoretical constructs in rhythmanalysis with mapping practices can result in new relations being observed; movement of the human body in space is integral in understanding the smellscape. Through a series of evanescent, dynamic mappings I have shown how the experienced qualitative smellscape might be understood as part of a meteorological phenomenon (for other ideas on visualizing non-visual landscape phenomena in this volume, see Siepmann et al. 2020).

References

Barkham, P. (2016). Being a beast by Charles Foster review—The man who ate worms like a badger. *The Guardian.* https://www.theguardian.com/books/2016/feb/03/being-beast-charles-foster-review-man-whoate-worms-like-badger.

Barrett, K. E., Brooks, H., Boitano, S., & Barman, S. (2012). *Ganong's review of medical physiology* (24 ed.). McGraw-Hill Medical.

Certeau, M. (2011). *The practice of everyday life* (3rd revised ed.). Berkeley: University of California Press.

Davies, B. J., Kooijman, D., & Ward, P. (2003). The sweet smell of success: Olfaction in retailing. *Journal of Marketing Management, 19*(5–6), 611–627.

Dodt, J., Bestgen, A.-K., & Edler, D. (2017). Ansätze der Erfassung und kartographischen Präsentation der olfaktorischen Dimension. *Kartographische Nachrichten, 67*(5), 245–256.

DNews. (2014). *Why cold air smells different.* Seeker—Science. World. Exploration. Seek for yourself. https://www.seeker.com/why-cold-air-smells-different-1768222636.html.

Drobnick, J. (2002). Toposmia: Art, scent, and interrogations of spatiality. *Angelaki, 7*(1), 31–48.

Edensor, T. (2010). Walking in rhythms: Place, regulation, style and the flow of experience. *Visual Studies, 25*(1), 69–79.

Edensor, T. (2012). The rhythms of tourism. In C. Minca & T. Oakes (Eds.), *Real tourism: Practice, care, and politics in contemporary travel culture* (pp. 54–71). Abingdon: Routledge.

Edensor, T. (2013). Rhythm and arrythmia. In P. Adey, D. Bissell, K. Hannam, P. Merriman, & M. Sheller (Eds.), *The Routledge handbook of mobilities* (pp. 163–171). London: Routledge.

Edwards, R., & Holland, J. (2013). *What is qualitative interviewing?* Bloomsbury.

Elden, S. (2013). Rhythmanalysis: An introduction. In H. Lefebvre (Ed.), *Rhythmanalysis: Space, Time and Everyday Life.* Bloomsbury Academic.

Engen, T. (1991). *Odor sensation and memory.* Greenwood Publishing Group.

Frayling, C. (1993). Research in art and design. *Royal College of Art Research Papers, 1*(1), 1–9.

Haraway, D. (1988). Situated knowledges: The science question in feminism and the privilege of partial perspective. *Feminist Studies, 14*(3), 575–599.

Henshaw, V. (2014). *Urban smellscapes: Understanding and designing city smell environments.* Routledge.

Henshaw, V., Medway, D., Warnaby, G., & Perkins, C. (2016). Marketing the 'city of smells.' *Marketing Theory, 16*(2), 153–170.

Hodson, H. (2013). *Smell-o-vision screens let you really smell the coffee.* New Scientist. https://www.newscientist.com/article/mg21729105-900-smell-o-vision-screens-let-you-really-smell-the-coffee/.

Horowitz, A. (2016). *Being a dog: Following the dog into a world of smell.* Scribner Book Company.

Internet Encyclopedia of Ukraine. (2001). *Pine.* https://www.encyclopediaofukraine.com/display.asp?linkpath=pages%5CP%5CI%5CPine.htm.

Keller, A. (2008). *Phenomenology of smell (Term paper).*

Keller, A. (2009). *Phenomenology of smells* [Masters Thesis]. The New School for Social Research.

Kephart, K. B., & Mikesell, R. E. (2000). *Manure odors* (pp. 1–11). Pennsylvannia State University.

Kitchin, R., & Dodge, M. (2007). Rethinking maps. *Progress in Human Geography, 31*(3), 331–344.

Kitchin, R., Gleeson, J., & Dodge, M. (2013). Unfolding mapping practices: A new epistemology for cartography. *Transactions of the Institute of British Geographers, 38*(3), 480–496.

Klee, P. (1961). *Notebooks, Volume 1: The Thinking Eye* (J. Spiller, Ed.; R. Manheim, Trans.). Lund Humphries.

Lamb, H. H. (1975). Our understanding of the global wind circulation and climatic variations. *Bird Study, 22*(3), 121–141.

Lefebvre, H. (2013). *Rhythmanalysis: Space, time and everyday life.* Bloomsbury Academic.

Lupton, E., & Lipps, A. (2018). *Senses: Design beyond vision.* Princeton Architectural Press.

McBride, M., & Nolan, J. (2017). Situating olfactory literacies. In V. Henshaw, D. Medway, K. McLean, C. Perkins, & G. Warnaby (Eds.), *Designing with smell: Practices, techniques and challenges* (pp. 187–195). Routledge.

McLean, K. (2017). Communicating and mediating Smellscapes: The design and exposition of olfactory mappings. In V. Henshaw, D. Medway, K. McLean, C. Perkins, & G. Warnaby (Eds.), *Designing with smell: Practices, techniques and challenges* (pp. 67–76). Routledge.

McLean, K. (2018). Mapping the invisible and the ephemeral. In A. J. Kent & P. Vujakovic (Eds.), *The Routledge handbook of mapping and cartography* (pp. 500–515). Routledge.

Moore, R. M. (2013). The beat of the city: Lefebvre and rhythmanalysis. *Situations: Project of the Radical Imagination, 5*(1).

Porteous, J. D. (1985). Smellscape. *Progress in Physical Geography, 9*(3), 356–378.

Porteous, J. D. (1990). *Landscapes of the mind: Worlds of sense and metaphor.* University of Toronto Press.

Rodaway, P. (1994). *Sensuous geographies: Body, sense and places.* Routledge.

Shcherbak, E. (2017). *Zhytniy market Kiev | Spotted by Locals.* Spotted by Locals Kiev. https://www.spottedbylocals.com/kiev/zhytniy-market/.

Siepmann, N., Edler, D., & Kühne, O. (2020). Soundscapes in cartographic media. In D. Edler, C. Jenal, & O. Kühne (Eds.), *Modern approaches to the visualization of landscapes* (pp. 247–263). Wiesbaden: Springer VS.

Simpson, P. (2012). Apprehending everyday rhythms: Rhythmanalysis, time-lapse photography, and the space-times of street performance. *Cultural Geographies, 19*(4), 423–445.

Vorotnikov, V. (2015). *Will Ukraine ever come clean?* https://www.recyclingwasteworld.co.uk/in-depth-article/will-ukraine-ever-come-clean/84996/.

Wilson, H. F. (2017). On geography and encounter: Bodies, borders, and difference. *Progress in Human Geography, 41*(4), 451–471.

Wunderlich, F. M. (2013). Place-temporality and urban place-rhythms in urban analysis and design: An aesthetic akin to music. *Journal of Urban Design, 18*(3), 383–408.

Dr Kate McLean works at the intersection of human-perceived smellscapes, cartography and the communication of 'eye-invisible' sensed data. To achieve this she leads public smellwalks internationally and translates the resulting data using digital design, watercolour, animation, scent diffusion and sculpture into smellscape mappings. Her recently-completed PhD, "Nose-first: practices of smell walking and smellscape mapping" from the Royal College of Art examines representation and communication of the smellscape as theorised by J. Douglas Porteous and activated by Victoria Henshaw, the research explores how social performative mapping might contribute to communication of non-visual sensory olfactory information. In so doing it tests existing theories to build a deeper understanding of the smellscape. Kate leads the BA Hon Graphic Design programme at Canterbury Christ Church university and undertakes international smellscape mapping commissions. Her work can be seen at https://sensorymaps.com/.

Soundscapes in Cartographic Media

Nils Siepmann, Dennis Edler and Olaf Kühne

Abstract

The acoustic dimension is a topic that has attracted cartographers for decades. In many examples of 2D and 3D cartography, sound-related matters have been used as map themes or/and as topics of cartographic communication. This article briefly summarizes relevant approaches that address the acoustic dimension in cartography. From the point of view of social constructivist landscape theory, there are several potentials associated with audiovisual stimuli: It deals with aspects of physical space that have so far been underrepresented in spatial research, it clarifies the contingency of spatial patterns of interpretation and evaluation, and it provides a basis for research into the social construction of the world, in this case landscape.

Keywords

Soundscape · Landscape · Audiovisual cartography · Social constructivism · Sound

N. Siepmann (✉) · D. Edler
Ruhr-Universität Bochum, Bochum, Germany
e-mail: nils.siepmann@ruhr-uni-bochum.de

D. Edler
e-mail: dennis.edler@ruhr-uni-bochum.de

O. Kühne
Eberhard Karls Universität Tübingen, Tübingen, Germany
e-mail: olaf.kuehne@uni-tuebingen.de

© Springer Fachmedien Wiesbaden GmbH, part of Springer Nature 2020
D. Edler et al. (eds.), *Modern Approaches to the Visualization of Landscapes*, RaumFragen: Stadt – Region – Landschaft,
https://doi.org/10.1007/978-3-658-30956-5_13

1   Introduction

Acoustic stimuli are of considerable importance for the individual and social construction of landscapes, especially in its native and commons-sense dimensions (Kühne
2018a, 2020a; in this volume: Kühne and Jenal 2020). The consideration of the non-
visual dimensions of landscape remained underrepresented in the academic discussion (Bischoff 2007; Kazig 2013, 2019; Winkler 2005, 2006[1995]; see in this volume:
McLean 2020). Volatile phenomena (such as sounds and odors) present challenges for
the detection, categorization, quantification and presentation. These challenges are much
less evident compared to visual and haptic phenomena (Edler and Kühne 2019; Winkler
2005). Although spatial sciences reveal a preference for visually perceptible matters (see
in this volume: Berr 2020), the audiovisual representation of geospatial information has
been a topic that has attracted cartographers for decades. (see, for example, recent publications by Ballatore et al. 2018; Edler 2020; Edler et al. 2019a; Lammert-Siepmann
et al. 2017; Schito and Fabrikant 2018). The establishment of computers as a mass media
tool, at home and at work, gave a 'fresh boost' to new digital cartographic solutions. The
1980s and 1990s are often associated with the rise of animation software which enabled
people to expand static media with interactive and animated elements (Dickmann 2018,
pp. 169–171; Kraak and Ormeling 2010, p. 155; Peterson 1995, p. 75). At the beginning of animation research in cartography, the projects were primarily focused on visual
components (e.g. DiBiase et al. 1992; Koussoulakou and Kraak 1992; Moellering 1980;
Monmonier 1990; Peterson 1993; see in this volume: Hochschild et al. 2020). However,
the research topic of animation in multimedia cartography had soon reached a multisensory stage. Map animation also included sound—in different variants (see Edler et al.
2019a)—to communicate spatial information.

   A pioneering article that emphasizes the potential of sound for the representation of
spatial information was authored by John B. Krygier (1994), who also presented first
examples of audiovisual maps at the 1993 Association of American Geographers (AAG)
Conference in Atlanta, Georgia. Krygier's article mainly deals with a set of "abstract
sound variables" that could be used to represent ordinal and nominal data in maps (see
also Schiewe 2015). The author also recommends the investigation of a more environmental and geographical perspective on sound which is bound to "day to day experience
with sound" (Krygier 1994, p. 150).

   The idea of using sound in maps to represent sound experiences in the real environment has been taken up in several cartographic examples over the last decades (e.g.
Aiello et al. 2016; Edler et al. 2015; Pulsifer et al. 2007; Müller et al. 2001). Such
examples refer to the concept of the "soundscape" which was introduced by R. Murray
Schafer in (1977). Based on sound recordings or computer-based sound simulations,
cartographic media can provide audiorealistic impressions of the sonic environment and
point to locational references at the same time (Laakso and Sarjakoski 2010). Obviously,
soundscape elements can have a high individual component and they can help to put

across a more realistic impression of a large-scale excerpt of landscapes. From a positivistic perspective, these impressions can have an intended 'standard message', often related to communicating quantities. Such audiorealistic sound elements in maps may also be used to individualize the interpretation of a specific landscape (excerpt). This rather corresponds to constructivist approaches of landscape perception and (individual and social) evaluation.

This article presents different approaches to using soundscapes in cartographic visualizations. The particular focus is on the potentials of soundscapes in modern 2D and 3D maps for the social construction of landscapes. The aspects of (recorded) language and speech as possible additional soundscape elements are not explicitly discussed in this article (but see the following papers for more information about the use and experimental investigation of language and speech in audiovisual cartography: Edler et al. 2012; Morrison and Ramirez 2001; Lammert-Siepmann et al. 2011; Siepmann et al. 2019, 2020).

2   Soundscapes in 2D Maps

The representation of soundscapes is not necessarily bound to the implementation of auditory map elements. Established examples of maps exist that represent soundscape phenomena based on map graphics, long before sound was used as cartographic design element to communicate spatial information (see examples in Harnapp and Noble 1987; Scharlach 2002). A prominent and widely known example is the cartographical representation of noise. In Europe, the mapping of noise has been an obligatory task by public authorities since 2002 and is often associated with the adoption of the "Environmental Noise Directive" (END—Directive 2002/49/EC). Noise maps represent sound pressure levels, and different intensities are represented by polygons. The borders between these polygons consist of isophones, linear map features indicating different noise intensities in the represented area. The noise maps and noise management action plans which are updated every 5 years are intended to inform the public about the current status of noise and noise pollution (Weninger 2014). Approaches to acquiring noise data also include modern sensors and data (models), such as smartphone-microphones (Maisonneuve et al. 2010; Zuo et al. 2016), big data sets derived from social media (Zheng et al. 2014) or user-oriented urban geo-data models, incl. sound quality indicators (Gomez et al. 2017; Lavandier et al. 2016; Ricciardi et al. 2015).

Apart from publicly administered and maintained 'visual soundscapes' in noise maps, other cartographic examples include graphical representations or verbal descriptions of sound, such as objects that compile urban soundscapes based on social media data (Aiello et al. 2016). In this example (see Fig. 1), the urban soundscape is visually communicated as a generalized and classified description of what could be acoustically experienced in the streets of London.

Fig. 1 "Chatty Map" of Central London (Aiello et al. 2016—with kind permission to print this Figure)

Maps representing areas on an individual/personal level can only hardly be found in the literature of cartography. In such examples, descriptions of objects producing (urban) soundscapes are much more specific and are closely related to constructivist approaches to landscape evaluation (see also Edler and Kühne 2019). Figure 2 gives an example of an individual perception of the Riga district "Bolderāja", incl. references to sound-emitting parks as well as river and road (transport) systems (c.f. Edler 2011).

Going beyond a solely graphical communication of soundscapes in 2D maps, cartographic researchers and practitioners have suggested many examples including sound features as cartographic elements of communication. In connection with rising multimedia development and processing tools on computers, such as professional and 'home-brew' animation software, audiovisual cartography has been established from the 1990s onwards. It is argued that the video gaming industry published several widely known examples incl. soundscape representations before (Edler and Dickmann 2017, 2016). These examples include sound sequences that underline the (usually) fictional landscape representations in the game story (see also Grimshaw 2014; Nitsche 2008, pp. 141–144).

In the development of audiovisual cartography, approaches have been made to transferring the (above mentioned) example of noise representation to the acoustic dimension. Depending on the position on the cursor in the map field, the soundscape is calculated, acoustically compiled and played with reference to different sound-emitting objects,

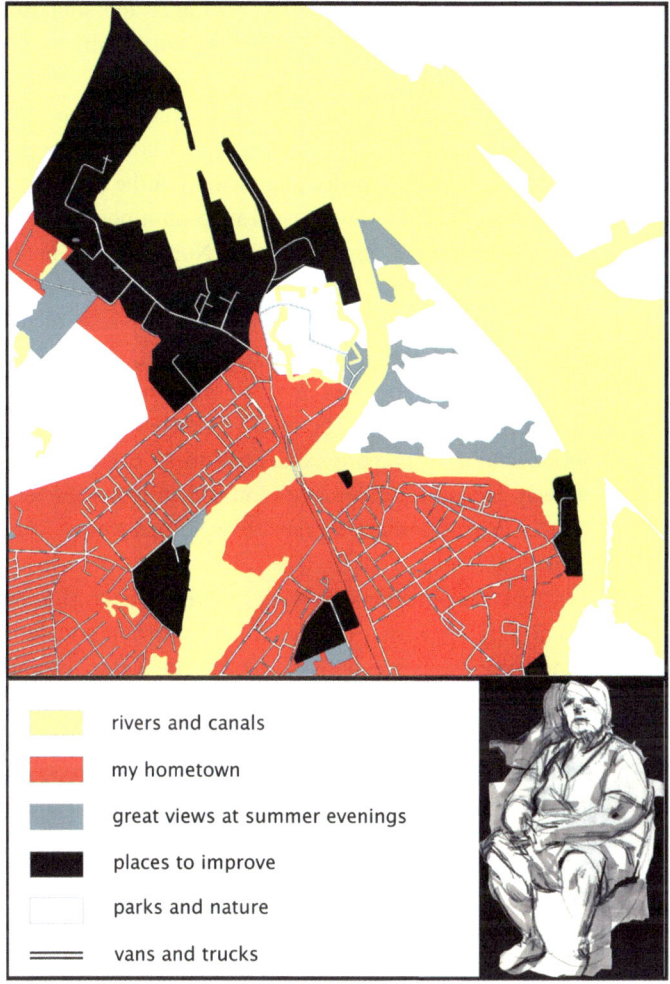

Fig. 2 The individual construction of Bolderāja by a Riga citizen (Edler 2011)

such as urban traffic noise and parks (see early examples by Müller and Scharlach 2001; Scharlach 2002; also Lavandier et al. 2016). Based on the location-dependent soundscape compilation and loudness, the map user can get a simulated auditory impression of the sound in the 'real urban landscape'.

Another popular category of multimedia cartographic applications that include sound features in interactive 2D maps are (audiovisual) tourist maps (for examples on the olfactory dimension in multimedia cartography, see Dodt et al. 2017; McLean 2016; Lauriault and Lindgaard 2006; see in this volume: McLean 2020). Visitors of the national park

"Nuuksio" (Finland), for example, are invited to experience the soundscape by using an audiovisual map (Laakso and Sarjakoski 2010). Selected highlights of the park are acoustically represented. By clicking on point symbols, sound recordings representing local sound sequences can be (repeatedly) played. The sequences can be heard while a 2D layout of the park area can be read. In this way, an audiovisual impression can be transferred, and visitors can get a multisensory imagination of the area.

A similar idea is behind a published audiovisual map on a post-industrial site in the Ruhr district (Germany), "Landscape Park Duisburg Nord" (Edler et al. 2015). This example is intended to put across sound imaginations of present usages in an area that still preserves the industrial past architecturally. The selected places of interests (POIs) highlight popular attractions in the park, such as a diving pool, skate park or a summertime cinema. By clicking on the pictograms of these attractions, sound sequences representing the present usage can be heard. The map was also suggested for applications in (geography) school education to prepare excursions to an example site of structural transformation in formerly industrial urban areas (related in this volume: Bagoly-Simó 2020; Papadimitriou 2020; Prisille and Ellerbrake 2020; Stintzing et al. 2020).

In addition to the two examples of audiovisual tourist maps described before, there are other map themes that cartographers have dealt with so far (for overviews, see Brauen 2014; Edler et al. 2019a). Based on modern web mapping techniques, such as leaflet-based web cartography (https://leafletjs.com/), several online examples of audiovisual maps have been designed and published to highlight the potentials, characteristics and peculiarities of different areas and their soundscapes (e.g. Cities and Memory Contributors 2020; Radicchi 2009; Tschaikner et al. 2013). These web applications invite audiophile citizens to participate by uploading and locating individual sound files. In this way, contributors around the globe can enrich and accentuate their individual impressions of places by making volunteered geographic information publicly available. Usually, this sound data is acquired with mobile devices that are equipped with good-quality sound sensors, such as smartphones.

These web projects indicate the public interest in mapping and sharing individualized and location-specific audiorealistic impressions, a social phenomenon which also occurs in social media platforms, video-sharing websites, open sound data bases and common messenger services (for more information on open spatial data, see in this volume: Edler et al. 2020; Kleber et al. 2020; Meyer-Heß 2020). Audiorealism establishes a high relationship between the signifying map element (sound sequence) and the signified object or concept, and this allows to underline the atmosphere and meaning of individuality in communication (see also Edler et al. 2019a; Papadimitriou et al. 2009). As the level of realism in the communication of spatial information can be increased in 3D environments, it is not surprising that simulations of soundscapes have found their ways into modern 3D cartographic media.

3   Soundscapes In 3D Maps

Some cartographic approaches show that the representation of soundscapes in 3D cartography could be reduced to the visual dimension (e.g. Herman and Řezník 2014; Kornfeld 2008; Stoter et al. 2008). The free availability of 3D gaming software, such as the game engines *Unreal Engine 4* (UE4) and *Unity 3d*, brought a whole new impetus to a highly realistic 3D visualization and immersive experience of the outcoming 3D environments (see also Edler et al. 2018a; Hruby et al. 2019; Kersten et al. 2018; Vetter 2019; see in this volume: Edler et al. 2020; Prisille and Ellerbrake 2020; Vetter 2020). These new technical opportunities also enable a 3D representation of soundscapes (Berger and Bill 2019; Hruby 2019, Edler et al. 2019b). Online available (open) sound databases, such as freesound.org (https://freesound.org/), offer an easy access of soundscape sequences gathered worldwide.

Before the new rise of VR visualization, sound sequences were already discussed and included in 3D cartography (e.g. Buchroithner 2005; Edler and Dodt 2010; Harding and Souleyrette 2009; Jobst and Germanchis 2007). Thanks to modern VR-based immersive usage possibilities, the three-dimensionality of sound is however stronger emphasized. Users can explore a 3D environment by real-time changes of their position. To create a highly realistic sound impression, the communication of sound also requires a real-time

Fig. 3   The 3D coverage of a sound element (audiorealistic sound sequence of singing birds) in a VR application (indicated by *orange circles* around a tree; own representation)

rendering and a positional (3D-dimensional) cue, which is supported by modern game engines (see also Edler et al. 2019a, Hruby 2019). Figure 3 gives an example that indicates the 3D experience of a located soundscape sequence in a UE4 VR application. Accordingly, the (real-time) immersion into 3D soundscapes based on modern VR techniques brings new potentials that are currently discussed in the cartographic literature.

For example, Kersten and colleagues (2017) suggest that sound, as one key feature of multimedia, adds value to the immersive experience of historical landscapes and virtual museums represented in VR. In the context of (simulated) historical landscapes, VR application may also serve as 'walkable 3D archives' that preserve historical sounds and contain educational media including soundscape impressions of the past (Edler et al. 2019b). Berger and Bill (2019) propose an immersive VR model which is intended to simulate (urban) noise differences and support procedures of public participation. The idea of using 3D cartographic media as support tool for public participation and planning procedures has been suggested several times (e.g. Lovett et al. 2015; Ma et al. 2020; Salter et al. 2009). The immersive component however allows a 3D sound perception at changing positions, which offers a new accuracy for participatory procedures.

Edler and colleagues (2018b) also discuss potentials of immersive VR applications for the individual and social level of evaluating represented landscapes. Highly realistic visual as well as auditory features located at specific places in a VR-based 3D model (suggesting a 1:1 scale) counteract traditional generalization processes in cartography and individualize a 3D model. This individualization of cartographic media may create new opportunities to analyse individual and social constructions of (realistically represented) landscapes (related in this volume: Al-Khanbashi 2020; Bellini and Leonardi 2020; Berr 2020; Fontaine 2020a, b; Kühne 2020b; Kühne and Jenal 2020; Linke 2020; Loda et al. 2020).

4   Potentials of Soundscapes in Cartography for the Exploration of the Individual and Social Construction of Landscape

As stated in the introduction, sounds have a special significance in the everyday construction of landscape. They are also strongly linked to specific spatio-temporal conditions: Certain sounds only occur at specific locations and times, both historically and at certain times of the year. For example, social and technical modernization has had a significant impact on the disappearance and emergence of soundscapes, although increasingly restrictive noise regulations are leading to an increasing silence of the landscape (Bernat and Hernik 2015; Kühne 2018a; Upton 2007). Accordingly, audiovisual cartography can make a significant contribution to the contingency of landscape constructions. But not only the physical foundations of landscape are subject to spatio-temporal variability, individual constructions and social constructions are also variable (among many:

Aschenbrand 2017; Bruns and Kühne 2015; Bruns and Paech 2015; Hunziker et al. 2008; Stotten 2015; in summary: Kühne 2018b, 2019): Individual landscape constructions are based on social conventions on the one hand, but also on an individual actualization which incorporates observations as well as the experience of physical space. Especially the 'normal domestic landscape' is strongly influenced by individual contributions, here especially in childhood. Here, sounds (but also smells) are accepted as 'normal' elements of an unquestioned given space; for some this is cowbells and the hum of insects, for others the sounds of a busy street in a metropolitan area. However, stereotypical perceptions of the landscape are not only associated with sounds (who would not expect noise on a beach?) but also their variability (who would expect noise-producing fans in a football stadium outside of matches?).

From the perspective of social constructivist landscape theory, the following potentials for the 2D and 3D representation of landscape contingencies and the diversity of landscape constructions arise:

1. Research aspects: It deals with aspects of physical space that have been underrepresented in spatial research, especially in landscape research.
2. Historical aspects: Audiovisual cartography can be used to document the emergence and disappearance of certain sounds for specific places and spaces (such as the advent of the steam railway, its replacement by diesel or electric traction vehicles).
3. Seasonal and diurnal aspects, especially in comparison: Here, specific sequences of sounds in the course of the year can be simulated for certain places and compared with them.
4. Aspects of spatial locations: The representation or simulation of characteristic noise arrangements in different parts of a city, its surroundings, up to intercontinental comparisons contributes to a visualization of spatial diversity.
5. Contingency of future sounds: Here the sound consequences of planning can be simulated.
6. Contingency of evaluation: with the help of sound simulations, different evaluation patterns of people, for example in dependence on ideological attitudes, can be recorded and examined (a drastic example: the owner of a historic Harley-Davidson with a weakly damped exhaust system will assess its contribution to the nightly soundscape differently than the resident looking for a night's rest after a stressful day).
7. Aspects of education: Due to the high importance of audiovisual cartography in everyday life and its low degree of canonization up to now, the subject is particularly suitable for the production of maps in school and university education, as it allows for own creative solutions (see e.g. Papadimitriou 2010).

A connection between social constructivist landscape research and audiovisual cartography is only just beginning, although the considerable potential of this connection can already be estimated.

5   Summary and Outlook

As explained in the previous section, there is considerable potential for using audiovisual cartographic media in social constructivist landscape research, but also for considering theoretical foundations of social constructivist landscape research in cartographic visualization. These potentials can be found in several dimensions.

1. Audiovisual cartography provides the opportunity for a social-constructivist-oriented landscape research to investigate and theoretically classify construction patterns of landscape. This is particularly true for intertemporal and interregional comparisons.
2. Social constructivist landscape research, or more generally social constructivist research, can provide audiovisual cartography with a basis for reflection, for example with regard to the question: What conventions of perception are created with the help of audiovisual cartography? Which stereotypes are updated and reinforced? What role does audiovisual cartography play in the interpretation of the world?
3. Landscape research with a social-constructivist orientation can provide audiovisual cartography with an orientation framework for the production of maps, for example, to illustrate contingencies in the physical foundations of landscape in order to point out different interpretations of space.

This article might have indicated that future research on audiovisual cartography could also leave the traditional positivistic framework of cartography. Modern techniques of visualizing and experiencing cartographic media increase the possibilities to individualize the representation of spatial contents. These modern cartographic media may serve as valuable materials for constructivist approaches of landscape evaluation, such as the social constructivist approach highlighted in this contribution.

References

Aiello, L. M., Schifanella, R., Quercia, D., & Aletta, F. (2016). Chatty maps: Constructing sound maps of urban areas from social media data. *Royal Society Open Science, 3*(3), 150690. https://doi.org/10.1098/rsos.150690.

Al-Khanbashi, M. (2020). Using matrix as a qualitative data display for landscape research and a reflection based on the social constructivist perspective. In D. Edler, C. Jenal, & O. Kühne (Eds.), *Modern approaches to the visualization of landscapes* (pp. 103–118). Wiesbaden: Springer VS.

Aschenbrand, E. (2017). *Die Landschaft des Tourismus. Wie Landschaft von Reiseveranstaltern inszeniert und von Touristen konsumiert wird.* Wiesbaden: Springer VS.

Bagoly-Simó, P. M. (2020). Landscape in geography textbooks. In D. Edler, C. Jenal, & O. Kühne (Eds.), *Modern approaches to the visualization of landscapes* (pp. 371–386). Wiesbaden: Springer VS.

Ballatore, A., Gordon, D., & Boone, A. P. (2018). Sonyfying data uncertainty with sound dimensions. *Cartography and Geographic Information Science, 46*(5), 385–400. https://doi.org/10.10 80/15230406.2018.1495103.

Bellini, A., & Leonardi, L. (2020). Prato: The social construction of an industrial city facing processes of cultural hybridization. In D. Edler, C. Jenal, & O. Kühne (Eds.), *Modern approaches to the visualization of landscapes* (pp. 549–572). Wiesbaden: Springer VS.

Berger, M., & Bill, R. (2019). Combining VR visualization and sonification for immersive exploration of urban noise standards. *Multimodal Technologies and Interaction, 3*(34), 3. https://doi. org/10.3390/mti3020034.

Bernat, S., & Hernik, J. (2015). Polnische Klanglandschaft um die Jahrhundertwende. In O. Kühne, K. Gawroński, & J. Hernik (Eds.), *Transformation und Landschaft. Die Folgen sozialer Wandlungsprozesse auf Landschaft* (pp. 247–267). Wiesbaden: Springer VS.

Berr, K. (2020). Visuality, aesthetics and landscape. For the enlightenment and self-enlightenment of constructivist landscape research. In D. Edler, C. Jenal, & O. Kühne (Eds.), *Modern approaches to the visualization of landscapes* (pp. 181–215). Wiesbaden: Springer VS.

Bischoff, W. (2007). *Nicht-visuelle Dimensionen des Städtischen. Olfaktorische Wahrnehmung in Frankfurt am Main, dargestellt an zwei Einzelstudien zum Frankfurter Westend und Ostend* (dissertation, Johann Wolfgang Goethe-Universität). Wahrnehmungsgeographische Studien, 23.

Brauen, G. (2014). Interactive audiovisual design for cartography: Survey, prospects, and example. In T. Lauriault & D. R. F. Taylor (Eds.), *Developments in the theory and practice of cybercartography. Applications and indigenous mapping* (2nd ed., pp. 141–159). Amsterdam: Elsevier.

Bruns, D., & Kühne, O. (2015). Zur kulturell differenzierten Konstruktion von Räumen und Landschaften als Herausforderungen für die räumliche Planung im Kontext von Globalisierung. In B. Nienaber, & U. Roos (Eds.), *Internationalisierung der Gesellschaft und die Auswirkungen auf die Raumentwicklung. Beispiele aus Hessen, Rheinland-Pfalz und dem Saarland* (Arbeitsberichte der ARL, vol. 13, pp. 18–29). Hannover: Selbstverlag.

Bruns, D., & Paech, F. (2015). „Interkulturell_real" in der räumlichen Entwicklung. Beispiele studentischer Arbeiten zur Wertschätzung städtischer Freiräume in Kassel. In B. Nienaber & U. Roos (Eds.), *Internationalisierung der Gesellschaft und die Auswirkungen auf die Raumentwicklung. Beispiele aus Hessen, Rheinland-Pfalz und dem Saarland* (Arbeitsberichte der ARL, vol. 13, pp. 54–71). Hannover: Selbstverlag.

Buchroithner, M. F. (2005). Interactive real-time VR cartography. *Proceedings of the 22nd International Cartographic Conference (ICC), 22.* https://www.cartesianos.com/geodoc/ icc2005/pdf/oral/TEMA15/Session%205/MANFRED%20BUCHROITHNER.pdf.

Cities and Memory Contributors (2020). *Cities and memory. A global, collaborative sound project. Remixing the world, one sound at a time.* https://citiesandmemory.com/.

DiBiase, D., MacEachren, A. M., Krygier, J. B., & Reeves, C. (1992). Animation and the role of map design in scientific visualization. *Cartography and Geographic Information Science, 19*(4), 201–214. https://doi.org/10.1559/152304092783721295.

Dickmann, F. (2018). *Kartographie*. Braunschweig: Westermann.

Dodt, J., Bestgen, A.-K., & Edler, D. (2017). Ansätze der Erfassung und kartographischen Präsentation der olfaktorischen Dimension. *KN—Journal of Cartography and Geographic Information, 67*(5), 245–256. https://doi.org/https://doi.org/10.1007/BF03545321.

Edler, D. (2020). Where spatial visualization meets landscape research and "Pinballology": examples of landscape construction in pinball games. *KN—Journal of Cartography and Geographic Information*, online first: https://doi.org/https://doi.org/10.1007/s42489-020-00044-1.

Edler, D., Keil, J., & Dickmann, F. (2020). From Na Pali to Earth—An 'unreal' engine for modern geodata? In D. Edler, C. Jenal, & O. Kühne (Eds.), *Modern approaches to the visualization of landscapes* (pp. 279–291). Wiesbaden: Springer VS.

Edler, D., & Dodt, J. (2010). Eine audio-visuelle Flash-Applikation zur Fremdsprachenvermittlung. *Kartographische Nachrichten, 60*(5), 276–278.

Edler, D., Husar, A., Keil, J., Vetter, M., & Dickmann, F. (2018a). Virtual reality (VR) and open source software: A workflow for constructing an interactive cartographic VR environment to explore urban landscapes. *KN—Journal of Cartography and Geographic Information, 68*(1), 3–11. https://doi.org/https://doi.org/10.1007/BF03545339.

Edler, D., Kühne, O., Jenal, C., Vetter, M., & Dickmann, F. (2018b). Potenziale der Raumvisualisierung in Virtual Reality (VR) für die sozialkonstruktivistische Landschaftsforschung. *KN—Journal of Cartography and Geographic Information, 68*(5), 245–254. https://doi.org/https://doi.org/10.1007/BF03545421.

Edler, D., & Dickmann, F. (2017). The impact of 1980s and 1990s video games on multimedia cartography. *Cartographica, 52*(2), 168–177. https://doi.org/10.3138/cart.52.2.3823.

Edler, D., & Dickmann, F. (2016). Interaktive Multimediakartographie in frühen Videospielwelten—Das Beispiel "Super Mario World". *KN—Journal of Cartography and Geographic Information, 66*(2), 51–58. https://doi.org/https://doi.org/10.1007/BF03545205.

Edler, D., Kühne, O., Keil, J., & Dickmann, F. (2019). Audiovisual cartography: Established and new multimedia approaches to represent soundscapes. *KN—Journal of Cartography and Geographic Information*, 69(1), 5–17. https://doi.org/https://doi.org/10.1007/s42489-019-00004-4.

Edler, D., Keil, J., Wiedenlübbert, T., Sossna, M., Kühne, O., & Dickmann, F. (2019). Immersive VR experience of redeveloped post-industrial sites: The example of "Zeche Holland" in Bochum-Wattenscheid. *KN—Journal of Cartography and Geographic Information, 69*(4), 267–284. https://doi.org/10.1007/s42489-019-00030-2.

Edler, D., & Kühne, O. (2019). Nicht-visuelle Landschaften. In O. Kühne, F. Weber, K. Berr, & C. Jenal (Eds.), *Handbuch Landschaft* (pp. 599–612). Wiesbaden: Springer VS.

Edler, D., Jebbink, K., & Dickmann, F. (2015). Einsatz audio-visueller Karten in der Schule—Eine Unterrichtsidee zum Strukturwandel im Ruhrgebiet. *KN—Journal of Cartography and Geographic Information, 65*(5), 259–265. https://doi.org/https://doi.org/10.1007/BF03545162.

Edler, D., Lammert-Siepmann, N., & Dodt, J. (2012). Die akustische Dimension in der Kartographie—Eine Übersicht. *KN—Journal of Cartography and Geographic Information, 62*(4), 185–195.

Edler, D. (2011). Visualising intimate spaces—The example of "Bolderāja inside-out(side)". *KN—Journal of Cartography and Geographic Information, 61*(4), 171–177. https://doi.org/https://doi.org/10.1007/BF03544495.

Fontaine, D. (2020a). Landscape in computer games—The examples of GTA V and watch dogs 2. In D. Edler, C. Jenal, & O. Kühne (Eds.), *Modern approaches to the visualization of landscapes* (pp. 293–306). Wiesbaden: Springer VS.

Fontaine, D. (2020b). Virtuality and landscape. In D. Edler, C. Jenal, & O. Kühne (Eds.), *Modern approaches to the visualization of landscapes* (pp. 267–278). Wiesbaden: Springer VS.

Gomez, S., Dominguès, C., Aumond, P., Lavandier, C., Palka, G., & Serrhini, K. (2017). Cartographic representation of soundscape: Proposals and assessment. In M. Leitner & J. J. Arsanjani (Eds.), *Citizen empowered mapping* (pp. 27–51). Cham: Springer.

Grimshaw, M. (2014). Sound. In M. J. P. Wolf & B. Perron (Eds.), *The Routledge companion to video game studies* (pp. 117–124). New York: Routledge.

Harding, C., & Souleyrette, R. R. (2009). Investigating the use of 3D graphics, haptics (touch), and sound for highway location planning. *Computer-Aided Civil and Infrastructure Engineering, 25*(1), 20–38. https://doi.org/10.1111/j.1467-8667.2008.00591.x.

Harnapp, V. R., & Noble, A. G. (1987). Noise pollution. *GeoJournal, 14*(2), 217–226. https://doi.org/10.1007/BF00435812.

Herman, L., & Řezník, T. (2014). Web 3D visualization of noise mapping for extended INSPIRE building models. In J. Hřebíček, G. Schimak, M. Kubásek & A. E. Rizzoli (Eds.), *Environmental software systems. Fostering information sharing.* (ISESS 2013. IFIP advances in information and communication technology, vol. 413, pp. 414–424). Berlin: Springer.

Hochschild, V., Braun, A., Sommer, C., Warth, G., & Omran, A. (2020). Visualizing landscapes by geospatial techniques. In D. Edler, C. Jenal, & O. Kühne (Eds.), *Modern approaches to the visualization of landscapes* (pp. 47–78). Wiesbaden: Springer VS.

Hruby, F. (2019). The sound of being there: Audiovisual cartography with immersive virtual environments. *KN—Journal of Cartography and Geographic Information, 69*(1), 19–28. https://doi.org/https://doi.org/10.1007/s42489-019-00003-5.

Hruby, F., Ressl, R., & de la Borbolla del Valle, G. (2019). Geovisualization with immersive virtual environments in theory and practice. *International Journal of Digital Earth, 12*(2), 123–136. https://doi.org/10.1080/17538947.2018.1501106.

Hunziker, M., Felber, P., Gehring, K., Buchecker, M., Bauer, N., & Kienast, F. (2008). Evaluation of landscape change by different social groups. Results of two empirical studies in Switzerland. *Mountain Research and Development, 28*(2), 140–147. https://doi.org/https://doi.org/10.1659/mrd.0952.

Jobst, M., & Germanchis, T. (2007). The employment of 3D in cartography—An overview. In W. Cartwright, M. P. Peterson, & G. Gartner (Eds.), *Multimedia cartography* (2nd ed., pp. 217–228). Berlin: Springer.

Kazig, R. (2019). Atmosphären und Landschaft. In O. Kühne, F. Weber, F. Berr, & C. Jenal (Eds.), *Handbuch Landschaft* (pp. 453–460). Wiesbaden: Springer VS.

Kazig, R. (2013). Landschaft mit allen Sinnen—Zum Wert des Atmosphärenbegriffs für die Landschaftsforschung. In D. Bruns & O. Kühne (Eds.), *Landschaften: Theorie, Praxis und internationale Bezüge* (pp. 221–232). Schwerin: Oceano.

Kersten, T. P., Tschirschwitz, F., & Deggim, S. (2017). Development of a virtual museum including a 4D presentation of building history in virtual reality. In D. Aguilera, A. Georgopoulos, T. Kersten, F. Remondino, & E. Stathopoulou, (Eds.), *The international archives of the photogrammetry, remote sensing and spatial information sciences* (Conference Paper, XLII-2/W3, 3D virtual reconstruction and visualization of complex architectures, pp. 361–367).

Kersten, T. P., Deggim, S., Tschirschwitz, F., Lindstaedt, M., & Hinrichsen, N. (2018). Segeberg 1600—Eine Stadtrekonstruktion in Virtual Reality. *KN—Journal of Cartography and Geographic Information, 68*(4), 183–191. https://doi.org/https://doi.org/10.1007/BF03545360.

Kleber, A., Edler, D., & Dickmann, F. (2020). Cartography and the sea: A javascript-based web mapping application for managing maritime shipping. In D. Edler, C. Jenal, & O. Kühne (Eds.), *Modern approaches to the visualization of landscapes* (pp. 173–186). Wiesbaden: Springer VS.

Kornfeld, A.-L. (2008). Die kartographische Visualisierung des akustischen Raums. *Kartographische Nachrichten, 58*(6), 294–301.

Koussoulakou, A., & Kraak, M.-J. (1992). Spatia-temporal maps and cartographic communication. *The Cartographic Journal, 29*(2), 101–108.

Kraak, M.-J., & Ormeling, F. (2010). *Cartography. visualization of spatial data* (3rd ed.). Pearson: The Guilford PUBN.

Krygier, J. B. (1994). Sound and geographic visualization. In A. M. MacEachren & D. R. F. Taylor (Eds.), *Visualization in modern cartography* (pp. 149–166). Oxford: Elsevier.

Kühne, O. (2018a). *Landschaft und Wandel. Zur Veränderlichkeit von Wahrnehmungen.* Wiesbaden: Springer VS.

Kühne, O. (2018b). *Landschaftstheorie und Landschaftspraxis. Eine Einführung aus sozialkonstruktivistischer Perspektive* (2., aktualisierte und überarbeitete Aufl.). Wiesbaden: Springer VS.

Kühne, O. (2019). *Landscape theories. A brief introduction.* Wiesbaden: Springer VS.

Kühne, O. (2020a). Landscape conflicts. A theoretical approach based on the three worlds theory of Karl Popper and the conflict theory of Ralf Dahrendorf, illustrated by the example of the energy system transformation in Germany. *Sustainability, 12*(17), 1–20. https://doi.org/10.3390/su12176772.

Kühne, O. (2020b). The social construction of space and landscape in internet videos. In D. Edler, C. Jenal, & O. Kühne (Eds.), *Modern approaches to the visualization of landscapes* (pp. 121–137). Wiesbaden: Springer VS.

Kühne, O., & Jenal, C. (2020). The threefold landscape dynamics—Basic considerations, conflicts and potentials of virtual landscape research. In D. Edler, C. Jenal, & O. Kühne (Eds.), *Modern Approaches to the visualization of landscapes* (pp. 389–402). Wiesbaden: Springer VS.

Laakso, M., & Sarjakoski, L. T. (2010). Sonic maps for hiking—Use of sound in enhancing the map use experience. *The Cartographic Journal, 47*(4), 300–307. https://doi.org/10.1179/00087 0410X12911298276237.

Lammert-Siepmann, N., Bestgen, A.-K., Edler, D., Kuchinke, L., & Dickmann, F. (2017). Audiovisual communication of object-names improves the spatial accuracy of recalled object-locations in topographic maps. *PLoS ONE, 12*(10), e0186065. https://doi.org/10.1371/journal.pone.0186065.

Lammert-Siepmann, N., Edler, D., Redecker, A. P., & Jürgens, C. (2011). Designing teaching units via WebGIS: Remotely sensed imagery in the language classroom. *EARSeL eProceedings, 10*(2), 149–158.

Lauriault, T. P., & Lindgaard, G. (2006). Scented cybercartography: Exploring possibilities. *Cartographica, 41*(1), 73–92. https://utpjournals.press/doi/https://doi.org/10.3138/W432-713U-3621-04N3.

Lavandier, C., Aumond, P., Gomez, S., & Dominguès, C. (2016). Urban soundscape maps modelled with geo-referenced data. *Noise Mapping, 3*(1), 278–294. https://doi.org/10.1515/noise-2016-0020.

Linke, S. (2020). Landscape in internet pictures. In D. Edler, C. Jenal, & O. Kühne (Eds.), *Modern approaches to the visualization of landscapes* (pp. 139–156). Wiesbaden: Springer VS.

Loda, M., Kühne, O., & Puttilli, M. (2020). The social construction of Tuscany in the German and English Speaking World—Presented by the analysis of internet images. In D. Edler, C. Jenal, & O. Kühne (Eds.), *Modern approaches to the visualization of landscapes* (pp. 157–171). Wiesbaden: Springer VS.

Lovett, A., Appleton, K., Warren-Kretzschmar, B., & von Haaren, C. (2015). Using 3D visualization methods in landscape planning: An evaluation of options and practical issues. *Landscape and Urban Planning, 142,* 85–94. https://doi.org/10.1016/j.landurbplan.2015.02.021.

Ma, Y., Wright, J., Gopal, S., & Phillips, N. (2020). Seeing the invisible: From imagined to virtual urban landscapes. *Cities, 98,* 102559. https://doi.org/10.1016/j.cities.2019.102559.

Maisonneuve, N., Stevens, M., & Ochab, B. (2010). Participatory noise pollution monitoring using mobile phones. *Information Polity, 15*(1/2), 51–71. https://doi.org/10.3233/IP-2010-0200.

McLean, K. (2020). Temporalities of the smellscape: Creative mapping as visual representation. In D. Edler, C. Jenal, & O. Kühne (Eds.), *Modern approaches to the visualization of landscapes* (pp. 217–246). Wiesbaden: Springer VS.

McLean, K. (2016). Mapping the city's smellscapes. In K. Harmon (Ed.), *You are here: NYC. Mapping the soul of the city* (pp. 144–147). New York: Princeton Architectural Press.

Meyer-Heß, F. (2020). Discovering forgotten landscapes. In D. Edler, C. Jenal, & O. Kühne (Eds.), *Modern approaches to the visualization of landscapes* (pp. 33–46). Wiesbaden: Springer VS.

Moellering, H. (1980). The real-time animation of three-dimensional maps. *the American Cartographer, 7*(1), 67–75. https://doi.org/10.1559/152304080784522892.

Monmonier, M. (1990). Strategies for the visualization of geographic time-series data. *Cartographica, 27*(1), 30–45. https://doi.org/10.3138/U558-H737-6577-8U31.

Morrison, J. L., & Ramirez, J. R. (2001). Integrating audio and user-controlled text to query digital databases and to present geographic names on digital maps and images. *Proceedings of the 20th International Cartographic Conference, Beijing, China.* https://icaci.org/files/documents/ICC_proceedings/ICC2001/icc2001/file/f12005.doc.

Müller, J.-C., Scharlach, H., & Jäger, M. (2001). Der Weg zu einer akustischen Kartographie. *Kartographische Nachrichten, 51*(1), 26–40.

Müller, J.-C., & Scharlach, H. (2001). Noise abatement planning—Using animated maps and sound to visualise traffic flows and noise pollution. *Proceedings of the 20th International Cartographic Conference (ICC), Beijing, China, 1*, 375–385.

Nitsche, M. (2008). *Video game spaces. Image, play, and structure in 3D worlds.* Cambridge: MIT Press.

Papadimitriou, F. (2010). A "neogeographical education"? The geospatial web, GIS and digital art in adult education. *International Research in Geographical and Environmental Education, 19*(1), 71–74. https://doi.org/10.1080/10382041003602969.

Papadimitriou, F. (2020). Visualization of future landscapes, postmodern cinema and geographical education. In D. Edler, C. Jenal, & O. Kühne (Eds.), *Modern approaches to the visualization of landscapes* (pp. 351–369). Wiesbaden: Springer VS.

Papadimitriou, K. D., Mazaris, A. D., Kallimanis, A. S., & Pantis, J. D. (2009). Cartographic representation of the sonic environment. *The Cartographic Journal, 46*(2), 126–135. https://doi.org/10.1179/000870409X459842.

Peterson, M. P. (1993). Interactive cartographic animation. *Cartography and Geographic Information Systems, 20*(1), 40–44. https://doi.org/10.1559/152304093782616724.

Peterson, M. P. (1995). *Interactive and animated cartography.* Upper Saddle River: Prentice Hall.

Prisille, C., & Ellerbrake, M. (2020). Virtual reality (VR) and geography education: Potentials of 360° 'experiences' in secondary schools. In D. Edler, C. Jenal, & O. Kühne (Eds.), *Modern approaches to the visualization of landscapes* (pp. 321–332). Wiesbaden: Springer VS.

Pulsifer, P. L., Caquard, S., & Taylor, D. R. F. (2007). Toward a new generation of community atlases—The cybercartographic atlas of Antarctica. In W. Cartwright, M. P. Peterson, & G. Gartner (Eds.), *Multimedia cartography* (2nd ed., pp. 195–216). Berlin: Springer.

Radicchi, A. (2009). *Firenze sound map.* https://www.firenzesoundmap.org/default.asp.

Ricciardi, P., Delaitre, P., Lavandier, C., Torchia, F., & Aumond, P. (2015). Sound quality indicators for urban places in Paris cross-validated by Milan data. *The Journal of the Acoustical Society of America, 138*(4), 2337–2348. https://doi.org/10.1121/1.4929747.

Salter, J. D., Campbell, C., Journeay, M., & Sheppard, S. R. J. (2009). The digital workshop: Exploring the use of interactive and immersive visualisation tools in participatory planning. *Journal of Environmental Management, 90*(6), 2090–2101. https://doi.org/10.1016/j.jenvman.2007.08.023.

Schafer, R. M. (1977). *The soundscape. Our sonic environment and the tuning of the world.* Rochester: Destiny Books.

Scharlach, H. (2002). *Lärmkarten. Kartographische Grundlagen und audiovisuelle Realisierung.* Dissertation, Ruhr-University Bochum.

Schiewe, J. (2015). Physiological and cognitive aspects of sound maps for representing quantitative data and changes in data. In J. Brus, A. Vondrakova, & V. Vozenilek (Eds.), *Modern trends in cartography. Selected papers of CARTOCON 2014* (pp. 315–324). Heidelberg: Springer.

Schito, J., & Fabrikant, S. I. (2018). Exploring maps by sounds: Using parameter mapping sonification to make digital elevation models audible. *International Journal of Geographical Information Science, 32*(5), 874–906. https://doi.org/10.1080/13658816.2017.1420192.

Siepmann, N., Edler, D., & Dickmann, F. (2019). A software tool for the experimental investigation of cognitive effects in audiovisual maps. *KN—Journal of Cartography and Geographic Information, 69*(1), 29–39. https://doi.org/https://doi.org/10.1007/s42489-019-00005-3.

Siepmann, N., Edler, D., Keil, J., Kuchinke, L., & Dickmann, F. (2020). The position of sound in audiovisual maps: An experimental study of performance in spatial memory. *Cartographica, 55*(2), 136–147. https://doi.org/10.3138/cart-2019-0008.

Stintzing, M., Pietsch, S., & Wardenga, U. (2020). How to teach "landscape" through games? In D. Edler, C. Jenal, & O. Kühne (Eds.), *Modern approaches to the visualization of landscapes* (pp. 333–349). Wiesbaden: Springer VS.

Stoter, J., de Kluijver, H., & Kurakula, V. (2008). 3D noise mapping in urban areas. *International Journal of Geographical Information Science, 22*(8), 907–924. https://doi.org/10.1080/13658810701739039.

Stotten, R. (2015). *Das Konstrukt der bäuerlichen Kulturlandschaft. Perspektiven von Landwirten im Schweizerischen Alpenraum* (alpine space—man & environment, vol. 15). Innsbruck: Innsbruck University Press.

Tschaikner, E., Scheib, C., Niedermayr, S., & Zimmer, F. (2013). *Personal soundscapes. Soundmap.* https://personal-soundscapes.mur.at/de/soundmap.

Upton, D. (2007). Sound as landscape. *Landscape Journal, 26*(1), 24–35. https://doi.org/10.3368/lj.26.1.24.

Vetter, M. (2020). Technical potentials for the visualization in virtual reality. In D. Edler, C. Jenal, & O. Kühne (Eds.), *Modern approaches to the visualization of landscapes* (pp. 307–317). Wiesbaden: Springer VS.

Vetter, M. (2019). 3D-Visualisierung von Landschaft—Ein Ausblick auf zukünftige Entwicklungen. In O. Kühne, F. Weber, K. Berr, & C. Jenal (Eds.), *Handbuch Landschaft* (pp. 199–214). Wiesbaden: Springer.

Weninger, B. (2014). A framework for color design of digital maps: An example of noise maps. In J. Brus, A. Vondrakova, & V. Vozenilek (Eds.), *Modern trends in cartography. Selected papers of CARTOCON 2014* (pp. 103–116), Heidelberg: Springer.

Winkler, J. (2005). Raumzeitphänomen Klanglandschaften. In V. Denzer, J. Hasse, K.-D. Kleefeld & U. Recker (Eds.), *Kulturlandschaft. Wahrnehmung—Inventarisation—regionale Beispiele* (Kulturlandschaft, vol. 14, pp. 77–88). Bonn: Habelt.

Winkler, J. (2006). *Klanglandschaften. Untersuchungen zur Konstitution der klanglichen Umwelt in der Wahrnehmungskultur ländlicher Orte in der Schweiz.* Basel: Akroama.

Zheng, Y., Liu, T., Wang, Y., Zhu, Y., Liu, Y., & Chang, E. (2014). Diagnosing New York city's noises with ubiquitous data. *Proceedings of the 2014 ACM International Joint Conference on Pervasive and Ubiquitous Computing, Seattle, WA, USA, 13–17 September*, 715–725.

Zuo, J., Xia, H., Liu, S., & Qiao, Y. (2016). Mapping urban environmental noise using smartphones. *Sensors, 16*(10), 1692. https://doi.org/10.3390/s16101692.

Nils Siepmann studied Geography (B.Sc., 2006–2009 and M.Sc., 2009–2011) at Ruhr-University Bochum, after a semester of psychology, philosophy and Scottish History at the University of Sterling. He is currently responsible for geodata management and cartographic design at the Archdiocese of Cologne and working on his doctoral thesis on audiovisual cartography. His regional focus is on the South Caucasus, where he took part in a study program at Ilia State University in Tbilisi, Georgia. Passionate about finding new ways to facilitate open data and free and open source software, he is also an associate lecturer at the Geography Department at Ruhr-University Bochum and a lecturer in vocational training in the field of Geomatics.

Dennis Edler studied English Philology and Geography (B.A., 2006–2009) at the Ruhr-University Bochum (RUB) and University College Cork (UCC). He also spent several study periods at other European universities and research institutes: Regent's Park College (Univ. of Oxford), Riga Technical University (RTU), Mediterranean Agronomic Institute of Chania (MAICh) and German Aerospace Center (DLR). As a master student (M.Sc., 2009–2011), he specialized in Cartography, GIS and Remote Sensing. In his doctoral thesis (2012–2015), he conducted experimental studies on the effects of artificial map elements, such as grids, on the accuracy of spatial memory. Since 2015, he holds a permanent staff position as a senior lecturer at the Ruhr-University Bochum (Geography Department). His current research and teaching activities involve modern analytical and visualization methods based on open source software and open geodata. Email: dennis.edler@rub.de

Olaf Kühne, born in 1973, is Professor of Urban and Regional Development at the University of Tübingen. He received his doctorate in geography (on an urban climatological topic) in Saarbrücken and sociology (on the social construction of landscape) in Hagen. His habilitation in Mainz dealt with ecological and social transformation processes in Poland. Today, he is particularly concerned with social issues of energy system transformation, landscape theory formation and urban transformation processes. These are the focus of his interest in feedback relationships between the material, individual and social world.

Part V
Modern Virtual Landscapes

Virtuality and Landscape

Dominique Fontaine

Abstract

The 21st century is marked by an increasing number of brilliant technological inventions, simulation in the context of virtuality being one of them. Nowadays, the entering of virtual landscapes in various domains of everyday life is quite substantial and that is why interactions between virtuality and landscape constitute both an interesting and legitimate field of research. While focusing on virtuality and landscape against the background of postmodernity and social constructivism, the complexity of the interplay between landscape, aesthetics and atmosphere becomes obvious. To what extend landscape can be artificially manipulated is shown in any kind of computer games. The link to escapism gives proof of social needs and pictures socio-cultural structures. The following article gives a synopsis about the interdependence of postmodernity and social constructivism with regard to landscape and draws conclusions on how cultural and social backgrounds are reflected in virtuality.

Keywords

Virtuality · Landscape · Aesthetics · Social constructivism · Atmosphere · Post-modernism · Escapism · Computer games

D. Fontaine (✉)
Warndt-Gymnasium Völklingen, Völklingen, Germany

© Springer Fachmedien Wiesbaden GmbH, part of Springer Nature 2020
D. Edler et al. (eds.), *Modern Approaches to the Visualization of Landscapes*, RaumFragen: Stadt – Region – Landschaft,
https://doi.org/10.1007/978-3-658-30956-5_14

1   Introduction

In times of global technological competition, virtuality—especially in the context of computer-based programs as well as computer games—gain high importance (see also in this volume: Berr 2020; Edler et al. 2020a, b; Kühne and Jenal 2020a; Papadimitriou 2020; Prisille and Ellerbrake 2020; Siepmann et al. 2020; Stintzing et al. 2020; Thomas 2020; Vetter 2020). The idea of landscape seems to be directly linked to this phenomenon: Virtual landscapes dominate our daily lives in many ways, as for example based on satellite images, providing information for safety, environment, politics etc. This shows that the scientific analysis of landscape is of prominent interest for several actors (e.g. politics, science etc.) and has in this sense a qualification for being highlighted in geographical studies, too. Different questions are of geographical as well as of sociologist interest, for example: In what way does virtuality affect our state of mind and perception? As current scientific issues show, there is a huge and ongoing discourse about landscape that is reflected against the background of virtuality (Hokema 2012, S. 231; Kühne et al. 2018, S. 22). The focus is on the simulation and the digitalization of data and landscape (Krecklau et al. 2014, S. 201) in order to serve several branches (e.g. tourism, urban planning, gaming etc.).

The own scientific approach and tradition relies on the social constructivism (see also in this volume: Al-Khanbashi 2020; Bellini and Leonardi 2020; Berr 2020; Fontaine 2020; Jenal 2020; Kühne and Jenal 2020b; Loda et al. 2020; Roßmeier 2020; Weber 2020). Space is—following this angle—seen as a construct of which one is internally aware and not kind of an external 'truth'. But how is space—and so on landscape—perceived by society? What kind of circumstances might influence the awareness and perception of landscape? Is there a link to ethnical backgrounds?

This article will give an overview of landscape both as a physical/geographical and social phenomenon against the background of its virtualization in postmodern times.

2   Theoretical Frame: The Importance of Landscape Research for Geographical Studies

Given that (social constructivist) geography examines space in all of its meanings (e.g. the constitution of space, differences throughout time and space according to dominating social structures) it seems necessary to have a closer look at this huge phenomenon giving a frame to all kind of life, called space. The social constructivist approach—on which is based this article—focuses on human influences on space, therefore called landscape.[1]

---

[1]In this article, landscape is defined as a social construct and as a derivative of space. Space is understood as the frame giving instance, while landscape is a socially constructed phenomenon within space. The process of construction is seen as a result of an experience of space which goes ahead. (Kühne 2018a, p. 15).

First of all, there is the question about the character of landscape: what is the exactly meaning of landscape? Who is to define what landscape is all about? Does its meaning differ from other perspectives according to different social or ethnic backgrounds? Such questions are an important part of geographically based space studies. The fact that social behavior goes hand in hand with the conditioning of landscape—and so has to be analyzed—was pointed out by Gimblett and Skov-Petersen (2008, p. 425) claiming that "there is a growing body of theoretical and applied research focused within the context of human–environment interactions." These so called "human–environment interactions" (Gimblett and Skov-Petersen 2008, p. 425) are reflected by landscape. Those can be seen as a physical mirror of human actions in combination with the physical environment.

2.1 The Historical Background of Landscape Research

In view of the historical development of landscape discourse in science, it has to be pointed out that the history of landscaping in a geographical way is rather a recent phenomenon as landscapes are often seen as exclusively physical and therefore not to question (Kühne 2018a, p. 135; see in this volume: Schenk 2020).

Landscape research analyzes how and why space is affected by humans. Research has shown that modifications patterns highly depend on social structures as well as historical events giving direction to what society needs. As a result, it can be concluded that space—and so landscape—has to be seen as a flexible human construct which can changed or manipulated any time. Of course, there is the question about the motivation to form or change landscape. First of all, landscape should respond to any human need. Besides, especially in the postmodern times, there is a frequent call for aesthetics. Objects as well as consequently landscape should not only satisfy basic human needs but they should fit socially defined patterns of aesthetics. That is why Wöbse (2002, p. 115) sees landscape as an aesthetic object nowadays.

Another postmodern approach consists in understanding landscape as a medium of power and potency (Kühne 2008, 2018b). By manipulating landscape (physically and virtually), the (technological and economic) power of human mankind becomes obvious. This scientific angle is rather a recent one and represents therefore a current topic in geographical studies, especially in social constructivism. Cosgrove (2008, p. 18) accentuates that "landscape emerge from specific geographical, social and cultural circumstances, that landscape is embedded in the practical uses of the physical world [...]. Cosgrove's wording emphasizes what the cultural turn pronates: landscaping is to be seen as a process that highly depends on social and cultural structures which leads directly to further thoughts about social constructivism in the following chapter.

2.2   Reflections About Landscape in Regard of Social Constructivism

As soon as landscape is discussed against a social constructivist background, it is seen as a product of human processes and actions. Landscapes are to be understood as social constructs of high variability reflecting social phenomenon as well as the society's system of values. This way of reading landscape today is put forward by Baralou and Shepherd (2008, p. 1) both referring to Morse (1998): "Virtuality is a socially constructed reality mediated by electronic media." The idea that landscape can be seen as a human product goes already back to the nineties when Hillis (1996, p. 64) mentioned: "Specific landscape forms represent specific cultural understandings constituted through specific social relations, times, and places."

The focus on such 'cultural understandings' is based on the 'cultural turn' (Dear 2001) and points out the social aspect in reading—and finally creating—landscape. With a closer look at the timescale (as Hillis' quotation shows), it seems logical that this social constructivist approach correlates with a new era of reading and understanding landscape (Kühne 2018b, 2019): postmodernity. Before the paradigm shift to postmodernity, a clear-cut understanding of what landscape has to be defined as ruled the landscape discourse. The modern times were emblematic for straight structures, for example in architecture (e.g. 'form follows function'), which went hand in hand to human influenced landscape. The following chapter will focus on the postmodern way to read landscape.

2.3   The Interpretation of Landscape in Postmodern Times

First of all, the terminus of postmodernity has to be explained in order to make clear in what way it influenced the recent scientific discourse about landscape. The postmodern times constitute in some sense an evolution of modern thoughts. Even if the paradigm shift towards postmodernity evokes the image of a kind of a rupture, it developed many modern ideas in a new direction. Postmodern keywords to mention are in this context diversity, flexibility, irony, fragmented space and simulation (Kühne 2012, S. 70; Baudrillard 1994; Bauman 1995).

Following Baudrillard's (1989, 1994) logic of postmodern simulation, the social constructivist approach gains a new dimension: starting from the point of view that humans might influence landscape by their actions and being aware of the fact that our actual century is strongly dominated by virtual simulations in any imaginable branch or domain of everyday life, one has to become aware of the undeniable link between simulation (of landscape) and postmodernity. Besides the technological progress that allows to create veritable 'virtual realities' or 'virtual landscapes', the freedom of thoughts and of aesthetics are of highest importance. There aren't any longer structural or ideological boundaries avoiding several shaping and designs of landscape. The guiding principle of the postmodern times is 'Form follows fun'—not exclusively in architectural issues. To speak in metaphors, landscape can be understood as a white piece of paper on which society prints and designs whatever they want to or need. The most obvious example for this kind of 'landscaping' is Disneyland (Fontaine 2017a, b). But next to those built landscapes in cities as Paris or Anaheim, there are lots of virtual spaces created according to social wishes. Computer games that come up with virtual landscapes allow users to do nearly whatever they want to and this without any consequence. In this sense, the postmodern landscape is a flexible, dynamic, ironic and iconic stage for human operations or storylines.

3   Virtuality: Another Approach to Paradise?

Today, the number of sold virtual reality classes and VR games increases steadily as a proof of high demand of virtuality in daily life and of course as a sign of cartographic and geographic works on 3D modeling techniques (Edler et al. 2018a, b; Vetter 2019; see in this volume: Edler et al. 2020b; Hochschild et al. 2020; Siepmann et al. 2020; Vetter 2020). Not surprisingly that we have to face the question why. Why do people want that much virtuality in their lives? Do they behave in another way as soon as they act in virtual settings? If so, why? As interviews with users (for example of GTA V) show (Fontaine 2017a), the fact of being anonymous and acting in a completely virtual world allows the users to act totally different compared to 'real' (in the sense of non-virtual) or 'everyday' life. Of course, actions in virtual games do not affect their 'real' personal lives (e.g. social status), because they won't be arrested for example for having murdered someone. Virtuality creates a space of freedom in which borders have to be newly marked. Qvortrup (2002, p. 6) refers to Heim's point of view (1993) and considers that virtual space "is a space existing in parallel to 'real space' with its own, specific laws of existence". These so-called specific laws of existence include—especially with regard to computer games—deviant behavior and even forbidden actions that would lead directly to sensitive punishment in everyday life. The fact that landscapes are artificially created in a very realistic way allows users to fell free trying to test new limits without running the risk to be arrested or punished by law for example. Still the question remains: why is there seemingly the necessity of breaking borders in the context of (virtual) games? Is it the unbreakable frame of everyday life where we all need to function without any fail, without any space for irony or fun? The pressure to perform seems to be omnipresent in huge parts of society. Nevertheless, there is a need for freedom, a ludic drive that wants to be satisfied as well. The pursuit of perfection and social recognition may lead to a loss of freedom. This is exactly the point where virtual reality and virtual landscapes offer kind of a 'safe stage' on which one can behave in unusual ways, try new patterns, maybe fail in doing something without any dramatic consequence. This is why it can be stated that the success of virtuality is linked to escapism.

3.1   Virtuality in the Context of Postmodernity

On one hand, virtuality reflects technological progress, on the other hand, it gives proof of social demands and needs. This is why Brown et al. (2002, p. 96) state:

> "The demand for information has escalated alongside both the capacity for the collection of data on landscapes, and the requirements of the end-user, potentially placing VR tools in a key position to aid in the description, explanation and communication of often sophisticated concepts to audiences of differing reference levels and capabilities with regard to landscape-related issues."

Following this logic, the creation of virtual spaces can be attributed to the postmodern times in which simulation and codes play a prominent role. While copying physical landscapes and transmitting them into virtuality, there is a process of simulation going on which is emblematic for postmodernism. Patterns and structures are simulated and obey to several codes. These codes are read and classified by society and that is the moment where aesthetics and atmospheres become relevant. Who is to decide whether an object or a landscape is beautiful/picturesque/ugly or sublime? The answer seems to be both simple and complex: it depends on social structures, ethnic and cultural backgrounds, moral point of views and familiar structures. The complexity is due to the huge variability of social and cultural backgrounds as well as personal perceptions. Therefore Luhmann, a radical constructivist, (2009, p. 11) pleads for "indexes" as means of description that help to decode our physical and social environment. Postmodernity is well known for its complexity and variety of meanings. As a consequence, it seems necessary to hold on to certain keys of lecture in order to read or literally decode systems—or even landscape. Kühne et al. (2019, p. 124) reference to Luhmann's approach while stating that the process of reading landscape in aesthetic codes goes back to "self-referential systems".

3.2   Virtual Effects: Creating 'Real' Atmospheres Artificially

By creating landscapes artificially or virtually, several skills come together and create kind of a symbiose. To mention are technological skills as well as scientific knowledge that are necessary to generate atmospheres. In addition to that, competencies in art are required to realize what phantasm created abstractly. It can thus be stated that there is an interface between different skills that go hand in hand when designing virtual landscapes and atmospheres.

The cultural turn[2] that is emblematic for postmodernity incorporates the idea that landscapes should create certain atmospheres. The postmodern plea consists in "an

---

[2]The term 'cultural turn' goes back to the presumption that social living is defined by culture and its influences (Susen 2015) and that there is a certain "vision of pluralism" which is emblematic for the postmodern times (Jameson 1998, p. 50).

'anything goes' pluralist approach to urban design" (Turner 1996, p. 8). This creative freedom allows to evoke special atmospheres that appeals to different ways of perception: watching, hearing, feeling, tasting and even the sense of balance (Weidinger 2018, p. 26). Hauskeller (1995) stresses the affective component of atmospheres in addition to the named characteristics. In the chart !, two reactions of GTA V-users[3] show to what extend virtual effects can thrill:

Chart 1: GTA V user's quotations about virtual effects within the game. (Source: own presentation)

User's name	Date of publication	Quotation
Susann	12.05.2015	" [...} The graphics are overwhealming. I find it realistic when they speak English, it gives a certain feeling to the game. The sound effects while having a phone call or the police radio via the controller amaze me. I also see a red and blue reflecting controller on the TV screen as soon as the policemen discover me. [...]"
Dnil.mjrn	18.12.2016	"The created virtual city is amazing!"

Taking into account those examples, the importance of successful virtual effects become obvious: only by convincing the user by realistic landscaping, atmospheres are created and experienced. Saggio (2007, p. 398) claims:

"In a word, 'mental landscape' refers to the fact that architects of the new generation are working to make an architecture that draws upon certain aspects and characteristics of the virtual world."

According to Saggio's logic, not only 'real' (in the sense of physical) landscapes give birth to virtual ones but also vice versa. Social needs that can easily be traduced into virtual landscapes should now be respected in architectural questions.

3.3   Motivation for Virtuality: A Plea for Escapism?

As radical as it might seem, there is one question to be responded: even though we live in an affluent society with uncountable options, is it possible that there a basal things missing so that the demand of virtual space where human beings can do whatever they want to doesn't stop to increase? Does everyday life with all its norms, stated values

---

[3]Source: https://www.amazon.de/Grand-Theft-Auto-Standard-PlayStation/product-reviews/B00K W2FKAQ/ref=cm_cr_getr_d_paging_btm_next_2?ie=UTF8&reviewerType=all_reviews&pageN umber=2&filterByStar=five_star (13.11.2019). *Notice: The orthograph has not been modified by the author.*

and laws encourage us to escape towards spheres of freedom? There might not be one single clear-cut answer to those questions, but apparently there is a need in society to escape from daily struggle (both in professional and private life) from time to time. Virtual landscapes provide an approach to a perfect world, a world of freedom and self-fulfillment. Starting from the social constructivist point of view that landscape is a social product and considering the technological progress nowadays, there might be the simple conclusion that virtual landscapes are created against the background of social needs and wishes (that might be missed in 'real' daily life). This is exactly the main idea of escapism: literally to escape from time and space and to dive into new spheres without any feared limit (see also Edler 2020). Of course, there is a wide range of scopes for design depending from the interest of creating landscapes. For example, a person living in a global city working all day might seek for calming and relaxing spaces, green oases that allow to take a breath from daily routine. If there is no possibility to have a walk in a park nearby or if there are physical problems avoiding taking a 'real' stroll, a virtual landscape on the screen might be a perfect alternative in such a case. As Zapatka (1995) puts forward, dense urban zones can be graded up by green oases, so why not use them virtually too?

While creating virtual places according to social needs and wishes, aesthetic rules have to be considered. As a characteristic of postmodernism, aesthetic freedom is of higher importance: even 'kitsch' is allowed as well as clashing with existent styles. Art and aesthetics are being democratized in a way according to Kaesbohrer (2010, p. 27). This emphasizes one more time the paradigm shift from modernity to postmodernity: the cultural turn gave birth to a new comprehension of aesthetics and its freedom. Even 'kitsch' or ugly objects are part of the existing and socially approved environment—and following this logic, of the surrounding landscape, too.

Another link can be drawn in this context: the appreciation and importance of the aesthetical can be considered as a characteristic of social constructivism. First of all, space—and therefore landscape—is marked and designed against the background of fundamental needs. In a second step, according to Hartmann (2010, p. 322) aesthetical values come into operation that put emphasis on the individual character of reading and understanding objects. Every object itself has a certain value, but the ways to read or judge them are various and complex. That is the moment that escapism comes into play again. By designing landscape not only useful but aesthetically worthy, society as well as individuals can profit to the top. The focus is not longer on the evaluation of space or landscape, but on the feelings towards so-called atmospheres,[4] claims Böhme (2013). By creating virtual landscapes, socially desired atmospheres can be produced and can thus favor escapism.

---

[4]According to Rauh (2012, p. 164) atmospheres are to be considered as the current situation of perception.

4   Conclusion

As this article shows, reflections about landscape and virtuality represent an interface between social studies, geographical studies and marketing studies as well as game designing. Each and every perspective has to be considered in the process of creating virtuality. Landscape is kind of a 'stage' on which social wishes and actions become obvious. Design patterns are linked to the attempt of creating special atmospheres that take into account aesthetic questions with the aim to promote escapism. As landscaping is an active and dynamic process, virtuality and landscapes are to be reflected against a social constructivist background.

It was stated that there exist interactions between society and space. In the context of the cultural and spatial turn, landscape is seen as something dynamic, a postmodern phenomenon that includes several protagonists and focuses on simulation. Space and landscape are perceived in atmospheres (Kazig 2019), the current situation decides on whether the impression is positive or negative. The attribution of codes helps to categorize landscape and therefore to reduce complexity, a typical postmodern characteristic. Further scientific examinations could turn the attention to intercultural differences of landscaping, pursuing the question in what way cultural or even ethnic backgrounds can influence the perception and—further on—the creation of (virtual) landscape.

References

Al-Khanbashi, M. (2020). Using matrix as a qualitative data display for landscape research and a reflection based on the social constructivist perspective. In D. Edler, C. Jenal, & O. Kühne (Eds.), *Modern approaches to the visualization of landscapes* (pp. 103–118). Wiesbaden: Springer VS.

Baralou, E., & Shepherd, J. (2008). Going virtual. In J. Kisielnicki (Ed.), *Virtual technologies: Concepts, methodologies, tools, and applications.* New York: Hershey.

Baudrillard, J. (1989). *America.* London: Verso.

Baudrillard, J. (1994). *Simulacra and simulation.* Ann Arbor: The University of Michigan Press.

Bauman, Z. (1995). *Ansichten der Postmoderne.* Hamburg: Argument.

Bellini, A., & Leonardi, L. (2020). Prato: The social construction of an industrial city facing processes of cultural hybridization. In D. Edler, C. Jenal, & O. Kühne (Eds.), *Modern approaches to the visualization of landscapes* (pp. 549–572). Wiesbaden: Springer VS.

Berr, K. (2020). Visuality, aesthetics and landscape. For the enlightenment and self-enlightenment of constructivist landscape research. In D. Edler, C. Jenal, & O. Kühne (Eds.), *Modern approaches to the visualization of landscapes* (pp. 189–215). Wiesbaden: Springer VS.

Böhme, G. (2013). *Atmosphäre: Essays zur neuen Ästhetik.* Berlin: Suhrkamp.

Brown, I. M., et al. (2002). Virtual landscapes. Introduction. In P. Fisher & D. Unwin (Eds.), *Virtual reality in geography.* London: Taylor and Francis.

Cosgrove, D. (2008). Introduction to social formation and symbolic landscape. In R. Z. De Lue & J. Elkins (Eds.) *Landscape theory.* New York: Routledge.

Dear, M. (2001). *The postmodern urban condition.* Hoboken: Wiley.

Edler, D. (2020). Where spatial visualization meets landscape research and "Pinballology": examples of landscape construction in pinball games. *KN—Journal of Cartography and Geographic Information*. https://doi.org/10.1007/s42489-020-00044-1.

Edler, D., Kühne, O., Jenal, C., Vetter, M., & Dickmann, F. (2018a). Potenziale der Raumvisualisierung in Virtual Reality (VR) für die sozialkonstruktivistische Landschaftsforschung. *Kartographische Nachrichten, 68*(5), 245–254 (Bonn: Kirschbaum Verlag).

Edler, D., Husar, A., Keil, J., Vetter, M., & Dickmann, F. (2018b). Virtual reality (VR) and open source software: A workflow for constructing an interactive cartographic VR environment to explore urban landscapes. *Kartographische Nachrichten, 68*(1), 3–11 (Bonn: Kirschbaum Verlag).

Edler, D., Jenal, C., & Kühne, O. (2020a). Modern approaches to the visualization of landscapes—An introduction. In D. Edler, C. Jenal, & O. Kühne (Eds.), *Modern approaches to the visualization of landscapes* (pp. 3–15). Wiesbaden: Springer VS.

Edler, D., Keil, J., & Dickmann, F. (2020b). From Na Pali to Earth—An 'unreal' engine for modern geodata? In D. Edler, C. Jenal, & O. Kühne (Eds.), *Modern approaches to the visualization of landscapes* (pp. 279–291). Wiesbaden: Springer VS.

Fontaine, D. (2017a). *Simulierte Landschaften in der Postmoderne. Reflexionen und Befunde zu Disneyland, Wolfersheim und GTA V*. Wiesbaden: Springer.

Fontaine, D. (2017b). Ästhetik simulierter Welten am Beispiel Disneylands. In O. Kühne, H. Megerle, & F. Weber (Eds.), *Landschaftsästhetik und Landschaftswandel*. Wiesbaden: Springer.

Fontaine, D. (2020). Landscape in computer games—The examples of GTA V and Watch Dogs 2. In D. Edler, C. Jenal, & O. Kühne (Eds.), *Modern approaches to the visualization of landscapes* (pp. 293–306). Wiesbaden: Springer VS.

Gimblett, R., & Skov-Petersen, H. (Eds.). (2008). *Monitoring, simulation, and management of visitor landscape*. Tucson: The University of Arizona Press.

Hartmann, N. (2010). *Ästhetik*. Berlin: De Gruyter.

Hauskeller, M. (1995). *Atmosphären erleben. Philosophische Untersuchungen zur Sinneswahrnehmung*. Berlin: Akademie.

Heim, M. (1993). *The metaphysics of virtual reality*. Oxford: Oxford University Press.

Hillis, K. (1996). *Geography, identity, and embodiment in virtual reality*. Madison: University of Wisconsin.

Hochschild, V., Braun, A., Sommer, C., Warth, G., & Omran, A. (2020). Visualizing landscapes by geospatial techniques. In D. Edler, C. Jenal, & O. Kühne (Eds.), *Modern approaches to the visualization of landscapes* (pp. 47–78). Wiesbaden: Springer VS.

Hokema, D. (2012). *Landschaft im Wandel? Zeitgenössische Landschaftsbegriffe in Wissenschaft, Planung und Alltag*. Wiesbaden: Springer.

Jameson, F. (1998). *The cultural turn. Selected writings on the postmodern 1983–1998*. London: Verso.

Jenal, C. (2020). Visualizations of 'landscape' in protest movements. On exclusive and inclusive patterns of vision and interpretation using the example of resistance to the expansion of the electricity grid in Germany. In D. Edler, C. Jenal, & O. Kühne (Eds.), *Modern approaches to the visualization of landscapes* (pp. 427–445). Wiesbaden: Springer VS.

Kaesbohrer, B. (2010). *Die sprechenden Räume. Ästhetisches Begreifen von Bühnenbildern der Postmoderne. Eine kunstpädagogische Betrachtung*. München: Herbert Utz.

Kazig, R. (2019). Atmosphären und Landschaft. In O. Kühne, et al. (Eds.), *Handbuch Landschaft*. Wiesbaden: Springer VS.

Krecklau, L., et al. (2014). Rekonstruktion urbaner Umgebungen und ihre Anwendungen. In S. Jeschke, L. Kobbelt, & A. Dröge (Eds.), *Exploring virtuality. Virtualität im interdisziplinären Diskurs*. Wiesbaden: Springer.

Kühne, O. (2008). *Distinktion—Macht—Landschaft: Zur sozialen Definition von Landschaft.* Wiesbaden: Springer.

Kühne, O. (2012). *Stadt-Landschaft-Hybridität. Ästhetische Bezüge im postmodernen Los Angeles mit seinen modernen Persistenzen.* Wiesbaden: Springer.

Kühne, O. (2018a). *Landschaftstheorie und Landschaftspraxis. Eine Einführung aus sozialkonstruktivistischer Perspektive.* Wiesbaden: Springer.

Kühne, O. (2018b). *Landscape and power in geographical space as a social-aesthetic construct.* Dordrecht: Springer.

Kühne, O. (2019). *Landscape theories. A brief introduction.* Wiesbaden: Springer.

Kühne, O., & Jenal, C. (2020a). The threefold landscape change—Basic considerations, conflicts and potentials of virtual landscape research. In D. Edler, C. Jenal, & O. Kühne (Eds.), *Modern approaches to the visualization of landscapes* (pp. 389–402). Wiesbaden: Springer VS.

Kühne, O., & Jenal, C. (2020b). The threefold landscape dynamics—Basic considerations, conflicts and potentials of virtual landscape research. In D. Edler, C. Jenal, & O. Kühne (Eds.), *Modern approaches to the visualization of landscapes* (pp. 389–402). Wiesbaden: Springer VS.

Kühne, O., Weber, F., & Jenal, C. (2018). *Neue Landschaftsgeographie: Ein Überblick.* Wiesbaden: Springer.

Kühne, O., Weber, F., & Jenal, C. (2019). Neue Landschaftsgeographie. In O. Kühne, et al. (Eds.), *Handbuch Landschaft.* Wiesbaden: Springer VS.

Loda, M., Kühne, O., & Puttilli, M. (2020). The social construction of Tuscany in the German and English speaking world—Presented by the analysis of internet images. In D. Edler, C. Jenal, & O. Kühne (Eds.), *Modern approaches to the visualization of landscapes* (pp. 157–171). Wiesbaden: Springer VS.

Luhmann, N. (2009). *Soziologische Aufklärung 5. Konstruktivistische Perspektiven.* Wiesbaden: VS Verlag für Sozialwissenschaften.

Morse, M. (1998). *Virtualities: Television, media art, and cyberculture.* Bloomington: Indiana University Press.

Papadimitriou, F. (2020). Visualization of future landscapes, postmodern cinema and geographical education. In D. Edler, C. Jenal, & O. Kühne (Eds.), *Modern approaches to the visualization of landscapes* (pp. 351–369). Wiesbaden: Springer VS.

Prisille, C., & Ellerbrake, M. (2020). Virtual reality (VR) and geography education: potentials of 360° 'experiences' in secondary schools. In D. Edler, C. Jenal, & O. Kühne (Eds.), *Modern approaches to the visualization of landscapes* (pp. 321–332). Wiesbaden: Springer VS.

Qvortrup, L. (Ed.). (2002). *Virtual space. Spatiality in virtual inhabited 3D worlds.* London: Springer.

Rauh, A. (2012). *Die besondere Atmosphäre. Ästhetische Feldforschungen.* Bielefeld: Transcript.

Roßmeier, A. (2020). Urban/rural hybridity in pictures—The creation of neighborhood images using the example of San Diego's urbanizing inner-ring suburbs East Village and Barrio Logan. In D. Edler, C. Jenal, & O. Kühne (Eds.), *Modern approaches to the visualization of landscapes* (pp. 479–498). Wiesbaden: Springer VS.

Saggio, A. (2007). The new mental landscape. Why games are important for architecture. In F. von Borries, S. P. Walz, & M. Böttger (Eds.), *Space, time, play. Computer games, architecture and urbanism: The next level.* Basel: Birkhäuser.

Schenk, W. (2020). Visualization of the fundamental dimensions of "landscape" in landscape paintings around 1500 A.D. In D. Edler, C. Jenal, & O. Kühne (Eds.), *Modern approaches to the visualization of landscapes* (pp. 19–32). Wiesbaden: Springer VS.

Siepmann, N., Edler, D., & Kühne, O. (2020). Soundscapes in cartographic media. In D. Edler, C. Jenal, & O. Kühne (Eds.), *Modern approaches to the visualization of landscapes* (pp. 247–263). Wiesbaden: Springer VS.

Stintzing, M., Pietsch, S., & Wardenga, U. (2020). How to teach "landscape" through games? In
D. Edler, C. Jenal, & O. Kühne (Eds.), *Modern approaches to the visualization of landscapes*
(pp. 333–349). Wiesbaden: Springer VS.

Susann, Dnil.mjrn. (2019). On Amazon (no title): Amazon: https://www.amazon.de/Grand-Theft-
Auto-Standard-PlayStation/product-reviews/B00KW2FKAQ/ref=cm_cr_getr_d_paging_
btm_next_2?ie=UTF8&reviewerType=all_reviews&pageNumber=2&filterByStar=five_star.
Accessed 13 Nov 2019.

Susen, S. (2015). *The 'postmodern turn' in the social sciences*. New York: Palgrave & Macmillan.

Thomas, P. M. (2020). The digitalizing society—Transformations and challenges. In D. Edler, C.
Jenal, & O. Kühne (Eds.), *Modern approaches to the visualization of landscapes* (pp. 447–
457). Wiesbaden: Springer VS.

Turner, T. (1996). *City as landscape: A post-postmodern view of design and planning*. London: E
& FN Spon.

Vetter, M. (2019). 3D-Visualisierung von Landschaft—Ein Ausblick auf zukünftige
Entwicklungen. In O. Kühne, F. Weber, K. Berr, & C. Jenal (Eds.), *Handbuch Landschaft*.
Wiesbaden: Springer VS.

Vetter, M. (2020). Technical potentials for the visualization in virtual reality. In D. Edler, C. Jenal,
& O. Kühne (Eds.), *Modern approaches to the visualization of landscapes* (pp. 307–317).
Wiesbaden: Springer VS.

Weber, F. (2020). Blurring the boundaries of landscape visualization: Welcome to fabulous Las
Vegas. In D. Edler, C. Jenal, & O. Kühne (Eds.), *Modern approaches to the visualization of
landscapes* (pp. 461–478). Wiesbaden: Springer VS.

Weidinger, J. (Ed.). (2018). *Atmosphären entwerfen*. Berlin: Universitätsverlag der TU Berlin.

Wöbse, H. H. (2002). *Landschaftsästhetik. Über das Wesen, die Bedeutung und den Umgang mit
landschaftlicher Schönheit*. Stuttgart: Ulmer.

Zapatka, C. (1995). *The American landscape*. New York: Princeton Architectural Press.

Dominique Fontaine studied French and Geography at the University of Lorraine and the
University of the Saarland. In 2016, she did a doctor's degree searching and writing about simulated postmodern landscapes. From 2016 to 2019, she worked as a teacher (French, bilingual geography) at the Robert-Schuman-Gymnasium Saarlouis. Since 2019, she has been working at the
Warndt-Gymnasium Völklingen.

From Na Pali to Earth—An 'Unreal' Engine for Modern Geodata?

Dennis Edler, Julian Keil and Frank Dickmann

Abstract

The free availability of powerful game engines, such as Unreal Engine 4 (UE4), has revolutionized the possibilities of 3D visualization. These engines include, amongst many other features, a simulation of a realistic physical system. Their early versions are often associated with first-person shooter games, and there are many more application scenarios. Based on the freely available UE4, people can create and share their own virtual landscapes which could be experienced through techniques of immersive VR, such as modern VR headsets. VR software and hardware are currently developing into affordable mass media systems. The new possibilities are explored by researchers of many disciplines, incl. cartographers, geographers and landscape researchers. Research teams dealing with spatial data and information have a high interest in the development of virtual landscapes which are based on highly accurate spatial representations. To achieve this high spatial accuracy, different geodata resources are available. These resources refer to rather traditional origins, such as official surveying departments. Beyond this, the Internet community offers many data sets of

D. Edler (✉) · J. Keil · F. Dickmann
Ruhr-Universität Bochum, Bochum, Germany
e-mail: dennis.edler@ruhr-uni-bochum.de

J. Keil
e-mail: julian.keil@ruhr-uni-bochum.de

F. Dickmann
e-mail: frank.dickmann@ruhr-uni-bochum.de

© Springer Fachmedien Wiesbaden GmbH, part of Springer Nature 2020
D. Edler et al. (eds.), *Modern Approaches to the Visualization of Landscapes*, RaumFragen: Stadt – Region – Landschaft,
https://doi.org/10.1007/978-3-658-30956-5_15

sophistically designed and animated 3D objects which could be suitable data sets for virtual landscapes. This article presents and discusses two examples indicating how geodata resources could be used in the creation of modern VR landscapes.

Keywords

Virtual reality · 3D cartography · Open data · Landscape research · Unreal engine

1   Introduction

Whoever was 'seriously' involved in video and computer gaming of the 1990s might have experienced the (fictional) plot of the first "Unreal" video game, created by Tom D. Sweeney and released by his company Epic Games in 1998: A fictional character called "Prisoner 849" is transported to a moon-based prison. Before the spacecraft (Vortex Rikers) reaches its destination, the voyage is disturbed and pulled to an unexplored planet called "Na Pali". This planet is ruled by technologically advanced reptilian humanoids (the Skaarj). After a crash landing on this planet, troops of the Skaarj begin to fight against the surviving passengers of the spacecraft. The only survivor is Prisoner 849, who is controlled by the player from a first-person perspective. The aim is to survive and to find a way to escape from the planet. This is bound to ordered game missions.

The game is mainly concentrated on fighting situations (3D shoot-'em-up) between the main character (player) and the supernatural antagonists, as it is in many examples of the first-person shooter (FPS) genre. In this article, it is not intended to go deeper into the game stories of FPS games and their—intensively investigated, discussed and criticized (see for e.g. Anderson and Bushman 2001; Salminen and Ravaja 2008; Voorhees 2014)—psychological and social impacts. From a technical perspective, the game Unreal can be regarded as a pioneering publication, as the game was built on a developer framework ("game engine") whose later version (Unreal Engine 4) has meanwhile become one of the most widespread freely available game engines. In 2014, Epic Games launched a new license model of the Unreal Engine 4, which made the game engine publicly available for free. The engine includes a physics engine that provides an (approximate) simulation of the physical system people experience on Earth. Based on this realistic physical simulation, the freely available game engine can be used by professionals and 'home-brewers' to create sophisticated virtual landscapes, incl. visual and auditory stimuli (see also Edler et al. 2019a; Hruby 2019; Jerald 2016). These virtual landscapes can further be coupled with modern hardware innovations, such as Virtual Reality (VR) headsets. Applying this, people can get an experience of being immersed into virtual realities (related in this volume: Prisille and Ellerbrake 2020; Stintzing et al. 2020; Vetter 2020). Based on several modern navigation and locomotion tools established by the gaming industry (see also Edler et al. 2019b; Kersten et al. 2018), people can move through these simulated virtual landscapes, as if they were a physical part of it. This also involves a deeper understanding of empirically investigated cognitive and usability issues (see Clarke et al. 2019; Lokka and Çöltekin 2020; Roth et al. 2017; Edsall and Larson 2009). Technological innovations and ongoing research, together with neighboring disciplines, unlock new potentials and opportunities for the user-oriented 3D visualization and simulated experience of geographic space (Çöltekin et al. 2019; Kersten and Edler 2020; Hruby et al. 2019; Keil et al. 2020; Walmsey and Kersten 2020). It also leads to new application scenarios for spatial sciences, such as urban geography and planning (Edler et al. 2018b; Jamei et al. 2017; Ma et al. 2020), cultural heritage and architecture (Büyüksalih et al. 2020; Kersten et al. 2018), hydrology (Lütjens et al. 2019), biodiversity (Hruby et al. 2019), noise pollution (Berger and Bill 2019), education (Carbonell-Carrera and Saorín 2017; Šašinka et al. 2019) and landscape research (Edler et al. 2018a; Vetter 2019).

The popular interest in creating and 'playing with' (individual) virtual 3D landscapes, the technological opportunities (with available game engines as a key software component), and modern communicational opportunities offered by the Internet are three crucial driving forces behind 3D visualization in the last decade. In conjunction with each other, these developments make it possible to build up new digital and open databases which can be filled with new kinds of geodata (related in this volume: Edler et al. 2020; Kleber et al. 2020; Siepmann et al. 2020, Stratmann et al. 2020). This geodata and their properties extend 'traditional' kinds of geodata, by introducing new data formats and higher levels of details. The possibility to apply a realistic simulation of known physical systems also allow cartographers to create 3D cartographic media which can be used to experience landscapes in a simulated 1/1 scale from a first-person perspective (related in this volume: Prisille and Ellerbrake 2020; Vetter 2020).

What might have started off in a fictional and merciless planet called Na Pali seems to bring a lot of real—not 'unreal'—public benefit 'back on Earth'. This article deals with examples of modern forms of geodata which can be processed in the game engine Unreal Engine 4 to simulate highly realistic virtual landscapes. Whereas some recent publications have already highlighted and discussed the potentials of modern virtual landscapes from a (social) constructivist point of view (Edler 2020a; Edler et al. 2019b, 2018a; in this volume: Bellini and Leonardi 2020; Berr 2020; Fontaine 2020a, 2020b; Kühne 2020; Kühne and Jenal 2020), the landscape concept in this article is reduced to a common positivistic understanding of cartography (c.f. Kühne 2019, 2018). Examples are shown that represent and translate the material world into a virtual form, and the idea behind is to achieve a high level of (measurable) congruence.

2   Examples of VR-Based Landscape Construction

The examples presented in the following take up two geodata resources. The first one originates in data acquisition and supply by official surveying departments. The second one is offered by the VR-gaming community through the Unreal Engine marketplace. Both examples are based on open data sets which can be publicly downloaded without any fees.

2.1   From Numbers to a VR-Compatible 3D Elevation Model

The year 2007 represents a starting point for a new orientation of geodata infrastructures in Europe. The introduction of the INSPIRE directive provides a basis for the establishment of an infrastructure for spatial information in Europe, with a specific focus on the support of environmental policies and related activities. The directive includes three appendices with 34 different geodata themes which refer to environmental applications. It also provides definitions of common terms and concepts which are important for building up, maintaining and using a joint spatial data infrastructure (SDI). Other aims are the establishment of interoperable standards for an efficient exchange of geodata and their metadata and the increase the simplicity of data access. This is in line with the idea of a highly transparent public services in Europe (see also Bartha and Kocsis 2011; Bernard 2005; Bielecka and Medyńska-Gulij 2015; Janssen 2010; Litwin and Rossa 2011; Tsiavos et al. 2012; in this volume: Hochschild et al. 2020; Meyer-Hess 2020).

Although the INSPIRE directive does not explicitly involve a mandatory open geodata policy, meaning that the data is made available without any fees, the directive is a milestone which has changed the access to—high quality—official geodata sets. Many contributing countries upload their latest geodata and metadata sets on a jointly created website that includes different data sets viewer: https://inspire-geoportal.ec.europa.eu/. Open governance websites that provide open geodata are also offered by, for e.g., the United States (https://www.data.gov/), Canada (https://open.canada.ca/en/open-data), Australia (https://data.gov.au/) and New Zealand (https://www.data.govt.nz/) [last access: 29 Feb 2020].

The INSPIRE geoportal shows that many data sets have been made available as open data. In Germany, for example, several responsible public authorities have delivered contributions. This not only refers to the national scale (managed by the German Federal Agency for Cartography and Geodesy—BKG), but also on regional scales. The surveying department of North Rhine-Westphalia ("Geobasis NRW"), for example, provides many geodata services and geodata, incl. thematic data, on their websites and open data download portal (https://open.nrw/open-data).

Among this data, people get open access to digital elevation data (1 m, spatial resolution) in different formats, such as xyz-text files (ASCII character set). The ASCII files contain numeric data on geographic Cartesian coordinates (UTM: easting and northing) and terrain elevation. A complete coverage of NRW is available, and the users can make individual spatial selections based on different data packages (sorted by the names of municipalities).

The open elevation data is not yet provided in game engine formats. In other words, users need to find ways to display the data as VR compatible 3D models in a modern game engine. A new promising approach which has recently been developed and released as a plugin for Unreal Engine 4 allows a processing of point clouds, such as data sets acquired from laser scanning devices (see Epic Games 2019, date of release: 10 Dec 2019). Moreover, elevation data can be opened, further processed and prepared by using tools implemented in Geographic Information System (GIS), such as QGIS (open source). An example of a detailed and applicable workflow based on xyz-text files and QGIS is suggested by Edler (2020b).

According to this workflow, a rasterized image (in PNG format, data type: uint16) can be used to visualize a 3D elevation model in Unreal Engine 4. This image file can be processed by the "landscape tool", and a texture ("landscape material") can be added in the same step. Figure 1 gives a visual impression of a 3D landscape based on open digital elevation data, from an oblique bird's eye perspective in UE4. The illustrated 'raw model' represents the (illuminated and simply textured) terrain of a walkable and revitalized mining heap in Gelsenkirchen ("Halde Rheinelbe"). In next steps, it could be further modulated and enriched by adding photorealistic textures and (audio-visually animated and/or interactive) 3D objects. Moreover, a virtual "character" can be added. This character is a virtual object (moveable camera) that makes it possible to experience a virtual environment through an ego-perspective. It enables to visualize the transfer of the user's sensor-tracked head and body movements to the virtual landscape. The physical system of reality is reproduced by the integrated game physics.

Figure 2 is a snapshot showing a character (switching to a first-person perspective is possible) who watches a post-industrial landscape (near the industrial heap used as example in Fig. 1). The snapshot includes several static and animated 3D objects, such as buildings, water bodies and vegetation. These objects are used in this 'VR landscape' to increase the realistic impression, richness in detail and the feeling of immersion. Another example is presented in Fig. 3. The elevation model is expanded by adding 3D

Fig. 1 3D elevation model in Unreal Engine 4 based on open geodata. (Source: own presentation)

Fig. 2 An elevation model becomes a detailed virtual landscape—new perspectives for urban planning and landscape architecture. (Source: own presentation)

Fig. 3 An elevation model becomes a detailed virtual landscape—planning new electricity lines and poles. (Source: own presentation)

objects simulating a (future) scenario for the electricity supply in rural areas. Both examples might indicate that VR-based landscapes could be a helpful cartographic media for planning applications which supports an individual, interactive and immersive experience in real-time (see also Jamei et al. 2017; Lovett et al. 2015; Virtanen et al. 2015). In such immersive VR environments, objects can also be consciously omitted. Examples are acoustic or visual elements that could disturb a realistic simulation scenario, such as planning simulations used for surveys or lab studies (see also Edler et al. 2018a).

2.2   From (Animated) 3D Objects to Open Geodata

The idea of placing additional (animated) 3D objects onto an accurate virtual terrain simulation to add details and improve realistic impressions is obvious. In terms of 3D visualization, the implementation of such 3D elements and effects could be a tough problem of a visualization project. The modelling of individual objects, no doubt, is possible—for example, by using open source 3D graphics software (e.g. Blender) and game engines. Depending on the level of detail and functionality, 3D modelling can however be a very time-consuming—sometimes even project- or budget-breaking—task.

Therefore, 3D landscape visualization in VR can benefit a lot from work which is done by many 3D artists worldwide. As 3D modelling software and game engines are available for free, many users have created their own (mostly fictional) spatial settings. These projects include 3D objects that can attract people and appear highly realistic. These data packages are often offered online, either on private websites or on established portals, like the UE4 marketplace (https://www.unrealengine.com/marketplace/en-US/store—last access: 29 Feb 2020). These data bases can also have commercial interests, but they also include data sets offered for free. These freely available can be compared to open datasets discussed before.

Game engines, in contrast to GIS, are not based on external geographic reference systems. They use an internal reference system to keep up spatial relations. Therefore, 3D objects created with game engines (and graphics software) differ from geographically referenced data, such as geodata stored in widespread GIS-compatible formats (e.g. Shapefile, GeoTIFF, GeoPackage and GeoJSON). In VR landscapes, 3D objects can represent specific geographical facts and information at clearly defined locations. An accumulation of 3D trees can represent a forest, and a concentration of high-rise buildings can shape out a central business district. Depending on the context information of the application, users understand which geographical location the forest or CBD belong to. Featured with this indirect and context-dependent geocoding, 3D objects made available as rather solitary products of computer arts can be used and considered as geodata. If these objects are offered as free downloads, they can be classified as open (geo-)data which extend the common geodata resources and the opportunities of cartography and geovisualization.

Fig. 4 A landscape consisting of 3D objects offered as freely available data (Source: own presentation)

Figure 4 gives an impression of (animated) 3D objects which can be downloaded as open data from the UE4 marketplace [last access: 29 Feb 2020]. The image shows a vivid forest landscape.

The trees and plants, for example, may serve as 3D objects which could represent vegetation on the floodplain of rivers. Of course, the use of such data involves the consideration whether the available object would be suitable for the purpose of the final application. If a VR landscape is concentrated on the discovery of specific plant species, the artistic design level of (open data) trees might not be an appropriate choice. If the trees are just used to indicate the existence of trees at a specific location, a lower degree of iconicity might well be enough and this (open data) could be valuable components for a VR landscape representation. Depending on the purpose of the VR application, these levels of detail can be adjusted.

3   Summary

This article might have indicated that freely available game engines, such as Unreal Engine 4 (UE4), are examples of modern geoprocessing software which serve society to create highly realistic 3D representations of landscapes. The immersive experience of virtual environments also involves a simulation of the physics system on Earth. This is the base for realistic animation and the involvement of the user's head and body movements.

The geodata used to create these virtual environments can come from different resources, such as data repertoires of official surveying departments and online data bases bringing together the examples and approaches developed by the web community, including companies and 'home-brewers'. The increasing opportunities for the creation and experience of 3D visualization offered by game engines bring together traditional and modern approaches of spatial data. It seems that a software whose 'ancestor' was developed to guide the users (players) through the fictional and dangerous landscape of "Na Pali" has now become an established tool to create highly realistic visualization products and 3D data 'on Earth'.

The new possibilities of landscape visualization in VR bears a lot of potentials for audio-visually exploring the planned landscapes. It enriches the amount of details to represent 3D visualizations which could be accessed in real-time, from an ego perspective and by using body movements. In this way, the level of detail and the feeling of immersion can be increased (see also Hruby et al. 2020). Immersive VR could also leave the 'traditional' positivistic view on cartography, as modern VR-based 3D visualizations may also serve as 'mediators' to evaluate (social) constructivist perspectives on landscapes, incl. individual and social meanings (Edler et al. 2018a).

Acknowledgements   The authors wish to thank Timo Wiedenlübbert and Marco Weissmann for their valuable assistance in creating the figures.

References

Anderson, C., & Bushman, B. J. (2001). Effects of violent video games on aggressive behavior, aggressive cognition, aggressive effect, psychological arousal, and prosocial behavior: A meta-analytic review of scientific literature. *Psychological Science, 12*(5), 353–359. https://doi.org/10.1111/1467-9280.00366

Bartha, G., & Kocsis, S. (2011). Standardization of geographic data. The European INSPIRE directive. *European Journal of Geography, 2*(2), 79–89.

Bellini, A., & Leonardi, L. (2020). Prato: The social construction of an industrial city facing processes of cultural hybridization. In D. Edler, C. Jenal, & O. Kühne (Eds.), *Modern approaches to the visualization of landscapes* (pp. 549–572). Wiesbaden: Springer VS.

Berger, M., & Bill, R. (2019). Combining VR visualization and sonification for immersive exploration of urban noise standards. *Multimodal Technologies and Interaction, 3*(34), 3. https://doi.org/10.3390/mti3020034.

Bernard, L. (2005). INSPIRE—Der Weg zu einer Europäischen Geodateninfrastruktur. *KN—Journal of Cartography and Geographic Information, 55*(5), 232–236. https://doi.org/https://doi.org/10.1007/BF03544018.

Berr, K. (2020). Visuality, aesthetics and landscape. For the enlightenment and self-enlightenment of constructivist landscape research. In D. Edler, C. Jenal, & O. Kühne (Eds.), *Modern approaches to the visualization of landscapes* (pp. 181–215). Wiesbaden: Springer VS.

Bielecka, E., & Medyńska-Gulij, B. (2015). Zur Geodateninfrastruktur in Polen. *KN—Journal of Cartography and Geographic Information, 65*(4), 201–208. https://doi.org/https://doi.org/10.1007/BF03545142.

Büyüksalih, G., Kan, T., Özkan, G. E., Meriç, M., Isın, L., & Kersten, T. P. (2020). Preserving the knowledge of the past through virtual visits: From 3D laser scanning to virtual reality visualisation at the Istanbul Çatalca İnceğiz caves. *PFG—Journal of Photogrammetry, Remote Sensing and Geoinformation Science*. https://doi.org/https://doi.org/10.1007/s41064-020-00091-3.

Carbonell-Carrera, C., & Saorín, J. L. (2017). Geospatial Google street view with virtual reality: A motivational approach for spatial training education. *ISPRS International Journal of Geo-Information, 6*(9), 261. https://doi.org/10.3390/ijgi6090261.

Clarke, K. C., Johnson, J. M., & Trainor, T. (2019). Contemporary American cartographic research: A review and prospective. *Cartography and Geographic Information Science, 46*(3), 196–209. https://doi.org/10.1080/15230406.2019.1571441.

Çöltekin, A., Oprean, D., Wallgrün, J. O., & Klippel, A. (2019). Where are we now? Re-visiting the digital Earth through human-centered virtual and augmented reality geovisualization environments. *International Journal of Digital Earth, 12*(2), 119–122. https://doi.org/10.1080/1753 8947.2018.1560986.

Edler, D. (2020a). Where spatial visualization meets landscape research and "pinballology": Examples of landscape construction in pinball games. *KN—Journal of Cartography and Geographic Information*. https://doi.org/10.1007/s42489-020-00044-1.

Edler, D. (2020b). VR ready? Ein methodischer Ansatz zur Erschließung und Weiterverarbeitung freier Geodaten (Open Data) für die 3D-Landschaftsvisualisierung in Game Engines. *Berichte. Geographie und Landeskunde* (accepted for publication).

Edler, D., Kühne, O., Jenal, C., Vetter, M., & Dickmann, F. (2018). Potenziale der Raumvisualisierung in Virtual Reality (VR) für die sozialkonstruktivistische Landschaftsforschung. *Kartographische Nachrichten, 68*(5), 245–254.

Edler, D., Husar, A., Keil, J., Vetter, M., & Dickmann, F. (2018). Virtual reality (VR) and open source software: A workflow for constructing an interactive cartographic VR environment to explore urban landscapes. *Kartographische Nachrichten, 68*(1), 3–11.

Edler, D., Kühne, O., Keil, J., & Dickmann, F. (2019a). Audiovisual cartography: Established and new multimedia approaches to represent soundscapes. *KN—Journal of Cartography and Geographic Information*, *69*(1), 5–17. https://doi.org/https://doi.org/10.1007/ s42489-019-00004-4.

Edler, D., Keil, J., Wiedenlübbert, T., Sossna, M., Kühne, O., & Dickmann, F. (2019). Immersive VR experience of redeveloped post-industrial sites: The example of "Zeche Holland" in Bochum-Wattenscheid. *KN—Journal of Cartography and Geographic Information, 69*(4), 267–284. https://doi.org/10.1007/s42489-019-00030-2.

Edler, D., Jenal, C., & Kühne, O. (2020). Modern approaches to the visualization of landscapes—An Introduction. In D. Edler, C. Jenal, & O. Kühne (Eds.), *Modern approaches to the visualization of landscapes* (pp. 3–15). Wiesbaden: Springer VS.

Edsall, R. M., & Larson, K. L. (2009). Effectiveness of a semi-immersive virtual environment in understanding human-environment interactions. *Cartography and Geographic Information Science, 36*(4), 367–384. https://doi.org/10.1559/152304009789786317.

Epic Games (2019). LiDAR point cloud. https://www.unrealengine.com/marketplace/en-US/ product/lidar-point-cloud#.

Fontaine, D. (2020a). Landscape in computer games—The examples of GTA V and watch dogs 2. In D. Edler, C. Jenal, & O. Kühne (Eds.), *Modern approaches to the visualization of landscapes* (pp. 293–306). Wiesbaden: Springer VS.

Fontaine, D. (2020). Virtuality and landscape. In D. Edler, C. Jenal, & O. Kühne (Eds.), *Modern approaches to the visualization of landscapes* (pp. 267–278). Wiesbaden: Springer VS.

Hochschild, V., Braun, A., Sommer, C., Warth, G., & Omran, A. (2020). Visualizing landscapes by geospatial techniques. In D. Edler, C. Jenal, & O. Kühne (Eds.), *Modern approaches to the visualization of landscapes* (pp. 47–78). Wiesbaden: Springer VS.

Hruby, F., Sánchez, L. F. Á., Ressl, R., & Escobar-Briones, E. G. (2020). An empirical study on spatial presence in immersive geo-environments. *PFG—Journal of Photogrammetry, Remote Sensing and Geoinformation Science.* https://doi.org/https://doi.org/10.1007/s41064-020-00107-y.

Hruby, F. (2019). The sound of being there: Audiovisual cartography with immersive virtual environments. *KN—Journal of Cartography and Geographic Information, 69*(1), 19–28. https://doi.org/https://doi.org/10.1007/s42489-019-00003-5.

Hruby, F., Ressl, R., & de la Borbolla del Valle, G. (2019). Geovisualization with immersive virtual environments in theory and practice. *International Journal of Digital Earth, 12*(2), 123–136. https://doi.org/https://doi.org/10.1080/17538947.2018.1501106.

Jamei, E., Mortimer, M., Seyedmahmoudian, M., Horan, B., & Stojcevski, A. (2017). Investigating the role of virtual reality in planning for sustainable smart cities. *Sustainability, 9*(11), 2006. https://doi.org/10.3390/su9112006.

Janssen, K. (2010). *The availability of spatial and environmental data in the European Union: At the crossroads between public and economic interests.* Austin: Kluwer Law International.

Jerald, J. (2016). *The VR book. Human-centered design for virtual reality.* San Rafael: ACM & Morgan & Claypool Publishers.

Keil, J., Korte, A., Ratmer, A., Edler, D., & Dickmann, F. (2020). Augmented Reality (AR) and spatial cognition: Effects of holographic grids on distance estimation and location memory in a 3D indoor scenario. *PFG—Journal of Photogrammetry, Remote Sensing and Geoinformation Science.* https://doi.org/https://doi.org/10.1007/s41064-020-00104-1.

Kersten, T. P., & Edler, D. (2020). Special Issue "Methods and applications of virtual and augmented reality in geo-information sciences". *PFG—Journal of Photogrammetry, Remote Sensing and Geoinformation Science.* https://doi.org/https://doi.org/10.1007/s41064-020-00109-w.

Kersten, T. P., Deggim, S., Tschirschwitz, F., Lindstaedt, M., & Hinrichsen, N. (2018). Segeberg 1600—Eine Stadtrekonstruktion in Virtual Reality. *KN—Journal of Cartography and Geographic Information, 68*(4), 183–191. https://doi.org/https://doi.org/10.1007/BF03545360.

Kleber, A., Edler, D., & Dickmann, F. (2020). Cartography and the sea: A javascript-based web mapping application for managing maritime shipping. In D. Edler, C. Jenal, & O. Kühne (Eds.), *Modern approaches to the visualization of landscapes* (pp. 173–186). Wiesbaden: Springer VS.

Kühne, O. (2018). *Landscape and power in geographical space as a social-aesthetic construct.* Cham: Springer.

Kühne, O. (2019). *Landscape theories. A brief introduction.* Wiesbaden: Springer VS.

Kühne, O. (2020). The social construction of space and landscape in internet videos. In D. Edler, C. Jenal, & O. Kühne (Eds.), *Modern approaches to the visualization of landscapes* (pp. 121–137). Wiesbaden: Springer VS.

Kühne, O., & Jenal, C. (2020). The threefold landscape dynamics—basic considerations, conflicts and potentials of virtual landscape research. In D. Edler, C. Jenal, & O. Kühne (Eds.), *Modern approaches to the visualization of landscapes* (pp. 389–402). Wiesbaden: Springer VS.

Litwin, L., & Rossa, M. (2011). *Geoinformation metadata in INSPIRE and SDI. Understanding. Editing. Publishing.* Heidelberg: Springer.

Lokka, I. E., & Çöltekin, A. (2020). Perspective switch and spatial knowledge acquisition: Effects of age, mental rotation ability and visuospatial memory capacity on route learning in virtual environments with different levels of realism. *Cartography and Geographic Information Science, 47*(1), 14–27. https://doi.org/10.1080/15230406.2019.1595151.

Lovett, A., Appleton, K., Warren-Kretzschmar, B., & von Haaren, C. (2015). Using 3D visualization methods in landscape planning: An evaluation of options and practical issues. *Landscape and Urban Planning, 142,* 85–94. https://doi.org/10.1016/j.landurbplan.2015.02.021.

Lütjens, M., Kersten, T. P., Dorschel, B., & Tschirschwitz, F. (2019). Virtual reality in cartography: Immersive 3D visualization of the Arctic Clyde Inlet (Canada) using digital elevation models and bathymetric data. *Multimodal Technologies and Interaction, 3*(1), 9. https://doi.org/10.3390/mti3010009.

Ma, Y., Wright, J., Gopal, S., & Phillips, N. (2020). Seeing the invisible: From imagined to virtual urban landscapes. *Cities, 98*, 102559. https://doi.org/10.1016/j.cities.2019.102559.

Meyer-Hess, F. (2020). Discovering forgotten landscapes. In D. Edler, C. Jenal, & O. Kühne (Eds.), *Modern approaches to the visualization of landscapes* (pp. 33–46). Wiesbaden: Springer VS.

Prisille, C., & Ellerbrake, M. (2020). Virtual reality (VR) and geography education: Potentials of 360° 'experiences' in secondary schools. In D. Edler, C. Jenal, & O. Kühne (Eds.), *Modern approaches to the visualization of landscapes* (pp. 321–332). Wiesbaden: Springer VS.

Roth, R. E., Çöltekin, A., Delazari, L., Fonseca Filho, H., Griffin, A. L., Hall, A., et al. (2017). User studies in cartography: Opportunities for empirical research on interactive maps and visualizations. *International Journal of Cartography, 3*(sup1), 61–89. https://doi.org/10.1080/2372 9333.2017.1288534.

Salminen, M., & Ravaja, N. (2008). Increased oscillatory theta activation evoked by violent digital game events. *Neuroscience Letters, 435*(1), 69–72. https://doi.org/10.1016/j.neulet.2008.02.009.

Siepmann, N., Edler, D., & Kühne, O. (2020). Soundscapes in cartographic media. In D. Edler, C. Jenal, & O. Kühne (Eds.), *Modern approaches to the visualization of landscapes* (pp. 247–263). Wiesbaden: Springer VS.

Stintzing, M., Pietsch, S., & Wardenga, U. (2020). How to teach "landscape" through games? In D. Edler, C. Jenal, & O. Kühne (Eds.), *Modern approaches to the visualization of landscapes* (pp. 333–349). Wiesbaden: Springer VS.

Stratmann, J., Ristea, A., Leitner, M., & Paulus, G. (2020). Exploring urban "blightscapes" applying spatial video technology and geographic information systems: A case study from Baton Rouge, USA. In D. Edler, C. Jenal, & O. Kühne (Eds.), *Modern approaches to the visualization of landscapes* (pp. 499–517). Wiesbaden: Springer VS.

Šašinka, Č., Stachoň, Z., Sedlák, M., Chmelík, J., Herman, L., Kubíček, P., Šašinková, A., Doležal, M., Tejkl, H., Urbánek, T., Svatoňová, H., Ugwitz, P., & Juřík, V. (2019). Collaborative immersive virtual environments for education in geography. *ISPRS International Journal of Geo-Information 2019, 8*(1), 3. https://doi.org/https://doi.org/10.3390/ijgi8010003.

Tsiavos, P., Pediaditi, K., Nedas, K., & Athanasiou, S. (2012). Cultivating open data ecologies: Lessons from the implementation of the INSPIRE directive. In K. Janssen & J. Crompvoets (Eds.), *Geographic data and the law. Defining new challenges* (pp. 37–52). Leuven: Leuven University Press.

Vetter, M. (2020). Technical potentials for the visualization in virtual reality. In D. Edler, C. Jenal, & O. Kühne (Eds.), *Modern approaches to the visualization of landscapes* (pp. 307–317). Wiesbaden: Springer VS.

Vetter, M. (2019). 3D-Visualisierung von Landschaft—Ein Ausblick auf zukünftige Entwicklungen. In O. Kühne, F. Weber, K. Berr, & C. Jenal (Eds.), *Handbuch Landschaft* (pp. 199–214). Wiesbaden: Springer.

Virtanen, J.-P., Hyyppä, H., Kämärainen, A., Hollström, T., Vastaranta, M., & Hyyppä, J. (2015). Intelligent open data 3D maps in a collaborative virtual world. *ISPRS Journal of Geo-Information, 4*(2), 837–857. https://doi.org/10.3390/ijgi4020837.

Voorhees, G. (2014). Shooting. In M. J. P. Wolf & B. Perron (Eds.), *The Routledge companion to video game studies* (pp. 251–258). Abingdon: Routledge.

Walmsey, A. P., & Kersten, T. (2020). The IMPERIAL Cathedral in Königslutter (Germany) as an immersive experience in virtual reality with integrated 360° panoramic photography. *Applied Sciences, 10*(4), 1517. https://doi.org/10.3390/app10041517.

Dennis Edler is a Senior Lecturer in the Institute of Geography at the Ruhr-University Bochum. In his research and teaching activities, he deals with open spatial data and modern approaches to visualizing 3D environments. He also chairs the joint commission "Virtual and Augmented Reality" of the German Cartographic Society and German Society for Photogrammetry, Remote Sensing and Geoinformation.

Julian Keil is a research fellow in the Institute of Geography at the Ruhr-University Bochum. His major research interests are user studies related to cognitive aspects of map use as well as 3D-modelling of virtual landscapes.

Frank Dickmann is Professor of Cartography and Geo-Information Science at the Ruhr-University Bochum. His main interests are 3D cartography, cognitive cartography, and exploring map efficiency using modern integrated media.

Landscape in Computer Games—The Examples of GTA V and Watch Dogs 2

Dominique Fontaine

Abstract

In view of the present and still ongoing virtual gaming boom, it seems appropriate to focus on the recent understanding of landscape. Based on observations of physical landscapes in 'real' life (in contrast to the virtual one), artificial landscapes are produced in order to make computer games as realistic as possible. Virtual effects in combination to a sophisticated storytelling (which is typical for the postmodern times) favor escapism and bring landscaping up to a new level. The chosen examples of GTA V and Watchdogs 2 are representative for computer games having an accent on the landscaping part. The simulations of Los Angeles and San Francisco show how urban landscape is perceived and in what way it can be manipulated in order to fulfill several actions in the game. From a scientific point of view, the active creation (and designing) of landscape goes back to the social constructivist approach pronating that landscape has to be understood as a dynamic sociocultural product. Social constructivism is rather a contemporary phenomenon rooted in the paradigm shift towards postmodernity.

Keywords

Landscape · Social constructivism · Postmodernity · Computer games · Aesthetics · Escapism

D. Fontaine (✉)
Warndt-Gymnasium Völklingen, Völklingen, Germany
e-mail: domiflo@gmx.de

© Springer Fachmedien Wiesbaden GmbH, part of Springer Nature 2020   293
D. Edler et al. (eds.), *Modern Approaches to the Visualization of Landscapes*, RaumFragen: Stadt – Region – Landschaft,
https://doi.org/10.1007/978-3-658-30956-5_16

1   Introduction

As soon as landscape is discussed against the background of computer games, there is one component that becomes evident: in computer games, landscape is always designed either by humans or by programs that were programmed and controlled by humans (see also in this volume: Edler et al. 2020a). This starting point leads us to the question in what way landscape is influenced by human nature and to what extend it is manipulated. In the context of scientific geographical research, this subject seems especially interesting to be reflected based on social constructivism, (e.g. Kühne 2019; Berr and Kühne 2020) a postmodern phenomenon closely related to the cultural turn (cf. Dear 2001; Susen 2015). There is no longer a clear cut between 'true' or 'false'. Figuratively spoken: there are many 'shades of grey' in opposition to 'black or white' attributions or labels during modernism (Susen 2015). The paradigm shift towards postmodern thoughts opened the minds for pluralism, simulation and irony as well as hybridity (as will be shown later on in this article). Nevertheless, where is the link between those postmodern approaches, landscape and computer games? Well, by visualizing and creating landscape artificially (e.g. in computer game designing), the active construction of landscape patterns follow several structures. This leads consequently to the necessity of defining the author's scientific point of view.

The own tradition of research relies on the social constructivist approach and will consequently run like a golden thread through this article. Central questions to be answered are: What does the term of 'landscape' mean from a social constructivist point of view? Can it be clearly defined and if so, how? How is the link between (simulated) landscapes and postmodernity to be explained? Last, not least remains the question about escapist motives behind computer games and their designing. During the author's researches on landscaping in computer games, two outstanding examples of detailed urban simulation as well as social interaction were found. Therefore, the examples of GTA V (picturing Los Angeles) and Watch Dogs 2 (simulating San Francisco) were chosen to demonstrate the landscaping in computer games.

2   The Concept Of Landscape From A Social Constructivist Perspective

Landscapes are a constant companion of society and thus both a component of everyday discussions and of scientific interest. Reflections on 'landscape images', 'beautiful' or even 'ugly' landscapes (cf. Rosenkranz 1853), anthropogenically 'overprinted' natural landscapes, 'industrial landscapes' or even 'landscapes of longing' are firmly anchored in everyday language, although 'landscape' is rather seen as a given matter of course and is hardly examined from a scientific point of view. The fact that the landscape of the twenty-first century is largely shaped by anthropogenic influences becomes obvious at the latest when looking at the lively discussions about the change of landscape in connection with the energy transition (cf. e.g. Eiselt 2012; Gailing 2013). Aschenbrand (2017, p.32) emphasises the construct character of landscape by stating that landscape is fundamentally subject to examination and comparison with the known:

> "The viewer assigns what he sees to a previously learned typification and in doing so integrates objects that are considered to fit the applied typification—i.e. typical—into his perception of landscape."

The perception of landscape is thus based on a referencing of landscape-forming objects with the aim of determining harmony or discrepancy and then to make a (subjective) assessment of landscape. Landscape is thus always also symbolically charged (cf. Cosgrove 1984). Citizens' initiatives which, for example, oppose the erection of wind farms fight for the so-called preservation of their surrounding landscape, i.e. a previously existing landscape image is perceived as *typical* and thus serves as a reference norm. Altered landscapes are often initially given a code of ugliness—this is due to the fact that landscape redesign leads to a break with the familiar, a guarantor of security and a carrier of identity. It is to be noted that landscape that has been reshaped by human hand is more polarized than ever before and is becoming the focus of social and scientific discussions (e.g. Kühne 2008). For this reason, a determined examination of the scientific and theoretical basis of the concept of landscape is necessary and at the same time inevitable (cf. Stemmer 2016, p. 51).

2.1 Social Constructivism As A Young Research Tradition Within Landscape Research

Within geography, landscape research from a social constructivist perspective falls under the domain of human geography. The central point of condensation in the scientific examination of the construct 'landscape' is the question of factors that shape space: How and for what reasons does man influence and shape landscape? What social needs are reflected in a landscape that has been shaped by man? These and similar questions are the focus of landscape research motivated by social constructivism, with particular attention being paid to the process character of landscape transformation (see in this volume: Al-Khanbashi 2020; Bellini and Leonardi 2020; Berr 2020; Edler et al. 2020b; Fontaine 2020; Jenal 2020; Kühne 2020; Kühne and Jenal 2020; Loda et al. 2020; Roßmeier 2020; Siepmann et al. 2020; Weber 2020). Economic as well as ecological, political and social parameters control the development of landscape. This is subject to a constant dynamic process of change. At the same time, however, in addition to landscape construction, the human perception of landscape is also variable over time: human perception of landscape can vary depending on a wide range of factors (cf. Kühne 2018a). For example, Kühne (2018a, p. 1) states that "rapid social changes" often imply "significant changes in physical spaces". This assumption can be demonstrated transparently using a concrete example: In the course of the aforementioned energy revolution, a sensitive change in landscape has taken place. The erection of wind farms (both *onshore* and *off-shore*) alternates a previously existing landscape. As explained above, political decisions (resolutions on energy system transformation), economic goals (financial viability of wind power plants), ecological considerations (orientation and localisation of wind farms, consideration of possible nature conservation areas) and social considerations (acceptance or rejection) all play a major role in the processual changeability of landscape. In addition to landscape change, the above-mentioned factors are also subject to possible change: political agreements can be replaced or even overthrown in the course of changes of government, the financial starting position can change massively, ecological expert opinions can possibly prevent wind farms in one place and make them possible in another instead, and the social desirability of large and possibly noisy wind turbines can experience a new acceptance and dimension in the course of habituation processes. In summary, it can be stated that landscape is a complex mosaic of different limiting factors and—as small as individual changes may seem—is in a continuous process of change (cf. Denecke 2005, p. 239; Fontaine 2017, p. 63).

2.2   The Plurality Of Meanings Of The Landscape Concept

The concept of landscape per se shows a certain degree of complexity (cf. Hokema 2012), which can be explained on the one hand by the fact that 'landscape' is an everyday construct, but on the other hand it is also the subject of scientific debate—and this across all disciplines. It seems obvious that the term should be discussed against a geographical (both physical-geographical and cultural-geographical) background, but also scientific disciplines such as spatial planning, art or even sociology are affected when it comes to design measures or even patterns of perception. Questions of the aesthetic are also raised (e.g. Kühne et al. 2017; Kühne 2018b, 2019), landscape is thus connoted and read in different ways. Postmodernism, which for its part is characterized by catchwords such as heterogeneity (cf. Vielhaber 2001, p. 10), plurality and tolerance, allows a multilayered view of the landscape construct. What on the one hand opens up the possibility of open interpretations and readings, at the same time poses the challenge of increased complexity and thus represents an attractive topic for scientific discourse.[1] The chronology of the etymology of the word 'landscape' can be traced back to the Middle Ages. In the course of the centuries, the concept of landscape has repeatedly acquired new connotations in the light of the respective times: for example, especially in the Middle Ages, landscape was considered a "concept of the spatial synopsis of social norms and customs", whereas since industrialization it has increasingly acquired the dimension of "physical manifestations of aesthetic and ethical norms" (cf. Kühne 2018a, p. 12). At

---

[1]The concept of discourse is based on the understanding of Laclau and Mouffe (2014, p. 175), according to which meaning is generated and terminated from several perspectives.

present, landscape is primarily read from a postmodern perspective and Kühne (2018a, p. 14) specifies this recent view of landscape by formulating:

> "In the context of postmodern concepts, the romantic understanding of science has been updated since the 1980s with the integration of cognitive, aesthetic and emotional interpretations of the world (...). With the development of postmodernism, especially its characteristic of not rejecting the historical as traditional, but rather appreciating it, objects that have shaped the industrial era of the Western world are being given an increasingly positive connotation."

As the above remarks show, the understanding of landscape and in particular the concept of landscape has changed significantly over the past centuries. Thus, a more narrowly defined medieval concept stands in contrast to the current, far more complex assumption that landscape is a social construct and can be read in a variety of ways against the background of various influencing factors.

3 The Importance of Simulation in Postmodernism

The chronology of postmodernism records its beginnings in the second half of the twentieth century and represents a striking paradigm shift from previous modernity (cf. Lyotard 1979; Welsch 2018), although not all principles have been completely dissolved, but rather alternated and adapted to postmodern concepts and ways of thinking. What is certain is that postmodernism has generated new interpretations of space and landscape, which will be explained below.

In principle, paradigmatic shifts often require new readings, which are essential for breaking down the new ways of thinking. The reading of postmodernism is that of codes that need to be decoded (cf. inter alia Venus 1997, p. 86). The fact that the decoding process is by all means a demanding one becomes clear when looking at common attributes of postmodernism such as "new complexity" or also "aberration of reason" (cf. Reich et al. 2005, p. 112). Bauman (1995, p. 182) also underlines—following on from Baudrillard's understanding of postmodernism—the increase in complexity that has taken hold in the course of postmodernization when he proclaims that "between simulation and truth, image and reality" no longer seem to have any clear boundaries. Metaphorically speaking: the black and white of modernity has been replaced by the grey zone of postmodernism.

3.1 Characteristics of Postmodernism

Postmodernism, which according to Kühne (2012, p. 164) considers itself as the "new emergence level of modernity", by no means breaks with all the maxims of modernity, but has been developing new readings and patterns of thought since the 1970s. It differs in particular from modernity in that it has a high tolerance for hybridity and plurality. This includes all domains and areas of life, such as architecture (cf. Welsch 2008, p. 18) and art, but also economic production methods. Thus, a tendency towards Fordism in modernism (cf. *Economies of scale*) is differentiated, which contrasts with the post-fordist, flexible and consumer-oriented tendencies of postmodernism (cf. *Economies of scope*; cf. Kühne 2012, p. 164). Geographical space is also undergoing a change of meaning: it is characterised by a higher degree of fragmentation, successively dissolving earlier densotomies of city and country (cf. Kühne 2012, p. 167) and reflecting socio-spatial patterns. Accordingly, space is no longer to be understood as something static, but is *lived* and actively shaped by people. In the course of postmodernization, which, as already mentioned, has committed itself to plurality and diversity, dichotomous boundaries, as they were typical of modernity, have become increasingly obsolete. Thus, once prominent transitions "between simulation and truth, image and reality" have become blurred (cf. Bauman 1995, p. 182). Best and Kellner (1997, p. 80) emphasize that space can henceforth be read and deciphered according to certain codes, by formulating—in reference to Baudrillard:

> "For Baudrillard, political economy and the era of production are finished, and we live in a new, dematerialized society of signs, images, and codes."

It becomes clear that materialistic features of a society successively give way to a complex, multilayered and emblematic world of life, which is a world of simulacra (cf. Baudrillard 1994). Baudrillard (1989, p. 28 f.) states that America functions as a prototype of simulated worlds, even though this seems to be the European perspective, since American citizens regard themselves and their world as *real* rather than *simulated:*

> "America is neither dream nor reality. It is hyperreality. It is a hyperreality because it is a utopia (…). Everything here is real and pragmatic, and yet it is all the stuff of dreams too. It may be that the truth of America can only be seen by a European, since he alone will discover here the perfect simulacrum – that of the immanence and material transcription of all values. The Americans, for their part, have no sense of simulation. They are themselves simulation in its most developed state, but they have no language in which to describe it, since they themselves are the model."

The scope of simulation in postmodernism is reflected in the local case study GTA V, which represents a simulated image of the city of Los Angeles. The extent to which simulation efforts go hand in hand with escapist motives will be outlined below.

3.2   Escapist References

Escapism, from an etymological point of view, comes from the English verb '*to escape*'—which means '*to escape* something'—and is therefore based on the assumption that people follow the need to escape their everyday world—at least temporarily. Inevitably, the question of motivational backgrounds arises. Probably factors such as a consumption-driven working world striving to maximize performance play a major role in a globally networked world characterized by competitive pressure and fast pace of life. In addition, the natural striving for recognition and the experience of one's own competence (so-called "secondary motives", see Faller and Lang 2010, p. 139) play a prominent role. Computer games often serve as an appropriate medium for satisfying escapist motives (see also Edler 2020). Why not slip into the role of a well-off person for a few hours and enjoy the *American Dream* with your own villa, fast cars and attractive women, as is the case in GTA V, for example? Although the examples listed may seem bold and stereotypical, they reflect the desire of many users of computer games, as expert interviews show (cf. Fontaine 2017).

Is the desire to slip into a different role and lead the life of another person for a limited period of time guiding the action, to reach for the game console? As the survey of GTA V users showed (cf. Fontaine 2017), the testing of legally prohibited and socially frowned upon actions with the aim of consciously and consistently crossing borders is of paramount importance. If you steal a car in GTA V and then disregard all traffic regulations, you will not face any legislative or social punishment. Similarly, visiting a prostitute remains without any consequence: social desirability and moral value systems are neglected, if not completely ignored, for the duration of the game and thus take on a background function. The combination of the feeling of apparently unrestricted freedom on the one hand and the innovativeness of deviant behavior on the other hand provides the basis for explaining the user behavior described above.

With regard to the characteristics of postmodernism, the final conclusion can be drawn that simulation plays a prominent role in the realization of escapist motives for action: By virtually depicting—in the context of computer games—actually existing landscapes and also simulating actions in detail, escapist motives can be served optimally (cf. Evans 2001). This is shown in the following chapter using GTA V and Watch Dogs 2 as examples.

4   Comparison of the Case Studies GTA V and Watch Dogs 2

In order to investigate the staging of landscape in computer games, it is initially advisable to select games that are reminiscent of existing locations—such as the city of Los Angeles in this case—since a certain degree of comparability can be applied as a benchmark for successful simulation. Following this logic, the two computer games GTA V and Watch Dogs 2 will be presented and compared below, both of which have created images of US cities as a game setting.

4.1   GTA V and the Simulated Image of Los Angeles As a Prototype of the Postmodern City

Participate in an illegal city car race after assaulting an annoying passer-by and robbing him of his sports car? GTA V makes all this possible. But without having to take responsibility for it in a way that could harm your own life. In Los Santos, a Californian L.A. based model, the user is presented with a considerable range of optional deviant behavior: whether it's muggings, visits to prostitutes or helicopter flights without the appropriate ticket—the user has the choice. Görig (2013[2]) points out:

> "You can drive through the big city by car, race from the highest mountain in the game on a mountain bike, fly over the desert in a plane, go diving. Or drive aimlessly through the landscape, listen to hip hop classics, dub or country radio. Be happy to have Bootsy Collins or Pam Grier host, or sit in front of the TV in his lair and watch "Republican Space Rangers". Can wonder how Rockstar manages to present completely different landscapes in a very short time and in a very small space without noticeable breaks in between. How they manage to do that particular California light."

The game also suggests that—if you have acted quickly and cunningly enough—you don't have to expect any consequences (such as an arrest). If one suffers a serious accident in connection with a speeding or a fight, the worst thing that happens to the player is an awakening in a clinic, which he can then leave completely regenerated and set off on new adventures. Thus, there is a complete decoupling of action and reaction, at least as far as real life is concerned. A feeling of freedom and escapism control the user behavior (cf. Fontaine 2017; Bowditch et al. 2019, p. 46) to a large extent and are particularly encouraged by a high degree of authenticity (scenic and interactive nature). With regard to the underlying constructivist background, this means that bonds are depicted so faithfully, as in the example of a city here, that the simulation hardly differs from the model template. In concrete terms, this means that if someone is playing the GTA V game and is also familiar with the city of Los Angeles, they will perceive a high level of detail in the game with regard to urban design. This may have an effect on the player's behavior. An exemplary thought scenario could therefore look as follows: The player recognizes a certain area with all infrastructures (street network, shops, bars …) of Los Angeles downtown and may decide to take an action, an action in the game, which he would normally never do in 'real contact' outside the computer (for example a theft).

The richness of detail as a mirror of technological and scientific progress is reflected on the one hand in the lifelike depiction of existing locations within Los Angeles and the exact imitation of e.g. wind and wave movements, and on the other hand in the communication between the individual actors in the game. Conversations and emotionality

---

[2]Online at: https://www.spiegel.de/netzwelt/games/grand-theft-auto-v-durchgespielt-a-923737.html (14.11.19).

are revealed, for example, in situations between the cheated husband and the wife who is planning to elope with the yoga teacher, and thus form a next higher level, namely that of socio-critical discourse (cf. Fontaine 2017, pp. 211, 277).

4.2   Watch Dogs 2 and the Simulacrum of San Francisco

Watch Dogs 2 is a computer game from 2016 that puts its user in the role of a hacker who is supposed to save society from total surveillance by the *ctOS security system* (*central operating system*). In the course of the game, corrupt machinations of the FBI and some companies based in Silicon Valley are also to be proven and made public. The scene of the action is San Francisco, which is simulated in detail and has a high affinity for computers due to its proximity to Silicon Valley. In this sense, it is a prime example of a living environment that is anthropogenically controlled in the extended sense, since it is electronically programmed. As in GTA V, moral and normative boundaries are crossed during the game, but deeds often go unpunished and the user is stylized as the saviour of society through this conscious crossing of boundaries. From this it can be concluded that social desirability in this case is achieved in particular by breaking with rule systems and that computer crime appears in Watch Dogs 2 as a new form of deviant behaviour. Fehrenbach (2016) states:

> "Holiday resort, anarchic playground, dystopia: virtual cities can be many things—and they are becoming more and more elaborate. Because games companies are investing millions in the construction of digital metropolises in which gamers can let off steam."

The above-mentioned 'digital metropolises' serve in this sense as a setting and platform for—at least in real life—unpunished deviance. Limits can be tested without danger or even exceeded. With a view to the ever-increasing demand for virtual game worlds, Fehrenbach (2016) continues:

> "A number of destinations have been added this autumn: a feverish sixties version of New Orleans in "Mafia III", for example, a port city with steampunk bonds in "Dishonored 2" and a detailed rebuilt San Francisco in "Watch Dogs 2", where hackers and large corporations duel each other. The power of game computers and consoles contributes to the digital construction boom."

The above-mentioned 'digital construction boom' is fed by technological advances on the one hand and a constantly growing demand on the world market on the other. But what exactly is the special appeal of a game like Watch Dogs 2 or GTA V? One user, Zkosan (in Schott, 2016), comments on this question when he states (18.11.2016, access 27.03.2019):

> "Here are a few things I have experienced (...) I overheard an argument of a couple, I hacked into the mobile phone of the guy and then a chat started, if I could not pretend to be his

mistress for a short time via chat. He showed it to his wife and she ran away crying. (...) A dog attacked another passer-by, the attacked person screamed and railed about why he couldn't control his dog and went after the owner. Then his dog defended him again. I called the police ... Then someone stood alone outside a restaurant with flowers and whispered to himself that he shouldn't be a coward. I watched the event for a longer time, shortly afterwards he went to the restaurant, went to a woman, she was totally happy and they hugged each other. Even every car standing at the side of the road is used by a passer-by at some point and doesn't just stand there stupidly forever. And that is a TIME small part that you experience, just by walking through the city. There is nothing scripted for the moment when you are there. The world LIVES! Even without us. And THIS is an EXTREMER progress compared to e.g. GTA V!"

Based on this evaluative commentary, it can be deduced that the enthusiasm and intrinsic motivation to choose just *this* computer game over an immense selection of other options seems to be correlated with authenticity. The more true-to-life the reproduction of a location, the reaction of other game characters (whether controlled by the user or the computer) or the interaction possibilities between self- and remote-controlled persons, the higher the game motivation (cf. Bostan and Altun 2016, p. 70). It can therefore be stated that a focus on detail is increasingly in demand in the course of the technological revolution and at the same time—with regard to Watch Dogs 2—still represents a kind of unique selling point, at least at present, although games such as GTA V already exhibit a high degree of simulation accuracy (especially the latest adaptations for the Playstation). Lange (1999, p. 52) and outlines the development and carries out:

> "Many of the somewhat more elaborately designed computer games already feature seemingly realistic virtual landscapes. This endeavor becomes more difficult if the goal is to create virtual landscapes that actually exist in reality and are thus anchored in the world of experience of every single viewer familiar with this landscape."

What was still considered a serious challenge at the end of the twentieth century has become common practice twenty years later thanks to the rapid technological development towards more powerful processors. Kupfer (2010, p. 11) speaks in this context of the "development towards ever greater realism". The users' living world is depicted as authentically as possible in the game setting: This circumstance allows maximum identification with their own environment and consequently generates familiarity (cf. Adriaanse and Enskat 1999, pp. 77, 399). This in turn suggests security and thus forms a starting point for—sometimes daring—actions in the game.

4.3 Comparison of Landscape References and Potential for Action

In summary and regarding the two case studies presented, it can be stated that the landscape created in both computer games functions as the stage for all actions. Depending on the setting, different perspectives of action arise in the game (such as yoga classes in the park or garden or a chase through the streets of Los Santos in GTA V or a romantic dinner in a restaurant or the manipulation of computer data in offices as in Watch Dogs 2). The landscape does not only provide a framework, it also implicitly animates possible actions in the game. The high level of detail and the enormous simulation density allow for maximum identification with the depicted landscape. Weather phenomena as well as fashion and music trends or advertising are considered in the design and visualization of the virtual landscape and once again emphasize the construct character of landscape. As in everyday life far away from the computer screen, anthropogenic actions are significantly influenced by a multitude of (unconscious or conscious) decisions: certain actions are carried out in 'good' weather or in 'favorable' traffic conditions, while for other actions the opposite might be explicitly advantageous (e.g. large traffic volumes or crowds of people). As both computer games take this complexity of the setting into account, a wider range of actions becomes possible—a fact that in turn confirms the high degree of simulation and closes the ring logic to the postmodern staging of simulated worlds.

5   Conclusion

Following the social constructivist perspective, on which is based the present article; landscape must be analyzed against the background of social and cultural circumstances. The cultural turn was emblematic for a new approach towards a higher esteem of the sociocultural frame that determinates how space—and landscape—is constructed and finally perceived. Even if a clear-cut definition of what landscape is seems to be hardly to achieve, it becomes evident that landscape is a dynamic product of human-environment interactions. Beyond doubt, this scientific field is still rather young in geographic and it offers various interfaces to other scientific disciplines: sociology, market research and so on. Besides, it has become clear in what way social behavior and social needs have an effect on landscaping. The socially desired is what pictures 'real' life—this stands in contrast to the claim of freedom that is satisfied in computer games. None of the virtual landscapes will demand for less fuel consumption while 'real' life with its politics does. None of the virtual landscapes (with all of its actors) will condemn one for having stolen something while justice and social rules in 'real' life tell you not to do so. This explains the immense success of computer games such as GTA V and Watch Dogs 2 that provide freedom in a perfectly simulated setting that conveys the impression of acting 'real'—just without any annoying consequences. Further research on landscape in the context of computer games might focus on the different representations of landscapes in international computer games so that sociocultural as well as ethnical differences may be revealed one day (see also Edler et al. 2019, 2018). In times of further globalization and the image of the world as a 'melting pot', the various understandings and designs of landscapes seem to be of high interest.

References

Adriaanse, H. J., & Enskat, R. (1999). *Fremdheit und Vertrautheit: Hermeneutik im europäischen Kontext*. Leuven: Peeters.

Al-Khanbashi, M. (2020). Using matrix as a qualitative data display for landscape research and a reflection based on the social constructivist perspective. In D. Edler, C. Jenal, & O. Kühne (Eds.), *Modern approaches to the visualization of landscapes* (pp. 103–118). Wiesbaden: Springer VS.

Aschenbrand, E. (2017). *Die Landschaft des Tourismus: Wie Landschaft von Reiseveranstaltern inszeniert und von Touristen konsumiert wird*. Wiesbaden: Springer.

Baudrillard, J. (1994). *Simulacra and simulation*. Ann Arbor: The University of Michigan Press.

Baudrillard, J. (1989). *America*. London: Verso.

Bauman, Z. (1995). *Ansichten der Postmoderne*. Hamburg: Argument.

Bellini, A., & Leonardi, L. (2020). Prato: The social construction of an industrial city facing processes of cultural hybridization. In D. Edler, C. Jenal, & O. Kühne (Eds.), *Modern approaches to the visualization of landscapes* (pp. 549–572). Wiesbaden: Springer VS.

Berr, K. (2020). Visuality, aesthetics and landscape. For the enlightenment and self-enlightenment of constructivist landscape research. In D. Edler, C. Jenal, & O. Kühne (Eds.), *Modern approaches to the visualization of landscapes* (pp. 181–215). Wiesbaden: Springer VS.

Berr, K., & Kühne, O. (2020). *„Und das ungeheure Bild der Landschaft …" The Genesis of Landscape Understanding in the German-speaking Regions*. Wiesbaden: Springer.

Best, S., & Kellner, D. (1997). *The postmodern turn*. New York: The Guilford Press.

Bostan, B., & Altun, S. (2016). Goal-directed player behavior in computer games. In Bostan, B. (ed.), *Gamer psychology and behavior*. Switzerland: Springer International.

Bowditch, L., Naweed, A., & Chapman, J. (2019). Escaping into a simulated environment: A preliminary investigation into how MMORPGs are used to cope with real life stressors. In Naweed, A., Bowditch, & L., Sprick, C. (ed.), *Intersections in Simulation and Gaming. Disruption and Balance*. Singapore: Springer Nature.

Cosgrove, D. (1984). *Social formation and symbolic landscape*. Madison: The University of Wisconsin Press.

Dear, M. (2001). *The postmodern urban condition*. Hoboken: Wiley.

Denecke, D. (2005). *Wege der Historischen Geographie und Kulturlandschaftsforschung*. Stuttgart: Franz Steiner.

Edler, D. (2020). Where spatial visualization meets landscape research and "pinballology": Examples of landscape construction in pinball games. *KN—Journal of Cartography and Geographic Information*. https://doi.org/https://doi.org/10.1007/s42489-020-00044-1.

Edler, D., Keil, J., & Dickmann, F. (2020). From Na Pali to Earth—An 'unreal' engine for modern geodata? In D. Edler, C. Jenal, & O. Kühne (Eds.), *Modern approaches to the visualization of landscapes* (pp. 279–291). Wiesbaden: Springer VS.

Edler, D., Jenal, C., & Kühne, O. (2020). Modern approaches to the visualization of landscapes—An introduction. In D. Edler, C. Jenal, & O. Kühne (Eds.), *Modern approaches to the visualization of landscapes* (pp. 3–15). Wiesbaden: Springer VS.

Edler, D., Keil, J., Wiedenlübbert, T., Sossna, M., Kühne, O., & Dickmann, F. (2019). Immersive VR experience of redeveloped post-industrial sites: The example of "Zeche Holland" in Bochum-Wattenscheid. *KN—Journal of Cartography and Geographic Information, 69*(4), 267–284. https://doi.org/10.1007/s42489-019-00030-2.

Edler, D., Kühne, O., Jenal, C., Vetter, M., & Dickmann, F. (2018). Potenziale der Raumvisualisierung in Virtual Reality (VR) für die sozialkonstruktivistische Landschaftsforschung. *KN—Journal of Cartography and Geographic Information, 68*(5), 245–254. https://doi.org/https://doi.org/10.1007/BF03545421.

Eiselt, J. (2012). *Dezentrale Energiewende. Chancen und Herausforderungen*. Wiesbaden: Springer Vieweg.

Evans, A. (2001). *This virtual life: Escapism and simulation in our media world*. Nashville: Fusion Press.

Faller, H., & Lang, H. (2010). *Medizinische Psychologie und Soziologie*. Berlin: Springer.

Fehrenbach, A. (2016). *Zwölf Videospielstätten, die immer noch faszinieren*. Erschienen bei Spiegel. https://www.spiegel.de/netzwelt/games/faszinierende-staedte-in-videospielen-hier-laesst-sich-s-digital-leben-a-1123068.html. Accessed March 2019.

Fontaine, D. (2017). *Simulierte Landschaften in der Postmoderne: Reflexionen und Befunde zu Disneyland, Wolfersheim und GTA V*. Wiesbaden: Springer.

Fontaine, D. (2020). Virtuality and landscape. In D. Edler, C. Jenal, & O. Kühne (Eds.), *Modern approaches to the visualization of landscapes* (pp. 267–278). Wiesbaden: Springer VS.

Gailing. L. (2013). Die Landschaften der Energiewende—Themen und Konsequenzen für die sozialwissenschaftliche Landschaftsforschung. In Gailing, L., Leibenath, M. (Hrsg.), *Neue Energielandschaften—Neue Perspektiven der Landschaftsforschung*. Wiesbaden: Springer.

Görig, C. (2013). *Gesellschaftskritik mit dem Holzhammer*. Erschienen bei: Spiegel. https://www.spiegel.de/netzwelt/games/grand-theft-auto-v-durchgespielt-a-923737.html. Accessed 14 Nov 2019.

Hokema, D. (2012). *Landschaft im Wandel? Zeitgenössische Landschaftsbegriffe in Wissenschaft, Planung und Alltag*. Wiesbaden: Springer.

Jenal, C. (2020). Visualizations of 'landscape' in protest movements. On exclusive and inclusive patterns of vision and interpretation using the example of resistance to the expansion of the electricity grid in Germany. In D. Edler, C. Jenal, & O. Kühne (Eds.), *Modern approaches to the visualization of landscapes* (pp. 427–445). Wiesbaden: Springer VS.

Kühne, O. (2008). *Distinktion— Macht—Landschaft. Zur sozialen Definition von Landschaft*. Wiesbaden: VS Verlag.

Kühne, O. (2012). *Stadt-Landschaft-Hybridität. Ästhetische Bezüge im postmodernen Los Angeles mit seinen modernen Persistenzen*. Wiesbaden: Springer.

Kühne, O. (2018a). *Landschaft und Wandel: Zur Veränderlichkeit von Wahrnehmungen*. Wiesbaden: Springer.

Kühne, O. (2018b). *Landscape and power in geographical space as a social-aesthetic construct*. Dordrecht: Springer International Publishing.

Kühne, O. (2019). *Landscape theories: A brief introduction*. Wiesbaden: Springer.

Kühne, O. (2020). The social construction of space and landscape in internet videos. In D. Edler, C. Jenal, & O. Kühne (Eds.), *Modern approaches to the visualization of landscapes* (pp. 121–137). Wiesbaden: Springer VS.

Kühne, O., & Jenal, C. (2020). The threefold landscape dynamics—Basic considerations, conflicts and potentials of virtual landscape research. In D. Edler, C. Jenal, & O. Kühne (Eds.), *Modern approaches to the visualization of landscapes* (pp. 389–402). Wiesbaden: Springer VS.

Kühne, O., Megerle, H., & Weber, F. (2017). *Landschaftsästhetik und Landschaftswandel*. Wiesbaden: Springer.

Kupfer, D. (2010). *Entwicklung und Implementierung einer Animationstechnik zur Simulation natürlicher Bewegungen in Computerspielen*. Hamburg: Diplomica.

Laclau, E., & Mouffe, C. (2014). *Hegemony and socialist strategy: Towards a radical democratic politics*. London: Verso.

Lange, E. (1999). *Realität und computergestützte visuelle Simulation. ORL-Bericht 106/1999*. Zürich: vdf Hochschulverlag AG.

Loda, M., Kühne, O., & Puttilli, M. (2020). The social construction of Tuscany in the German and English speaking world—Presented by the analysis of internet images. In D. Edler, C. Jenal,

& O. Kühne (Eds.), *Modern approaches to the visualization of landscapes* (pp. 157–171). Wiesbaden: Springer VS.

Lyotard, J.-F. (1979). *La condition postmoderne. Rapport sur le savoir.* Paris: Les Éditions de Minuit.

Reich, K., Sehnbruch, L., & Wild, R. (2005). *Medien und Konstruktivismus. Eine Einführung in die Simulation als Kommunikation.* Münster: Waxmann.

Rosenkranz, K. (1853). *Ästhetik des Häßlichen.* Königsberg: Gebr. Bornträger.

Roßmeier, A. (2020). Urban/Rural hybridity in pictures—The creation of neighborhood images using the example of San Diego's urbanizing inner-ring suburbs east village and Barrio Logan. In D. Edler, C. Jenal, & O. Kühne (Eds.), *Modern approaches to the visualization of landscapes* (pp. 479–498). Wiesbaden: Springer VS.

Schott, D. (2016). *Watch Dogs 2—Die Spielwelt ist so wunderbar, denn wir sind ihr egal.* Erschienen bei Gamepro online. https://www.gamepro.de/artikel/watch-dogs-2-die-spielwelt-ist-so-wunderbar-denn-wir-sind-ihr-egal,3305445.html. Accessed 26 March 2019.

Siepmann, N., Edler, D., & Kühne, O. (2020). Soundscapes in cartographic media. In D. Edler, C. Jenal, & O. Kühne (Eds.), *Modern approaches to the visualization of landscapes* (pp. 247–263). Wiesbaden: Springer VS.

Stemmer, B. (2016). *Kooperative Landschaftsbewertung in der räumlichen Planung. Sozialkonstruktivistische Analyse der Landschaftswahrnehmung der Öffentlichkeit.* Wiesbaden: Springer.

Susen, S. (2015). *The 'postmodern turn' in the social sciences.* New York: Palgrave Macmillan.

Venus, J. (1997). *Referenzlose Simulation?: Argumentationsstrukturen postmoderner Medientheorie am Beispiel von Jean Baudrillard.* Würzburg: Königshausen und Neumann.

Vielhaber, C. (2001). *Die Präfixe der Postmoderne oder: Wie man mit dem Mikroskop philosophiert.* Münster: LIT.

Weber, F. (2020). Blurring the boundaries of landscape visualization: Welcome to fabulous Las Vegas. In D. Edler, C. Jenal, & O. Kühne (Eds.), *Modern approaches to the visualization of landscapes* (pp. 261–278). Wiesbaden: Springer VS.

Welsch, W. (2008). *Unsere postmoderne Moderne.* Berlin: Akademie.

Welsch, W. (2018). *Wege aus der Moderne: Schlüsseltexte der Postmoderne-Diskussion.* Berlin: Akademie.

Dominique Fontaine studied French and Geography at the University of Lorraine and the University of the Saarland. In 2016, she did a doctor's degree searching and writing about simulated postmodern landscapes. From 2016 to 2019, she worked as a teacher (French, bilingual geography) at the Robert-Schuman-Gymnasium Saarlouis. Since 2019, she has been working at the Warndt-Gymnasium Völklingen.

Technical Potentials for the Visualization in Virtual Reality

Mark Vetter

Abstract

Virtual Reality (VR) is becoming a tool that is more often used in various types of activities, including the investigation of landscape visualizations. For the creation of VR environments, it is necessary to explore the existing technical potentials. In this paper, the author will mainly focus on the possibilities of creating 3D worlds and the interaction possibilities in these worlds. This also includes special considerations to produce effective real-time capabilities of VR environments. In addition, this article addresses the creation of visual effects, such as daylight or nightlight simulations.

Keywords

3D-visualisation · Visual effects · Geo-VR · Game engine · Geodata · Lighting

1 Introduction

VR technology has developed rapidly in recent years. More and more subjects and industries are discovering the potential that lies in these methods, tools and techniques. The high expectations of this technology face great technical challenges and necessary theoretical considerations (see in this volume: Fontaine 2020). Although the focus of this paper is on the visual effects of VR, VR also aims at other senses: Aristoteles (after Jütte 2000) knew theses 5 senses:

M. Vetter (✉)
Hochschule Für Angewandte Wissenschaften Würzburg-Schweinfurt, Würzburg, Germany
e-mail: mark.vetter@fhws.de

© Springer Fachmedien Wiesbaden GmbH, part of Springer Nature 2020   307
D. Edler et al. (eds.), *Modern Approaches to the Visualization of Landscapes*, RaumFragen: Stadt – Region – Landschaft,
https://doi.org/10.1007/978-3-658-30956-5_17

- Hearing (auditory perception with the ears, hearing)
- Smelling (olfactory perception with the nose, smell)
- Tasting (gustatory perception with the tongue, taste)
- Vision (visual perception with the eyes, visual perception)
- Touch (tactile perception with the skin, feeling)

The way people perceive things crucially depends on their point of view, their previous knowledge and their experience. As to VR, it is important to consider exactly these aspects. A material spatial structure, or what people think it is, is an artificial creation in VR. It is enriched with artificial details to make the presumed reality more associative and accessible (see in this volume: Hochschild et al. 2020; Prisille and Ellerbrake 2020; Siepmann et al. 2020).

A VR system should therefore be able to evoke the above-mentioned sensory triggers in humans. It is therefore a computer system consisting of suitable hardware and software to provide the imagination of virtual reality. A virtual world is therefore the represented content in this system. It comprises 3D models of objects, their description of behavior for the simulation model and their arrangement in space. A virtual environment is the virtual world experienced by a user in a VR system (Dörner et al. 2019).

Even though VR technology is of particular scientific interest, we can also ascribe an enormous economic potential to the technology. According to the Gartner Hype Cycle (gartner.com 2017) for the market use of innovative technologies in 2017, the technology "Virtual Reality" is at the bottom of the valley of tears. After this low point, it can generally be assumed that the technology is about to become increasingly important for marketing in the coming years. The time span for the "breakthrough" is approximately 5–10 years. Hence, we expect the marketability of the technology in the period of the years 2022–2027. For the use of geodata in connection with VR, there is still a great need for investigation for the presentation and evaluation of application scenarios or different workflows.

The first aim of this article is to show which technical and conceptual challenges exist. A further objective is to describe the potentials of VR technology for geovisualization (including landscape visualization) and the entire GI-Science.

From the point of view of the author, the following potentials arise when considering the technical potential for the (geo)-Visualization of VR:

1. Technical potential to create virtual 3D worlds that do not (yet) correspond to a known, material spatial structure
2. The potential to put oneself in these worlds and interact with them, including the objects within them, using locomotion techniques, such as teleportation
3. Use of a head-mounted display to protect the user from disturbing everyday influences. Intensification of the immersive experience through VR glasses. A more intensive/real experience through the perspective and the haptic experience with the controllers.

4. Creation of physical or optical 3D effects such as light, shadow, shine, illumination, colour effects
5. Generation of specific acoustic effects that can be perceived differently in the simulated 3D space (taking into account the distance or shielding from certain acoustic sources)
6. Consideration of physical or kinetic effects for a quasi-real behaviour of objects in 3D space (e.g. reaction of objects to wind or other forms with acceleration force)
7. Potential to apply cartographic requirements in the 3D world (highlighting/relegating objects to the background, application of interactive 3D symbols, support in pathfinding, provision, spatial context-dependent, additional information etc.)

In order not to go beyond the scope of this article, only the aspects of VR potentials concerning visualization follow in more detail below (i.e. the above-mentioned points 1–4). Point 7 (potential for cartography) should be given even more attention in research, in addition to work that has already been done on this area (e.g. Edler et al. 2019; Edler and Vetter 2019).

2   Basics of Virtual Reality

2.1   The Three „I " in Virtual Reality

For a better understanding of the potentials of this technology, some essential basics of Virtual Reality are addressed in a first step, which are indispensable for the understanding of VR applications. Three terms for the evaluation of VR experiences are crucial (Fig. 1). A short explanation follows (Akenine-Möller et al. 2018).

Fig. 1 The 3 I of virtual reality. (Source: own presentation)

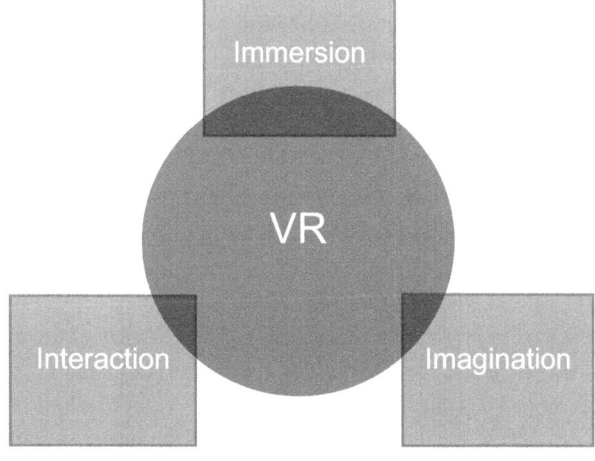

Immersion
The quality of virtual perception depends on the degree of immersion. This means the intensity of immersion in the virtual environment. The aim is to achieve the most realistic possible response of human senses (visual, acoustic, haptic, olfactory), which should give users the feeling of being integrated and present in the virtual worlds (Slater 2009).

Interaction
This term describes the reciprocal action, i.e. the extent to which the user interacts with the virtual environment and the virtual objects occurring in it through real-world actions. Thus, the user can change or use objects in the virtual world. The higher the possibilities and degrees of freedom for interaction, the higher the degree of immersion.

Imagination
Imagination means power of imagination or descriptive thinking. Hence, the user's imagination creating the experience is part of a virtual world. Thus, the imagination is highly dependent on the quality of the immersion and interaction.

2.2   Virtual Worlds

The term virtual world covers the virtual contents of a VR environment. This content primarily consists of 3D graphic objects, but there is much more related to the concept of a virtual world. The 3D objects must have the ability to respond dynamically to user input or manipulation. Additionally, in order to complement these worlds, further addition of abstract, invisible objects is required. These objects serve to support the virtual world by simulations or by representations. This also includes light and sound sources, virtual cameras or certain objects in the "replace"-position for collision checks or the application of physical laws of nature (Hartley and Zisserman 2003).

VR—in contrast to complex renderings without great dynamic requirements for particularly demanding visualizations—must, above all, consider real-time capability for a high-performance VR experience. The three I's of VR are especially negatively influenced if this real-time capability is not or only partially available. The acceptance for the user during a VR experience decreases immediately if the image has interference or is even "frozen" for a short time. Ideally, the user should not perceive any delay during interactions with the VR world compared to movement in everyday, material spatial structures. Per certain unit of time in a VR experience, simultaneous query of possible user inputs, processing of these signals, the calculation of the user's field of view including any changes in the field of view and the output on a display must take place in the background (Dörner et al. 2019). The calculation of changes in the field of view may lead to further operations for the creation/change of computer graphics including rendering. Thus, the biggest challenge of visualizations in the context of VR experiences is mainly the calculation of the visible computer graphics. Sophisticated algorithms and especially high graphic card performance with a high processor power are therefore pre-requisites for a real-time VR experience. In fact, deficits due to less performant hardware are currently still the bottleneck for immersive VR experiences.

The more the so-called visual realism in the displayed computer graphics is considered, the higher are the demands on the graphics performance of the VR computer. However, the real-time capability is more important than computer visualizations that are as close to reality as possible. Furthermore, due to the Uncanny Valley Effect (Mori et al. 2012), reality fidelity is not necessarily to be aspired to.

The need for powerful computer hardware, especially in terms of memory and processing power, also becomes clear when the highest possible degree of interactivity is relevant. The geometrically simplest possible 3D body should be adapted to this requirement. This means that usually cuboids or spheres, or modified forms of these bodies are used. This is especially necessary to perform simple collision calculations with high performance.

3 Aspects for Implementation in VR-Systems

For an evaluation of the technical potentials, we need to know the general workflow required to create a VR simulation.

3.1 3D Modelling

As a first step of a VR experience, especially if it is created for geographic research purposes, 3D modelling of objects for usage in the virtual world is necessary. Many application scenarios are based on existing, material spatial structures or on simulations of these scenarios. Therefore, the focus is on the 3D modelling of everyday objects, such as buildings, landscape elements, technical infrastructure, etc. If the objects to be modelled are as close to nature as possible, the main focus is not only on the 3D construction of the body, but also on the texturing (Remondino and El-Hakim 2006).

Users who frequently work with 3D data in cities or for infrastructure objects for visualization in landscapes usually use 3D modelling software or CAD systems (e.g. AUTOCAD, 3DS-Max, Cinema 4D, Blender etc.). Some of these software products allow the creation of animations, but it is not very efficient to set up the animations in the 3D software product, as these settings cannot usually be exported later for further processing. The 3D models, often of high complexity, must be geometrically simplified before importing them; otherwise, the above-mentioned real-time capability cannot be achieved. We can use procedural modelling, especially for a larger constructed areal by buildings. Even "natural" objects, such as rock formations or vegetation (Fig. 2) can be modelled procedurally (Deussen and Lintermann 2005).

In both cases, the algorithms are structured in such a way that frequently recurring physiognomies are modelled according to the same pattern using the same materials.

Fig. 2 A flower how it looks like in the material editor. (Source: own presentation based on the Unreal Engine Material Editor)

Based on geodetic methods for the acquisition of spatial data (e.g. 3D laser scanning), information on objects can also be obtained by considering surface conditions. If necessary, however, extensive post-processing may be necessary. This includes the filling of acquisitional gaps or the simulation of hidden areas that cannot be captured by the scanner, such as casting shadows. Under certain circumstances, it may be more efficient to model the structures from the beginning than to use any scans that require extensive post-processing.

3.2 Preparation of 3D-Objects for VR

For the preparation of 3D objects for the usage in virtual worlds, an adaptation of the object geometries and the conversion into importable file formats is necessary. Essentially, a reduction of the number of polygons of a 3D object takes place, which leads to a minimized memory requirement and thus higher performance when using these objects. Automated procedures in the 3D software make it possible to reduce the number of polygons. It is also possible to use very simple geometries of the objects and to use more detailed, more realistic visualization results by using shape-matched textures. This is also referred to as "texture baking" (Spini et al. 2016). In cartography and geovisualization, the observation of objects at different scales plays a major role. Therefore, in this sub-step, one should think of providing several instances of the objects in different resolution accuracies, for use in different scale ranges. During the runtime in the VR experiment, a corresponding selection of the objects depends on the distance of the observer (Level of Detail). Finally, the objects would have to be provided in a format which is supported by the software for further use in the VR environment. The proprietary *.fbx format (Autodesk) can usually be read, but also other formats, such as *.obj. Non-proprietary standards, such as those defined by the Web 3D Consortium (*Web3D Consortium|Open Standards for Real-Time 3D Communication* 2020), also have the advantage that the use of open source software projects is supported. Here, for example, COLLADA (*.dae) or X3D (*.x3d) should be mentioned. X3D is the successor of VRML (Virtual Reality Markup Language). The advantage is that not only 3D models but also instructions for animations in 3D scenes are supported (for other studies considering open initiatives, see in this volume: Kleber et al. 2020; Meyer-Heß 2020; Stratmann et al. 2020; Thomas 2020).

3.3 Implementation of 3D Objects in VR Environments

In order to make a VR experience work, an appropriate software environment is required so that a transfer of 3D objects into real-time interaction with the user is possible. Game engines are ideal platforms for this purpose (see in this volume: Edler et al. 2020). Here, adding collision geometries, but also virtual cameras, light sources, audio sources and backgrounds (e.g. virtual walls, horizon, celestial phenomena etc.) for possible interaction with the objects is a simple process.

4 Visualization and Interaction Potentials in VR

4.1 Scene Graphs

A technical potential of VR is certainly the possibility of interaction with objects in the VR world. Using this potential is feasible if the corresponding internal structure for the arrangement of objects and their possible actions are defined in the VR system. These conditions are usually arranged in a scene. The "stage directions" for the arrangement of the objects and the visibility or interaction possibilities in a scene are described by scene graphs. Based on graph theory, these graphs are directed and acyclic and allow the construction of hierarchically structured scenes (Kamat and Martinez 2002).

4.2 Physiognomic Characteristics

Geometric bodies describe the inner structure of 3D objects, while textures and materials describe the appearance. The selection of an appropriate material defines the visual appearance. It considers corresponding material properties. In this context, properties are also defined that are decisive for the appearance in the rendered scene, namely reflection or transparency properties of the material. With the corresponding lighting models, which are defined for an entire scene, physico-optical lighting conditions should be created as close to reality as possible.

Textures, on the other hand, try to imitate the structures of surfaces such as wood, stone and metal as realistically as possible. Textures are stored as raster images, whereby texture coordinates represent the assignment of the individual pixels of the images to the 3D object (Fig. 3).

Shaders are used to create shading effects. These shaders start at the surface of the objects. They consider the materials or textures, and they change the appearance of a pixel on this surface, including a shadow effect. This means that usually only colour fragments of the original 3D objects are displayed. In the calculation process of this representation, other objects are also considered if necessary, such as semi-transparent objects in front of the object (window glass).

4.3   Optimization Techniques for 3D Objects

In order to realize the technical requirements of real-time capability, optimization techniques for 3D objects must be applied. First, this includes the simplification of 3D polygon meshes in order to reduce the number of polygons that are displayed, in a first step. The provision of 3D objects in different levels of detail is a necessity for the creation of high-performance virtual worlds. A reduction of the number of polygons for the representation of complex surfaces can also be achieved by texture baking. Here, the memory space of the polygon geometries is reduced to the visible part only and only the appearance of these geometries is reproduced as texture. Often the use of billboards is an efficient option. These billboards consist of simple geometries, quasi of textured squares. For example, a two-dimensional surface in a virtual world displays the image of a facade of a building. The disadvantage is that these objects can only be viewed from one side, i.e. the objects should be placed at a greater distance so that the details cannot be seen.

Fig. 3 Wood similiar material as texture for 3D-objects. (Source: own presentation based on the Unreal Engine Material Editor)

4.4   Animation and Object Behaviour

For geovisualizations, e.g. in thematic 3D cartography, it is necessary to change the displayed 3D objects in terms of colour, size, position, orientation or geometry. The animation of rigid bodies should obey the laws of physics even in the virtual world, i.e. certain physical properties such as mass, in case of a kinetic impulse, such as the effect of speed (also in case of rotation of objects), material damping parameters such as inertia or friction, but also elasticity should be taken into account accordingly when moving these objects or even in case of collisions. The system reacts if a user in a VR-experiment approaches an object, like a market stall (Fig. 4). As all these processes have only an indirect influence on the visualization, no further explanations will follow in this context.

4.5   Illumination

The lighting conditions in a 3D scene are of crucial importance for visualization. In addition to the material properties described above, which have a decisive influence on the visual impression of an object, it is of course also decisive which artificial or quasi-natural light sources are used to illuminate a virtual world, or which effects are supposed to be achieved in this context. The position in the 3D scene of the light sources is also of decisive importance. This bears a high potential in virtual worlds, as visualizations can be created that evoke special moods or reactions in the VR users. At the push of a button, a completely different illumination situation can be created; suddenly a scene with daytime lighting conditions is transformed into night-time conditions (Fig. 5).

Fig. 4 A middle age setting prepared for the interaction with the user. (Source: own presentation)

Fig. 5 Lighting-Simulation. *Left side* for 10 AM, *Right side* for 6 PM. (© Figures provided by Helge Olberding)

5   Conclusion

The creation of virtual worlds confronts us with the great challenge of achieving a sensory experience that is as authentic as possible, while it achieves reasonable real-time capability at the same time. For the next few years, therefore, we can expect a high technical and empirical research demand to make 3D models leaner and to effectively limit animation paths and reuse geometries several times.

References

Akenine-Möller, T., Haines, E., & Hoffman, N. (2018). *Real-time rendering* (4th ed.). Milton: Chapman and Hall/CRC.

Deussen, O., & Lintermann, B. (2005). *Digital design of nature. Computer generated plants and organics* (X.media.publishing). Berlin: Springer.

Dörner, R., Broll, W., Grimm, P., & Jung, B. (Eds.). (2019). *Virtual und augmented reality (VR/AR). Grundlagen und Methoden der Virtuellen und Augmentierten Realität* (2nd ed.). Berlin: Springer Berlin Heidelberg & Heidelberg: Springer Vieweg.

Edler, D., Keil, J., & Dickmann, F. (2020). From Na Pali to Earth—An 'unreal' engine for modern geodata? In D. Edler, C. Jenal, & O. Kühne (Eds.), *Modern approaches to the visualization of landscapes* (pp. 279–291). Wiesbaden: Springer VS.

Edler, D., Kühne, O., Keil, J., & Dickmann, F. (2019). Audiovisual cartography: Established and new multimedia approaches to represent soundscapes. *KN—Journal of Cartography and Geographic Information, 69*(1), 5–17. doi:https://doi.org/10.1007/s42489-019-00004-4.

Edler, D., & Vetter, M. (2019). The simplicity of modern audiovisual web cartography: An example with the open-source javascript library leaflet.js. KN - Journal of Cartography and Geographic Information. doi:https://doi.org/10.1007/s42489-019-00006-2.

Fontaine, D. (2020). Virtuality and landscape. In D. Edler, C. Jenal, & O. Kühne (Eds.), *Modern approaches to the visualization of landscapes* (pp. 267–278). Wiesbaden: Springer VS.

gartner.com. (2017). Top trends in the gartner hype cycle for emerging technologies, 2017. https://www.gartner.com/smarterwithgartner/top-trends-in-the-gartner-hype-cycle-for-emerging-technologies-2017/. Accessed 25 April 2020.

Hartley, R., & Zisserman, A. (2003). *Multiple view geometry in computer vision* (2nd ed.). Cambridge: Cambridge University Press.

Hochschild, V., Braun, A., Sommer, C., Warth, G., & Omran, A. (2020). Visualizing landscapes by geospatial techniques. In D. Edler, C. Jenal, & O. Kühne (Eds.), *Modern approaches to the visualization of landscapes* (pp. 47–78). Wiesbaden: Springer VS.

Jütte, R. (2000). *Geschichte der Sinne. Von der Antike bis zum Cyperspace*. München: Beck.

Kamat, V. R., & Martinez, J. C. (2002). Scene Graph and Frame Update Algorithms for Smooth and Scalable 3D Visualization of Simulated Construction Operations. *Computer-Aided Civil and Infrastructure Engineering, 17*(4), 228–245. doi:https://doi.org/10.1111/1467-8667.00272.

Kleber, A., Edler, D., & Dickmann, F. (2020). Cartography and the sea: A javascript-based web mapping application for managing maritime shipping. In D. Edler, C. Jenal, & O. Kühne (Eds.), *Modern approaches to the visualization of landscapes* (pp. 173–186). Wiesbaden: Springer VS.

Meyer-Heß, F. (2020). Discovering forgotten landscapes. In D. Edler, C. Jenal, & O. Kühne (Eds.), *Modern approaches to the visualization of landscapes* (pp. 33–46). Wiesbaden: Springer VS.

Mori, M., MacDorman, K., & Kageki, N. (2012). The Uncanny Valley [From the Field]. *IEEE Robotics & Automation Magazine, 19*(2), 98–100. doi:https://doi.org/10.1109/MRA.2012.2192811.

Prisille, C., & Ellerbrake, M. (2020). Virtual reality (VR) and geography education: Potentials of 360° 'experiences' in secondary schools. In D. Edler, C. Jenal, & O. Kühne (Eds.), *Modern approaches to the visualization of landscapes* (pp. 321–332). Wiesbaden: Springer VS.

Remondino, F., & El-Hakim, S. (2006). Image-based 3D modelling: A review. *The Photogrammetric Record, 21*(115), 269–291. doi:https://doi.org/10.1111/j.1477-9730.2006.00383.x.

Siepmann, N., Edler, D., & Kühne, O. (2020). Soundscapes in cartographic media. In D. Edler, C. Jenal, & O. Kühne (Eds.), *Modern approaches to the visualization of landscapes* (pp. 247–263). Wiesbaden: Springer VS.

Slater, M. (2009). Place illusion and plausibility can lead to realistic behaviour in immersive virtual environments. *Philosophical transactions of the Royal Society of London. Series B, Biological sciences, 364*(1535), 3549–3557. doi:https://doi.org/10.1098/rstb.2009.0138.

Spini, F., Marino, E., D'Antimi, M., Carra, E., & Paoluzzi, A. (2016). Web 3D indoor authoring and VR exploration via texture baking service. In I. Techniques (Ed.), *Proceedings, Web3D '16. The 21st International Conference on Web 3D Technology: Anaheim, California, July 22–24* (pp. 151–154). New York: ACM.

Stratmann, J., Ristea, A., Leitner, M., & Paulus, G. (2020). Exploring urban "blightscapes" applying spatial video technology and geographic information system: A case study from Baton Rouge, USA. In D. Edler, C. Jenal, & O. Kühne (Eds.), *Modern approaches to the visualization of Landscapes* (pp. 499–517). Wiesbaden: Springer VS.

Thomas, P. M. (2020). The digitalizing society—Transformations and challenges. In D. Edler, C. Jenal, & O. Kühne (Eds.), *Modern approaches to the visualization of landscapes* (pp. 447–457). Wiesbaden: Springer VS.

Web3D Consortium|Open Standards for Real-Time 3D Communication. https://www.web3d.org/. Accessed 14 May 2020.

Professor Dr. Mark Vetter is a geographer and geoinformation scientist at the Würzburg University of Applied Sciences. His research focuses on "geo-virtual" reality and climate modelling.

Part VI

Landscape Visualization in Education

Virtual Reality (VR) and Geography Education: Potentials of 360° 'Experiences' in Secondary Schools

Christopher Prisille and Marko Ellerbrake

Abstract

This article focuses on different options to integrate VR into geography education. At first, digital media in general and its role for geography are discussed. There are numerous digital media which rely on geolocation. As students are in contact with digital media outside of school, it is important that they are used and addressed in class as well. Therefore, many media literacy models have emerged in recent years which display the skills students are supposed to acquire. Virtual Reality (VR) is a digital medium which makes it possible to develop media literacy on the one hand. On the other hand, it allows to experience remote locations in geography classes among many other benefits. Constructivist theories are particularly suitable as a theoretical background and foundation. However, the different potentials pointed out are mostly subject to future research and yet to be confirmed.

Keywords

VR · Geography · Education · 360-degree · Constructivism · Problem-based learning · Head-mounted displays

C. Prisille (✉) · M. Ellerbrake
Ruhr-Universität Bochum, Bochum, Germany
e-mail: christopher.prisille@rub.de

M. Ellerbrake
e-mail: marko.ellerbrake@rub.de

© Springer Fachmedien Wiesbaden GmbH, part of Springer Nature 2020
D. Edler et al. (eds.), *Modern Approaches to the Visualization of Landscapes*, RaumFragen: Stadt – Region – Landschaft,
https://doi.org/10.1007/978-3-658-30956-5_18

1   Introduction

Nowadays, digital media affect our everyday life (Michel et al. 2011, p. 4; Quade and Felgenhauer 2013, p. 262; see in this volume: Edler et al. 2020a; Fontaine 2020; Kühne and Jenal 2020; Thomas 2020). According to the JIM-study (mpfs 2018, p. 8), 97% of all adolescents in Germany own a smartphone. In the case of eighteen to nineteen-year-olds, this percentage is even higher (99%; mpfs 2018, p. 10). Furthermore, nine out of ten adolescents are on the internet on a daily basis (mpfs 2018, p. 13). A lot of the software on computers and apps on mobile devices rely on geolocation and are therefore of geographic interest (e. g. Google Maps, public transport navigation apps, geocaching, Pokemon Go; Kanwischer 2014, p. 12).

Digital media used for geographic purposes are continuously evolving in the same way (Barnikel and Vetter 2011, p. 5). As a result, there are various new options for teaching geography in school. Hence, geographically related digital media are an essential part of twenty-first century education and evenly important for a contemporary geography education (Lindner-Fally and Zwartjes 2012, p. 272; see in this volume: Papadimitriou 2020; Stintzing et al. 2020).

There are three sections in this article. In the first section, questions about the role of Geography Education in a digitalized world and how it can contribute to media literacy are discussed. In the next section, potentials of VR in geography classes and the way constructivism can serve as a suitable learning theory are pointed out. In the final section, research gaps which are still lingering and challenges to an efficient utilization of VR are shown to conclude the article.

2   Geography Education in a Digital World

While using navigation devices, apps or during a GPS-based search for the closest and top-rated pizza delivery service, users get in contact with geo-information in the course of their daily contact with digital media. By rating restaurants, cinemas or similar places and activities, users are not only consumers of geo-information but also take part in the creation of geo-information. Thus, users become producers of geo-information (Kanwischer 2014, p. 13). Users need to be able to act responsibly and appropriately in the digital world, which is why school education should facilitate media literacy.

Part of these competences is not only the skillset to handle hardware and software (see Baacke 1997; Selwyn 2007; Tulodziecki 2011). Media literacy is also about searching and processing information, as well as communicating with help of digital media and about digital media. In addition, students should have the ability to create their own contents while keeping certain aspects like addressees and intentions in mind and to reflect on their own actions with regard to digital media. There are also other important matters which come across while using the internet and smartphones every day, like questions about the safety of personal data and protection against dangers (computer virus, addiction, insults; KMK 2017, pp. 16–19).

But how can geography education specifically take part in teaching media literacy? (related in this volume: Bagoly-Simó 2020; Papadimitriou 2020; Stintzing et al. 2020) There are already numerous examples: Chatel and Falk (2017, p. 155) suggest that smartphones can be integrated into learning environments to give additional access to information that support the process of learning and teaching. In doing so, learners become aware of the use of smartphones as educational media and reflect on their user behaviour in everyday life. Moreover, the use of smartphones leads to an individualisation and reinforcement of student-centred classes (Chatel and Falk 2017, p. 156).

All over the world, the use of GIS has been adopted into many curricula (Düren and Bartoschek 2013, p. 388; see approaches to the GIS-based visualization in this volume in: Kleber et al. 2020; Poplin et al. 2020; Stratmann et al. 2020). Yet, GIS is still used very rarely due to technical obstacles and a lack of support by teachers (Düren and Bartoschek 2013, pp. 388–389). The most notable technical obstacle in question is the complexity of the software which emphasises the importance of controlling and operating of digital media. The same applies to smartphones. Despite the fact that students use smartphones almost as fluent and naturally as their mother tongue, teachers who do not own a smartphone are often not well informed about the possibilities to use digital media in educational settings (Stojšic et al. 2016, p. 84).

360-degree pictures and 360-degree videos also belong to digital media with spatial reference students get in touch with on the internet. There is a huge variety of digital media online which ranges from 360-degree videos on YouTube and pictures on social media like Facebook or Instagram to interactive 360-degree content like a 360-degree discovery tour of the Brandenburg biosphere consisting of swamps, forests and meadows on the website wilde-welten.de and is expanding steadily (Fig. 1). Stereoscopic head-mounted displays (HMD) which use sensors to transmit user movements into the VR of the respective 360-degree content, present the best experience of 360-degree media. In the following, this paper focuses on Virtual Reality (VR) as a medium which creates the opportunity to experience 360-degree content.

Wilde-welten.de offers interactive 360-degree contents of meadow, forest and swamp biotopes. By taking a look around, learners can access countless information, depicted in different ways. A cursor, located in the centre of the field of view, can be used to activate content by aiming at different icons.

3 Potentials of VR for Geography Classes in School

In recent years, VR has taken the spotlight because of the increased usage of so-called head-mounted displays, such as HTC Vive, Oculus Rift and Playstation VR in the field of gaming (related in this volume: Edler et al. 2020b; Hochschild et al. 2020; Vetter 2020). Smartphone apps (e. g. YouTube, Google Expeditions, Cardboard) which are capable of

Fig. 1 Sample of the 360-degree environment of wilde-welten.de. (Stiftung NaturSchutzFonds Brandenburg, n.d.)

displaying 360-degree content are equally on the rise. A meta-study by Merchant et al. (2014, p. 29) indicates that games, simulations and VR can have a positive impact on knowledge gains. The same applies to education in schools. The implementation of VR in classrooms goes along with learning online and with computers and smartphones (Stojšic et al. 2016, p. 85). In similar fashion to the methods of generalisation and selection of attributes which are common in cartography (Hruby et al. 2019, p. 129) and similar to every other medium, lesson content and VR have to be chosen carefully and adapted to each other in order to ensure it benefits the comprehension of the learning content (Stojšic et al. 2016, p. 90). Because of the level of reality learners experience in VR settings which is only excelled by experiences first hand and the special way of learning in the virtual reality, it is unmatched by any other learning environments used in geography classes. Thus, VR in geography education offers the opportunity to promote media literacy such as the ability to change perspectives or the ability to critically assess the virtual world. This kind of assessment can be achieved by discussing the experiences made in class, for example.

As mentioned before, there is particular interest in immersive VR which has enjoyed recent popularity in the entertainment industry and especially gaming. In general, VR can be defined as a realistic simulation of a three-dimensional space which can be experienced and controlled by movements of the user (Deggim et al. 2017, p. 87). While non-immersive VR displays spaces and gives access to them through a computer, immersive VR creates the feeling of being present in a simulated environment with the help of head-mounted displays (Stojšic et al. 2016, p. 85).

By the use of head-mounted displays, spaces are visualised in form of a 360-degree setting and create a VR. This type of visualisation has different advantages over the conventional two-dimensional visualisations of maps and GIS, as it likely improves the learner's map-reading competence (Edler et al. 2018, p. 5). Furthermore, the three-dimensionality replicates the land configuration more realistically and opens up a different way of analysing topographic maps. When transferring the round shape of planet earth onto a two-dimensional map, there are restrictions which in turn do not apply to VR (Mikropoulos 1996).

3D-objects and VR are therefore able to explain and visualise contents that texts fail to convey (Stojšic et al. 2016, p. 92) and additionally able to add georeferenced data. Moreover, they can help to portray several complex issues at once (Havenith et al. 2019, p. 182), mostly without using symbols like a map (Mikropoulos 1996). Another way of utilising both VR and maps is to make learners connect the symbols of a map to structures in VR as a matter of improving their orientation.

Depending on the nature of the VR environment, the level of the immersion produced can vary (Vetter 2019, p. 562). To achieve a high immersion in geography classes, the virtual setting has to be as realistic as possible (Hruby et al. 2019, p. 129). Through head-mounted displays and the transmission of movements into the VR a high level of immersion is reached and therefore a significantly stronger feeling of being a part of the VR (Hruby et al. 2019, p. 128; Sánchez et al. 2000, p. 349). Generating a physical and spatial presence in the virtual world, results in an extremely realistic perception of the displayed space.

The immersion created by the VR is especially interesting if there is no way to put field trips into practice. There are different reasons why this might be the case: First of, the group size could be too large to go on a field trip or the location of the field trip might be out of reach due to various issues (e. g. distance, costs, logistics, time) (Kingston et al. 2012, p. 1282; Mikropoulos 1996). Through the virtual embodiment in the VR, learners first get in touch with visualisations of a specific area and can make first experiences and interpretations (Sánchez et al. 2000, p. 348) which is particularly useful when preparing a field trip. That way the time outside the classroom can be utilised much more effectively (Kingston et al. 2012, p. 1282). According to Mikropoulos (1996), there are no deficits in the students' motivation when working with models or VR compared to working with authentic objects or to first-hand experiences.

In the following, this article is going to focus on the possibilities for students have to create their own VR. There are free tools which can be used to create a user-friendly and appealing display of 3D-landscapes (Vetter 2019, p. 564). Especially so-called game engines like Unreal Engine, Unity3D and Cry Engine, which are used to programme video games, are worth mentioning (Vetter 2019, p. 564). Similar to the case of GIS, the future of game engines in classrooms settings depends on whether they are easy enough to handle. However, the experience of creating an own VR-world can be a great benefit for learners if the software allows them to visualise, edit and combine geo-information (Mikropoulos 1996). When creating a VR-world, learners need to make different decisions that are comparable to creating a poster or a flyer. Among them the addressee of the product, the fashion in which it is designed, the information given, as well as cartographic ones like the scale or the level of detail. Yet, it has to be noted that the participation and action of a VR-environment are unique characteristics and separate VR from any other learning product (Mikropoulos 1996).

Side Note: Learning According to Constructivist Theories and the Role of Preconceptions

The term constructivism stands for a learning theory which is the foundation of many didactic approaches. According to the constructivist paradigm, humans construct their own reality. Therefore, the knowledge acquired only exists in the mind of the creator and cannot be transferred or, metaphorically speaking, handed over like a book. However, knowledge is in fact a product of an active construction process which is based on individual experiences and preconceptions (Otto 2012, p. 47).

As social factors like specific settings are not considered important in conventional constructivist theories, so-called moderate constructivist theories, which in turn emphasise the influence of learning environments for the construction of knowledge, are nowadays widely accepted as the scientifically correct way of knowledge acquisition.

As a result, teachers have to create appropriate classroom settings in order to facilitate the learning process (Rinschede and Siegmund 2020, p. 155). Problem-based learning (PBL) as well as discovery learning are two approaches that are especially suited to implement moderate constructivism in class because learners participate actively and they rely on situated learning and authentic problems learners have to solve in different settings and contexts which are relatively close to their everyday life (Otto and Schuler 2012, pp. 137–138).

As mentioned before, according to constructivist theories, present preconceptions, often labelled misconceptions although they mostly prove reasonable in the context of their creation, are highly important for the construction of new knowledge because they function as a foundation. Yet, they can also emerge as potential obstacles towards scientifically accurate conceptions if new conceptions and information are rejected. There is even the possibility that material or information are not interpreted in the way originally intended by the teacher and learners might not be able to grasp them. This is also the case with landscapes which also represent social constructs and are therefore perceived differently by every individual (Kühne 2006, p. 146). Either way, it is essential to be aware of learners' preconceptions to optimise the learning process and hence strive for a conceptual change in class - a continuous change from everyday conceptions to scientific ones.

There are different reasons why constructivism presents the best theoretical approach to implement VR into teaching (Stojšic et al. 2016, p. 86). One of them is that the 5 principles for problem-based learning which are part of the constructivist didactics (Otto and Schuler 2012, p. 143) can be applied to VR (Fig. 2). Once a question is posed, learners face an authentic problem and situated learning with a social situation similar to the first principle (1). During the situated learning in the VR, a social situation is created through the mentioned feeling of spatial presence and it is authentic because the high level of immersion leads to a feeling of being physically present. The knowledge gained can be methodically applied to other 360-degree environments and ensures that the learning process takes place in multiple contexts (2). Therefore, the topics which are discussed have to be selected according to their potential of transferring the knowledge acquired to different topics. The criterion of multiple perspectives (3) is also accomplished when using VR in class. Depending on the type of VR, learners can even choose between

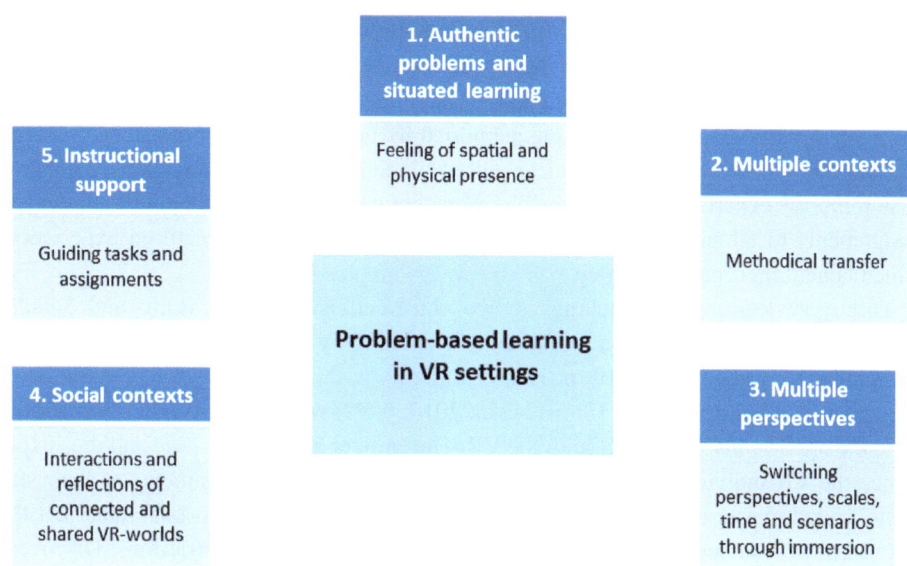

Fig. 2 The 5 principles of problem-based learning (Otto and Schuler 2012) in VR settings

several perspectives. However, it is important to keep in mind that every medium is a subjective display of reality. Although, unlike an ordinary 2D-picture, there is no focus on a particular situation or a selected object from a certain perspective, a 360-degree picture is still based on a preselected space and a specific form of representation, both determined by the creators' objective. This subjectivity should be discussed critically in the aftermath of the 360-degree experience. On the one hand, there are VR-environments where users have a stationary point-of-view and on the other hand, there are those where users can navigate through the VR-environment (Sánchez et al. 2000, p. 349). Learners appreciate the individualised approach of a subject and the fact that they can dictate their own pace (Kingston et al. 2012, p. 1283). Only a first-person-perspective offers the possibility of immersion and a relatively realistic navigation through the VR and encounter with the VR while facilitating a certain point-of-view and thereby aiming at a subjectivity that can be the basis of a critical discussion (Sánchez et al. 2000, p. 350). The geographic scale can change from a ratio of 1:1 (Hruby et al. 2019, p. 128) to a different one (Sánchez et al. 2000, p. 350). Especially topics which cannot be visualised or explored without a change in scale are suited for this type of approach (Mikropoulos 1996). Not only changes in scale but also between a variety of scenarios or in time can create multiple perspectives in the sense that learners get to know several realities possible (Havenith et al. 2019, p. 186). Learning in social contexts (4) takes place in the form of connected and shared VR-worlds which means, learners can meet each other and interact in the VR (Sánchez et al. 2000, p. 350). Additionally, there is the option of working in groups by using VR as a medium. Students can explore the VR individually and afterwards complete tasks and exchange their experiences in groups. With the help of this approach, a debate is encouraged and media literacy in the fields of communication about media, cooperation and reflection of media and the use of media is promoted. The final principle of problem-based learning is learning supported by instruction (5). If assistance is missing and the VR does not follow a clear purpose it can be overwhelming. This is also the case if the VR experience is too long. Teachers therefore need to create guiding tasks and assignments to allow their students to have customised but yet structured experiences which benefit their media literacy.

Discovery learning also belongs to the constructivist didactics (Otto and Schuler 2012, p. 139). VR can contribute to discovery learning by allowing to discover a subject interactively (Hruby et al. 2019, p. 128). Furthermore, this active way of learning helps to increase the learners' focus (Stojšic et al. 2016, p. 92) which is why every user should have the opportunity to interact with the VR (Havenith et al. 2019, p. 187) and thus experience the VR individually. To do so, teachers can again give assignments to their students, called quests in the field of gaming, they have to complete to guarantee that the VR world is discovered independently if there are no movement restrictions. Discovery learning and learning in the VR share even more features in the sense that learners have an active role, are independent, are able to learn individually and gain insight into complex issues (Otto and Schuler 2012, p. 139).

An important strategy of discovery learning is to create so-called cognitive conflicts (Otto and Schuler 2012, p. 140) by confronting students with information which contradicts their present preconceptions (see side note). VR can also serve as a medium to create such conflicts. A common example for a preconception which differs from scientific ones is that there are only deserts consisting of sand. In this case, 360-degree pictures and 360-degree videos of different types of deserts can initiate a cognitive conflict. While students are moving independently through a VR-world where they are able to gather own experiences, quests and instructional support can provide help even after cognitive conflict by imparting scientific concepts and triggering a conceptual change.

Not only preconceptions and VR, but also foreknowledge and VR offer a potential for geography education. Edler et al. (2019) identify audiovisual maps called soundscapes. As the auditory dimension is most important next to the visual dimension, these maps are expanded by sounds, melodies, language, music and audiorealistic recordings (Edler et al. 2019, p. 5; see in this volume: Siepmann et al. 2020). Auditory impulses can be used to activate foreknowledge if for example a 360-degree picture of an urban environment is accompanied by the sound of a busy street (Edler et al. 2019, p. 12). The learners subsequently link their impressions to experiences they have already made and recall their foreknowledge about being next to a busy street which may also include known smells and emotions among other things (Edler et al. 2019, p. 10).

4 Conclusion

This paper has pointed out many options and reasons to integrate VR into geography education. On the one hand media literacy can be promoted and on the other hand common topics and typical methods of geography education can be taught. An immersive VR-environment provides an opportunity to experience spaces which are out of reach for a class due to a variety of reasons. Methods used on field trips can be adopted as well as quests and software normally used on behalf of the gaming industry. The various ways of using VR and new media in class and the numerous potentials shown are mostly yet to be confirmed empirically but helped to illustrate how VR and new media can and should be a part of geography education in general. For the most part, the theoretical foundation (e. g. constructivist didactics, cartography) is already in place and can be transferred to the use of VR in class.

However, there are still obstacles to overcome. First off, the equipment in school is important, as most schools do not own head-mounted displays to facilitate VR-experiences. Although cardboards have proofed to be an inexpensive alternative, smartphones are necessary and the scope is limited compared to head-mounted displays and motion control on a computer. After all, also motion sickness is another problem of immersive VR-experiences due to the perceived movements in the VR while the body is not actually moving. Furthermore, there is only a limited range of interactive VR-contents so far which are suitable for geography education and it is unclear who could fill this void. Most teachers rarely have the technical skills to create an own VR for their lessons. Teachers need further training regarding media education and didactics to identify and to utilise the potential of VR in class.

If these challenges are met, VR-experiences bear a tremendous potential for constructivist learning settings because of their support of individual construction processes. The five principles of problem-based learning in VR-settings can serve as a guidance for the development of new VR content. By allowing learners to perceive numerous perspectives of a topic with less restrictions than conventional learning settings due to the level of participation, authenticity and freedom, geography classes could eventually be transformed into more efficient learning environments regarding every topic.

References

Baacke, D. (1997). Medienpädagogik. In E. Straßner (Ed.), *Grundlagen der Medienkommunikation 1*. Tübingen: De Gruyter.

Bagoly-Simó, P. M. (2020). Landscape in geography textbooks. In D. Edler, C. Jenal, & O. Kühne (Eds.), *Modern approaches to the visualization of landscapes* (pp. 371–386). Wiesbaden: Springer VS.

Barnikel, F., & Vetter, M. (2011). Digitale Medien im Geographieunterricht—Nie war Unterrichten schöner! *Diercke 360°, 2011*(02), 4–5.

Chatel, A., & Falk, G. (2017). Smartgeo—mobile learning in geography education. *European Journal of Geography, 8*(2), 153–165.

Deggim, S., Kersten, T. P., Tschirschwitz, F., & Hinrichsen, N. (2017). Segeberg 1600— Reconstructing a historic town for virtual reality visualisation as an immersive experience. *The International Archives of the Photogrammetry, Remote Sensing and Spatial Information Sciences, XLII-2/W8*, 87–94.

Düren, M., & Bartoschek, T. (2013). Assessing the usability of WebGIS for Schools. In Jekel, T., Car, A., Strobl, J., & Griesebner, G. (Eds.), *GI_Forum 2013. Creating the GISociety* (pp. 388–398) Berlin: Herbert Wichmann.

Edler, D., Husar, A., Keil, J., Vetter, M., & Dickmann, F. (2018). Virtual reality (VR) and open source software: A workflow for constructing an interactive cartographic VR environment to explore urban landscapes. *KN—Journal of Cartography and Geographic Information, 68*(1), 5–13.

Edler, D., Kühne, O., Keil, J., & Dickmann, F. (2019). Audiovisual cartography: Established and new multimedia approaches to represent soundscapes. *KN—Journal of Cartography and Geographic Information, 69*(1), 5–17.

Edler, D., Jenal, C., & Kühne, O. (2020). Modern Approaches to the visualization of landscapes— An introduction. In D. Edler, C. Jenal, & O. Kühne (Eds.), *Modern approaches to the visualization of landscapes* (pp. 3–15). Wiesbaden: Springer VS.

Edler, D., Keil, J., & Dickmann, F. (2020). From Na Pali to Earth—An 'unreal' engine for modern geodata? In D. Edler, C. Jenal, & O. Kühne (Eds.), *Modern approaches to the visualization of landscapes* (pp. 279–291). Wiesbaden: Springer VS.

Fontaine, D. (2020). Landscape in computer games—the examples of GTA V and watch dogs 2. In D. Edler, C. Jenal, & O. Kühne (Eds.), *Modern approaches to the visualization of landscapes* (pp. 293–306). Wiesbaden: Springer VS.

Havenith, H.-B., Cerfontaine, P., & Mreyen, A.-S. (2019). How virtual reality can help visualise and assess geohazards. *International Journal of Digital Earth, 12*(2), 173–189.

Hochschild, V., Braun, A., Sommer, C., Warth, G., & Omran, A. (2020). Visualizing landscapes by geospatial techniques. In D. Edler, C. Jenal, & O. Kühne (Eds.), *Modern approaches to the visualization of landscapes* (pp. 47–78). Wiesbaden: Springer VS.

Hruby, F., Ressl, R., & de la Borbolla del Valle, G. (2019). Geovisualization with immersive virtual environments in theory and practice. *International Journal of Digital Earth, 12*(2), 123–136.

Kanwischer, D. (2014). Digitale Geomedien und Gesellschaft. *Geographische Rundschau, 2014*(6), 12–17.

Kingston, D. G., Eastwood, W. J., Jones, P. I., Johnson, R., Marshall, S., & Hannah, D. M. (2012). Experiences of using mobile technologies and virtual field tours in physical geography: Implications for hydrology education. *Hydrology and Earth System Sciences, 2012*(16), 1281–1286.

Kleber, A., Edler, D., & Dickmann, F. (2020). Cartography and the sea: A javaScript-based web mapping application for managing maritime shipping. In D. Edler, C. Jenal, & O. Kühne (Eds.), *Modern approaches to the visualization of landscapes* (pp. 173–186). Wiesbaden: Springer VS.

KMK (2017). *Strategie der Kultusministerkonferenz „Bildung in der digitalen Welt".* Beschluss der Kultusministerkonferenz vom 08.12.2016 in der Fassung vom 07.12.2017.

Kühne, O. (2006). Landschaft und ihre Konstruktion—Theoretische Überlegungen und empirische Befunde. *Naturschutz Und Landschaftsplanung, 2006*(5), 146–152.

Kühne, O., & Jenal, C. (2020). The threefold landscape dynamics—Basic considerations, conflicts and potentials of virtual landscape research. In D. Edler, C. Jenal, & O. Kühne (Eds.), *Modern approaches to the visualization of landscapes* (pp. 389–402). Wiesbaden: Springer VS.

Lindner-Fally, M., & Zwartjes, L. (2012). Learning and teaching with Digital Earth–Teacher training and education in Europe. In T. Jekel, A. Car, J. Strobl, & G. Griesebner (Eds.), *GI_Forum*

2012: Geovisualization, Society and Learning (pp. 272–282). Berlin: Herbert Wichmann Verlag.

Medienpädagogischer Forschungsverbund Südwest (mpfs) (2018). *JIM-Studie 2018.* https://www.mpfs.de/studien/jim-studie/2018/. Accessed 10 Sep 2019.

Merchant, Z., Goetz, E. T., Cifuentes, L., Keeney-Kennicutt, W., & Davis, T. J. (2014). Effectiveness of virtual reality-based instruction on students' learning outcomes in K-12 and higher education: A meta-analysis. *Computers & Education, 70,* 29–40.

Michel, U., Siegmund, A., & Volz, D. (2011). Digitale Revolution im Klassenzimmer. *Praxis Geographie, 2011*(11), 4–9.

Mikropoulos, T. A. (1996). Virtual geography. *VR in the Schools* 2(2). https://www.academia.edu/24151342/Virtual_Geography. Accessed 4 Nov 2020.

Otto, K.-H. (2012). Didaktische Modelle und Prinzipien. In J.-B. Haversath (Moderator), *Geographiedidaktik. Theorie—Themen—Forschung* (pp. 37–55). Braunschweig: Westermann.

Otto, K.-H./Schuler, S. (2012). Pädagogisch-psychologische Ansätze. In: J.-B. Haversath (Moderator), *Geographiedidaktik. Theorie—Themen—Forschung* (pp. 133–164). Braunschweig: Westermann.

Papadimitriou, F. (2020). Visualization of future landscapes, postmodern cinema and geographical education. In D. Edler, C. Jenal, & O. Kühne (Eds.), *Modern approaches to the visualization of landscapes* (pp. 351–369). Wiesbaden: Springer VS.

Poplin, A., de Andrade, B., & Mahmud, S. (2020). Exploring tangible and intangible landscapes of evocative places: Case study of the City of Vitória in Brazil. In D. Edler, C. Jenal, & O. Kühne (Eds.), *Modern approaches to the visualization of landscapes* (pp. 519–547). Wiesbaden: Springer VS.

Quade, D., & Felgenhauer, T. (2013). Section Editorial: Geoinformation and society: Practising and comprehending geomedia. In T. Jekel, A. Car, J. Strobl, & G. Griesebner (Eds.), *GI_Forum 2013. Creating the GISociety* (pp. 262–271). Berlin: Herbert Wichmann.

Rinschede, G., & Siegmund, A. (2020). *Geographiedidaktik* (4., völlig neu bearbeitete und erweiterte Aufl.). Paderborn: Schöningh.

Sánchez, Á., Barreiro, J. M., & Maojo, V. (2000). Design of virtual reality systems for education: A cognitive approach. *Education and Information Technologies, 5*(4), 345–362.

Selwyn, N. (2007): Dealing with digital inequality: Refocusing our approach towards young people, technology, and social exklusion. In Kompetenzzentrum Informelle Bildung (Ed.), *Grenzenlose Cyberwelt? Zum Verhältnis von digitaler Ungleichheit und neuen Bildungszugängen für Jugendliche* (pp. 31–44). Wiesbaden: VS Verlag.

Siepmann, N., Edler, D., & Kühne, O. (2020). Soundscapes in cartographic media. In D. Edler, C. Jenal, & O. Kühne (Eds.), *Modern approaches to the visualization of landscapes* (pp. 247–263). Wiesbaden: Springer VS.

Stiftung NaturSchutzFonds Brandenburg. (n.d.). *Expedition Wilde-welten.de.* https://www.expedition-wilde-welten.de. Accessed 4 Nov 2020.

Stintzing, M., Pietsch, S., & Wardenga, U. (2020). How to teach "landscape" through games? In D. Edler, C. Jenal, & O. Kühne (Eds.), *Modern approaches to the visualization of landscapes* (pp. 333–349). Wiesbaden: Springer VS.

Stratmann, J., Ristea, A., Leitner, M., & Paulus, G. (2020). Exploring urban "blightscapes" applying spatial video technology and geographic information system: A case study from Baton Rouge, USA. In D. Edler, C. Jenal, & O. Kühne (Eds.), *Modern approaches to the visualization of landscapes* (pp. 499–517). Wiesbaden: Springer VS.

Stojšic, I., Džigurski, A. I., Maričić, O., Bibić, L. I., & Vučković, S. Đ. (2016). Possible application of virtual reality in geography teaching. *Journal of Subject Didactics, 2*(1), 83–96.

Thomas, P. M. (2020). The digitalizing society—Transformations and challenges. In D. Edler, C. Jenal, & O. Kühne (Eds.), *Modern approaches to the visualization of landscapes* (pp. 447–457). Wiesbaden: Springer VS.

Tulodziecki, G. (2011): Handeln und Lernen in einer von Medien mitgestalteten Welt—Konsequenzen für Erziehung und Bildung. In C. Albers, J. Magenheim, & D. M. Meister (Eds.), *Schule in der digitalen Welt. Medienpädagogische Ansätze und Schulforschungsperspektiven* (pp. 43–64). Wiesbaden: VS Verlag.

Vetter, M. (2019). 3D-Visualisierung von Landschaft—Ein Ausblick auf zukünftige Entwicklungen. In O. Kühne, F. Weber, K. Berr, & C. Jenal (Eds.), *Handbuch Landschaft* (pp. 559–573). Tübingen: Springer.

Vetter, M. (2020). Technical potentials for the visualization in virtual reality. In D. Edler, C. Jenal, & O. Kühne (Eds.), *Modern approaches to the visualization of landscapes* (pp. 307–317). Wiesbaden: Springer VS.

Christopher Prisille is a researcher at Ruhr-Universität Bochum in Germany. He has earned an M. Ed. degree in Mathematics and Geography in 2017. In his research, he specialised in media literacy of students and media literacy education of geography teachers. At university, he teaches classes about digital media and VR for upcoming geography teachers.

Marko Ellerbrake is a researcher at Ruhr-Universität Bochum in Germany. He has earned an M. Ed. degree in English and Geography in 2018. In his research, he specialised in preconceptions about climate change and adaption with a focus on heatwaves. At university, he teaches upcoming geography teachers.

How to Teach "Landscape" Through Games?

Maximilian Stintzing, Stephan Pietsch and Ute Wardenga

Abstract

Augmented Reality (AR) adds virtual features to the physical, real world that support the latter's exploration. A learning environment is enriched with location-based digital learning resources such as texts, images, maps and audio-visual presentations on certain subjects. In an ongoing project, we have applied AR to teach today's pupils the profound change of a cultural landscape—the former drain field area of Hobrechtsfelde north of Berlin, today a part of Barnim Nature Park. The paper presents the design of the digital excursion game "The Hunt in Hobrechtsfelde Forest", and how it has worked in first tests. The game explains the landscape's development through retrospective methods from Historical Geography, and illustrates it for players "in situ" through digital data on regional geography. The didactic basis consists in cross sections of time through which players have to pass. In the location-based game approach, the GPS coordinates contain the relevant, digitally prepared information on the landscape after the draining fields. Players interact, in a 'mixed reality', with physical as well as virtual objects, which cognitively activates the knowledge that they have possessed before, and/or what they are given during the excursion. At the same time, the AR game produces emotions and experiences in the context of the subject 'landscape', which further supports the transfer of knowledge.

M. Stintzing (✉) · S. Pietsch · U. Wardenga
Leibniz-Institut für Länderkunde, Leipzig, Germany
e-mail: m_Stintzing@ifl-leipzig.de

S. Pietsch
e-mail: S_Pietsch@ifl-leipzig.de

U. Wardenga
e-mail: U_Wardenga@ifl-leipzig.de

© Springer Fachmedien Wiesbaden GmbH, part of Springer Nature 2020
D. Edler et al. (eds.), *Modern Approaches to the Visualization of Landscapes*, RaumFragen: Stadt – Region – Landschaft,
https://doi.org/10.1007/978-3-658-30956-5_19

Keywords

Landscape · Augmented reality · Education · Location-based gaming · Serious games · Knowledge transfer through games · Digitization.

1 Introduction

This paper deals with the question how to teach today's pupils the profound change of a cultural landscape (German: *Kulturlandschaft*) through games; by means of new media (in this volume, connections between games and modern media are also addressed by Edler et al. 2020a; Fontaine 2020; Vetter 2020). Our example is the former drain field (German: *Rieselfeld*) area around Hobrechtsfelde, north of Berlin. Since 1998, it has been part of Barnim Nature Park. The Federal Agency for Nature Conversation (BfN) defines it as a historic cultural landscape (Schwarzer et al. 2018, p. 429).

In this article, we understand landscape as a perception of physical space constructed by humans, i.e. as an individual and social construction (Kühne 2018; see in this volume: Edler et al. 2020b; Kühne 2020; Kühne and Jenal 2020). The environment is thus viewed from a socially learnt perspective, modified by subjective primary and secondary experiences, to form a certain landscape, whereby the physical elements and structures are each given different meanings (Kühne 2019). Landscape is thus subject to constant change as the physical foundations change (Kühne 2018): A former sewage farm landscape with the function of sewage treatment can be transformed into a recreational landscape, whereby other elements and structures overlay the historical relics for the viewer and disappear from the landscape in his field of vision. On the other hand, society itself is also subject to constant change, which means that the perception of landscape features, for example aesthetic ideas of a beautiful area, also leads to a changed ideal of landscape (Schenk 2006).

At the paper's beginning, we describe, based on findings from Historical Geography, the landscape's change, which can also be understood as a change of relationships between humans and nature. Then, we outline the preconditions for the development of educational games. They include, first, a sufficient interest of teachers to convey to pupils basic notions of ever-changing landscapes by means of out-of-school learning and didactics based on media and games. Second, a gamification process, through location-based digital games, allows pupils to look behind what is directly visible; to turn that façade into an "Augmented Reality" and thus make it comprehensible as a result of dynamic change. Third, in order to translate the theoretical approach into a practical game, we need a convincing game story, which is embedded in the context of the teaching and lets the landscape crossed during the game become a new kind of experience.

2   The Former Drain Field Area of Hobrechtsfelde: History and Geography

The drain fields are located in the plain tracts of Lietzengraben and Panke rivers (Gärtner 2015) and form the western part of Barnim, a glacial plateau northeast of Berlin whose surface was shaped by the Saale ice age (ca. 186,000 years ago; Liedtke 2001). After the glaciers of the Weichsel ice age (ca. 18,000 years ago) retreated, the gains in the Barnim plateau, which originated in the Saale ice age, collected sand and gravel carried by meltwater (Bussemer et al. 2001). This gave rise to extensive outwash plains upon which a forest of birches and pines emerged. Expanding human settlement gradually supplanted it, creating a clear landscape with pockets of forest and a few linear villages by the late eighteenth century (Fig. 1).

Before the nineteenth century, the sand soils were sparsely populated because they held little water for agriculture. Industrialization, however, assigned a new purpose to them. In Berlin, the population grew so much that faeces could no longer be dumped into the city's canals and River Spree: it led to regular outbreaks of cholera, typhoid and dysentery in the German capital, since drinking water was taken from those watercourses (Koch 2015).

Fig. 1 Adapted extract from "Schmettau Maps", sheet no. 64: Bernau. (Source: Staatsbibliothek-Berlin/preussischer Kulturbesitz, dl-de/by-2-0)

Therefore, a new infrastructure for sewage disposal had to be set up. In England, large cities were having success with disposing of their sewages in drain fields. The example was followed by Berlin. Between 1862 and 1873, according steps were directed by urban planner James Hobrecht (1825–1902) and medical scientist Rudolf Virchow (1821–1902). As drain fields, the City purchased the sand soils in the north (Hobrecht 1884). They could serve as filters thanks to their high permeability to water. In addition, the disposal process would yield fertilizers for local agriculture (Gärtner 2015).

At the time, the drain fields were still located outside of the city. The sewage was brought there by 12 pump stations. It flowed through basins where larger suspended solids sank to the ground; they were then used as fertilizers. The pre-cleaned water was distributed on "draining tables" (German: *Rieseltafeln*): rectangular basins separated from each other by dams up to 100 cm tall. While draining into the sandy soil, the water was mechanically cleaned several times. Then, it was collected by underground pipes, led into ditches and returned to the general water cycle via fish ponds (a natural indicator of its quality) and river systems (Hobrecht 1884). After a round of draining, agriculture was possible on the draining tables, producing grains and vegetables for the Berlin metropolis (Gärtner 2015, p. 4).

The cultivation of the draining fields started in 1906. In 1908, the City's property of Hobrechtsfelde received its name in honour of James Hobrecht, the originator of the dual-use concept. In the following decades, its productive agriculture expanded to the point of employing over 200 people (Schulze 2015). After World War II, Berlin's sewage output soared again due to growing industry and population. At the same time, sewage came to be contaminated with industrial waste such as heavy metals and other inorganic materials. As a consequence, large parts of the draining fields were switched to intense filtering, that is, the originally flat draining tables were upgraded to tall basins. Here, the heavily contaminated sewage constantly stood at several centimetres and contaminated the soil, so that agriculture ceased to be possible. Intense filtering continued through GDR times until 1986, when a modern sewage treatment plant was constructed in neighbouring Schönerlinde (Senatsverwaltung für Stadtentwicklung Berlin 1992, p. 6).

In Hobrechtsfelde, attempts were then made at reforestation and renaturation; the draining fields were levelled. At first, success was limited: due to the draining system, the sandy soil was extremely permeable to water, so that up to 60% of new plants dried up. At the same time, the ground was heavily contaminated with pollutants that had to be bound through various measures in the ground. In the process, the former draining fields were gradually transformed into a recreational area for Berlin (Kappel 2015).

Today, it can be experienced as a seemingly natural, half-open forest with a great diversity of species. Even so, the area continues to require human management and cultivation in order to prevent, for example, an excessive natural reforestation or the drying-up of places as pre-cleaned water is discharged by the Schönerlinde sewage treatment plant (Kappel 2015).

3   "Landscape" Taught in School

The term "landscape" (German: *Landschaft*), once central for school teaching, has fallen into disrepute since the 1970s (Schultz 1980). In times enthusiastic about modernization (Wardenga 2019), the historical approach associated with the "landscape" term seemed outdated and not lending itself to preparing mature citizens for the future of a more and more globalized world. In school teaching, however, some perspectives and practices have survived that have belonged to the foundation of geography teaching since the late nineteenth century and that now go by the more innocuous term "space" (Schlottmann and Wintzer 2019; Wardenga 2002). One is, for example, the core skill of "orientation in space". It is supposed to enable pupils to act geographically, "make well-founded decisions in their everyday world and participate in the democractic development of the society" (Landesinstitut für Schule und Medien Berlin-Brandenburg 2015b, p. 5).

The teaching of that skill depends on concrete places of learning that allow "spaces" to be read and experienced as (historically changing) complex systems of interactions between humans and nature, and thus as "landscape". Interdisciplinary skills such as "accepting diversity" and "recognizing and appreciating (also historically) diverse life concepts" can be developed "on site" more than anywhere else (Landesinstitut für Schule und Medien Berlin-Brandenburg 2015a, p. 25), through activity-oriented learning (Schreiber 2016, p. 101). Encouraged by experiencing self-efficacy, pupils can thus develop abilities to "participate responsibly in societal and political opinion formation and decision processes"; to "negotiate their own intentions, tolerate diverse interests and find democratic solutions in conflicts" (Landesinstitut für Schule und Medien Berlin-Brandenburg 2015a, p. 26).

In Germany, first efforts have been made in recent years to adopt the "outdoor school" concept (German: *Draußenschule*), which has been successfully implemented in Scandinavia (*Udeskole*) since the 1930s (Bentsen 2016). From 2014 to 2016, the Pedagogical Institute of Johannes Gutenberg University Mainz (JGU) and the German Hiking Association (DWV) conducted the joint project "School hiking. Experience outdoors. Discover diversity. Move people" (*Schulwandern. Draußen erleben. Vielfalt entdecken. Menschen bewegen*), which included an experimental "outdoor school" (e.g. Gräfe et al. 2015, 2016a; b; Harring 2015, 2016). In Schleswig-Holstein and Hamburg, the agency "Landscape Adventure" (www.landschaftsabenteuer.de) has operated an "Outdoor School—Environmental education practical and close to life" (*Draußenschule—Umweltbildung praktisch und lebensnah*) since 2008, supported by the State of Schleswig-Holstein (Education for Sustainable Development [BNE]—Certificate "School fo the Future in Schleswig-Holstein" [*Zukunftsschule.SH*]). Both projects were accompanied by scientists and aimed at developing "best-practice models" to complement the everyday abstract teaching of knowledge in schools with out-of-school places of learning.

However, such activities, subsumed under the term "outdoor education" (von Au und Gade 2016), have so far remained exceptions: Germany in this regard still belongs to the "developing countries" (von Au 2016, p. 14) when compared to Denmark (Bentsen 2016), Scotland (Telford et al. 2016) or part of the US (Morrison 2016). In German schools, the teaching of processes related to cultural landscapes still takes place in a predominantly abstract way, with little reference to concrete views. In order to make subjects of regional geography accessible to the generation of digital natives, we need to develop attractive methods that cater to the specific needs of that group, provide links to their everyday world, and use new media for that purpose.

Taking up that task, the Federal Ministry of Education and Research (BMBF) funds the project "Spaces to Play: Lanscape as a Space for Discovery and Experience" (*SpielRäume: Entdeckungs- und Erlebnisraum Landschaft*; launched in 2018), which aims at making structures of landscape and concepts of regional geography accessible to pupils through games in out-of-school places of learning; the testing ground is Barnim Nature Park. In this framework, the authors of this paper develop digital excursion games in cooperation with the Nature Park. They are location-based games based on the application "Wherigo". Each game contains a storyline and elements of "Augmented Reality" (AR). Our tests of a game related to the drain field area of Hobrechtsfelde have shown that it both heightened pupils' motivation to look into subjects of regional geography and their ability to recap such a subject.

4   Location-Based Gaming and Gamification

Games are an abstraction of reality. They feature arcs of suspense as well as their own time and space; they are repeatable and played voluntarily (Sailer 2016, p. 20). In the process of *gamification*, elements of the real world are mixed with mechanics and elements of game so that the game maintains and increases users' motivation to carry out the game's assignments.

For today's smartphones or tablets, it is easy to recognize, visualize and anaylze spatial objects, structures and phenomena. Their User interface is constantly improved, simplified and adapted Today, almost every end device possesses numerous technical instruments that can collect space-related and geographic data and can thus be used for the geographical teaching of landscape, especially outdoors: camera, altimeter, GPS, compass, voice and sound recorders, etc. (Feulner and Kremer 2014).

Our project uses the integrated application "Wherigo". It allows the creation of GPS-based games for specific participants outdoors. For example, we can create interactive excursions or city tours including a storyline. To start the respective "game", the excursion's leader just needs to import it into the app as a so-called cartridge (that is, a programmed course of the game). The "Wherigo Builder" translates/transforms the cartridges into the respective programming language (in this case: Lua). The available options are manifold. For example, audio-visual data (such as images, audio files, texts) can be imported that users see only in certain areas. Various actions, assignments and requests to enter things can be created through which developers can equip the excursion with elements of game (Teamer und sein Trupp 2019). Places in the physical environment are defined that players have to visit. The application assigns each place to a so-called zone. After a zone has been activated, certain events are triggered (signalled also by sounds) when the zone is approached or entered.

Wherigo has advantages also for beta versions to test methods of teaching subjects of regional geography through games. The software works without internet connection after the game file has been downloaded to the tablet or smartphone; navigation works through GPS and map systems available offline. Around Wherigo, various types of assignments and cases can be created by each developer according to the respective needs and goals (Pánek et al. 2018). Furthermore, the respective contents can be digitally prepared and, by means of "Augmented Reality" (AR), made available on the respective end device in a way that adequately complements the physical environment with virtual information.

4.1   Teaching of "Landscape" and Augmented Reality (AR)

AR is any technology that lets physical reality co-exist with virtual, two- or three-dimensional objects (Joan 2015). AR bridges the gap between the two worlds by enhancing and mixing reality with digital materials (Bacca et al. 2014; Hsin-Kai et al. 2013). That "mixed reality" (Milgram et al. 1994, p. 283; van Krevelen and Poelman 2010) ties (at least) people more closely to their surroundings (Klopfer and Sheldon 2010). It creates immersive moments and experiences that stir motivation (Lindner et al 2019) and activate existing knowledge (Hsin-Kai et al. 2013).

Therefore, AR is highly useful for education, because the concrete place of learning in the real world can be enriched with digital educational resources such as texts, images, maps and audio-visual presentations. AR thus allows pupils to experience in many possible ways phenomena that remain hidden from the eye in the real world since they are only, or can only be, "told" (Joan 2015). This shows, "[…] that AR can make educational environments more productive, pleasurable and interactive […]." (Lee 2012, p. 19) and is generally speaking, able to improve the the quality of the respective information.

AR needs to be regarded not as a mere technology, but as a very concept of teaching where AR glasses are only one component (Hsin-Kai et al. 2013). If AR concepts are incorporated into game designs and mechanics in a reflected manner, they can help players interact with the place of learning/playing.

To this end, we can for example develop virtual personalities whom players need in order to find a real, physical object in the landscape (Hsin-Kai et al. 2013). Thereby, different stages of a place's historical evolution can be made comprehensible, or the surroundings can be charged with various meanings (Hsin-Kai et al. 2013; Klopfer and Sheldon 2010). Or, the game assigns problems that can be solved only if one or several cooperating pupils combine digital information with such from the analogue/physical surroundings. When solving the problem, new places and new virtual data, that is, new insights for pupils are unlocked gradually according to the principle of "cascading information". This principle prevents the game from overwhelming pupils with too much input at once (Joan 2015).

Moreover, AR location-based games allow pupils to organize themselves along digital/virtual information, which deepens their immersion in the subjects to be learnt. Finally, by encountering problems visually, pupils will have less difficulty solving them (Joan 2015).

4.2   Storytelling

Video games tell stories. Most of them feature narrative introductions that attach meaning to the actions expected from players. Besides, "many games have quest structures, and most games have protagonists" (Juul 2001). The event of the game is, to be sure, each time an individually constructed event. However, elements such as plot, game design and assignment structure are constitutive, as they define the framework of the subjective experience. Therefore, in video games, we should distinguish two types of interactivity: narrative situations and gaming situations. The latter offer the freedom to choose from several options (according to the game design) to solve the game's problems. In narrative situations, by contrast, the plot is driven forward in a *certain* way. In this case, it is thus impossible to adapt the plot to a player's individuality. Instead, the plot may be tied to non-player characters (NPCs; Simons 2007). In any case, to be well received by its target group, a game requires a good balance between free/gaming and predetermined/narrative situations (Lochner 2014, p. 144).

A coherent game plot, whose narrative is based on characters (Juul 2001), requires a script that plausibly combines the contents to be conveyed. From the beginning, the challenge is to meet players in their everyday world, because "every newly created world [in a game], whether set in reality or fictional, needs to have things in common with humans' surroundings" (Lochner 2014, p. 43).

For the fun of the game, it is also decisive that players first make themselves familiar with the rules of the game. This can be supported by a tutorial that is, as in our game, part of the very game, being represented by the virtual head of the Nature Park. He welcomes the players on site and makes them familiar with basic game control through little tasks of orientation.

The rest of the excursion game, which we present in the next section, follows the dramatic "three-act structure" of exposition, confrontation and resolution (Lochner 2014, p. 81). After the introduction, players are confronted with an inciting incident that causes them to embark on a journey.

5 The Drain Field Game: Hunting the Ghost of James Hobrecht

The "trigger" for our game in Hobrechtsfelde is the ghost of James Hobrecht who fears for his legacy and seeks to reactivate the drain field area. The goal of the game is to prevent him from that. In the vein of an adventure game, players, united in groups of three or four people, have to solve puzzles (such as triangulation exercises) or find objects (such as a ghost trap) that drive the plot forward (Lochner 2014, p. 147). A narrative climax is reached when players try to catch the ghost, but he narrowly escapes. Afterwards, there is a final chance to save today's Nature Park—or fail—by passing one last test.

When we tested the journey's simple version, the pupils (five times one class of grades 7 and 8) who played it gave it an average rating between "very good" (32.3%) and "good" (37%). Their satisfaction may spring from the fact that they had subconsciously been familiar with our classic way of narrating, since it is used in Hollywood films or fairytales as well (Kinateder 2012, p. 36).

A core feature of a coherent story design is the convincing design of the characters. In accordance with our goal to pass through different periods of the landscape around Hobrechtsfelde, we selected characters who represent the respective period through both their outward appearance and their way of speaking. For example, Count Schmettau speaks like an eighteenth-century Prussian nobleman, while the 1950s drain field attendant tells his work experiences with the post-war accent of Berlin. In fact, designing a game is generally not so much an academic but an artistic process. In particular, writing dialogues for the different scenes bears resemblance to composing a film script or a theatre play—including the invention of characters' personal backgrounds. Characters need to seem authentic enough to be embraced by the players.

While designing the game, we focussed on the antagonist James Hobrecht, who acts as the main teacher on the subject of drain field operation. The main challenge was to honour him, on the one hand, for his highly innovative idea of sewage treatment in the late nineteenth century, and on the other hand to make clear to players why that method is no longer viable today. We seem to have solved that challenge: 60 out of 127 pupils indicated that they especially liked Hobrecht's character.

Besides, 52 pupils stated to have learnt something about landscape development in Hobrechtsfelde—including its history as well as the functioning of its drain fields and of drain fields in general. We assume that the game was kept in motion above all by the non-player characters (NPCs) devised as antagonists. They drove the plot forward and thus proved suitable for teaching certain contents.

5.1 Conveying Time Sections

To let players retrace the historical evolution of Hobrechtsfelde Forest, we have identified characteristic cross sections of time and interrelated them with one another, resulting in a longitudinal section of time (Plöger 2003). This is a method from Historical Geography. The cross sections are from four points of the past for which cartographic visualizations are available, each of a different mode (Schramm 2009, p. 9): the eighteenth century with the Schmettau maps; the year 1953 with an aerial picture of the drain fields; the year 2008 with an aerial picture of today's recreational area; and the year 2019 with a digital model of the area's soil structures. To deepen the immersion, the stages of development were "filled" with contemporary characters and objects. Players can experience them at a total of 16 learning stations.

5.2 The Eighteenth Century

After the brief introduction to the game's mechanics as well as plot by the virtual head of the Nature Park (Fig. 2), the appearance of the landscape in 1780 is presented through a geo-referenced excerpt of the Schmettau maps, which players see as virtual object. Their own location is indicated on the historical map and can be synchronized with a current map of the area, in order to maximize the sense of place (Klopfer and Sheldon 2010, p. 89; Pánek et al. 2018). Then, players meet the historical character of Count Schmettau (1743–1806), who is surveying the area with a military detachment.

Players are confronted with the time's language/etiquette as well as its surveying devices and methods. The devices appear as virtual objects along with optional

Fig. 2 *Left*: P.G., virtual head of Barnim Nature Park; *right*: James Hobrecht, originator of the drain fields. (Source: Project SpielRäume)

information on their use. Players are assigned a triangulation exercise, equally inspired by the period, on the region's oldest road, the Bernau Military Road, which today serves as a gravel path for walking and biking. After the exercise, players are given two more hints, concerning, first, the landscape's topography at the time, and second, the whereabouts of Hobrecht's ghost.

In sum, the real surroundings are enhanced with digital elements so that players immerse themselves in the time period, remember it later and anticipate the following periods (Hsin-Kai et al. 2013, p. 45; Pánek et al. 2018).

5.3   Operating the Drain Fields in 1953

Searching for Hobrecht's ghost, players are led to the next cross section of time. A worker who has been attending Berlin's drain fields in Hobrechtsfelde is debating with the ghost. The latter then disappears after opening a sewage pipe, which players have to close under time pressure. The grateful worker explains to them the operating mode and eventual problem of the drain fields. Virtual objects are provided, such as a technical plan of the drain fields and a pipe with optional extra information (unlike real objects or characters, virtual ones do not depend on external providers, so that preparation is accordingly easier). Players are assigned exercises on Berlin's demographic growth and sewage contamination.

Further information on the landscape's evolution is given to players through data on virtual objects such as a geo-referenced 1953 aerial picture showing a treeless area altered by humans. This is the players' first visual contact with the rectangular, today almost invisible structures of a drain field area as well as with the overflowing basins in times of intense filtering, which contaminated the soil. Today, the landscape at the respective game stations seems very natural, with players standing in the middle of a forest. When they are asked to compare their own location with the aerial picture, they sense a paradox. The drain field attendant deepens it with photographs of the attempts at reforestation and renaturation in 1986, which were thwarted by insufficient planning. At the same time, he points out lasting landscape features such as the Bernau Military Road and River Lietzengraben, which has to be crossed and is visible in all maps.

5.4   The Recreational Area in 2008

Still searching for the ghost, players reach a viewing platform. The virtual character of a professor at the University for Sustainable Development in nearby Eberswalde appears and explains the methods and problems of the area's renaturation. In a geo-referenced aerial picture from 2008, players see that the area's management has at last been successful, especially when they look at the half-open forest, with its structure of small parcels received from the rectangular draining tables (Fig. 3).

Fig. 3 Adapted extract of the drain field area of Hobrechtsfelde; *top*: 1953; *bottom*: 2009. (Source: Geoportal Berlin [Luftbilder 1953], dl-de/by-2-0; GeoBasis-DE/LGB, dl.de/by-2-0)

In the run-up to the final encounter with the ghost of the draining fields, players have to find physical boards explaining methods of re-watering and the drainage problem of a former draining field area (for other methodological approaches based on remotely sensed imagery in this volume, see: Hochschild et al. 2020; Meyer-Heß 2020). Through visual interaction with the virtual professor, the "physical" data are put in the right context. The complexity of the subject "landscape development" is alleviated by a slow, step-by-step flow of information.

5.5   The 2015 Digital Area Model

In the next and final cross section of time, a landscape planner gives players a digital model of the region, in which vegetation has been removed to carve out soil strcutures. The digital image, again geo-referenced, clearly shows the rectangular stuctures that remain from the operation of the drain fields. Moreover, explanations are provided on the lasting contamination from certain drain field products such as heavy metals in the ground; they make it necessary to constantly monitor water and soil qualities. Equipped with that knowledge, players can now confront Hobrecht's ghost, to convince him that today's landscape concept is more suitable.

6   Conclusion

The location-based digital game "The Hunt in Hobrechtsfelde Forest" conveys the genesis of a current recreational area. The landscape's development is explained through retrospective methods from Historical Geography (Plöger 2003, p. 18), and illustrated for players "in situ" through digital data on regional geography. The didactic basis consists in cross sections of time through which players have to pass; they are supported by maps and represent different stages of the development (related in this volume: Prisille and Ellerbrake 2020). Thus, the landscape appears in different perspectives of the past and the present, is given new meanings and turns out to reflect the changing relationships between human and nature (Klopfer and Sheldon 2010, p. 89). AR concepts help players experience and explore their real surroundings in an authentic manner, for example when the area of a present-day forest is shown, in an aerial picture from 30 years before, without trees, and the change is explained by virtual information or characters (Hsin-Kai et al. 2013).

In the location-based game approach, the GPS coordinates contain the relevant, digitally prepared information on the landscape after the draining fields. Players interact, in a "mixed reality", with physical as well as virtual objects, which cognitively activates the knowledge that they have possessed before, and/or what they are given during the excursion (Hsin-Kai et al. 2013, pp. 45–46). At the same time, the AR game seems to produce emotions and experiences in the context of the subject "landscape", which further supports the transfer of knowledge.

Translated from german by Maximilian Georg, Leipzig

References

Bacca, J., Baldiris, S., Fabregat, R., Graf, S., & Kinshuk. (2014). Augmented reality trends in education: A systematic review of research and applications. *Educational Technology & Society, 17*(4), 133–149.

Bentsen, P. (2016). „Udeskole" in Dänemark. Von einer „Bottom-Up-" zu einer „Top-Down-Bewegung". In von Au & Gade 2016 (pp. 50–63).

Bussemer, S., Gärtner, P., & Thieke, H. (2001). Jungquartäre Reliefentwicklung auf der Hochfläche des Barnims (NE-Brandenburg). In S. Bussemer (Ed.), *Das Erbe der Eiszeit* (pp. 135–147). Langenweißbach: Beier & Beran.

Edler, D., Keil, J., & Dickmann, F. (2020). From Na Pali to Earth—An 'unreal' engine for modern geodata? In D. Edler, C. Jenal, & O. Kühne (Eds.), *Modern approaches to the visualization of landscapes* (pp. 279–291). Wiesbaden: Springer VS.

Edler, D., Jenal, C., & Kühne, O. (2020). Modern approaches to the visualization of landscapes—An introduction. In D. Edler, C. Jenal, & O. Kühne (Eds.), *Modern approaches to the visualization of landscapes* (pp. 3–15). Wiesbaden: Springer VS.

Feulner, B., & Kremer, D. (2014). Using geogames to foster spatial thinking. In R. Vogler, A. Car, J. Strobl, & G. Griesebner (Eds.), *GI_Forum 2014: Geospatial Innovation for Society* (pp. 344–347). Berlin: Herbert Wichmann & Vienna: Austrian Academy of Sciences Press.

Fontaine, D. (2020). Landscape in computer games—The examples of GTA V and watch dogs 2. In D. Edler, C. Jenal, & O. Kühne (Eds.), *Modern approaches to the visualization of landscapes* (Eds.), *Modern approaches to the visualization of landscapes* (pp. 293–306). Wiesbaden: Springer VS.

Gärtner, P. (2015). Vorwort—Naturpark Barnim. In S. Stoll-Kleemann (2015) (Ed.), *Wahrnehmung und Akzeptanz des bundesländerübergreifenden Naturparks Barnim* (pp. 1–6). Greifswald: Institut für Geographie und Geologie der Ernst-Moritz-Arndt Universität Greifswald.

Gräfe, R., Gillessen, C., Harring, M., Sahrakhiz, S., & Witte, M.D. (2016a). Bildungsräume anders denken. Das Modellprojekt Draußenschule. In von Au & Gade 2016 (pp. 70–78).

Gräfe, R., Gillessen, C., Harring, M., Sahrakhiz, S., & Witte, M.D. (2016b). Einmal wöchentlich draußen unterrichten? Eine qualitativ-empirische Studie zur Draußenschule aus der Perspektive von Grundschullehrerinnen. In von Au & Gade 2016 (pp. 79–95).

Gräfe, R., Harring, M., & Witte, M. D. (Eds.). (2015). *Körper und Bewegung in der Jugendbildung. Interdisziplinäre Perspektiven auf schulische und außerschulische Bildungsprozesse.* Baltmannsweiler: Schneider Hohengehren.

Harring, M. (2015). Schulische Jugendbildung? Ausgangslagen und Bedingungen für eine ganzheitliche Bildung im Kontext von Schule. In Gräfe et al. 2015 (pp. 75–91).

Harring, M. (2016). Freizeit und informelles Lernen. In M. Harring, M.D. Witte, & T. Burger (Eds.), *Handbuch informelles Lernen. Interdisziplinäre und internationale Perspektiven* (pp. 416–438). Weinheim: Beltz Juventa.

Hobrecht, J. (1884). *Die Canalisation von Berlin. Im Auftrage des Magistrats der Königl. Haupt- und Residenzstadt Berlin.* Berlin: Ernst & Korn.

Hochschild, V., Braun, A., Sommer, C., Warth, G., & Omran, A. (2020). Visualizing landscapes by geospatial techniques. In D. Edler, C. Jenal, & O. Kühne (Eds.), *Modern approaches to the visualization of landscapes* (pp. 47–78). Wiesbaden: Springer VS.

Hsin-Kai, W., Wen-Yu Lee, S., Hsin-Yi, C., & Jyh-Chong, L. (2013). Current status, opportunities and challenges of augmented reality in education. *Computers and Education, 62,* 41–49.

Joan, D. R. R. (2015). Enhancing education through mobile augmented reality. *Journal of Educational Technology, 11*(4), 8–14.

Juul, J. (2001). Games telling stories? A brief note on games and narratives. *Game studies: The International Journal of Computer Game Research, 1*(1). https://www.gamestudies.org/0101/juul-gts/. Accessed 06 Oct. 2019.

Kappel, R. (2015). Ein Erholungswald für Berlin. In Förderverein Naturpark Barnim (Ed.), *Rieselfeldlandschaft Hobrechtsfelde. Nutzung, Umgestaltung und Entwicklung einer intensiv von Menschen geprägten Landschaft im Norden Berlins* (pp. 52–57). Wandlitz: Förderverein Naturpark Barnim.

Kinateder, B. (2012). Klassische Erzählformen. *Televizion, 25*(2), 34–35.

Klopfer, E., & Sheldon, J. (2010). Augmenting your own reality: Student authoring of science-based augmented reality games. *New Directions for Youth Development, 128,* 85–94.

Koch, K. (2015). James Hobrecht und die Berliner Stadtentwässerung. In Förderverein Naturpark Barnim (Ed.), *Rieselfeldlandschaft Hobrechtsfelde. Nutzung, Umgestaltung und Entwicklung einer intensiv von Menschen geprägten Landschaft im Norden Berlins* (pp. 25–29). Wandlitz: Förderverein Naturpark Barnim.

Kühne, O. (2018). *Landschaft und Wandel. Zur Veränderlichkeit von Wahrnehmungen.* Wiesbaden: Springer VS.

Kühne, O. (2019). *Landscape theories: A brief introduction.* Wiesbaden: Springer VS.

Kühne, O. (2020). The social construction of space and landscape in internet videos. In D. Edler, C. Jenal, & O. Kühne (Eds.), *Modern approaches to the visualization of landscapes* (pp. 121–137). Wiesbaden: Springer VS.

Kühne, O., & Jenal, C. (2020). The threefold landscape dynamics—Basic considerations, conflicts and potentials of virtual landscape research. In D. Edler, C. Jenal, & O. Kühne (Eds.), *Modern approaches to the visualization of landscapes* (pp. 389–402). Wiesbaden: Springer VS.

Landesinstitut für Schule und Medien Berlin-Brandenburg (2015a). Rahmenlehrplan Jahrgangsstufen 1–10, Teil B: Fächerübergreifende Kompetenzentwicklung. https://bildungsserver.berlin-brandenburg.de/fileadmin/bbb/unterricht/rahmenlehrplaene/Rahmenlehrplanprojekt/amtliche_Fassung/Teil_B_2015_11_10_WEB.pdf. Accessed 05 Dec. 2019.

Landesinstitut für Schule und Medien Berlin-Brandenburg (2015b). Rahmenlehrplan Jahrgangsstufen 1–10, Teil C. Geografie, Jahrgangsstufen 7–10. https://bildungsserver.berlin-brandenburg.de/fileadmin/bbb/unterricht/rahmenlehrplaene/Rahmenlehrplanprojekt/amtliche_Fassung/Teil_C_Geografie_2015_11_10_WEB.pdf. Accessed 05 Dec. 2019.

Lee, K. (2012). Augmented reality in education and training. *Tech-Trends, 56*(2), 13–21.

Liedtke, H. (2001). Das nordöstliche Brandenburg während der Weichseleiszeit. In S. Bussemer (Ed.), *Das Erbe der Eiszeit* (pp. 119–133). Langenweißbach: Beier & Beran.

Lindner, C., Rienow, A., & Jürgens, C. (2019). Augmented reality applications as digital experiments for education—An example in the Earth-Moon system. *Acta Astronautica, 161,* 66–74.

Lochner, D. (2014). Storytelling in virtuellen Welten. Konstanz: UVK.

Meyer-Heß, F. (2020). Discovering forgotten landscapes. In D. Edler, C. Jenal, & O. Kühne (Eds.), *Modern approaches to the visualization of landscapes* (pp. 33–46). Wiesbaden: Springer VS.

Milgram, P., Takemura, H., Utsumi, A., & Kishino, F. (1994). Augmented reality: A class of displays on the reality-virtuality continuum. *Proceedings of SPIE (Society of Photo-Optical Instrumentation Engineers), 2351,* 282–292.

Morrison, L. L. (2016). Outdoor education in Iowa, United States. Eine kurze Übersicht. In von Au & Gade 2016 (pp. 64–70).

Pánek, J., Gekker, A., Hind, S., Wendler, J., Perkins, C., & Lammes, S. (2018). Encountering place: Mapping and location-based games in interdisciplinary education. *Cartographic Journal, 55*(3), 285–297.

Plöger, R. (2003). *Inventarisation der Kulturlandschaft mit Hilfe von Geographischen Informationssystemen (GIS). Methodische Untersuchungen für historisch-geographische Forschungsaufgaben und für ein Kulturlandschaftskataster.* PhD thesis, Rheinische Friedrich-Wilhelms-Universität Bonn. https://hss.ulb.uni-bonn.de/2003/0156/0156.pdf. Accessed 20 Jan. 2020.

Prisille, C., & Ellerbrake, M. (2020). Virtual reality (VR) and geography education: Potentials of 360° 'experiences' in secondary schools. In D. Edler, C. Jenal, & O. Kühne (Eds.), *Modern approaches to the visualization of landscapes* (pp. 321–332). Wiesbaden: Springer VS.

Sailer, M. (2016). *Die Wirkung von Gamification auf Motivation und Leistung. Empirische Studien im Kontext manueller Arbeitsprozesse.* Wiesbaden: Springer.

Schenk, W. (2006). Der Terminus "gewachsene Kulturlandschaft" im Kontext öffentlicher und raumwissenschaftlicher Diskurse zu "Landschaft" und "Kulturlandschaft" . In U. Matthiesen, R. Danielzyk, S. Heiland, & S. Tzschaschel (Eds.), *Kulturlandschaften als Herausforderung für die Raumplanung: Verständnisse—Erfahrungen Perspektiven* (pp. 9–21). Hannover: Verlag der ARL.

Schlottmann, A., & Wintzer, J. (2019). *Weltbildwechsel. Ideengeschichten geographischen Denkens und Handelns.* Bern: Haupt.

Schramm, M. (2009). *Digitale Landschaften.* Stuttgart: Franz Steiner.

Schreiber, J.-R. (2016). Kompetenzen, Themen, Anforderungen, Unterrichtsgestaltung und Curricula. In J.-R. Schreiber & H. Siege (Eds.), *Orientierungsrahmen für den Lernbereich Globale Entwicklung im Rahmen einer Bildung für nachhaltige Entwicklung* (2nd ed., pp. 84–110). Bonn: Engagement Global.

Schultz, H.-D. (1980). *Die deutschsprachige Geographie von 1800 bis 1970. Ein Beitrag zur Geschichte ihrer Methodologie.* Berlin: Geographisches Institut der Freien Universität Berlin.

Schulze, A. (2015). Das Berliner Stadtgut Hobrechtsfelde und die Rieselfeldbewirtschaftung. In Förderverein Naturpark Barnim (Ed.), *Rieselfeldlandschaft Hobrechtsfelde. Nutzung, Umgestaltung und Entwicklung einer intensiv von Menschen geprägten Landschaft im Norden Berlins* (pp. 30–38). Wandlitz: Förderverein Naturpark Barnim.

Schwarzer, M., Mengel, A., Konold, W., Reppin, N., Mertelmeyer, L., Jansen, M., Gaudry, K.-H., & Oelke, M. (2018). *Bedeutsame Landschaften in Deutschland. Gutachtliche Empfehlungen für eine Raumauswahl, Bd. 1: Schleswig-Holstein und Hamburg, Niedersachsen und Bremen, Mecklenburg-Vorpommern, Nordrhein-Westfalen, Sachsen-Anhalt, Brandenburg und Berlin.* Bonn-Bad Godesberg: Bundesministerium für Umwelt, Naturschutz und nukleare Sicherheit (BMU).

Senatsverwaltung für Stadtentwicklung und Wohnen Berlin (1992). *Umweltatlas Berlin, 01.10: Rieselfelder (Ausgabe 1992).* https://www.stadtentwicklung.berlin.de/umwelt/umweltatlas/e_text/k110.pdf. Accessed 20 Jan. 2020.

Simons, J. (2007). Narrative, games, and theory. *Game Studies: The International Journal of Computer Game Research, 7*(1). https://gamestudies.org/0701/articles/simons. Accessed 06 Dec. 2019.

Teamer und sein Trupp (2019). *Wherigo-Tutorial.* https://www.geocaching-dresden.de/?s=wherigo. Accessed 10 Dec. 2019.

Telford, J., Beames, S., & Christie, B. (2016). „Outdoor Learning" in Schottland. Ein Überblick. In von Au & Gade 2016 (pp. 42–49).

Vetter, M. (2020). Technical potentials for the visualization in virtual reality. In D. Edler, C. Jenal, & O. Kühne (Eds.), *Modern approaches to the visualization of landscapes* (pp. 307–317). Wiesbaden: Springer VS.

van Krevelen, D. W. F., & Poelman, R. (2010). A survey of augmented reality technologies, applications and limitations. *International Journal of Virtual Reality, 9*(2), 1–20.

von Au, J. (2016). Einführung und Überblick. In von Au & Gade 2016 (pp. 13–39).

von Au, J., & Gade, U. (Eds.). (2016). *„Raus aus dem Klassenzimmer". Outdoor Education als Unterrichtskonzept.* Beltz Juventa: Weinheim & Basel.

Wardenga, U. (2002). Alte und neue Raumkonzepte für den Geographieunterricht. *Geographie Heute, 23*(200), 8–11.

Wardenga, U. (2019). Vergangene Zukünfte—oder: Die Verhandlung neuer Möglichkeitsräume in der Geographie. *Geographische Zeitschrift (Online first).* https://elibrary.steiner-verlag.de/article/https://doi.org/10.25162/gz-2019-0009. Accessed 10 Dec. 2019.

Maximilian Stintzing was a research assistant at the Leibniz Institute for Regional Geography, Leipzig (IfL) from 2017 to 2020. Since then, he works at the Bavarian State Library, Munich. He studied geography (Bachelor) and historical geography (Master) at Bamberg University. His research focuses on colonialism and spatial development, Geoanalysis and agricultural geography.

Stephan Pietsch has been a research assistant at the Leibniz Institute for Regional Geography, Leipzig (IfL) since 2013. He studied Modern and Contemporary History at Dresden Technical University, Geography (Bachelor) at Jena University, and Human Geography: Globalization, Media and Culture (Master) at Mainz University. His main research interests are spatial images, political geography, media geography, and the history of geography and cartography.

Professor Ute Wardenga has been working at the Leibniz Institute for Regional Geography, Leipzig (IfL) since 1996. She heads the department "Theory, Methodology and History of Regional Geography" and coordinates the research group "Historical Geographies".

Visualization of Future Landscapes, Postmodern Cinema and Geographical Education

Fivos Papadimitriou

Abstract

This study fuses geographical education with film studies, by addressing the question of how might university students of Geography visualize a future urban landscape on the basis of key elements of Riddley Scott's classic postmodern film "Bladerunner" (1982). Despite the fact that postmodern urbanism is a standard topic in university geography curricula, the literature is still short of studies relating to teaching postmodern geography. This paper contributes with an empirical study in this field, conducted on university students who had been taught postmodern urbanization. The students were asked to rate some of the main characteristics of postmodernity from with respect to their relevance to Athens and to predict the time by which Athens would show signs of the "Bladerunner" city. They identified areas of Athens resembling most to postmodern traits, predicted that Athens will resemble the Bladerunner city sometime before 2040, and that the anticipated changes will first become felt visually, aesthetically and socially and then, gradually, by behavioral/psychological characteristics. A map of future postmodern Athens areas was also produced from their responses.

F. Papadimitriou (✉)
Eberhard Karls Universität Tübingen, Tübingen, Germany
e-mail: fivos.papadimitriou@mnf.uni-tuebingen.de

© Springer Fachmedien Wiesbaden GmbH, part of Springer Nature 2020   351
D. Edler et al. (eds.), *Modern Approaches to the Visualization of Landscapes*, RaumFragen: Stadt – Region – Landschaft,
https://doi.org/10.1007/978-3-658-30956-5_20

Keywords

Landscape   visualization · Postmodern   landscapes · Urban   landscape · Athens ·
Bladerunner · Landscape complexity · Postmodern cinema · Geographical education ·
Film and landscape · Film and education

1   Introduction

This paper, which heavily draws on a previous publication by the author (Papadimitriou
2019), contributes to the simultaneous exploration of three different, but interwoven
needs: (a) To use film and cinema studies in geography (and geographical education in
particular); (b) To use students' informed opinions to visualize future urban changes; (c)
To explore the increasing relevance of postmodern realities for geography and geographical education.

1.1   Postmodern Landscapes

Postmodern urbanization is a recurrent theme in urban geography and town planning
(Harvey 1989; Hannigan 1995, 1998; Boyer 1994; Dalby 1999; Newman and Paasi
1998; Dear and Flusty 1998; Jameson 1991; Dear 2000; Soja 2000; Brooker 2007; for
studies addressing postmodernism in this volume, see: Fontaine 2020a, 2020b, Kühne
2020; Weber 2020) and has been considered as a sign of late capitalism, or a characteristic of cities of the developing world (Murray 2004). While studying postmodern urbanization therefore, the emphasis may be on design (Besteliu and Doevendans 2002), on
innovation and industrialization (von Tunzelmann 1997), on economic restructuring, or
on broader social and cultural contexts. Within the framework of the rising interest in
landscapes of postmodernity (Kühne 2012, 2018, 2019), typically postmodern spaces are
examined such as hybridization, polyvalency of spaces, pastiche and other ones.

These characteristics become evident as postmodern urban transformations melt the
social with the cultural, the traditional with the modern, eventually forming some kind
of peculiar "cultural citizenships" (Stevenson 1997). The emphasis of postmodernism is
on transient styles and modes of life, on fluidity, uncertainty and incoherence, as well
as on anthropocentric perceptions of reality. Unpredictability and unexpectedness lurk
in every step of a postmodern city, as do personal worldviews and multiculturalism. The
deconstruction of hierarchies and narratives appears in tandem with assertion of minorities and varieties instead of majorities and established groups, highlighting the importance of locality. These signs adopt to urban geographies at local scales, so it is probably
erroneous to claim that an entire city is "postmodern", but it would probably be safer to
say that certain sites, neighborhoods and locales are such.

Among the most widely accepted characteristics for delineating urban postmodernity are strong social diversity, ubiquity of signs of globalization, lack of local firms, extensive cultural fragmentation. These may occur along with excessive social control, even within areas of anarchy or gentrification, alongside with hybrid architectural forms and high technologies.

1.2   Postmodern Cinema

Media and cinema are prominent technological and conceptual vehicles for conveying concepts of postmodernity (Masirevic 2010). The film "Bladerunner" (by Ridley Scott, released in 1982) is the standard archetype (and forerunner) of postmodern urbanization and has been considered as such repeatedly in the scientific literature (Bruno 1987; Bull 1997; Davis 1998; Hefner 2002). This film will be used as a conceptual framework for the present research, although beyond "Bladerunner", the "postmodern cinema" can furnish other (possibly less symbolic) examples (see Neuser 2012), such as Stanley Kubrick's "Clockwork Orange" (1971) and "2001 A Space Odyssey" (1968).

Bladerunner is the cinematic metaphor of Philip Dick's novel "Do Androids dream of Electric Sheep?" by Warner Bros, featuring Harrison Ford and Darryl Hannah. The story takes place in Los Angeles in 2019. The central hero, Rick Deckard, is the "blade runner"; a hunter of genetically engineered replicant humans who are designed for labor in extraterrestrial colonies. These replicants illegally intermingle with non-engineered humans, in an attempt to extent their lifespan. Deckard chases the replicants Roy Batty, Zhora, Pris and Leon around the city, and eventually down to the Tyrell Corporation, a gigantic establishment with headquarters in pyramid-shaped buildings in which replicants are created. Replicants try to hide themselves and conceal their true identity, because they know they will be "retired" (terminated) when found. After many adventures, Deckard eventually ends up by being together with one of the replicants, Rachel, who has no termination date. Besides hybrid identities (replicants intermingling with ordinary humans), individuals in Bladerunner have no strong sense of personal identity (see characteristic 24) while replicants conceal their identities. But even nowadays, our "identity-less and place-less modes of existence" (Papadimitriou 2009a, p. 1331) tend to reshape cyber-identities.

In the context of the movie "Bladerunner", it is also interesting to notice that "Where in modernity the (as pure as possible) model of enlightenment and reason was valid, postmodernism aestheticizes" (Kühne 2019, p. 120). Indeed, aesthetics is central in understanding postmodern in explaining urban representations in cinema in contrast to the characteristics of the real space (Brandt 2009; Araujo 2015).

1.3   Visualizing Future Postmodern Urban Landscapes

The role of landscape in cinema has been studied repeatedly (Steimatsky 1995; Arecco 2001; Azevedo 2007, 2012; Harper and Rayner 2010), while some reseaerchers have examined issues of postmodernity in cinema (Azevedo 2015; Alvarado Duque and Escobar Ramirez 2019). For instance, desert landscapes or old warehouses in the films of Quentin Tarantino are characteristic landscapes (rural and urban respectively) of Tarantino's typically postmodern settings.

The concept of postmodern city has also been associated with futuristic anticipations of a fictional metropolis, with a strong focus on aesthetic parameters (Brandt 2009), which, at times, can be provocative or unusual to say the least. However, although it appears important to define visual/aesthetic criteria for the attribution of "postmodern" as a character of a city, no precise such criteria exist, and, consequently, some more subtle geographical and even psychological criteria may be needed (Parker 1996). For instance, as Brandt (2009, pp. 553) wrote, postmodern cities may resemble "a labyrinth that leaves protagonists and readers in a state of disorientation, fragmentation, and constant decentering" and this state "in favor of the concepts of transition and ambiguity". Further, as biotechnology constitutes a major field in "futures research" (Parker and Zilberman 1995), postmodern urbanities may be inhabited even by bio-engineered bodies in the distant future (in a more techno-favorable scenario), as, in fact, rather early in time, the film "Bladerunner" has already suggested.

1.4   Cinema and Geographical Education

Despite the fact that postmodern geography is a standard theme in all university curricula, the challenges related to teaching postmodern geography have hitherto been disregarded. Further, geographical education using films is rather uncommon, with the exception of few studies only (i.e. Madsen 2014; Sigler and Albandoz 2014). It is therefore the aim of this paper to explore ways by which urban postmodern landscapes can be perceived and assessed by students of human geography.

The fusion of technologies with everyday life creates new modes of living and therefore new cultural mindsets, which current cultural geography should examine (Papadimitriou 2006). While geospatial technologies have already penetrated the domain of geographical education and thus aid in its development (Papadimitriou 2010a, 2010b; see in this volume: Bagoly-Simó 2020; Prisille and Ellerbrake 2020; Stintzing et al. 2020), the realm of fantasy imperceptibly sometimes encroaches upon fields of geographical thinking. The result of this strange fusion of advanced technology, fantasy and geography can be a source of inspiration of students and tutors alike (Lastoria and Papadimitriou 2012; Beneker and van der Schee 2015), as geographers strive to explore visualization of future spaces, whether with the use of technologies or without them (Papadimitriou 2012a; see in this volume: Edler et al. 2020a, 2020b, Kleber et al. 2020; McLean 2020; Siepmann et al. 2020; Stratmann et al. 2020).

2 Methods

The subjects of this study have been a group of 2nd year students of Human Geography of the Department of European Culture of the Hellenic Open University, in May 2011. All students were citizens of Athens and their average ages ranged in between 30–35 years old (this public university enrols students of various ages).

Before participating in the research project, the students had already attended the annual course of General Geography, Human Geography and Civilization of Europe (equivalent to three semesters of study) and volunteered to participate in this research after the end of their course. Before doing so, they had also attended a 4-h long tutorial on postmodern urban geography. This covered topics on the theory of postmodernism, postmodern geography, postmodern urbanism, as well as examples of postmodern geographies from Europe and Greece. They also had to write a 2500 words-long essay of on the transition from modernity to postmodernity in European cities. Only those students who succeeded in this assignment (and hence had an adequately clear knowledge of postmodernity) participated in this research project.

Consequently, upon the successful completion of their assignment, they were given a brief description of characteristics of postmodernity from the film "Bladerunner" (translated into english as shown in the Appendix) and were asked to:

(1) Evaluate the characteristics of postmodernity as they were presented in the brief description of the movie, from 1 (most important) to 5 (least important), by underlining words and expressions in the text and by noticing a number from 1 to 5, next to the underlined words (according to the significance they assigned to each characteristic).

(2) Estimate the year by which Athens (the city they all live in) would resemble some of the descriptions given in the "Bladerunner" city (that is the time by which they anticipated that Athens would have assimilated some of the qualities of the "Bladerunner" city).

(3) Identify areas of Athens that already resemble some of the descriptions of the "Bladerunner" city.

As the characteristics of the "Bladerunner" city were several, they were classified into criteria, according to Table 1, for the purpose of analysis and the average rank per characteristic was calculated.

Table 1 Classification of the characteristics of the "Bladerunner" city, by category type (own depiction)

Type of characteristic	Code numbers of characteristics
Economic	2,12
Behavioral	13,14,15,16,21,22,23,24,25
Visual/Aesthetic	1,3,4,6,11,18,19,20
Technological	5,17
Socio-demographic	7,8,9,10

3   Results

3.1   Rank per Characteristic in the Text

The results of the average rank (scale 1 = highest to 5 = lowest) which students assigned to each characteristic of the Bladerunner city is shown in Table 2 and the number of students who chose each characteristic is given in Table 3.

3.2   Rank per Type of Characteristic

The ranks per characteristic are given in Table 4. Spearman correlation of ranks gave a correlation equal to 0.1, so the number of students and the rank of each type of characteristic are poorly correlated. This shows that the economic type of characteristics concentrated the highest rank among the responses, although the type that most respondents checked was the visual/aesthetic.

3.3   Visualizing Future Athens as a "Bladerunner City"

With N = number of students and t = years ahead of present time (counting as from 2011), the students' responses are shown in Fig. 1. The plot of N(t) can be taken as a probability function (the higher the number of N, the higher the probability of occurrence in time t), expressing the probability that Athens will resemble the "Bladerunner" city in as many years as those deemed more appropriate by the number of students N, whose responses correspond to each future year.

On the average, the students predicted that Athens will resemble the "Bladerunner" city after 28.3 years, that is in (approximately) the year 2040.

However, this does not mean that the all traits of postmodernity are expected to appear in the same time in future. Hence, by counting the average number of years anticipated per characteristic, the relationship between them and the average anticipated time of conversion to the state of "Bladerunner" city was established, as shown in Table 5 and Fig. 2 .

Table 2 Average rank (scale 1=highest to 5=lowest) per characteristic of postmodernity, as assigned by the University students of geography (own depiction)

Average rank	Codes of characteristics	Brief description of characteristics of the Bladerunner city
High (2–2.5)		
2.17	2	The city has the economic structure of late capitalism
2.18	12	Oriental merchants in the city
2.22	14	People of extreme attitudes or attires, punks
2.29	24,21,25	"Clones", people without strong sense of identity, only with a strong feeling of living in the present
2.4	1	Futurist post-industrial city
2.43	22	Memory-less people
2.5	23,15	People without sensitivity to history and past
Medium (2.5–3)		
2.53	3	Skyscrapers together with ruins
2.54	13	Impersonal masses
2.75	10	High population density in the city centre
2.89	16	Language mix
2.93	18	A sense of 3d world within the 1st world
2.94	19,20	Pastiche, Mixture of architectural and design styles
Low (3–4)		
3.1	6	Litter and waste
3.13	4	Sense of decomposition
3.14	5	High technology
3.33	8	Immigrants in the city centre
3.4	7	"Recyclers" working to recycle waste
3.5	9	Bourgeois in the suburbs
3.57	17	"Japanese simulacrum"
3.89	11	Abandoned buildings

Three main stages of future developments emerge from this table:

(a) In the first 20 years, there will be a mixture of types of postmodernity in Athens, although the "visual-aesthetic" characteristics will prevail.
(b) In the next 21–25 years, the "visual" characteristics will dominate the postmodern scenery of the city, with the "behavioural" such following second. Eventually,
(c) in the subsequent 26–40 years, the "behavioural" characteristics will prevail over all other characteristics of postmodern city.

Table 3 Numbers of students who checked each characteristic of postmodernity (own depiction)

Number of students who checked a characteristic	Codes of characteristics	Brief description of characteristics of the Bladerunner city
High		
19	3	Skyscrapers together with ruins
18	16	Language mix
17	10	High population density in the city centre
16	20	Mixture of architectural and design styles
14	5,18	High technology, A sense of 3d world within the 1st world
11	12,13	Oriental merchants in the city, Impersonal masses
Medium		
10	1,5,6	Futurist post-industrial city, High technology everywhere, Litter and waste
9	11,14,15	Abandoned buildings, People of extreme attitudes or attires
8	4	Sense of decomposition
7	21,22,24,17	"Memory-less" and "identity-less" people, Japanese simulacrum
Low		
6	2,8,9,23	The city has the economic structure of late capitalism, Immigrants in the city center, Bourgeois in the suburbs, People without sense of history and past
5	7	"Recyclers" working to recycle waste

Table 4 Ranks per characteristic type (own depiction)

Type of characteristic	Average rank in responses	Rank	Average rank in numbers of responses	Rank
Economic	2.175	1	8.5	4
Behavioral	2.431	2	9.33	3
Visual/Aesthetic	2.983	3	12.88	1
Technological	3.355	4	10.5	2
Socio-demographic	3.425	5	5.25	5

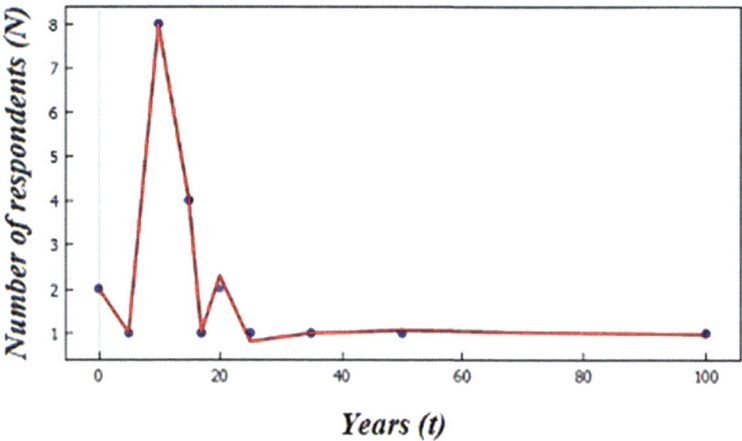

Fig. 1 The number of students (vertical axis) who forecasted that Athens might look like the "Bladerunner" city after as many years as those shown on the horizontal axis (as from present) (own depiction)

Table 5 Types of characteristics and years anticipated until Athens fully acquires the qualities of the "Bladerunner" city: Although the "visual" type characteristics dominate, the "behavioural" characteristics emerge and gain ground with the progress of time, until they eventually dominate the postmodern scene of the city (own depiction)

Time in the future (counting from 2011)	Codes of characteristics	Type of characteristics	Interpretation
0–20 years	1,3,5,6,7,8,9,11,12,13,16,17	4V,3S, 2B, 2 T, E	Visual mainly, Social secondly
21–25 years	2,4,10,14,15,	4V, 2B, E,S	Visual mainly, Behavioural secondly
26–40 years	16,30,33,34,38	5B	Behavioural only

How should these findings be interpreted? The students implicitly anticipated that there will be a remarkable shift in the perception of postmodern look of future Athens. This shift will be from the outer, physical and visible (visual/aesthetic and social) traces of postmoderntiy to the inner/psychological/ behavioural.

Otherwise stated, while in the first 20 years the changes of the city towards postmodernity will be *seen*, in the next 20 years that will follow (20–40) the changes will be *felt* (the changes will have become internalised by the population and thus will have become behavioural and psychological).

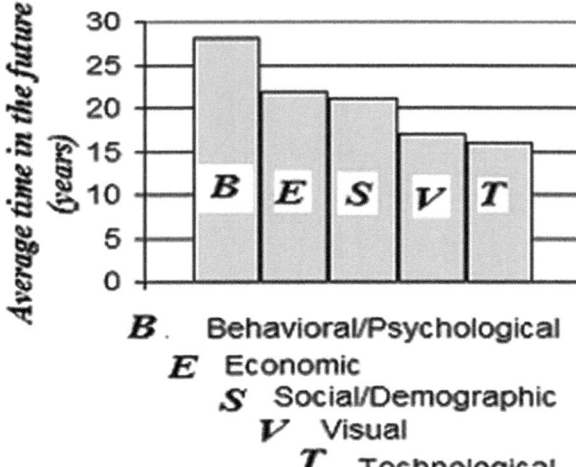

Fig. 2 Predicting the appearance of postmodern characteristics in Athens, as the city approaches the set of characteristics of the "Bladerunner" city: frequency of responses (horizontal axis) against time (average number of years in the future, vertical axis) (own depiction)

4 Visualizing Urban Landscapes of Athens Resembling the "Bladerunner" City: A Collective "Mental Map"

Grouping the students' responses, it was possible to draw a map of regions of Athens that students perceived that resembles the "Bladerunner" city most (for other empirical studies visualizing individual landscape perceptions, see in this volume: McLean 2020; Poplin et al. 2020).

The majority of students considered areas of central Athens (Omonia square for instance) to be the predominantly postmodern area of the city. Other areas of Athens frequently mentioned by the respondents were Psyrri, Gazi, Pireos, Panepistimiou (in fact, all areas around Omonia square). Yet, some students mentioned areas outside the city centre, such as Melissia/Vrilissia, Glyfada, The Mall Athens, Maroussi, Pireas, Elliniko, but these were isolated responses only, without confirmation by other responses, so they were excluded from the mapping procedure that subsequently followed. Counting these responses however, it was possible to map out the areas that (according to their opinion) resembled most to postmodern futures (Fig. 3).

5 Discussion

This research focused on a foresight-and-educational experiment within the context of a university-level human geography course. The triple combination of foresight ("futures studies") with geographical education and film studies makes it a completely novel approach to both education and foresight analyses, as well as an alternative to

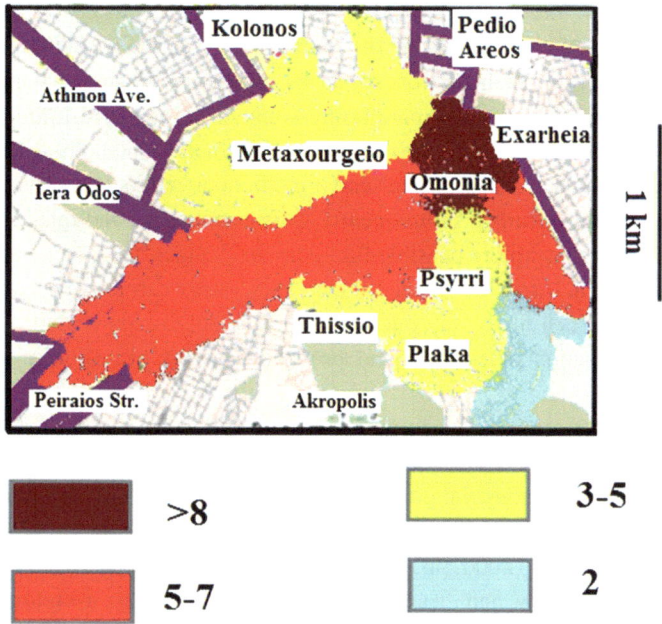

Fig. 3 A collective mental map visualizing areas of Athens resembling the "Bladerunner city" most, according to the students' responses. Omonia square was indicated by most respondents, while next most characteristic "postmodern" regions of the city were imagined to spread within the city centre also [the numbers below show how many students have indicated each area (own depiction)]

computer-generated and/or GIS-based landscape visualization techniques which have been used already (i.e. Edler et al. 2018, 2019).

In forecasting future changes however, several different methods can be applied (Schwartz 1991; Ringland 1998; Van der Heijden 2000). Particularly for forecasting future landscapes, while some authors (i.e. Antrop 2006) have considered social beliefs, perceptions and values in anticipating future landscape changes, others have focused on environmental and political aspects of future landscapes (Rotmans et al. 2000) and others still, have focused on anticipated spatial-environmental conflicts (Papadimitriou and Mairota 1996; Tress and Tress 2003). Expectedly, each foresight experiment is organized differently, has different aims and therefore different methods. Besides, "since we live in between predictability and unpredictability, it may be better to handle such experiments, as the particular circumstances require" (Rescher 1998).

To date however, the scientific literature is short of studies focusing on the junctions between postmodernism, geographical education and future studies. And it is in this particular field that this study aimed to contribute, because it goes beyond the established methods for future prediction, as it based on a fusion of reality and specialized tutoring (about postmodernism) with art (cinema).

Methodologically, the research papers published so far, related to modelling and computer simulation for predicting landscape changes (i.e. Godet 2000; Papadimitriou 2012a) and qualitative methods for scenarios of future changes (Barbarente et al. 2002). Asking participants to anticipate future situations (Jones and Emmelin 1995), or interviewing subjects (Van Dijk 1990) are other alternative methods towards this end (although such methods related to qualitative predictions without reference to aesthetic terms, such as postmodernity). Consequently, both quantitative and qualitative data had to be treated in order to explore possible future urban changes.

The mapping which resulted from students' responses can also be considered as a "Situational map". Such maps "lay out the major human, discursive and other elements in the research situation of concern and provoke analyses of relations among them" (Clarke 2003, pp. 554).

Interestingly, the region of Psyrri (which was referred to by several students) has been singled out by Gospodini (2006, pp. 325) as an interesting area of Athens bearing the characteristics of a post-industrial city. Other areas of formerly industrial or manufacturing areas of Athens were also indicated by the students. At this point, it is worth remembering that Sir Peter Hall (2000, pp. 640) referred to the transition "from manufacturing to informational economy and from informational economy to a cultural economy", while it is also worth noticing that "culture is now seen as the new magic substitute for all the lost factories and warehouses" (op. cit.).

It is precisely at those regions of Athens that the students indicated, where leisure sites are, interspersed with conservation projects of heritage buildings, avant-garde small theatres, alongside with mixtures of land uses appealing to different functions and different tastes. The multifunctional, multicultural and multinational character is strong in these areas and the concentration of such a variety of land uses is one of the hallmarks of transition from modernity to postmodernity in Athens (Gospodini 2004, 2006). Indeed, neglected sites (warehouses, abandoned buildings etc.) constitute one of the prominent characteristics of postmodern cities (Norcliffe et al. 1996) and such derelict buildings abound in the surroundings of Omonia square.

The landscape of Bladerunner city is clearly an exemplary complex landscape (skyscrapers mingle with abandoned buildings, high-tech rooms with ruins etc.). The term "landscape complexity" entails three different dimensions (defined by Papadimitriou 2010c): spatial/structural complexity (that is the complexity of the landscape as measured from its cartographic representation), functional complexity (the complexity of energy and mass flows in the landscape) and qualitative complexity (the complexity of semantics, connotations, affects, meanings etc. non-quantitative characters of the landscape). For the measurement of spatial landscape complexity the reader is referred to Papadimitriou (2009b, 2012a, 2020), of functional landscape complexity to Papadimitriou (2012c, 2013) and for the qualitative complexity to Papadimitriou (2012b).

The high spatial complexity of the urban landscape of the Bladerunner city is evident in the many different types of urban land uses in the city, which also entail high functional complexities (dense and efficient information, communication and traffic networks), as well as high qualitative complexity (i.e. mixture of architectural styles resembling Greek, Chinese, Egyptian styles, elevators made of stone but with inserted video screens in them, and other pastiche postmodern mixtures). In turn, high landscape complexity (in all its three dimensions) means also high difficulty to decipher, to decode, to fully understand that city (Papadimitriou 2012a). Plausibly, if it were any less difficult to decipher, it won't be as fascinating as it is and visualizations of its possible futures won't be as interesting either.

Besides the architectural and social context, technology is of paramount importance also in this account of geographical postmodern futures. Indeed, technology (ICTs, or biotechnology, as is presented in "Bladerunner") manages "to bring alternative worlds into being" (Hefner 2002, pp. 655) as futuristic cities are imagined in movies (Bull 1997, pp. 145). These alternative fictitious postmodern worlds entail contradictory situations (Wyn Davies 2011; Sandar 2010), as much as the real worlds do.

Yet, aside of technology and geography, in any experiment for the perception of postmodernity we should remember that the interplay between different time periods is an essential part of postmodernism (Rosenstone 2004).

And this is so, not only in architectural and stylistic terms, but also in terms of feeling time and perceiving spatial qualities through time. For this reason, reflecting on the "selflessness", or "ephemerality" of being as felt by the Bladerunner citizens, it is important to highlight the external (visual) postmodern forms, as well as the inner forms of profound psychological changes, that eventually develop among the people of postmodern cities.

The gradual shift from the visual and apparent to the inner and psychological (as indicated by the students' responses) is reminiscent of the "perpetual spatial present" that was proposed by Frederic Jameson in his explanation of postmodernism (Stephanson and Jameson 1989, p. 6). This may explain the final stage of "postmodernisation", whereas psychological changes may occur, with central to them being the sense of lack of memory and lack of sense of personal or collective history.

Such forms may give the impression of "unnatural beings" (Csicsery-Ronay 2002), or, more generally, as genetically-engineered and transformed human beings (as the "Bladerunner" movie suggests the "replicants" are). Besides, cyberbodies are, to a certain extent, "no bodies" (Abbott 2010). Further, both material and immaterial collections of heterogeneous qualities directly point to Haraway's "cyborg philosophy" (Haraway 1991; Salleh 2009) and post-humanist approaches of which the "Bladerunner" city is so much reminiscent of.

6 Conclusion

This study explored the perception of postmodernity in Athens among adult University students of geography. From the analysis of their responses, it was found that there are different groups of characteristics of postmodernity in the way they perceive it and that these groups are associated with different rankings and different anticipations for the future of Athens. While predicting that the city will show signs of the "Bladerunner city" after approximately 28 years, they also anticipated the changes to be first visual/aesthetic and social, then behavioral/psychological.

According to their estimates, the areas in downtown Athens, and particularly around Omonia square and the old commercial center of the city, attract the highest possibilities of becoming increasingly "postmodern" in the sense of the "Bladerunner movie" by that time. Evidently, similar foresight and visualization experiments can be organized for other cities, in other geographical and cultural settings, and can be useful in geographical education and/or film studies, for helping to elucidate how adult learners perceive and forecast changes in geographical settings which they are already familiar with.

Appendix

Translation of the text given to students, briefly presenting some of the main characteristics of the Bladerunner city (the numbering refers to the codes of characteristics of postmodern, which appear in the tables).

1. The Bladerunner city is a futuristic post-industrial city.
2. Its economic structures are characteristic of late capitalism.
3. It is a city of mainly skyscrapers and ruins.
4. There is a sense of decomposition everywhere.
5. High technology pervades everything and everyone.
6. There is litter and waste scattered all over.
7. The "recyclers" are those who deal specifically with waste recycling.
8. The center of the city is occupied by migrants.
9. The bourgeois live in the suburbs.
10. Population density is higher in older areas of the city.
11. Many abandoned buildings are in the city center, which nevertheless house advanced information and communication technologies.
12. Many oriental traders dwell in the streets.
13. "Impersonal masses" populate the city.
14. People of extreme attitudes or attires are common, such as punks and harekrishnas.
15. People appear to have no sensitivity to their own history of past.
16. The languages spoken are mixtures. One such is a mixture of Japanese and Spanish, with words from other languages. The city appears as a huge market place, dominated by easterners.
17. The center of the city is "monitored" from above by the "Japanese simulacrum" (this being a huge advertisement showing a Japanese face which turns into the shape of a Coca-Cola bottle).

18. The city gives the impression of "third world" emerging from within a "first world" city.
19. A Chinatown in Los Angeles in a "pastiche"-type architectural style.
20. The building interiors display mixtures of styles: Greek columns, Chinese dragons, Egyptian pyramids, elevators made of stone, yet with video screens.
21. The residents of the city are "replicants" (copies/clones).
22. They have only short-term memory.
23. They seem to not know of or not care about their past.
24. They have no strong sense of personal identity.
25. They only "live for the present".

References

Abbott, Ph. (2010). Should utopians have perfect bodies? *Futures, 42*, 874–881.

Alvarado Duque, C. F., & Escobar Ramirez, J. W. (2019). Metafora y metonimia: Estrategias retoricas de organizacion narrativa. Analisis de caso en el cine clasico y posmoderno. *Revista Signa, 28*, 373–399.

Antrop, M. (2005). Why landscapes of the past are important for the future. *Landscape and Urban Planning, 70*, 21–34.

Araujo, D. C. (2015). Memory, rhizome and postmodern sensitivity: Wong Kar-Wai and Brazilian Films. *East Asian Comparative Literature and Culture, 6*, 280–297.

Arecco, S. (2001). *Il paesaggio del cinema. Dieci studi da Ford a Almodóvar.* [The landscape of Cinema. Ten studies from Ford to Almodóvar]. Genova: Le Mani.

Azevedo, A. F. (2007). *Geografia e Cinema. Representaçoes culturais de paisagem, espaço e lugar na cinematografia portuguesa.* [Geography and Cinema. Cultural representations of landscape, space and place in Portuguese cinematography]. Braga: Repositorium da Universidade do Minho.

Azevedo, A. F. (2015). Políticas de pós-memória e paisagem cinematográfica como categoria epistémica. Um lugar rugoso da experiência. [Politics of post-memory and Cinematographic Landscape as an epistemic category. A rough place of experience]. In: A .F. Azevedo, R. Cerarols Ramirez, & W. Machado de Oliveira Jr. (Eds.) "*Intervalo II: Entre Geografias e Cinemas*" (pp. 81–95). Department of Geography of the University of Minho.

Azevedo, A.F. (2012). Cinema, arte e comunicação: o ser inteiro da paisagem e o fabrico dialógico da experiência. [Cinema, Art and Communication: the internal existence of landscape and the dialogical fabrication of experience]. In H. Pires & T. Mora (Eds.) „*Encontro de Paisagens*" (pp. 27–46). Minho: Centro de Estudos de Comunicação e Sociedade/Centro de Investigação em Ciências Sociais da Universidade do Minho.

Bagoly-Simó, P. M. (2020). Landscape in geography textbooks. In D. Edler, C. Jenal, & O. Kühne (Eds.), *Modern approaches to the visualization of landscapes* (p. 371–385). Wiesbaden: Springer VS.

Beneker, T., & van der Schee, J. (2015). Future geographies and geography education. *International Research in Geographical and Environmental Education, 24*(4), 287–293.

Besteliu, I., & Doevendans, K. (2002). Planning, design and the post-modernity of cities. *Design Studies, 23*, 233–244.

Boyer, M. C. (1994). *The city of collective memory: Its historical imaginary and architectural entertainments*. Cambridge: MIT Press.

Brandt, S. L. (2009). The city as liminal space: Urban visuality and aesthetic experience in postmodern US literature and Cinema. *Amerikastudien, 54*(4), 553–581.

Brooker, W. (2007). Everywhere and nowhere: Vancouver, fan pilgrimage and the urban imaginary. *International Journal of Cultural Studies, 10*(4), 423–444.

Bruno, G. (1987). Ramble City: Postmodernism and "Blade Runner". *October 41*, 61–74.

Bull, C. (1997). City: Repository of dreams – Realm of illusion – Experience of reality. *Urban Design International, 2*(3), 145–153.

Clarke, A. E. (2003). Situational analyses: Grounded theory mapping after the postmodern term. *Symbolic Interaction, 26*(4), 553–576.

Csicsery-Ronay, I., Jr. (2002). On the grotesque in science fiction. *Science-Fiction Studies, 29*, 71–99.

Dalby, S. (1999). Globalization or global apartheid? Boundaries and knowledge in postmodern times. In D. Newman (Ed.), *Boundaries, territory and postmodernity* (pp. 132–150). Portland: Frank Cass.

Davis, M. (1998). *Ecology of fear. Los Angeles and the imagination of disaster*. New York: Metropolitan Books.

Dear, M. (2000). *The postmodern urban condition*. New York: Blackwell.

Dear, M., & Flusty, S. (1998). Postmodern urbanism. *Annals of the Association of American Geographers, 1*, 50–72.

Edler, D., Kühne, O., Jenal, C., & Vetter, M. (2018). Potenziale der Raumvisualisierung in Virtual Reality (VR) für die sozialkonstruktivistische Landschaftsforschung [The Potentials of Spatial Visualization in Virtual Reality (VR) for the Social Constructivist Landscape Research]. *Kartographische Nachrichten, 5*, 245–254.

Edler, D., Keil, J., Wiedenlübbert, T., Sossna, M., Kühne, O., & Dickmann, F. (2019). Immersive VR experience of redeveloped post-industrial sites: The example of "Zeche Holland" in Bochum-Wattenscheid. *KN – Journal of Cartography and Geographic Information, 69*(4), 267–284.

Edler, D., Jenal, C., & Kühne, O. (2020a). Modern approaches to the visualization of landscapes—An introduction. In D. Edler, C. Jenal, & O. Kühne (Eds.), *Modern approaches to the visualization of landscapes* (p. 3–15). Wiesbaden: Springer VS.

Edler, D., Keil, J., & Dickmann, F. (2020b). From Na Pali to Earth—An 'Unreal' engine for modern geodata? In D. Edler, C. Jenal, & O. Kühne (Eds.), *Modern approaches to the visualization of landscapes* (pp. 279–291). Wiesbaden: Springer VS.

Fontaine, D. (2020a). Landscape in computer games—The examples of GTA V and Watch Dogs 2. In D. Edler, C. Jenal, & O. Kühne (Eds.), *Modern approaches to the visualization of landscapes* (pp. 293–306). Wiesbaden: Springer VS.

Fontaine, D. (2020b). Virtuality and landscape. In D. Edler, C. Jenal, & O. Kühne (Eds.), *Modern approaches to the visualization of landscapes* (p. 267–278). Wiesbaden: Springer VS.

Godet, M. (2000). The art of scenarios and strategic planning: Tools and pitfalls. *Technological Forecasting and Social Change, 65*, 3–22.

Gospodini, A. (2006). Portraying, classifying and understanding the emerging landscapes in the post-industrial city. *Cities, 23*(5), 311–330.

Gospodini, A. (2004). Glocalising urban landscapes: Athens and the 2004 Olympics. *Cities, 21*(3), 187–202.

Hannigan, J. (1995). The Postmodern City: A New Urbanization? *Current Sociology, 43*(1), 155–214.

Hannigan, J. (1998). *Fantasy city: Pleasure and profit in the postmodern metropolis.* New York: Routledge.

Hall, P. (2000). Creative cities and economic development. *Urban Studies, 37*(4), 639–649.

Haraway, D. (1991). *Simians, cyborgs and women.* London: Routledge.

Harper, G., & Rayner, J. (Eds.). (2010). *Cinema and landscape.* Chicago: The University of Chicago Press.

Harvey, D. (1989). *The condition of postmodernity: An enquiry into the origins of cultural change.* Oxford: Blackwell.

Hefner, P. (2002). Technology and human becoming. *Zygon, 37*(3), 655–665.

Jameson, F. (1991). *Postmodernism or the cultural logic of late capitalism.* New York and London: Verso.

Jones, M. & Emmelin, L. (1995). Scenarios for the visual impact of agricultural policies in two Norway landscapes. In J. F. Th. Schoute, P. A. Finke, F. R. Veeneklaas, & H. P. Wolfert (Eds.),*Scenario studies for the rural environment* (pp. 405–416). Dordrecht: Kluwer Academic Publishers.

Kleber, A., Edler, D., & Dickmann, F. (2020). Cartography and the sea: A javascript-based web mapping application for managing maritime shipping. In D. Edler, C. Jenal, & O. Kühne (Eds.), *Modern approaches to the visualization of landscapes* (pp. 173–186). Wiesbaden: Springer VS.

Kühne, O. (2012). Stadt – Landschaft – Hybridität. Ästhetische Bezüge im postmodernen Los Angeles mit seinen modernen Persistenzen [City-Landscape-Hybridity. Aesthetic references in postmodern Los Angeles with its modern persistence]. Wiesbaden: Springer VS.

Kühne, O. (2018). *Landscape and power in geographical space as a social-aesthetic construct.* Dordrecht: Springer International Publishing.

Kühne, O. (2019). *Landscape theories. A brief Introduction.* Wiesbaden: Springer VS.

Kühne, O. (2020). The social construction of space and landscape in internet videos. In D. Edler, C. Jenal, & O. Kühne (Eds.), *Modern approaches to the visualization of landscapes* (pp. 121–137). Wiesbaden: Springer VS.

Lastoria, A., & Papadimitriou, F. (2012). Geographical educaiton in Brazil: Past and present in "The country of the Future." *International Research in Geographical and Environmental Education, 21*(4), 327–335.

Madsen, K. D. (2014). Blue Indians: Teaching the political geography of imperialism with fictional film. *Journal of Geography, 113*(2), 47–57.

Masirevic, L. (2010). Media and postmodern reality. *Sociologija, 52*(2), 127–140.

McLean, K. (2020). Temporalities of the smellscape: Creative mapping as visual representation. In D. Edler, C. Jenal, & O. Kühne (Eds.), *Modern approaches to the visualization of landscapes* (pp. 217–246). Wiesbaden: Springer VS.

Murray, M.J. (2004). The spatial dynamics of postmodern urbanism: Social polarisation and fragmentation in Sao Paulo and Johannesburg. *Journal of Contemporary African Studies 22, 2,* May.

Neuser, M. (2012). *Postmodernes Kino; Stanley Kubricks Filmasthetik jenseits den Konventionen [Postmodern Cinema: Stanley Kubrick's Film Aeshetics beyond conventions].* Hamburg: Diplomica Verlag.

Newman, D., & Paasi, A. (1998). Fences and neighbors in the postmodern world: Boundary narratives in political geography. *Progress in Human Geography, 22*(2), 186–207.

Norcliffe, G., Bassett, K. & Hoare, T. (1996). The emergence of Postmodernism on the urban waterfront. Geographical perspectives on changing relationships. *Journal of Transport Geography 4(2),* 123–134.

Papadimitriou, F. (2006). The culture of cyberspace: examples from Greece. In J. Lidstone (Ed.), *Cultural issues of our time* (pp. 178–186). Cambridge: Cambridge University Press.

Papadimitriou, F. (2009a). A nexus of cyber-geography and cyber-psychology: Topos/"Notopia" and identity in hacking. *Computers in Human Behavior, 25*(6), 1331–1334.

Papadimitriou, F. (2009b). Modelling Spatial Landscape Complexity using the Levenshtein Algorithm. *Ecological Informatics, 4,* 48–55.

Papadimitriou, F. (2010a). Introduction to the complex geospatial web in geographical education. *International Research in Geographical and Environmental Education, 19*(1), 53–56.

Papadimitriou, F. (2010b). A "Neogeographical Education"? The geospatial web, GIS and digital art in adult education. *International Research in Geographical and Environmental Education, 19*(1), 71–74.

Papadimitriou, F. (2010c). Conceptual modelling of landscape complexity. *Landscape Research, 35*(5), 563–570.

Papadimitriou, F. (2012a). The algorithmic complexity of landscapes. *Landscape Research, 37*(5), 599–611.

Papadimitriou, F. (2012b). Artificial intelligence in modelling the complexity of mediterranan landcsape transformations. *Computers and Electronics in Agriculture, 81,* 87–96.

Papadimitriou, F. (2012c). Modelling landscape complexity for land management in Rio de Janeiro. *Brazil. Land Use Policy, 29*(4), 855–861.

Papadimitriou, F. (2013). Mathematical modelling of land use and landscape complexity with ultrametric topology. *Journal of Land Use Science, 8*(2), 234–254.

Papadimitriou, F. (2019). *Postmodern urban futures, film and geographical education.* Athens.

Papadimitriou, F. (2020). *Spatial Complexity. Theory, mathematical methods and applications.* Springer.

Papadimitriou, F., & Mairota, P. (1996). Spatial scale – Dependent policy planning for land management in Southern Europe. *Environmental Monitoring and Assessment, 39,* 49–60.

Parker, D. D., & Zilberman, D. (1995). Biotechnology and the future of agriculture and natural resources – An overview. *Technological Forecasting and Social Change, 50*(1), 1–7.

Parker, I. (1996). Psychology, science fiction and postmodern space. *South African Journal of Psychology, 26*(3), 143–149.

Poplin, A., Andrade, B. de, & Mahmud, S. (2020). Exploring tangible and intangible landscapes of evocative places: Case study of the City of Vitória in Brazil. In D. Edler, C. Jenal, & O. Kühne (Eds.), *Modern approaches to the visualization of landscapes* (pp. 519–547). Wiesbaden: Springer VS.

Prisille, C., & Ellerbrake, M. (2020). Virtual Reality (VR) and geography education: Potentials of 360° 'Experiences' in secondary schools. In D. Edler, C. Jenal, & O. Kühne (Eds.), *Modern approaches to the visualization of landscapes* (pp. 321–332). Wiesbaden: Springer VS.

Rescher, N. (1998). Predicting the future. An introduction to the theory of forecasting. New York: State University of New York Press.

Ringland, G. (1998). Scenario planning: Managing for the FUTURE. New York: Wiley.

Rosenstone, R. A. (2004). Confessions of a postmodern historian. *Rethinking History, 8*(1), 149–166.

Salleh, A. (2009). The dystopia of technoscience: An ecofeminist critique of postmodern reason. *Futures, 41,* 201–209.

Sandar, Z. (2010). Welcome to postmodern times. *Futures, 42,* 435–444.

Schwartz, P. (1991). The art of long view. New York: Doubleday.

Siepmann, N., Edler, D., & Kühne, O. (2020). Soundscapes in cartographic media. In D. Edler, C. Jenal, & O. Kühne (Eds.), *Modern approaches to the visualization of landscapes* (pp. 247–263). Wiesbaden: Springer VS.

Sigler, T., & Albandoz, R. I. (2014). Beyond representation: Film as a pedagogical tool in urban geography. *Journal of Geography, 113*(2), 58–67.

Soja, E. (2000). Postmetropolis: Critical studies of cities and regions. Malden, Massachusetts: Blackwell.

Steimatsky, N. (1995). The earth figured. An exploration of landscapes in Italian Cinema. New York University, Graduate School of Arts and Science: Dissertation submitted for the degree of Doctor of Philosophy. 594 p.

Stephanson, A., & Jameson, F. (1989). Regarding postmodernism – A conversation with Frederic Jameson. *Social Text, 21,* 3–30.

Stevenson, N. (1997). Globalization, national cultures and cultural citizenship. *Sociological Quarterly, 38*(1), 41–66.

Stintzing, M., Pietsch, S., & Wardenga, U. (2020). How to teach "Landscape" through games? In D. Edler, C. Jenal, & O. Kühne (Eds.), Modern approaches to the visualization of landscapes (pp. 333–349). Wiesbaden: Springer VS.

Stratmann, J., Ristea, A., Leitner, M., & Paulus, G. (2020). Exploring urban "Blightscapes" applying spatial video technology and geographic information system: A case study from Baton Rouge, USA. In D. Edler, C. Jenal, & O. Kühne (Eds.), *Modern approaches to the visualization of landscapes* (pp. 499–517). Wiesbaden: Springer VS.

Tress, B., & Tress, G. (2003). Scenario visualization for participatory landscape planning – A study from Denmark. *Landscape and Urban Planning, 64*(3), 161–178.

Van Der Heijden, K. (2000). Scenarios and forecasting: Two perspectives. *Technological Forecasting and Social Change, 65,* 31–36.

van Dijk, J. (1990). Delphi questionnaires versus individual and group interviews. *Technological Forecasting and Social Change, 37,* 293–304.

Von Tunzelmann, G. N. (1997). Innovation and industrialization: A long-term comparison. *Technological Forecasting and Social Change, 56*(1), 1–23.

Weber, F. (2020). Blurring the boundaries of landscape visualization: Welcome to Fabulous Las Vegas. In D. Edler, C. Jenal, & O. Kühne (Eds.), *Modern approaches to the visualization of landscapes* (pp. 461–478). Wiesbaden: Springer VS.

Wyn Davies, M. (2011). Postmodern times: Are we there yet? *Futures, 43,* 136–141

Dr. Fivos Papadimitriou studied geology (B.Sc.), physics (M.Sc.), environmental resources (M.Sc.) and education (M.Ed.), and gained a doctorate in Geography from the University of Budapest (Ph.D.) and another one from the University of Oxford (D.Phi.Oxon.). He has taught at Universities for several years and has accomplished cooperations or field researches in several countries. He is member of the Editorial Boards of ISI-listed journals, and has received numerous prizes, grants, awards, fellowships and distinctions. His main contributions to science consist in the creation of new algebraic, algorithmic and topological mathematical models and formulas for landscape complexity analysis. His papers have been cited by scientists from sixty-five countries. Aside of these, he also maintains vivid research interests in geographical education and cyber-geography.

Landscapes in Geography Textbooks

Péter Bagoly-Simó

Abstract

Landscapes have shaped both academic and school Geographies over the centuries. Research in both Physical and Human Geography contributed to an ongoing re-conceptualization of landscapes, turning it into a more open and inclusive concept. The particularities of national and regional geographical discourse colored the concept diverse. School Geographies—located at the interface of scientific progress and educational goals—distilled specific understandings of the concept of landscape. This chapter aims to explore two less examined case studies of lower secondary Geographies: Hungary and Venezuela. Both case studies stand for particular (school) geographical traditions and fall back on a particular perspectives on landscapes. The textbook analysis served to uncover, on the one hand, the way curricular requirements translated into textbooks, and, on the other hand, what kind of geographical and cross-curricular knowledge acquisition the landscape concept supports. The computer-assisted content analysis revealed that landscapes remain outside the key concepts of both school Geographies. Moreover, the textbooks showed a rather traditional perspective in terms of Physical Geography, and merely first steps were taken towards a Human Geography interested in the role of the individuals producing landscapes.

Keywords

Landscape · Geography · Textbook · Secondary school · Hungary · Venezuela

P. Bagoly-Simó (✉)
Humboldt-Universität zu Berlin, Berlin, Germany
e-mail: peter.bagoly-simo@geo.hu-berlin.de

© Springer Fachmedien Wiesbaden GmbH, part of Springer Nature 2020   371
D. Edler et al. (eds.), *Modern Approaches to the Visualization
of Landscapes*, RaumFragen: Stadt – Region – Landschaft,
https://doi.org/10.1007/978-3-658-30956-5_21

1   Introduction

School Geographies and landscapes are as tightly connected in the public understanding of the subject as the overview of all continents and selected countries. While this view of school Geographies is undoubtedly outdated, a glance at textbooks may mislead the reader into thinking that colorful maps and pictures still aim at familiarizing students with the world and its regions. A second look, however, raises several questions: Are school Geographies still about landscapes? How do the textbooks discuss landscapes? More importantly, what is the role of the landscapes in knowledge acquisition through Geography in contemporary schools? This chapter seeks to answer these questions by looking at selected Geography textbooks for lower secondary education. After a brief theoretical overview, the chapter describes the research methods along with the sample to subsequently introduce selected results. A brief discussion and some final thoughts conclude the chapter.

2   Theoretical Background

Landscapes are at the heart of geographical thought for centuries. Nonetheless, national and regional discourses in academic Geography attributed different importance to the concept of landscape. Similarly, school Geographies distilled particular ways of dealing with the landscapes as a result of the complex interplay between academic discourse, societal relevance, and political priorities in a given space and at a given time (in this volume: Edler et al. 2020a, Kühne and Jenal 2020).

The international discourse on landscapes looks back on a rich tradition both in Physical and Human Geography. When reviewing geographical research and thought on landscapes in Physical Geography, Gray (2008) distinguishes between primary and secondary layers. Each of the three primary landscape layers encompasses several secondary layers. The geological layer represents, in terms of geological time, the oldest layer and entails both landscape structures (i.e., rocks, sediments, soils, landforms) and processes (physical processes). The biological layer constitutes the second-oldest layer and comprises all non-human forms of life (e.g., biomes, megafauna). Finally, the cultural layer entails both the human impact on the landscape (e.g., land-use in the past and presents in terms of economic spaces, human settlements) and the experiences and associations humans have concerning the landscape.

Landscapes played, according to Gray (2008), quite different roles in physical-geographical research over time. The geomorphological tradition investigated both structures and processes of the geodiversity that shapes the landscapes. Thereby, genetic aspects of land-forms and landscapes played an equally important role as endo- and exogenous processes leading to landscape dynamics. The second half of the twentieth century marked the beginning of a turn that redefined the role of landscape in Physical Geography by focusing on landscape ecology and cultural landscapes. Paramount for this turn was the role of research for landscape conservation and restoration—both requiring an integrative perspective involving the human agent as well.

The primary interest of human geographers in the landscape, according to Morin (2008), is to comprehend its power when subverting or challenging social orders. In her reading, criticism from Human Geography targets two central aspects of the way geographers did landscape studies. On the one hand, Marxist geographers stressed the false dichotomy between morphology and landscape representation and suggested to shift the focus from landscapes to those inhabiting, and, ultimately, producing them. On the other hand, feminist approaches drew attention to the historical masculinism in geographical landscape studies (e.g., wars, representations, spatial and landscape planning). Both directions stressed the necessity to overcome paradigms that focused on cultural differences read through morphological landscape patterns and, instead, view the landscape as an intersection of competing discourses, interpretations, knowledge, and authority.

Both Gray (2008) and Morin (2008) use a language that claims the global validity of *the* role of *the* landscape in *Geography*. In reality, their work is limited to the Anglo perspective—a fact they reflect in a rather superficial manner. However, regional and national Geographies around the globe dealt with the concept of landscape in quite different ways. For example, Schultz (1971, 2014), Hard (1991), Hasse (1993,2011), and Kühne (2018, 2019) summarize both the development and challenges of the landscape tradition in the German-speaking countries and academic tradition. Similarly, Fernández Álvarez (2020) offers an overview of the role of landscapes in Spanish Geographies.

At the intersection of geographical relevance, social expectations concerning education, and (educational) policy agenda, school Geographies distilled a multitude of ways to include the concept of the landscape into formal education. While in some countries Geography prescribes the landscape as a content element, others view it as one of their key concepts that define the very core of the subject or even contextualize it within overarching educational objectives, such as Education for Sustainable Development (ESD). Despite their heterogeneity, school Geographies remain closely tied to the traditions of the academic Geographies they mirror.

In conceptual and curricular terms, British school Geographies repeatedly listed the concept of landscape among their key concepts. For example, both the Geography Advisors' and Inspectors' Network, as well as Clifford et al. (2008), placed *landscape and environment* on their list of key concepts along with space, place, time, scale, social formations, and physical systems. Subsequent conceptualizations turned toward overarching concepts, such as human—environment systems and cultural understanding—both of which require the consideration of landscapes, however, in a more implicit manner.

The German federal educational system allows state curricula to set individual accents when it comes to aims, basic concepts, and content. In recent years, the gradual implementation (cf. Schöps 2017) of the national Educational Standards in Geography for the Intermediate School Certificate (DGfG [2]2012– first published in 2006) led to a slow unification across state borders and school types. Concerning school Geography's basic concepts *(Basiskonzepte)*, the Standards—and, implicitly, all state curricula implementing them—adopt a systemic perspective that rests, on the one hand, on system components (i.e., structure, process, function) and, on the other hand, on scale (local, regional, national, international, and global). The landscape concept, thereby, plays a marginal role and does not belong to the key concepts.

Spanish school Geographies across the federal units *(Comunidades Autónomas)* look back on a long tradition in discussing the concept of landscape. In an extensive curricular analysis, Fernández Álvarez (2020) found significant differences between primary and lower secondary education regarding the presence of the landscape concept. His results confirmed the decreasing importance of the landscape concept while traveling through formal education. Moreover, the federal units dedicate differing attention to landscapes, with Aragón and La Rioja scoring lowest and Catalonia highest.

Landscapes play a rather marginal role in Geography Education research. Overall, international literature explores landscapes under two significant aspects. On the one hand, the way students as young individuals perceive and construct landscapes remains central for Geography Education. Based on a sample of 21 East Anglian schools, Robertson et al. (2003) found that personal experience decisively influenced the values individuals attributed to landscapes. The authors reflected on the outcome of their study in terms of the place concept in school Geography. Measham (2007) also found that experiential learning in different landscapes during childhood significantly influenced children's primary landscapes. Following Gayton's (1996) conceptualization, the authors defined primary landscapes as particular environments young humans are confronted with and grow up in that, subsequently, serve as points of comparison and reference for any other landscapes. Landscape perception was the main research question of empirical work carried out in Greece on agricultural terraces (Klonari et al. 2011). Both teachers and students perceived agricultural terraces—central landscape elements of the Mediterranean landscape in the authors' reading—as less important (see for the construction of Mediterranean landscapes also Loda et al. 2020 in this volume). The differences between urban and rural respondents were limited, which raised the question of the nature of the experiential learning of rural youth in agricultural environments. The authors explained the limited knowledge of teachers with the weak stand of Greek school Geography and the number of non-specialist teachers teaching the subject in schools across the country.

On the other hand, research in Geography Education also focused on landscapes as content in both Geography curricula and textbooks. While Klonari et al. (2011) proved the mandatory nature of landscapes as content in the Greek curriculum as well as in textbooks, Bagoly-Simó (2017) explored the way German textbooks for lower secondary education in the federal state of Berlin introduced landscapes. Both physical and human geographical perspectives were outdated and served primarily the acquisition of skills that lie outside the core of school Geography. Regardless of their aim, all studies discussed landscapes in relation to Environmental Education (EE) or ESD and referred to it as a critical topic in achieving the goals of these cross-curricular educational objectives. Particularly concerning recommendations for the role of landscape in school Geographies, EE/ESD enjoys a higher priority than the acquisition of geographical knowledge.

This chapter aims to explore the representation of the concept of landscape in selected Geography textbooks for lower secondary education—the most widely used educational media of school Geographies (for additional studies addressing educational media for schools in this volume, see Prisille and Ellerbrake 2020; Stintzing et al. 2020). Hungary and Venezuela, two countries with a stable and traditional school Geography, serve as examples to compare Central European and Latin American perspectives on landscape representation in lower secondary geographical education.

3   Method and Sample

Content analysis served to explore both the representation and the education objectives connected to the concept of landscape in textbooks. In the first step, software-assisted lexical analysis helped to identify all segments dedicated to landscapes. Thereby, discontinuous text elements remained unconsidered. The second step consisted of semantic disambiguation. Step three focused on the separation of tasks from all other content elements. The last step involved the separate analysis of both sub-samples. Task analysis followed pre-defined criteria that rest on three performance levels (PFL), namely reproduction (PFL1), reorganization and transfer (PFL2), and reflection and problem-solving (PFL3). The validation was based on parallel coding by two independent individuals and subsequent comparison.

The sample consisted of two Hungarian and two Venezuelan Geography textbooks for lower secondary education. The sampling considers, on the one hand, two countries absent from most studies centered on European countries. On the other hand, lower secondary education is mandatory in both case studies. Besides, upper secondary Geographies often adopt a strong propaedeutic perspective that may reflect more adequately the academic discourse revolving around the concept of landscape. However, Geography in upper secondary education is often an elective and fails to reach students leaving school after having concluded mandatory formal education.

4   Results

Hungarian and Venezuelan Geography textbooks paint different pictures of the way the concept of landscape fosters the objectives of school Geographies in specific terms, and cross-curricular educational objectives, such as ESD, in more general terms. Following the introduction of selected results according to countries, their discussion adopts a comparative glance (cf. Sect. 5).

4.1   Hungary

In grades 7 and 8 (students aged 14–15) of Hungarian lower secondary education, Geography is an independent subject. The subject combines two approaches to teaching Geography as it begins with a thematic introduction to Physical Geography and proceeds to offer an overview of regional entities at different scales.

The thematic units of *seventh-grade* Geography link back to basics of Climatology and guide students through the major geographical zones both along latitude and altitude. Subsequently, an additional unit summarizes the geomorphological content of the present (endogenous and exogenous processes) and in geological time. The regional-thematic units cover all extra-European continents as well as the global water bodies. In doing so, they focus on selected issues of both Physical and Human Geography rather than giving an exhaustive description of the continents.

Following semantic disambiguation, 73 segments referencing landscape remained in the sample. Content analysis revealed four significant ways the seventh-grade textbook deals with landscapes.

First, the textbook relies solely on previous knowledge when requiring students to work with the concept of landscape. A (re-introduced) definition of the concept is missing from the textbook. Furthermore, a certain conceptual unclarity surrounds the concepts of landscape and region.

Second, most segments address landscapes in a static, and often descriptive, manner. Each landscape's main characteristics primarily rest on geomorphology. However, students also explore other elements of the geosystem, such as the atmosphere, hydrosphere, and the biosphere. For example, the identification of Mediterranean landscapes relies mainly on the vegetation, but also on the climate. This pattern also applies to underwater landscapes that result from the interplay of geomorphology, characteristics of the water bodies, and the elements of flora and fauna inhabiting it. Also, the textbook considers human societies as part of the landscape. However, a division into original (or natural) and cultural landscapes is missing from the continuous text. Human elements described as parts of the landscape are often material cultural artifacts, such as cities and traces of economic activities. Nevertheless, the textbook prefers to refer to ethnographically relevant features, such as yurts in Central Asia and Bedouin tents in Northern Africa rather than skyscrapers in downtowns.

Third, the textbook adds a dynamic layer to the landscape concept in two significant ways. On the one hand, different forces—both natural and anthropogenic—induce a change in all landscapes. Along with erosion and change in geological time (e.g., the development of major African landscapes in time), students also learn about the impact of human societies on landscapes. For example, fertile soils and abundant water resources enabled societies to develop economic activities that ultimately changed the original features of the landscape. The textbook exemplifies these processes by referring to both permanent and temporary human settlements on several continents. In doing so, it highlights the interdependence of the main features of human settlements and economic activities with the main characteristics of the natural landscape. For example, the availability of building materials had as much of an impact on the settlement morphology as did the climate (e.g., white houses with flat roofs in hot and arid regions). While the textbook shows the historical development of landscapes, it emphasizes the interplay between nature and society and discards geodeterminism. On the other hand, seventh-grade Geography also introduces the dynamic observation of landscapes. Along with description, students learn to trace how the landscape changes with latitude and altitude. At the heart of this dynamics representation lies the observer's movement in space.

Fourth, 3.38% of all tasks featured in the textbook involves the concept of landscape. The tasks target all performance levels, ranging from landscape description (PFL1 and PFL2) through comparison (PFL2) to evaluation (PFL3). Thereby, the description relies strongly on geomorphological features, as shown in the case of Asia (Kusztor et al. 2017a, p. 163). Tasks dedicated to PFL2 often require information retrieval from both continuous and discontinuous text targeting their subsequent comparison. Such tasks may ask students to locate selected landscapes on an atlas map or match pictures of landscapes with their geographic position on maps. Finally, PFL3 assists students in acquiring evaluative skills. Some of the tasks featured in the seventh-grade textbook instruct students to match a list of North-American landscapes with pictures of their South-American counterparts. Overall, the tasks display a progression from lower to higher performance levels. Tasks also foster information retrieval from both continuous and (sometimes several) discontinuous text elements. Particularly maps and pictures remain at the heart of these tasks. Regarding map skills, students work both with maps in the sense of location exercises as well as with map production by sketching maps and simple, schematic cartographic representations of landscapes. Similarly, working with pictures ties theoretical concepts and descriptions to selected spatial examples that foster observation and description skills whenever direct observation in the field is impossible.

The *eighth-grade* Hungarian Geography textbook explores in a regional-thematic manner the Carpathian Basin, Hungary, and Europe. With 176 segments following semantic disambiguation, it refers to the concept of landscape more than twice as much as the textbook for grade seven. The content analysis led to five significant features that describe the way the eighth-grade textbook deals with the concept of landscape.

First, the landscape stands, according to the authors (Kusztor et al. 2017b, p. 3), in the center of the eighth-grade Geography textbook: "Let us explore the beautiful landscapes of our continent and its interesting countries! Let us get to know Europe!".

Second, in grade eight, students work with three concepts of landscape. In addition to the generic landscape concept, the textbook also introduces cultural landscapes as well as ethnographic landscapes. The generic landscape concept primarily covers descriptions based on geomorphologic and, in selected cases, additional physical-geographical features. One example is the identification of Spanish landscapes based on prevailing conifers. In contrast, cultural landscapes carry the fingerprints of human action. Thereby, the quality and quantity of societal impact vary. The textbook emphasizes the interplay between the original landscape and societies. Overall, students explore cause-effect relations in landscape development based on resources, their usage for different human (economic) activities, and the concurrent alteration of the original landscape. In historical terms, the textbook links raw materials and economic activities based on them with settlement development and patterns. In essence, students acquire the ability to recognize the main contemporary and past economic activities based on the landscape. Some examples are the former agricultural coastal areas subsequently developed into a resort town and the transition from (heavy) industry in Germany and England to the third economic sector. In both cases, the landscape carries not only traces of economic activities past but also the result of creative repurposing of space. Along with this development, the textbook also points out the consequences of political decisions as visualized by mining activities that can alter the landscape to the extent that excludes tourism from possible future activities. When introducing the third landscape concept (ethnographic landscapes), the textbook matches the natural and anthropogenic elements of a landscape in historical terms. In doing so, it explains the links between certain settlement types and landscapes and how the culture of human groups carries the features of a particular landscape. As ethnographic landscapes, such spaces reflect raw materials, their usage, as well as their presence in the language, mythology, and religion of people in selected landscapes. Thereby, the example of Hungarian ethnic minorities in the Carpathian Basin serves as an example.

Third, the representation of the landscapes remains mainly static. Change over time is an implicit dimension that serves to visualize the human impact on landscapes and supports students in reading current landscapes in light of their economic and social history. Also, it serves to highlight various relations between the environment and human societies.

Fourth, the textbook addresses the individual perception and value of the landscape for each individual. On the one hand, students learn to express their feelings concerning the aesthetic value of a given landscape for the individual. Thereby, one essential aspect is to learn how to handle their fellow students' possibly differing values and feelings. On the other hand, the textbook tackles the role of the landscape in spatial identity and belonging. Using poetry as an example, students reflect on the role of landscapes when thinking about home: "I cannot know what these landscapes may mean to others—to me, they are home" (Kusztor et al. 2017b, p. 72).

Fifth, 2.33% of all tasks featured in the textbook involves the concept of landscape. The tasks target all performance levels, ranging from landscape description (PFL1 and PFL2) through comparison (PFL2) to evaluation (PFL3). Some of the tasks fostering skill development in reproduction (PFL1) target the location of selected landscapes on maps, listing geographical entities (i.e., rivers, mountains), and descriptions of climates based on landscapes. The textbook provides students with a blueprint of landscape analysis along with the following categories: total surface (and Hungary's share); geographic location (including borders); sub-divisions; development in time (based on geomorphological features) (Kusztor et al. 2017b, p. 44). PFL2 requires students to reorganize their knowledge and transfer it to other thematic and regional entities. The eighth-grade textbook emphasizes the comparison of different landscapes based on prescribed criteria. The interplay between nature and society plays an essential role in these comparative tasks that often rest on discontinuous text (mainly maps and pictures). Finally, the PFL3 requires students to evaluate landscape based on personal criteria of aesthetics, and solve complex problems, such as "What type of landscape lies behind us, if we look at the landscape shown on picture 3.5?" (Kusztor et al. 2017b, p. 155). Overall, the tasks dealing with the concept of landscape actively contribute to maps skills and picture analysis skills. In terms of map skills, students locate landscapes, extract their subdivisions, and draw simple map sketches. Similarly, textbooks tasks require students to match pictures with maps, landscape descriptions and identify traces of landscape change in time based on pictures.

4.2   Venezuela

The Venezuelan lower secondary curriculum prescribes Geography as a mandatory subject in the first (students aged 12–13) and third grade (students aged 14–15). The overall approach is thematic. However, third-grade Geography explores six topics on the national scale.

The textbook for the *first grade* introduces students to Geography as a scientific discipline to subsequently turn to Physical and Human Geography. Regional Geography appears as a conglomerate of Political Geography and continental units. Finally, environmental challenges conclude the introductory course of secondary Geography.

Lexical retrieval identified 71 segments dedicated to the concept of landscape, all of which were semantically appropriate. The content analysis enabled the delimitation of five significant features concerning the way first-grade Geography deals with landscapes.

First, the textbook dedicates not only a separate thematic topic but also an additional space to introduce the concept of landscape. Topic nine of the physical-geographical unit defines geographical landscapes as "[…] a space with particular features defined by the elements that formed it. Physical events (relief, climate, hydrography, and soils) shape certain environmental parameters that allow different plant and animal communities to thrive (biogeographic realm) that shape each landscape. Humans also introduced cultural elements to improve their living conditions" (Rodríguez Requeña 2013, p. 64). Along with the definition, the textbook also provides a classification of landscapes into natural and cultural landscapes. While natural landscapes consist entirely of physical and natural components, cultural landscape emerged under human influence. An additional discontinuous text lists the physical (relief, climate, hydrography, soils, fauna, and flora) and cultural elements (settlements, infrastructure, agriculture, industry, and human populations) of the landscape. Subsequently, topic nine briefly introduces a total of eight landscapes: tropical rainforest, savanna, desert, Mediterranean shrublands, grasslands, forests, taiga, and polar. Accompanied by a map and a picture, a brief continuous text describes the physical features of each landscape and adds selected information on the cultural elements. Within the unit dedicated to Human Geography, the textbook returns to the concept of landscape and describes the agricultural landscapes composed of parcels, settlements, irrigation systems, cropping systems, and agricultural production systems. Within the subsequent units, the textbook mentions some of the main landscapes of each continent and describes hydropower as a clean source of energy that requires significant alterations of the landscape and changes in the lifestyle of the affected communities by large hydropower projects. Also, there are scattered mentioning of urban and industrial landscapes, reconstructed landscapes on dumps, and legal means to protect coastal landscapes.

Second, the textbook describes the main features of different types of natural and cultural landscapes statically. However, the remaining segments that frame these short fragments dedicated primarily to landscapes emphasize their dynamic nature both in the present (e.g., earthquakes, landslides, human impact) and in the geological past.

Third, landscape plays only a secondary role when introducing students to the regional structures on the continental scale.

Fourth, the textbook introduces landscapes as objectively defined structures. Any reference to subjective interpretation and perception is missing from the text. Even at the local scale, observation and description follow the objective criteria of landscape analysis.

Fifth, geographical skill acquisition encompasses all three performance levels. A total of 6.84% of all tasks deal with the concept of landscape. Tasks that target reproduction (PFL1) instruct students to list elements of selected landscapes or describe selected landscapes. Comparison is the main activity within PFL2, where students reorganize and transfer their knowledge to other content and regional examples. One task, for example, confronts students with an imaginary person's description of a landscape to require them to identify the elements of the agricultural landscape. Another task instructs students to apply their knowledge on landscape alteration through exogenous forces to Venezuelan case studies. Other tasks require informational retrieval from both continuous and discontinuous (pictures, never maps) text. Finally, a few tasks assist students in developing their critical thinking and problem-solving skills by asking them to design information material for specific audiences on a case study based on all their previous knowledge of landscapes.

Third-grade secondary Geography (Rodríguez Requeña 2012) discusses thematic content on the national scale of Venezuela. Units explore the country's geographical location, its Physical Geography, population, economy, regionalization, and environment. The 27 segments identified during lexical retrieval were unambiguous in semantical terms. Along with scattered references to the landscape, most segments belong to the topic four of the unit dedicated to Physical Geography that discusses the landscape along with the climate, flora, and fauna. The content analysis produced four main features of the representation of landscape in the third-grade textbook.

First, the main thematic topic dedicated to landscapes replaces the concept of the landscape by biogeographic realms and refers to unique combinations of climate, flora, and fauna. Indeed, the description of the eleven Venezuelan realms excludes information on the geomorphological features and solely focuses on altitude, climate, flora, and fauna. A map visualizes the spatial distribution of these realms (see other examples of cartographic results in this volume: Edler et al. 2020b; Kleber et al. 2020; McLean 2020; Meyer-Heß 2020; Poplin et al. 2020; Siepmann et al. 2020; Stratmann et al. 2020; Vetter 2020). Nevertheless, the lesson dedicated to the central geomorphological units of the country lists three landscapes of the Guiana Shield: tepui, peneplain, and the Great Savanna. All other segments also refer to landscapes instead of biogeographic realms.

Second, the third-grade textbook repeatedly stresses that landscapes are a result of constant change induced by both natural and human agents. Still, the description of landscapes within the country's geomorphology, as well as the introduction of the eleven biogeographical realms, offers a rather static image.

Third, units concerned with the population and the economy dedicate little attention to the matters of landscapes. The role of landscape in series in general and tourism, in particular, represents the only exception.

Fourth, the three tasks (1 % of all tasks) merely contribute to skill acquisition at PFL1, requiring students to reproduce content, look up information online, and list features of landscapes.

5   Discussion and Conclusions

Against the background of the substantial change that the concept of landscape experienced over the last decades, this chapter followed two aims. On the one hand, it explored the ways two school Geographies at the periphery of global attention introduced and represented the concept of landscape. On the other hand, it focused on the educational objectives tied to landscapes.

Regarding the initial question, whether school Geographies were still about landscapes, the results showed two different perspectives. While the Hungarian sample repeatedly and broadly used landscapes as primary descriptors and, often, as the main access to geographical space, Venezuelan textbooks introduced landscapes as one content element of Physical Geography. In consequence, Hungarian students still gain access to the world and its regions employing landscapes. In Venezuela, school Geography uses several approaches to introduce the world to the students. Moreover, the two case studies stand for two quite different ways of implementing the thematic, regional, and regional-thematic approaches prescribed in the curricula. While Venezuelan thematic units introduce landscapes along with other physical-geographical elements of the geosystem, their

Hungarian counterparts proceed to work with the concept without prior introduction in terms of definition and classification. Similarly, Hungarian regional units rest on major landscapes, while Venezuelan landscapes are merely one of the many possibilities to regionalize the country.

Concerning the second question targeting the ways lower secondary Geography textbooks discussed space, the two case studies displayed several similarities. First, all textbooks work with a landscape concept that is deeply rooted in the geomorphological tradition and, therefore, stand for the geological layer (Gray 2008). In addition, selected units add the biological, and, in some cases, even the cultural layer (Gray 2008) to the geological substrate. Nevertheless, a close reading of the cultural reading is yet to be achieved, as the textbooks merely entail dispersed references to the role of human societies for both landscape alteration and conservation/restoration. Second, the textbooks paint an overall static image of the landscape. Landscape dynamics as a process of change often remains at a declarative mentioning or constitute loose pieces scattered across the textbook (e.g., erosion, mining, and tourism). The request to view landscapes more dynamically and based on the perspectives of those who inhabit and produce them (Morin 2008) is another required future improvement. Third, all four textbooks mainly emphasize the existence of landscapes that can be identified based on criteria. Both case-studies present sets of criteria to categorize landscapes that exclude the individual perception. The Hungarian textbooks feature a few tasks that target individual experiences and perception, even values connected to selected landscapes, including the primary landscapes (Measham 2007). As both geographers (Morin 2008) and Geography educators (Robertson et al. 2003; Measham 2007; Klonari et al. 2011) stress, experiential learning is intimately tied to understanding both the subjective perception of landscapes and their constructed nature.

Finally, the third question targeted the role of the landscape concept in knowledge acquisition within school Geographies. The results uncovered, first, two very different strategies. On the one hand, Hungarian textbooks use the landscape concept as a guiding element of geographical learning. Nevertheless, definitions and classifications, for example, in a separate unit, are missing from both textbooks. This is surprising given the leading role of the concept for content structure. In consequence, landscapes may be omnipresent; however, they remain outside the core of Geography's conceptual learning. On the other hand, both Venezuelan textbooks dedicated a separate unit to the concept of landscape. However, the remaining units and chapters rest on other concepts that turn landscapes into one of the many content elements of Venezuelan school Geographies. Second, the results also show that landscapes—unlike in other case studies (see Sect. 1)– primarily serve the purposes of geographical knowledge acquisition instead of fostering cross-curricular objectives, such as EE and ESD. Nevertheless, both Hungarian and Venezuelan textbooks make references to the role protection and sustainability play. Finally, the tasks support geographical knowledge acquisition at all three PFLs, however, to a different extent. The Hungarian textbooks rest on an implicit spiral curriculum that involves an episodic return to the concept of landscape both in the continuous text and in the tasks. The Venezuelan textbooks entail few tasks that, nonetheless, cover all PFLs. In contrast to their Venezuelan counterparts, tasks in Hungarian textbooks also contribute to map skills, numeric literacy, and media education.

Summing up, the results of this chapter offered insight into both the content and objectives of two school Geographies located outside of the usual sampling (see Sect. 1). While the results differ in many ways from previous findings, some limitations apply. First, an in-depth analysis of both continuous and discontinuous text elements would complement the findings of this study. Second, interviews with textbook authors and editors may offer the required background information to better comprehend both content- and objective-related decisions. Third, the analysis of an overall conceptual structure of all textbooks would offer more precise information on the exact role of the landscape concept. Future work could complement this study, also addressing these shortcomings.

References

Bagoly-Simó, P. (2017). Przemilczany człowiekń krajobrazy kulturowe w berlińskich podręcznikach szkolnych. In R. Traba, V. Julkowska, & T. Stryjakiewicz (Hrsg.), *Krajobrazy kulturowe. Sposoby konstruowania i narracje* (S. 459–486). Warsaw/Berlin: Wydawnictwo Neriton.

Clifford, N., Holloway, S., Rice, S. P., & Valentine, G. (Eds.). (2008). *Key concepts in geography.* London: SAGE.

DGfG (Deutsche Gesellschaft für Geographie; German Geographical Association) (2012). *Educational standards in geography for the intermediate school certificate.* DGfG: Bonn.

Edler, D., Jenal, C., & Kühne, O. (2020a). Modern approaches to the visualization of landscapes—An introduction. In D. Edler, C. Jenal, & O. Kühne (Eds.), *Modern approaches to the visualization of landscapes* (pp. 3–15). Wiesbaden: Springer VS.

Edler, D., Keil, J., & Dickmann, F. (2020b). From Na Pali to Earth—An 'Unreal' engine for modern geodata? In D. Edler, C. Jenal, & O. Kühne (Eds.), *Modern approaches to the visualization of landscapes* (pp. 279–291). Wiesbaden: Springer VS.

Fernández Álvarez, R. (2020). The landscape in the teaching of geography: Analysis of the curricula contents of primary education and secondary education. *Zeitschrift für Geographiedidaktik\Journal of Geography Education* (accepted).

Gayton, D. (1996). *Landscapes of the interior: Re-explorations of nature and the human spirit.* Gabriola Island: New Society Publishers.

Gray, M. (2008). Landscape: The physical layer. In N. Clifford, S. Holloway, S. P. Rice, & G. Valentine (Eds.), *Key concepts in geography* (pp. 265–284). London: SAGE.

Hard, G. (1991). Landschaft als professionelles Idol. *Garten und Landschaft, 3*, 13–18.

Hasse, J. (1993). *Heimat und Landschaft – Über Gartenzwerge, Center Parcs und andere Ästhetisierungen.* Wien: Passagen Verlag.

Hasse. J. (2011). Zur mythischen Funktion deklarierter Natur-Landschaften. Das Beispiel des „Weltnaturerbes" Wattenmeer. In L. Fischer & K. Reise (Hrsg.), *Küstenmentalität und Klimawandel. Küstenwandel als kulturelle und soziale Herausforderung* (S. 97–113), München: oekom.

Kleber, A., Edler, D., & Dickmann, F. (2020). Cartography and the sea: A javascript-based web mapping application for managing maritime shipping. In D. Edler, C. Jenal, & O. Kühne (Eds.), *Modern approaches to the visualization of landscapes* (pp. 173–186). Wiesbaden: Springer VS.

Klonari, A., Dalaka, A., & Petanidou, T. (2011). How Evident is the apparent? Students' and teachers' perceptions of the terraced landscape. *International Research in Geographical and Environmental Education, 20*(1), 5–20.

Kühne, O. (2018). Macht, Herrschaft und Landschaft: Landschaftskonflikte zwischen Dysfunktionalität und Potenzial. Eine Betrachtung aus der Perspektive der Konflikttheorie Ralf Dahrendorfs. In K. Berr (Hrsg.), *Transdisziplinäre Landschaftsforschung: Grundlagen und Perspektiven* (S. 155–170). Wiesbaden: Springer VS.

Kühne, O. (2019). *Landscape theories: A brief introduction.* Wiesbaden: Springer VS.

Kühne, O., & Jenal, C. (2020). The threefold landscape dynamics—Basic considerations, conflicts and potentials of virtual landscape research. In D. Edler, C. Jenal, & O. Kühne (Eds.), *Modern approaches to the visualization of landscapes* (pp. 389–402). Wiesbaden: Springer VS.

Kusztor, A., Makádi, M., Pokk, P., & Szöllőssy, L. (2017a). *Földrajz 7 (Geography Textbook for Grade 7).* Eger: Eszterházy Károly Egyetem.

Kusztor, A., Makádi, M., Pokk, P., & Szöllőssy, L. (2017b). *Földrajz 8 (Geography Textbook for Grade 8).* Eger: Eszterházy Károly Egyetem.

Loda, M., Kühne, O., & Puttilli, M. (2020). The social construction of Tuscany in the German and English speaking world—Presented by the analysis of internet images. In D. Edler, C. Jenal, & O. Kühne (Eds.), *Modern approaches to the visualization of landscapes* (pp. 157–171). Wiesbaden: Springer VS.

McLean, K. (2020). Temporalities of the smellscape: Creative mapping as visual representation. In D. Edler, C. Jenal, & O. Kühne (Eds.), *Modern approaches to the visualization of landscapes* (pp. 217–246). Wiesbaden: Springer VS.

Measham, T. G. (2007). Primal landscapes: Insights for education from empirical research on ways of learning about environments. *International Research in Geographical and Environmental Education, 16*(4), 339–350.

Meyer-Heß, F. (2020). Discovering forgotten landscapes. In D. Edler, C. Jenal, & O. Kühne (Eds.), *Modern approaches to the visualization of landscapes* (pp. 33–45). Wiesbaden: Springer VS.

Morin, K. M. (2008). Landscape: Representing and interpreting the world. In N. Clifford, S. Holloway, S. P. Rice, & G. Valentine (Eds.), *Key concepts in geography* (pp. 265–284). London: SAGE.

Poplin, A., Andrade, B. de, & Mahmud, S. (2020). Exploring tangible and intangible landscapes of evocative places: Case study of the City of Vitória in Brazil. In D. Edler, C. Jenal, & O. Kühne (Eds.), *Modern approaches to the visualization of landscapes* (pp. 519–547). Wiesbaden: Springer VS.

Prisille, C., & Ellerbrake, M. (2020). Virtual Reality (VR) and geography education: Potentials of 360° 'Experiences' in secondary schools. In D. Edler, C. Jenal, & O. Kühne (Eds.), *Modern approaches to the visualization of landscapes* (pp. 321–332). Wiesbaden: Springer VS.

Robertson, M., Walford, R., & Fox, A. (2003). Landscape meanings and personal identities: Some perspectives of East Anglian Children. *International Research in Geographical and Environmental Education, 12*(1), 32–48.

Rodríguez Requeña, J. M. (2012). *Geografía de Venezuela. 3er año (Geography of Venezuela. Third Grade).* Caracas: Santillana.

Rodríguez Requeña, J. M. (2013). *Geografía General. 1er año (General Geography. First Grade).* Caracas: Santillana.

Schöps, A. (2017). The paper implementation of the German Educational Standards in Geography for the Intermediate School Certificate in the German Federal States. *Review of International Geographical Education Online (RIGEO), 7*(1), 94–117.

Schultz, H.-D. (1971). Versuch einer ideologiekritischen Skizze zum Landschaftskonzept. *Geografiker, 6*, 1–12

Schultz, H.-D. (2014). "Wie das Land, so das Volk, wie das Volk, so das Land": Landschafts- und Länderkunde (die klassische Geographie) auf Abwegen. In N. M. Franke & U. Pfennig (Eds.), *Kontinuitäten im Naturschutz* (pp. 23–79). Baden-Baden: Nomas.

Siepmann, N., Edler, D., & Kühne, O. (2020). Soundscapes in cartographic media. In D. Edler, C. Jenal, & O. Kühne (Eds.), *Modern approaches to the visualization of landscapes* (pp. 247–263). Wiesbaden: Springer VS.

Stintzing, M., Pietsch, S., & Wardenga, U. (2020). How to teach "Landscape" through games? In D. Edler, C. Jenal, & O. Kühne (Eds.), Modern approaches to the visualization of landscapes (pp. 333–349). Wiesbaden: Springer VS.

Stratmann, J., Ristea, A., Leitner, M., & Paulus, G. (2020). Exploring urban "Blightscapes" applying spatial video technology and geographic information system: A case study from Baton Rouge, USA. In D. Edler, C. Jenal, & O. Kühne (Eds.), *Modern approaches to the visualization of landscapes* (pp. 499–517). Wiesbaden: Springer VS.

Vetter, M. (2020). Technical potentials for the visualization in virtual reality. In D. Edler, C. Jenal, & O. Kühne (Eds.), *Modern approaches to the visualization of landscapes* (p. 307–317). Wiesbaden: Springer VS.

Péter Bagoly-Simó is full professor and chair of Geography Education with the Geography Department of Humboldt-Universität zu Berlin. His previous work focused on curriculum studies, Education for Sustainable Development, and educational media in the teaching and learning of Geography at K-12 level.

Part VII
Landscape Visualization and Conflicts

The Threefold Landscape Dynamics: Basic Considerations, Conflicts, and Potentials of Virtual Landscape Research

Olaf Kühne and Corinna Jenal

Abstract

The threefold landscape dynamic describes, firstly, the changes in societal conceptions of landscape, which, secondly, provide the framework for individual conceptions of landscape. Thirdly, it describes changes in the physical space which 'landscape' is projected onto. As a result of social differentiation processes, this threefold landscape dynamic is accelerating. With increasing contrasts in society, the interpretations and evaluations of 'landscape', as well as the normative conceptions associated with it, also differ. This leads to an increase in the number of landscape conflicts. These can certainly take on a productive function in society—when they are regulated. The influence of virtualizations of landscape on the threefold landscape dynamic and the related conflicts is manifold, but only rudimentarily considered in research: They form a component of the construction of social and individual landscapes, but they can also be applied in the context of landscape conflicts.

Keywords

Landscape · Constructivism · Virtuality · Social constructivism · Threefold landscape dynamic · Landscapes conflicts

O. Kühne (✉) · C. Jenal
Eberhard Karls Universität Tübingen, Tübingen, Germany
e-mail: olaf.kuehne@uni-tuebingen.de

C. Jenal e-mail: corinna.jenal@uni-tuebingen.de

© Springer Fachmedien Wiesbaden GmbH, part of Springer Nature 2020
D. Edler et al. (eds.), *Modern Approaches to the Visualization of Landscapes*, RaumFragen: Stadt – Region – Landschaft,
https://doi.org/10.1007/978-3-658-30956-5_22

1   Introduction

What landscape is for us is very diverse: view, insight, homeland, scenery. It is associated with many adjectives, such as beautiful, historical, industrial, picturesque, sublime, intact, typical, low mountain range, sometimes even kitschy or ugly, diverse, and much more (among many: Hard 1969; Hokema 2013; Penning-Rowsell and Lowenthal 1986). However and above all, landscape is very changeable. This changeability does not only affect the material world, in which we generally see landscape. It also affects our social conventions, which we may describe and not least evaluate as landscape without loss of social recognition. This changeability also affects our personal preferences, attachments, and rejections. In short—landscape is subject to a threefold dynamic (Kühne 2020a, b).

This threefold dynamic of landscape is definitely accelerating in many regions of the world: Societal demands on material spaces are accelerating with the increasing speed of social change, demands on infrastructures, the energy revolution, housing for a growing population, the desire for recreation, but also anthropogenic climate change with its biotic and abiotic consequences are inscribed in physical spaces. Societal patterns of construction, interpretation and evaluation of landscape change, for example, through the examination of changing physical structures (for example, through old industrialization, see Jenal 2019a; Kühne 2007; Wood 2012), additionally, they are also transformed by intercultural exchange processes (see for example Bruns and Kühne 2015; Bruns and Paech 2015), through educational processes (Kühne 2008b), and other things. Individual approaches to landscape are also subject to accelerated change with 'broken spatial biographies', for example through spatial shifts of centers of gravity, a conscious examination of 'landscape', an individually experienced access to 'landscape', individualized educational processes, etc. (Kühne and Schönwald 2015; Lehmann 1996; Münderlein et al. 2019; Sautter 2018). The fact that with the expansion of the world external to consciousness through virtual worlds, the change in the individual and social constructions of landscape will be further accelerated and can now be regarded as certain (Edler et al. 2018a, 2018b; Fontaine 2017; Kühne 2018; Lange 2001).

2   The Threefold Landscape Dynamic – Some Basic Considerations

When talking about landscape change, this usually refers to a change in the physical world, such as the expansion of renewable energies, agricultural structural change, the extraction of raw materials, the growth of cities and much more. Such a view is very presuppositional. On the one hand, it assumes that landscape is a material object. On the other hand, it assumes that there is just the *one* landscape. Although such a view is still widespread in large parts of the natural and planning sciences, it has been questioned in the social and cultural sciences and spatial sciences in recent decades. This is a reason the Swiss landscape researcher Lucius Burckhardt once formulated 'landscape is created in the mind'. (Burckhardt 2006). This expresses that landscape is not a clearly given material object that can be (more or less) unambiguously determined by empirical methods and whose structures and functions can be modelled. From this point of view, landscape is created by combining sensory impressions within our consciousness (see in this volume: McLean 2020; Siepmann et al. 2020). This combination is called 'landscape'. It is often bound to adjectives, like 'beautiful', 'romantic', 'industrial', or 'ugly'. However, these attributions do not arise from within ourselves. They are based on social conventions that we learn and that we (can) rub up against. If we have enough social capital (in the sense of Bourdieu 1989), we can also change social conventions (in this case in relation to landscape).

If such a constructivist perspective on landscape is adopted (as in landscape research, already suggested by: Cosgrove 1984; Duncan 1990; Greider and Garkovich 1994; Micheel 2012; Stotten 2015), the constitutive level for 'landscape' is not to be sought in the material world. It rather lies in the recursive relationship of social patterns of interpretation, the evaluation of landscape with their individual actualization, and in the comparison with experienced and reflected aspects. (Gergen and Gergen 2009; Jenal 2019b; Knorr-Cetina 1989). An essential 'hinge' between the individual and society is the socialization of landscape; this process will be briefly explained below.

3   The Socialization of Landscape

Like all complex and abstract concepts, the concept of landscape must be learned. What we can call a 'landscape' (preferably with an adjective preceding) in which social and spatial contexts are without loss of social recognition, we learn over the course of our lives. Three different modes of learning landscape can be distinguished: the 'native normal landscape', the 'stereotypical landscape', and—if we decide to deal with the topic of landscape professionally—'expert special knowledge landscape'. (Kühne 2019b).

The 'normal domestic landscape' is developed primarily in childhood. This particularly happens through personal experience, but also through the mediation of parents, grandparents, siblings, etc. The environment of the parental residence is explored. What is encountered is deemed 'normal' and is not critically questioned, 'homelike ties' to the environment are created. In contrast, the 'stereotypical landscape' is conveyed more cognitively, especially through school, movies, books, the internet, etc. (in this volume e.g.: Fontaine 2020; Kühne 2020c; Linke 2020, Loda et al. 2020; Papadimitriou 2020, 2021). Through these media, ideas are brought to the individual, such as what a 'beautiful', 'natural', 'ugly', etc., landscape should look like. In this context, landscape is no longer what I experience, but what I observe and judge. This is where a common sense understanding of landscape emerges. Even more distanced is the consideration of what we call 'landscape' based on 'expert special knowledge', which is usually acquired through scientific studies related to landscape—and which is strongly deficit-oriented. Landscape thus becomes the 'object' of optimization efforts. However, what is regarded as an optimum is very different from one discipline to another. For example, a landscape planner has a different (partial) idea of an optimum landscape than a tourism expert or an agricultural scientist (see Aschenbrand 2016; Nissen 1998; Stotten 2013; in this volume: Bagoly-Simó 2020).

But not only the 'expert special knowledge' is differentiated, but also the common sense understanding of landscape. Those former are more technical (landscape planners, agricultural scientists, human geographers, etc.) or paradigmatic (constructivists, essentialists, positivists, phenomenologists etc.; see Kühne et al. 2018), and the differences in common sense understanding are particularly pronounced with regard to linguistic characteristics (Bruns and Münderlein 2019; Döhla 2019; Drexler 2009; Makhzoumi 2015). This includes social differences (such as formal education level, age, and user interests; among many others: Herzog et al. 2000; Kearney and Bradley 2011; Kühne 2006a; Stotten 2019) or also with regard to the assignment to different ideological basic positions, such as conservatism, liberalism, and socialism (among others: Kirchhoff 2019; Kühne 2015; Vicenzotti 2011). These different constructions of landscape often do not tend towards a peaceful coexistence, but they are in conflict over the sovereignty of interpretation and evaluation of landscape. This is discussed in the following.

4   Landscape Conflicts and the Threefold Landscape Dynamic

Even in a world without the threefold dynamic of landscape, conflicts are inevitable, since the views on what we call landscape are very diverse, not only between experts of different disciplines, but also between these and the 'stereotypical' ideas of landscape. So, the ideas of a 'stereotypically beautiful landscape' do not necessarily coincide with the ideas of an 'economically productive' or 'natural' landscape as well with the 'normal landscape of the home country'. At the latest with this, the topic of 'change' becomes virulent, as spaces that are experienced in the mode of the 'normal domestic landscape' do not have to be stereotypically beautiful or valuable in terms of nature conservation (alternatively: economic, infrastructural,…; see in this volume: Berr 2020). They are rather familiar and stable. Changes in physical objects (such as the construction of a wind farm) are thus seen here as a threat to 'home'. To what extent the change of objects under the mode of the 'stereotypical landscape' is rejected depends on whether this change is perceived at all (this also applies to the mode of the 'homeborn-normal landscape'). If so, it is interpreted as 'positive' or 'negative'. If the change takes place gradually and is below the threshold of perception (e.g. the consequences of 'insect mortality'), it is not considered as problematic. If the change corresponds to the ideas of a 'stereotypically beautiful landscape', it is welcomed—if not rejected and sometimes even fought against. The assessment of change under the mode of 'expert special knowledge' is highly dependent on the technical background. Therefore, an agricultural economist will view land use extensification more critically than a landscape planner (among many: Breukers and Wolsink 2007; Gailing 2013; Hoeft et al. 2017; Kühne and Weber 2019; Pasqualetti et al. 2002).

However, it is not only the physical foundations of landscape that change: The 'normal landscape of the home' is dependent on the state of the spaces in childhood, which can change from generation to generation. The 'stereotypical landscape' is also undergoing a change: While 40 years ago, the mining and iron industry facilities were considered 'ugly', they have now become magnets for tourism, and they are considered 'interesting' (Jenal 2019a; Kühne 2018; in this volume: Weber 2020). Expert stocks of 'special knowledge' are, anyway, subject to constant change in science (in summary: Kühne 2019b; Winchester et al. 2003; Wylie 2007): Supporters of the various paradigms of landscape struggle for the sovereignty of interpretation over the scientific approach to 'landscape'. In the meantime, it is much more than about reputation within the specialist discourse. It is not just funding for further research that depends on the paradigm, but also on the impact of one's own approach in political and, especially, administrative practice (Burckhardt 2004; Kühne 2008b; Poerting and Marquardt 2019).

Conflicts (or at least contradictions) in relation to 'landscape' arise not only between persons, groups of persons, the adherents of different paradigms, etc., but they also arise between individual persons, especially those who deal more intensively with the topic of 'landscape' (Kühne 2006a). For example, for persons with 'expert special knowledge', the technical view (e.g. of process protection) of stereotypical ideas (e.g. an aesthetic preference of a 'Tuscan Landscape') can deviate significantly from the 'normal domestic landscape' of an old industrial area. The ways of dealing with these divergences and conflicts can vary greatly, from despair over the inconsistency of one's own understanding of landscape, to the attempts of adapting it to one another, and to the acceptance of its diversity (Bruns and Kühne 2013; Kühne 2006b, 2008a; Wojtkiewicz and Heiland 2012).

In the end, therefore, it can be summarized – 'Landscape is conflict!' This conflictual nature of and within landscape increases with the increasing social, economic, political, and cultural differentiation of society. Finally, the interpretations and evaluations of landscape also differentiate themselves, as do the demands on physical spaces and normative allowances (see for example Al-Khanbashi 2019; Kühne and Weber 2018 [online first 2017]; Kühne, Weber, and Berr 2019). However, conflicts in general and landscape conflicts in particular do not necessarily need to be understood as dysfunctional (as is the case with Parsons 1991 [1951]). According to the sociologist Ralf Dahrendorf, conflicts can be understood as 'social normality' (Dahrendorf 1961, 1972). For him, they can be understood as an essential driving force of social change, in which people exercise power over other people (quite differentiated – both direct and indirect as well as mutual). Nevertheless, individual people are also able to change social conditions. Thus, conflicts can certainly be attributed to productivity. However, conflicts can only achieve this productivity if they are settled. A solution does not seem possible as the social causes of conflicts would have to be resolved. They are often rooted in a differentiated distribution of power which is, in turn, inherent in all societies. The suppression of conflicts is also not sustainable, for Dahrendorf, as it only led to the further development of social contradictions, which ultimately erupted violently. Essential for a settlement of conflicts is the observance of four aspects (Dahrendorf 1972):

1. The conflictual contradictions must be recognized as a legitimate dimension of normality, not as a state contrary to the norm.
2. Conflict regulation refers to the manifestations of the conflict, not to its causes, since these lie outside the conflict framework.
3. If the efficiency of conflict resolution can be positively influenced by the degree of organization of the conflict parties, the better organized the conflict parties are, the higher the probability that a conflict resolution accepted by both sides will be achieved.
4. The success of conflict resolution depends on compliance with certain rules. These rules must not favor any of the conflicting parties, i.e. the conflicting parties must be regarded as equal (i.e. certain pre-defined procedural rules must be followed).

The complexity of landscape conflicts is increased by the fact that different levels of conflict can be identified: distribution conflicts, procedural conflicts, location or land use conflicts, identity conflicts and technological conflicts (Becker and Naumann 2018). In this respect, the need to create a regulated framework for the settlement of landscape conflicts becomes clear. This applies, for example, to sites for wind power plants, a very present subject of landscape conflicts currently (see inter alia Otto 2019; Roßmeier and Weber 2018; Weber 2018). If a settlement is to be reached, it seems sensible to relate the spatial framework as far as possible to the site and its immediate surroundings (not to address the meaning of the energy system transformation as a whole), and to organize conflict parties as completely as possible so that negotiators with a corresponding mandate can negotiate with each other. In this conflict debate, both parties must accept that the arguments of the other side are legitimate. Accordingly, moralizations of the conflict by defining one's own position as 'good' and the other as 'morally reprehensible' (e.g. as "climate destroyer" and "home wrecker") are of little help to a consensual settlement of the conflict; Bues and Gailing 2016; Eichenauer et al. 2018; Kühne 2019a).

5   Potentials of Virtual Landscape Research

Virtuality offers the possibility of generating arrangements understood as 'landscapes' beyond their social complexity, which 'materialize' socially undesirable perspectives on 'landscape' in virtual space on the basis of all its 'illusion capacities' (Braun and Friess 2019). Thus, it opens up new perspectives on 'landscape'. Through technical innovations, it is possible to combine visual influences through a broader spectrum of sensory impressions (Edler et al. 2019; see in this volume: Edler et al. 2020a; McLean 2020;

Siepmann et al. 2020) and thus to get closer to a synaesthetic experience of space similar to an 'analog' access to the world (Kazig 2007, 2013) than landscape visualizations of the past, as those in illustrations and moving pictures.

Even though the basic technical idea was initially developed according to Heilig (1961) and the apparatus of a *Sensorama simulator* "to stimulate the senses of an individual to simulate an actual experience realistically" by "the cooperative effects of the breeze, the odor, the visual images, and binaural sound that stimulate a desired sensation in the senses of an observer" (Heilig 1961, Chap. 1) the development of virtual realities has progressed considerably (see in this volume: Edler et al. 2020a, 2020b, Fontaine 2020, Hochschild et al. 2020, Prisille and Ellerbrake 2020, Vetter 2020). There are still technical restrictions, for example in still deviating graphics, the realization of combined sensory impressions such as smell, temperature fluctuations, etc., similar to the experience of analog worlds or the so-called cybersickness as well as a strong dependence upon expert special knowledge for their creation (vgl. u.a. Braun and Friess 2019; Edler et al. 2018a, 2018b; McCauley and Sharkey 1992; Rebenitsch and Owen 2016).

Nevertheless, especially against the background of a further technical development, it is necessary to examine to what extent VR methods can be used to extend the set of methods for the investigation of the processes of social construction of landscape, landscape-related conflicts, and also in the context of a secondary education to 'landscape' (see Edler et al. 2018a, 2018b). If the existing technical restrictions were to be overcome, the landscape-related virtual realities thus generated could represent a similar milestone in the production, attribution, and establishment of visual patterns and the visual habits of landscape stereotypes, as it was once assigned to landscape paintings (vgl. dazu u.a. Berr and Schenk 2019; Büttner 2006, 2019; Kortländer 1977; Stiens 2009).

6 Conclusion

If landscape—in the constructivist tradition of thought—is not taken as an object, but as an individual and social understanding, there are three essential consequences for dealing with the threefold landscape dynamic:

1. Landscape change is normal. This normality arises both for the individual construction of landscape (the understanding and experience of landscape changes in the course of one's life, or at least there exists the possibility of doing so), and for the social construction (for example, economic transformations such as deindustrialization or new interpretations of landscape by individuals, such as landscape architects or painters). This normality also applies to physical spaces that are adapted according to changing social needs.
2. The number of perspectives on landscape also increases along with social differentiation. This is especially true when people come from other cultural circles, some of whom have no concept of landscape (e.g. in Arabic) or have a more differentiated

(e.g. in Chinese) understanding of landscape than in Central Europe. But this also applies in relation to different milieus that attribute very different meanings to 'landscape'. For example, the interpretations and evaluations of and normative ideas about 'landscape' clearly differ between urban hedonists, family-oriented suburbanites, and rural farmers.

3. Conflicts over landscape are normal. On the one hand, this results from the finiteness of physical space as well as overlapping and contradictory interests of use. It also applies to different interpretations, evaluations, and normative ideas of and about landscape. Landscape conflicts can also be socially productive if they are carefully managed. This includes, for example, the acknowledgement that the position of the other conflict party is legitimate and that its moral condemnation is therefore of little use. Such a conflict settlement can, among other things, offer sensitivity to various understandings of landscape.

The significance of virtuality in the threefold landscape dynamic is manifold and has so far only been investigated in a rudimentary form. Virtual landscapes contribute to the creation of individual landscape constructs. In the majority of cases, social landscape stereotypes are represented in an idealized form, and the increasing importance of virtual landscapes compared to the reference to physical-material spaces has already begun. Virtual simulations of 'landscape changes' in landscape conflicts (e.g. around the erection of wind turbines, but also power grids, motorways, nature reserves, logistics centers, etc.) can contribute to their objectification. They can also affect an emotionalization, depending on the perspective, atmosphere, and, especially, the discursive framing. In this respect, the 'classical field' of research into the social construction of landscape is extended by a further method. Nevertheless, virtualizations can also be used as a method to capture and classify individual and social construction patterns of landscape, beyond disturbances in the field and beyond disturbances in the field of de-dynamization which occur, for example, in photographs (Edler et al. 2019a, 2019b).

References

Al-Khanbashi, M. (2019). Urban/Rural Hybrids and Conflicts: New Research Perspective in Jeddah, Saudi Arabia. In K. Berr & C. Jenal (Eds.), *Landschaftskonflikte* (pp. 607–625). Wiesbaden: Springer VS.

Aschenbrand, E. (2016). Einsamkeit im Paradies. Touristische Distinktionspraktiken bei der Aneignung von Landschaft. *Berichte. Geographie und Landeskunde, 90*(3), 219–234.

Bagoly-Simó, P. M. (2020). Landscape in geography textbooks. In D. Edler, C. Jenal, & O. Kühne (Eds.), *Modern approaches to the visualization of landscapes* (p. 371–385). Wiesbaden: Springer VS.

Becker, S., & Naumann, M. (2018). Energiekonflikte erkennen und nutzen. In O. Kühne & F. Weber (Eds.), *Bausteine der Energiewende* (pp. 509–522). Wiesbaden: Springer VS.

Berr, K., & Schenk, W. (2019). Begriffsgeschichte. In O. Kühne, F. Weber, K. Berr, & C. Jenal (Eds.), *Handbuch Landschaft* (pp. 23–38). Wiesbaden: Springer VS.

Berr, K. (2020). Visuality, Aesthetics and Landscape. For the enlightenment and self-enlightenment of constructivist landscape research. In D. Edler, C. Jenal, & O. Kühne (Eds.), *Modern approaches to the visualization of landscapes* (pp. 189–215). Wiesbaden: Springer VS.

Bourdieu, P. (1989). Social space and symbolic power. *Sociological theory, 7*(1), 14–25.

Braun, H., & Friess, R. (2019). Empirische Zugänge zur Virtual Reality. Heterogenes Netzwerk, Diskurs und Wahrnehmungsform. In D. Kasprowicz, & S. Rieger (Eds.), *Handbuch Virtualität* (pp. 1–21). Wiesbaden: Springer VS.

Breukers, S., & Wolsink, M. (2007). Wind power implementation in changing institutional landscapes: An international comparison. *Energy Policy, 35*(5), 2737–2750. https://doi.org/10.1016/j.enpol.2006.12.004

Bruns, D., & Münderlein, D. (2019). Interkulturelle Konstruktion. In K. Berr, C. Jenal, O. Kühne, & F. Weber (Eds.), *Handbuch Landschaft* (pp. 313–319). Wiesbaden: Springer VS.

Bruns, D., & Kühne, O. (2013). Landschaft im Diskurs. Konstruktivistische Landschaftstheorie als Perspektive für künftigen Umgang mit Landschaft. *Naturschutz und Landschaftsplanung, 45*(3), 83–88.

Bruns, D., & Kühne, O. (2015). Zur kulturell differenzierten Konstruktion von Räumen und Landschaften als Herausforderungen für die räumliche Planung im Kontext von Globalisierung. In B. Nienaber, & U. Roos (Eds.), *Internationalisierung der Gesellschaft und die Auswirkungen auf die Raumentwicklung. Beispiele aus Hessen, Rheinland-Pfalz und dem Saarland* (Arbeitsberichte der ARL, vol. 13, pp. 18–29). Hannover: Selbstverlag. https://shop.arl-net.de/media/direct/pdf/ab/ab_013/ab_013_02.pdf. Accessed: 26 November 2018.

Bruns, D., & Paech, F. (2015). „Interkulturell_real" in der räumlichen Entwicklung. Beispiele studentischer Arbeiten zur Wertschätzung städtischer Freiräume in Kassel. In B. Nienaber, & U. Roos (Eds.), *Internationalisierung der Gesellschaft und die Auswirkungen auf die Raumentwicklung. Beispiele aus Hessen, Rheinland-Pfalz und dem Saarland* (Arbeitsberichte der ARL, vol. 13, pp. 54–71). Hannover: Selbstverlag. https://shop.arl-net.de/media/direct/pdf/ab/ab_013/ab_013_05.pdf. Accessed: 26 November 2018.

Bues, A., & Gailing, L. (2016). Energy transitions and power: Between governmentality and depoliticization. In L. Gailing, & T. Moss (Eds.), *Conceptualizing Germany's energy transition. Institutions, materiality, power, space* (pp. 69–91). London: Palgrave Macmillan.

Burckhardt, L. (2004). *Wer plant die Planung? Architektur, Politik und Mensch.* Berlin: Martin Schmitz Verlag.

Burckhardt, L. (2006). *Warum ist Landschaft schön? Die Spaziergangswissenschaft.* Kassel: Martin Schmitz Verlag.

Büttner, N. (2006). *Geschichte der Landschaftsmalerei.* München: Hirmer.

Büttner, N. (2019). Landschaftsmalerei. In O. Kühne, F. Weber, K. Berr, & C. Jenal (Eds.), *Handbuch Landschaft* (pp. 577–584). Wiesbaden: Springer VS.

Cosgrove, D. E. (1984). *Social Formation and Symbolic Landscape.* London: University of Wisconsin Press.

Dahrendorf, R. (1961). *Gesellschaft und Freiheit. Zur soziologischen Analyse der Gegenwart.* München: Piper.

Dahrendorf, R. (1972). *Konflikt und Freiheit. Auf dem Weg zur Dienstklassengesellschaft.* München: Piper.

Drexler, D. (2009). Kulturelle Differenzen der Landschaftswahrnehmung in England, Frankreich, Deutschland und Ungarn. In T. Kirchhoff & L. Trepl (Eds.), *Vieldeutige Natur. Landschaft, Wildnis und Ökosystem als kulturgeschichtliche Phänomene* (Sozialtheorie, pp. 119–136). Bielefeld: transcript.

Duncan, J. S. (1990). *The city as text: The politics of landscape interpretation in the Kandyan Kingdom.* Cambridge: Cambridge University Press.

Döhla, H.-J. (2019). Sprache und Landschaft. In O. Kühne, F. Weber, K. Berr, & C. Jenal (Eds.), *Handbuch Landschaft* (pp. 429–440). Wiesbaden: Springer VS.

Edler, D., Jenal, C., & Kühne, O. (2020a). Modern approaches to the visualization of landscapes—An Introduction. In D. Edler, C. Jenal, & O. Kühne (Eds.), *Modern approaches to the visualization of landscapes* (pp. 3–15). Wiesbaden: Springer VS.

Edler, D., Husar, A., Keil, J., Vetter, M., & Dickmann, F. (2018a): Virtual Reality (VR) and open source software: A Workflow for constructing an interactive cartographic VR environment to explore urban landscapes. *KN – Journal of Cartography and Geographic Information, 68*(1) 3–11.

Edler, D., Keil, J., & Dickmann, F. (2020b). From Na Pali to Earth—An 'Unreal' engine for modern geodata? In D. Edler, C. Jenal, & O. Kühne (Eds.), *Modern approaches to the visualization of landscapes* (pp. 279–291). Wiesbaden: Springer VS.

Edler, D., Kühne, O., Keil, J., & Dickmann, F. (2019a). Audiovisual cartography: Established and new multimedia approaches to represent soundscapes. *KN – Journal of Cartography and Geographic Information, 69* 5–17. https://doi.org/10.1007/s42489-019-00004-4

Edler, D., Keil, J., Wiedenlübbert, T., Sossna, M., Kühne, O. and Dickmann, F. (2019b): Immersive VR experience of redeveloped post-industrial sites: The example of "Zeche Holland" in Bochum-Wattenscheid. *KN – Journal of Cartography and Geographic Information, 69*(4), 267–284.

Edler, D., Kühne, O., Jenal, C., Vetter, M., & Dickmann, F. (2018b). Potenziale der Raumvisualisierung in Virtual Reality (VR) für die sozialkonstruktivistische Landschaftsforschung. *Kartographische Nachrichten, 68*(5), 245–254.

Eichenauer, E., Reusswig, F., Meyer-Ohlendorf, L., & Lass, W. (2018). Bürgerinitiativen gegen Windkraftanlagen und der Aufschwung rechtspopulistischer Bewegungen. In O. Kühne & F. Weber (Eds.), *Bausteine der Energiewende* (pp. 633–651). Wiesbaden: Springer VS.

Fontaine, D. (2017). *Simulierte Landschaften in der Postmoderne. Reflexionen und Befunde zu Disneyland, Wolfersheim und GTA V.* Wiesbaden: Springer VS.

Fontaine, D. (2020). Virtuality and Landscape. In D. Edler, C. Jenal, & O. Kühne (Eds.), *Modern approaches to the visualization of landscapes* (pp. 267–276). Wiesbaden: Springer VS.

Gailing, L. (2013). Die Landschaften der Energiewende – Themen und Konsequenzen für die sozialwissenschaftliche Landschaftsforschung. In L. Gailing & M. Leibenath (Eds.), *Neue Energielandschaften – Neue Perspektiven der Landschaftsforschung* (pp. 207–215). Wiesbaden: Springer VS.

Gergen, K. J., & Gergen, M. (2009). *Einführung in den sozialen Konstruktionismus.* Heidelberg: Carl-Auer-Systeme Verlag.

Greider, T., & Garkovich, L. (1994). Landscapes: The Social Construction of Nature and the Environment. *Rural Sociology, 59*(1), 1–24. https://doi.org/10.1111/j.1549-0831.1994.tb00519.x.

Hard, G. (1969). Das Wort Landschaft und sein semantischer Hof. Zur Methode und Ergebnis eines linguistischen Tests. *Wirkendes Wort, 19*, (3–14).

Heilig, M. L. (1961). Heilig, Morton L. (Anmelder), US81864AUSA.

Herzog, T. R., Herbert, E. J., Kaplan, R., & Crooks, C. L. (2000). Cultural and developmental comparisons of landscape perceptions and preferences. *Environment and Behavior, 32*(3), 323–346. https://doi.org/10.1177/0013916500323002.

Hochschild, V., Braun, A., Sommer, C., Warth, G., & Omran, A. (2020). Visualizing landscapes by geospatial techniques. In D. Edler, C. Jenal, & O. Kühne (Eds.), *Modern approaches to the visualization of landscapes* (pp. 47–78). Wiesbaden: Springer VS.

Hoeft, C., Messinger-Zimmer, S., & Zilles, J. (Eds.). (2017). *Bürgerproteste in Zeiten der Energiewende. Lokale Konflikte um Windkraft, Stromtrassen und Fracking.* Bielefeld: transcript.

Hokema, D. (2013). *Landschaft im Wandel? Zeitgenössische Landschaftsbegriffe in Wissenschaft, Planung und Alltag.* Wiesbaden: Springer VS.

Jenal, C. (2019a). (Alt)Industrielandschaften. In O. Kühne, F. Weber, K. Berr, & C. Jenal (Eds.), *Handbuch Landschaft* (pp. 831–841). Wiesbaden: Springer VS.

Jenal, C. (2019b). *„Das ist kein Wald, Ihr Pappnasen!"* – Zur sozialen Konstruktion von Wald. *Perspektiven von Landschaftstheorie und Landschaftspraxis.* Wiesbaden: Springer VS.

Kazig, R. (2007). Atmosphären – Konzept für einen nicht repräsentationellen Zugang zum Raum. In C. Berndt, & R. Pütz (Eds.), *Kulturelle Geographien. Zur Beschäftigung mit Raum und Ort nach dem Cultural Turn* (pp. 167–187). Bielefeld: transcript.

Kazig, R. (2013). Landschaft mit allen Sinnen – Zum Wert des Atmosphärenbegriffs für die Landschaftsforschung. In D. Bruns, & O. Kühne (Eds.), *Landschaften: Theorie, Praxis und internationale Bezüge. Impulse zum Landschaftsbegriff mit seinen ästhetischen, ökonomischen, sozialen und philosophischen Bezügen mit dem Ziel, die Verbindung von Theorie und Planungspraxis zu stärken* (pp. 221–232). Schwerin: Oceano Verlag.

Kearney, A. R., & Bradley, G. A. (2011). The effects of viewer attributes on preference for forest scenes. Contributions of attitudes, knowledge, demographic factors, and stakeholder group membership. *Environment and Behavior, 43*(2), 147–181. https://doi.org/10.1177/0013916509353523

Kirchhoff, T. (2019). Politische Weltanschauungen und Landschaft. In O. Kühne, F. Weber, K. Berr, & C. Jenal (Eds.), *Handbuch Landschaft* (pp. 383–396). Wiesbaden: Springer VS.

Knorr-Cetina, K. (1989). Spielarten des Konstruktivismus. Einige Notizen und Anmerkungen. *Soziale Welt, 40*(1/2), 86–96.

Kortländer, B. (1977). Die Landschaft in der Literatur des ausgehenden 18. und beginnenden 19. Jahrhunderts. In A. Hartlieb von Wallthor, & H. Quirin (Eds.), *„Landschaft" als interdisziplinäres Forschungsproblem. Vorträge und Diskussionen des Kolloquiums am 7./8. November 1975 in Münster.* Münster: Aschendorff.

Kühne, O. (2006a). *Landschaft in der Postmoderne. Das Beispiel des Saarlandes.* Wiesbaden: DUV.

Kühne, O. (2018). *Landschaft und Wandel. Zur Veränderlichkeit von Wahrnehmungen.* Wiesbaden: Springer VS.

Kühne, O. (2019a). Die Produktivität von Landschaftskonflikten – Möglichkeiten und Grenzen auf Grundlage der Konflikttheorie Ralf Dahrendorfs. In K. Berr & C. Jenal (Eds.), *Landschaftskonflikte* (pp. 37–49). Wiesbaden: Springer VS.

Kühne, O. (2006b). Soziale Distinktion und Landschaft. Eine landschaftssoziologische Betrachtung. *Stadt+Grün* (12), 42–45.

Kühne, O. (2007). Soziale Akzeptanz und Perspektiven der Altindustrielandschaft. Ergebnisse einer empirischen Untersuchung im Saarland. *RaumPlanung* (132/133), 156–160.

Kühne, O. (2008a). Die Sozialisation von Landschaft – sozialkonstruktivistische Überlegungen, empirische Befunde und Konsequenzen für den Umgang mit dem Thema Landschaft in Geographie und räumlicher Planung. *Geographische Zeitschrift, 96*(4), 189–206.

Kühne, O. (2008b). *Distinktion – Macht – Landschaft. Zur sozialen Definition von Landschaft.* Wiesbaden: VS Verlag für Sozialwissenschaften.

Kühne, O. (2015). Weltanschauungen in regionalentwickelndem Handeln – die Beispiele liberaler und konservativer Ideensysteme. In O. Kühne & F. Weber (Eds.), *Bausteine der Regionalentwicklung* (pp. 55–69). Wiesbaden: Springer VS.

Kühne, O. (2019b). *Landscape theories. A brief introduction.* Wiesbaden: Springer VS.

Kühne, O. (2020a). The social construction of space and landscape in internet videos. In D. Edler, C. Jenal, & O. Kühne (Eds.), *Modern approaches to the visualization of landscapes* (pp. 121–137). Wiesbaden: Springer VS.

Kühne, O. (2020b): Die Landschaften 1, 2 und 3 und ihr Wandel. Perspektiven für die Landschaftsforschung in der Geographie – 50 Jahre nach Kiel. *Berichte. Geographie und Landeskunde, 92*(3–4) 327–231.

Kühne, O. (2020c). Landscape conflicts. A theoretical approach based on the three worlds theory of Karl Popper and the conflict theory of Ralf Dahrendorf, illustrated by the example of the energy system transformation in Germany. *Sustainability 12(17)*, 1–20. https://doi.org/10.3390/su12176772.

Kühne, O., & Schönwald, A. (2015). *San Diego. Eigenlogiken, Widersprüche und Hybriditäten in und von ‚America's finest city'*. Wiesbaden: Springer VS.

Kühne, O., & Weber, F. (2018 [online first 2017]. Conflicts and negotiation processes in the course of power grid extension in Germany. *Landscape Research, 43*(4), 529–541. https://doi.org/10.1080/01426397.2017.1300639

Kühne, O., & Weber, F. (2019). Landschaft und Heimat – argumentative Verknüpfungen durch Bürgerinitiativen im Kontext des Stromnetz- und des Windkraftausbaus. In M. Hülz, O. Kühne, & F. Weber (Eds.), *Heimat. Ein vielfältiges Konstrukt* (pp. 163–178). Wiesbaden: Springer VS.

Kühne, O., Weber, F., & Jenal, C. (2018). *Neue Landschaftsgeographie. Ein Überblick* (Essentials). Wiesbaden: Springer VS.

Kühne, O., Weber, F., & Berr, K. (2019). The productive potential and limits of landscape conflicts in light of Ralf Dahrendorf's conflict theory. *Società Mutamento Politica, 10*(19), 77–90.

Lange, E. (2001). The limits of realism: perceptions of virtual landscapes. *Landscape and Urban Planning, 54*(1–4), 163–182. https://doi.org/10.1016/S0169-2046(01)00134-7

Lehmann, A. (1996). Wald als „Lebensstichwort". Zur biographischen Bedeutung der Landschaft, des Naturerlebnisses und des Naturbewußtseins. *BIOS, 9*(2), 143–154.

Linke, S. (2020). Landscape in internet pictures. In D. Edler, C. Jenal, & O. Kühne (Eds.), *Modern approaches to the visualization of landscapes* (pp. 139–156). Wiesbaden: Springer VS.

Loda, M., Kühne, O., & Puttilli, M. (2020). The social construction of Tuscany in the German and English speaking world—Presented by the analysis of internet images. In D. Edler, C. Jenal, & O. Kühne (Eds.), *Modern approaches to the visualization of landscapes* (pp. 157–171). Wiesbaden: Springer VS.

Makhzoumi, J. M. (2015). Borrowed or rooted? The discourse of 'Landscape' in the Arab Middle East. In D. Bruns, O. Kühne, A. Schönwald, & S. Theile (Eds.), *Landscape culture – Culturing landscapes. The differentiated construction of landscapes* (pp. 111–126). Wiesbaden: Springer VS.

McCauley, M. E., & Sharkey, T. J. (1992). Cybersickness: Perception of self-motion in virtual environments. *Presence: Teleoperators and Virtual Environments, 1,*(3), 311–318. https://doi.org/10.1162/pres.1992.1.3.311

McLean, K. (2020). Temporalities of the smellscape: Creative mapping as visual representation. In D. Edler, C. Jenal, & O. Kühne (Eds.), *Modern approaches to the visualization of landscapes* (pp. 217–246). Wiesbaden: Springer VS.

Micheel, M. (2012). Alltagsweltliche Konstruktionen von Kulturlandschaft. *Raumforschung und Raumordnung, 70,*(2), 107–117. https://doi.org/10.1007/s13147-011-0143-x

Münderlein, D., Kühne, O., & Weber, F. (2019). Mobile Methoden und fotobasierte Forschung zur Rekonstruktion von Landschaft(sbiographien). In O. Kühne, F. Weber, K. Berr, & C. Jenal (Eds.), *Handbuch Landschaft* (pp. 517–534). Wiesbaden: Springer VS.

Nissen, U. (1998). *Kindheit, Geschlecht und Raum. Sozialisationstheoretische Zusammenhänge geschlechtsspezifischer Raumaneignung*. Weinheim: Beltz Juventa.

Otto, A. (2019). Landschaft und der Ausbau der Windenergie. In O. Kühne, F. Weber, K. Berr, & C. Jenal (Eds.), *Handbuch Landschaft* (pp. 859–869). Wiesbaden: Springer VS.

Papadimitriou, F. (2020). Visualization of future landscapes, postmodern cinema and geographical education. In D. Edler, C. Jenal, & O. Kühne (Eds.), *Modern approaches to the visualization of landscapes* (pp. 351–369). Wiesbaden: Springer VS.

Papadimitriou, F. (2021). *Spatial Complexity. Theory, mathematical methods and applications.* Cham: Springer.

Parsons, T. (1991 [1951]). *The Social System.* London: Routledge.

Pasqualetti, M. J., Gipe, P., & Righter, R. W. (Eds.). (2002). *Wind power in view: energy landscapes in a crowded world.* San Diego: Academic Press.

Penning-Rowsell, E. C., & Lowenthal, D. (Eds.). (1986). *Landscape meanings and values.* London: Allen and Unwin.

Poerting, J., & Marquardt, N. (2019). Kritisch-geographische Perspektiven auf Landschaft. In O. Kühne, F. Weber, K. Berr, & C. Jenal (Eds.), *Handbuch Landschaft* (pp. 145–152). Wiesbaden: Springer VS.

Rebenitsch, L., & Owen, C. (2016). Review on cybersickness in applications and visual displays. *Virtual Reality, 20*(2) 101–125. https://doi.org/10.1007/s10055-016-0285-9

Roßmeier, A., & Weber, F. (2018). Stürmische Zeiten. Bürgerschaftliches Engagement beim Windkraftausbau zwischen Befürwortung und Ablehnung. In A. Stefansky, & A. Göb (Eds.), *„Bitte wenden Sie!" – Herausforderungen und Chancen der Energiewende* (Arbeitsberichte der ARL, vol. 22, 52–79). Hannover: Selbstverlag.

Sautter, T. (2018). *Zwischen Hier und Dort – Gebrochene Raumbiographien von Geflüchteten.* Tübingen: Masterarbeit im Studiengang Humangeographie/Global Studies an der Eberhard Karls Universität Tübingen.

Siepmann, N., Edler, D., & Kühne, O. (2020). Soundscapes in cartographic media. In D. Edler, C. Jenal, & O. Kühne (Eds.), *Modern approaches to the visualization of landscapes* (pp. 247–263). Wiesbaden: Springer VS.

Stiens, G. (2009). *Gegen den Verfall lebensweltlicher Landschaften* (Beiträge zur Sozialästhetik, vol. 9). Bochum: Projekt-Verlag.

Stotten, R. (2013). Kulturlandschaft gemeinsam verstehen – Praktische Beispiele der Landschaftssozialisation aus dem Schweizer Alpenraum. *Geographica Helvetica, 68*(2), 117–127. https://doi.org/10.5194/gh-68-117-2013

Stotten, R. (2015). *Das Konstrukt der bäuerlichen Kulturlandschaft. Perspektiven von Landwirten im Schweizerischen Alpenraum* (alpine space – man & environment, vol. 15). Innsbruck: Innsbruck University Press.

Stotten, R. (2019). Kulturlandschaft als Ausdruck von Heimat der bäuerlichen Gesellschaft. In M. Hülz, O. Kühne, & F. Weber (Eds.), *Heimat. Ein vielfältiges Konstrukt* (pp. 149–162). Wiesbaden: Springer VS.

Vetter, M. (2020). Technical potentials for the visualization in virtual reality. In D. Edler, C. Jenal, & O. Kühne (Eds.), *Modern approaches to the visualization of landscapes* (pp. 307–317). Wiesbaden: Springer VS.

Vicenzotti, V. (2011). *Der „Zwischenstadt"-Diskurs. Eine Analyse zwischen Wildnis, Kulturlandschaft und Stadt.* Bielefeld: transcript.

Weber, F. (2018). *Konflikte um die Energiewende. Vom Diskurs zur Praxis.* Wiesbaden: Springer VS.

Weber, F. (2020). Blurring the boundaries of landscape visualization: Welcome to Fabulous Las Vegas. In D. Edler, C. Jenal, & O. Kühne (Eds.), *Modern approaches to the visualization of landscapes* (pp. 461–478). Wiesbaden: Springer VS.

Winchester, H. P. M., Kong, L., & Dunn, K. (2003). *Landscapes. Ways of imagining the world.* London: Routledge.

Wojtkiewicz, W., & Heiland, S. (2012). Landschaftsverständnisse in der Landschaftsplanung. Eine semantische Analyse der Verwendung des Wortes „Landschaft" in kommunalen Landschaftsplänen. *Raumforschung und Raumordnung, 70,* (2, 133–145). doi:https://doi.org/10.1007/s13147-011-0138-7

Wood, G. (2012). Zur Bedeutung von Images in der Entwicklung von Altindustrieregionen. Das Beispiel Ruhrgebiet. *Metropolis und Region, 8,* (129–140).

Wylie, J. (2007). *Landscape.* Abingdon: Routledge.

Olaf Kühne studied geography, modern history, economics, and geology at Saarland University and received his doctorate in geography and sociology there and at the Open University of Hagen. After working in various Saarland state authorities and at Saarland University, he was Professor of Rural Development/Regional Management at Weihenstephan-Triesdorf University of Applied Sciences from 2013 to autumn 2016 and Associate Professor of Geography at Saarland University in Saarbrücken. Since autumn 2016, he has been a professor in the Department of Geography at the Chair of Urban and Regional Development at the Eberhard Karls University of Tübingen. His research interests include landscape and discourse theory, social acceptance of landscape change, sustainable development, transformation processes in Southern California and the Southern States of the USA, regional development, and urban and landscape ecology.

Corinna Jenal studied German language and literature, political science, and philosophy at the University of Trier and completed the "Sustainability Certificate" at Saarland University at the Endowed Chair for Sustainable Development. At Saarland University and the Weihenstephan-Triesdorf University of Applied Sciences she worked on various research projects. Since autumn 2016 and summer 2019, respectively, she has been working as a research assistant and academic councilor in the research area of geography at the Chair of Urban and Regional Development at the Eberhard Karls University of Tübingen, where she received her doctorate in 2019 on the social construction of forests. Her research focuses on landscape research, energy system transformation, urban-rural hybrids, old industry, and social construction and negotiation processes of nature and forest as their associated part.

Linking Socio-Scientific Landscape Research with the Ecosystem Services Approach to Analyze Conflicts About Protected Area Management—The Case of the Bavarian Forest National Park

Erik Aschenbrand and Thomas Michler

Abstract

Landscape changes as a result of large-scale windthrows have sparked long-lasting conflicts about protected area management in Bavarian Forest National Park, Germany. We assess tourists, locals, and nature conservation professionals' attitudes to these landscape changes. Methodologically we look at social media, local newspapers and the National Parks press releases and combine these findings with existing results from socio-scientific landscape research that we transfer to the field of ecosystem service assessments. We find that locals rate the value of landscape based on familiarity, while for tourists the connectivity to touristic landscape stereotypes is important. Nature conservation professionals assess the value of landscape based on biodiversity and especially value the appearance of rare species. As all groups use different measures to assess the value of a landscape, their evaluations of the changes to the national parks landscape differ clearly. We argue that when the value of cultural ecosystem services is assessed, the social context needs to be specified in order to improve the accuracy of any cultural ecosystem services assessment.

E. Aschenbrand (✉)
Biosühärenreservat Mittelelbe, Schollene, Germany
e-mail: erik.aschenbrand@gmx.de

T. Michler
Nationalpark Bayerischer Wald, Grafenau, Germany
e-mail: thomas.michler@npv-bw.bayern.de

© Springer Fachmedien Wiesbaden GmbH, part of Springer Nature 2020   403
D. Edler et al. (eds.), *Modern Approaches to the Visualization of Landscapes*, RaumFragen: Stadt – Region – Landschaft,
https://doi.org/10.1007/978-3-658-30956-5_23

Keywords

Protected area · National park · Nature conservation · Cultural ecosystem services ·
Landscape · Landscape change · Tourism

1   Introduction: Conflicts and Cultural Ecosystem Services

In this article we apply a social constructivist perspective on landscape to cultural ecosystem services research. We explore how a conflict about protected area management
reveals differing valuations of landscape that can be interpreted as different valuations
of cultural ecosystem services. We do not calculate the value of cultural ecosystem services but focus on the mechanisms of value attribution instead. Looking at conflicts
appears helpful from our perspective. This can contribute to the understanding of cultural ecosystem services provision. As conflicts reveal the existence of differing valuations, they highlight the social aspect of cultural ecosystem services provision. Conflicts
are understood by Dahrendorf (1958) as basically productive situations that are necessary for social change. Democratic societies should from this perspective not suppress
conflict but find ways of conflict regulation which includes the freedom of speech to
allow the articulation of problems (Kühne 2017b). We want to take advantage of the fact
that openly articulated conflicts make the existence of differing opinions explicit and
offer a possibility to analyze social processes that can explain changes in public opinion. Specifically, we investigate how cultural ecosystem services can change along with
social attitudes towards nature in a National Park. First, we discuss the ecosystem services approach. Then we introduce the social-constructivist perspective on landscape and
give some introductory remarks about the Bavarian Forest National Park before coming
to the methods and results chapters.

The ecosystem services framework as published by the millennium ecosystems
assessment was characterized by Daniel et al. (2012, p. 8812) as an attempt "to reflect
and guide human attitudes and actions toward the natural environment". Ecosystem services are defined in the same report as the benefits that people receive from ecosystems
(Millennium Ecosystem Assessment 2005, p. 40). The ecosystem services approach has
been very successful as it "has attracted tremendous attention from policymaking, planning and interdisciplinary sciences" (Kühne and Duttmann 2020, p. 1). The approach has
been characterized as anthropocentric because it focusses on the services that ecosystems
offer to human societies as opposed to the idea of an intrinsic or God-given value of
nature (Kirchhoff 2019). Jax et al. (2013) challenge such a contrasting juxtaposition by
arguing that the "ecosystem service concept neither necessarily excludes the consideration of other than economic values nor does it capture the whole array of values which
people connect with nature" (Jax et al. 2013, p. 266). Jax et al. (2013) have the intention
to overcome the weakness of the ecosystem services approach in assessing cultural ecosystem services for example by demanding more transparency regarding intentions of

ecosystem services assessments while (Kirchhoff 2018, p. 18) concludes that "intrinsic, non-instrumental, aesthetic, symbolic and moral values" should not be integrated in the ecosystem services approach and consequently demands that the idea of cultural ecosystems should be rejected.

The results of ecosystem services research are supposed to address the problem of negative economic externalities and to strengthen arguments for nature conservation because they illustrate that ecosystem services are important not only for nature lovers but for the society as a whole (Chan et al. 2017). Ecosystem services can hence work as a translation and offer a new way to address ecological issues beside moral, political or legal communication (Kühne and Duttmann 2020). The ecosystem services concept has been criticized for generally conceptualizing nature as an ecosystem (Kirchhoff 2019) and thereby simplifying human-nature relationships (Kühne 2014; Kühne and Duttmann 2020)—an argument that gets more important the more influential the ecosystem services approach becomes. Payments for ecosystem services have been critically assessed by Chan et al. (2017) for potentially creating new externalities, misplacing rights and responsibilities, crowding out existing motivations and limited applicability among other reasons. The assessment of cultural ecosystem services has proven to be especially difficult (Daniel et al. 2012) because the aesthetic or symbolic qualities of natural features cannot be explained by analyzing ecosystems alone (Kirchhoff 2019). An artist's innovative visualization of a landscape may spark general interest that later leads to touristic demand (related in this volume: McLean 2020). Looking only at ecosystems to explain touristic demand could be considered as an ecological reductionism. The dependence on social processes and hence on culture does not only mean that cultural ecosystem services are potentially valued differently on different continents, as for example Hunziker et al. (2008) or Kühne (2017a) have shown that landscape changes are evaluated differently by different social groups within one society. In this article we also want to look at cultural differences within a given society that bear different perspectives on cultural ecosystem services and hence create management challenges for protected areas.

For the Bavarian Forest National Park, a major management challenge is finding a balance between conservation, tourism and acceptance with the local population—that is to balance different and often mutually exclusive ecosystem services. National Parks usually bring restrictions in use rights which can affect the local population (Cernea and Schmidt-Soltau 2006). On the other hand, National Parks attract visitors and can increase revenue from tourism because they are associated with beautiful landscape and offer touristic infrastructure like information centers, hiking trails and guided tours (Lacy and Whitmore 2006). National Parks can also lead to changes in the landscape if land use practices are changed by the National Park management. This is exactly what happened in the Bavarian Forest National Park which was established in an area where forestry has been the dominant land use form for centuries. Conflicts about National Park management make differing attitudes towards nature and divergent valuations of cultural ecosystem services explicit—this is why we think that analyzing conflicts can be useful in order to understand the provision of cultural ecosystem services.

1.1   Linking a Social Constructivist Perspective on Landscape to Cultural Ecosystem Services Research

Results from landscape research can enhance the understanding of cultural ecosystem services as Schaich et al. (2010) have argued. In this chapter we want to link a specific approach of landscape research—the social constructivist perspective on landscape—to the cultural ecosystem services approach. Social constructivist research brings with it a focus on language and communication (Kühne 2019) and we want to explore how it can help understanding cultural ecosystem services.

If the aesthetic value of an area is discussed, the word *landscape* is frequently used to describe the subject (Kühne 2019). We reconstruct the differing attributions of value to the National Parks forest landscape from locals, tourists and conservation scientists by using a social constructivist perspective on landscape and on ecosystem services showing how the value of cultural ecosystem services is being negotiated and also newly created in social processes.

When dealing with social phenomena, language becomes especially important, but translation can never transmit all aspects of meaning. The German word *Landschaft* for example, usually is translated to *landscape*. There are however historic differences in how both terms developed their meaning and in the connotations that they carry (Kühne 2018). The same is true for words like *nature* (German: Natur) or *wilderness* (German: Wildnis) (Cronon 1996). We still think that substantial aspects of the conflicts around protected areas, cultural ecosystem services and landscape are comparable to situations in other parts of the world.

From a social constructivist perspective, landscape is understood as a social construct (see in this volume: Al-Khanbashi 2020; Bellini and Leonardi 2020; Berr 2020; Edler et al. 2020; Fontaine 2020a, b; Jenal 2020; Kühne and Jenal 2020; Linke 2020; Loda et al. 2020; Roßmeier 2020; Siepmann et al. 2020; Weber 2020). This perspective does not question the existence of physical things nor their meaning for society. It asks how these meanings come into being and how they are communicated and changed (Kühne 2018, 2019; see). From a social constructivist perspective, landscape is not an object that can be defined and divided from its surrounding by physical criteria. It is rather a way of looking at a constellation of objects and attributing value to them (Cosgrove 1988). If landscape is no existing object but rather a learned and unconsciously performed way of looking at a constellation of objects, it follows that landscape recognitions depend on previously learned aspects. Things like mountains, trees, forests, lakes and waterfalls are themselves social categories (Berger and Luckmann 1966; Kahneman 2012) and at the same time they are the physical basis of landscape. But landscape is different from its physical basis because the word landscape expresses the contemplator's interpretation of a given constellation of physical objects which is created unconsciously on the basis of knowledge and socialization (Kühne 2019). This means that knowledge and socialization are as important in the provision of cultural ecosystem services as plants, animals, mountains and other physical features.

1.2   The Bavarian Forest National Park and its Bark Beetle Crisis

After severe wind throws in the 1980s, the National Park management was confronted with an epidemic of the European Spruce bark beetle *Ips typographus*. This large-scale disturbance killed mature spruce trees on an area of 6000 hectares and therefore lead to substantial changes in the appearance of the landscape. The bark beetle as a forest pest was not new to the region. Neither were large losses of timber resulting from bark beetle epidemics. New was however the way the national park management reacted to the pest. They coined the term 'Natur Natur sein lassen' (let nature be nature) to emphasize the natural character of this disturbance and argued to accept the bark beetle epidemic as part of natural dynamics. Implicitly this argumentation gave the natural dynamic priority over the familiar appearance of the landscape.

The Bavarian Forest National Park management's decision to interpret the bark beetles killing of thousands a living green trees as a natural phenomenon and hence to reject intervention like salvage logging challenged a centuries-old traditional attitude shared by many locals that demanded to fight the bark beetle as a pest that destroys the forest (Müller 2011). Neither the dimension of change that this decision brought to the landscape nor the impact it had on the acceptance of the National Park with the local population were anticipated by the National Park management at that time.

Local politicians as well as citizen`s initiatives argued that the bark beetle epidemic destroys cultural landscape and expressed fears that this would threaten the region's value for recreation and tourism. The local protesters made it clear that the familiar appearance of the forest plays a crucial role for regional identity. Conservation scientists from the National Park administration called the bark beetle epidemic a natural process whose undisturbed activity would bring about more stable natural forests. Later the conservation scientists' arguments focused on the appearance of rare species as witnesses of an intact forest ecosystem. In ecosystem services terminology the question was: which type of forest management provides more valuable cultural ecosystem services? Critics of the Bavarian forest national park claim until today that the forest was being destroyed by the bark beetle with negative effects on regional identity and tourism (Bürgerbewegung zum Schutz des Bayerischen Waldes 2018). From their perspective, the national park management has decreased the value of the Bavarian forests' cultural ecosystem services. Not all locals share this opinion (Liebecke et al. 2011) but the critics frequently managed to gain media attention and influence on local politics. This forced the National Park management to deal with the topic. Mainly the provision of cultural ecosystem services was put forward by the critics, namely the aesthetic value of the forest and its ability to provide regional identity. Other Ecosystem services—like the changing forests ability to buffer enough drinking water—were raised but soon proven to be irrelevant (Beudert et al. 2015) and have never been prominent in the discussion.

2   Methods

Many surveys have been conducted in cultural ecosystem services research and a wide variety of different methods have been applied as Milcu et al. (2013) have shown in their literature review. Measuring cultural ecosystem services has many times proven to be difficult (Daniel et al. 2012) and if it is conducted on a local scale with methods like community mapping it is also highly time-consuming (e. g. Plieninger et al. 2013). Our intention is to explore the potential of non-reactive methods like media- and literature-analysis for understanding cultural ecosystem services in addition to surveys. With their quantitative analyses of social media photos, Oteros-Rozas et al. (2018) and Sonter et al. (2016) have given a promising outlook on the possibilities of non-reactive methods in cultural ecosystem services research on a broader scale. As we are focusing on conflicting perspectives in a single protected area we decided to use a qualitative approach that allows us to look at the articulated arguments in detail in order to understand the opposing positions at depth. We think that important aspects of the cultural meaning of ecosystems are revealed in cultural artefacts like newspapers, magazines, scientific publications, tourism advertisements, social media, film, literature or visual arts. Analyzing visualizations of landscapes and statements concerning landscapes (for example in newspapers) and the reactions to them (for example in letters to the editor) can help understanding the provision of cultural ecosystems. This approach can benefit from integrating social media because publishing visualizations and statements is much easier in social media as is reacting to them. For this study we collected local and national newspaper articles, social media posts and scientific results and identified the arguments that were made concerning the natural disturbance event in Bavarian forest National Park.

We reconstruct the perspectives of locals, tourists and conservation scientists. To understand the local National Park critics perspective, we analyze the online content and publications of a local National Park-critic citizens' initiative as well as articles from local newspapers including letters to the editor. We also took into consideration research about local history as the local culture has a specific connection to forestry that can provide valuable context for certain arguments that were made in the conflict. We included the tourism-perspective in the analysis because tourism is an often-quoted cultural ecosystem service and as such it was prominent in the discussion. Critics argued the natural disturbance that they called "destruction of landscape" damages the regional economy by scaring off tourists. To estimate whether the cultural ecosystem service "tourism" has decreased due to the natural disturbance event, we could rely on surveys that already investigated this issue with a focus on visitor numbers and visitor satisfaction concluding that no effects on tourism could be found. To understand why tourism has not suffered from the natural disturbance we investigated the mechanisms of tourism marketing in general. To analyze the conservationist's perspective, we analyzed the National Parks press releases and publications from conservation science.

In the results we show that attitudes toward changes in landscape differ between social groups as different mechanism are at work in every social group, leading their members to divergent valuations of the same forests cultural ecosystem services. Among the main conflicting parties were of course the National Park administration on one side. On the other side were various privately acting and organized National Park critics that share one characteristic: their self-description as locals. Tourists were part of the conflict more indirectly as their feared absence due to a destroyed landscape was used as an argument by the local critics. Later tourists became more active affirming the National Parks position in social media.

A protected area can be tourist destination (tourist perspective), investigation area (conservation scientists' perspective) or simply home (locals' perspective) at the same time. Each of these three groups (tourists, conservation scientists and locals) applies distinctive knowledge and expectations when thinking about the National Parks changing forest landscape. This does not mean however that being a local would lead automatically to a certain mindset concerning landscape and conservation. We therefore strongly reject any determinism in the development of attitudes. The following results rather show ideal types that are meant to illustrate social mechanisms. Realizing the meaning of social mechanisms in the provision of cultural ecosystems can—and this is our hope—in the best-case lead to mutual understanding between conflicting parties.

In the following chapter we reconstruct the locals, tourists and conservation scientists' perspectives on the National Parks changing landscape. On the basis of these reconstructed ideal–typical perspectives we estimate for each group if cultural ecosystem services have increased, decreased or remain unchanged.

3   Results and Discussion

3.1   Familiar Landscapes—Locals Like Stability

If areas are referred to as cultural landscape, this usually means that changes in the landscape due to natural disturbance events are reduced to a minimum. This results in stability of landscape features, e.g. forest landscapes dominated by green mature trees. Stability results in familiarity which has proven to be most important for a positive attitude towards one's home surrounding area because the qualities of the landscape where one lives his daily life are hardly thought of or questioned as long as they do not change (Kühne 2018). The Bavarian Forest National Park has put this stability of landscape into question with its claim for natural dynamics. While the National Park management's approach of approving the bark beetle epidemic as a legitimate part of natural dynamics was certainly innovative for the region, it was also insensitive to the cultural heritage aspect and the local's relation to their landscape. Cultures create specific ways of constructing landscape that result in specific forms of place-attachment. In the following part we will have to describe briefly the locals place-attachment in the Bavarian Forest in order to understand the forests role for the local's perception of their home region.

Many local critics of the National Park use the word "Heimat" to describe their place attachment. "Heimat" translates to home or homeland. In the German language the word has specific connotations which why we decided to introduce the German concept of "Heimat" and then to use the term without translation. Weber et al. (2019) call "Heimat" a polyvalent construct: The word has a long history of positive and negative connotations (Scharnowski 2019). In everyday life the word "Heimat" is used for referring to the living environment, to social relations, to familiar landscapes and foremost to describe a feeling of rootedness and belonging to a place. A film genre called "Heimatfilm" is characterized by Moltke (2005, p. 3) as circling "obsessively around questions of home and away, tradition and change, belonging and difference inscribed in the German term *Heimat*". "Heimat" is strongly connected to identity and in this respect it is a political term that became prominent in political discourse in times of social or political changes (Scharnowski 2019). We will now describe the construction of "Heimat" in the Bavarian forest.

"Bavarian Forest", see Fig. 1, is not just the name of the National Park but of a whole region that is difficult to circumvent but that is certainly much larger than the protected area. Within the region, the forest is seen as the dominant landscape feature and as a defining element for local culture and identity. Many people call this region their "Heimat" and call themselves "Waidler" or "Waldler" which is local dialect and can be translated as the "people from the forest". Many of these locals value their descendance from lumberjacks (Müller 2011), which means that ancestors have earned their living

Fig. 1 The Bavarian Forest. *Source* Bavarian Forest National Park/Frank Bielau (2020)

from forestry, have spent their professional life in the forest and have shaped its appearance. Like almost anywhere in Western Europe, the economic importance of forestry has decreased over the decades but here the forest remained important as a focal point for regional identity. Attributions from the outside may also have helped to (self-)romanticize the locals living with the forest as many documentaries about the Bavarian forest until today show the forest as a mysterious place and underline the excellence of traditional artisanry thus painting the picture of a traditional society not far away from the "noble savage"-stereotype. Binder (2000) traces these stereotypes of a traditional living with the forest and the staging of what he calls *picturesque poverty* back to the 1930s and shows how these descriptions of the Bavarian Forest have fit into the Nazi propaganda.

The development of regional identities is difficult to reconstruct but what is important for the understanding of the following process is that the forest was integrated in the self-description of the local people as "Waidler" (Binder 2000; Müller 2011). While the word *forest* can have many different meanings from a tropical rainforest to an urban forest it was clear to the locals that in their context *forest* would mean one specific forest with its familiar appearance: the green rolling hills of the Bavarian forest with its large spruce trees, a forest that was managed and shaped by ancestors over centuries. A forest, whose timber was the material for many traditional objects from wooden shoes to artisanal glass production.

3.1.1 Local Impact of the National Parks Bark Beetle Crisis

Beginning in the mid-1980s the National Park management had to deal with largescale windthrows and resulting bark beetle epidemics. Until 2012 the bark beetle epidemics caused 6000 hectares of spruce forest to die. These dead trees remained in the landscape thus changing its appearance dramatically as some observers pointed out. This has caused a lot of public attention throughout Germany. The post-disturbance landscape was called "tree-graveyard" and "death zone" in a large German newspaper (Metzner and Meffert 1997). Critics doubted that forest would ever grow again in these areas and assumed that where once the green roof of Europe has been only a Tundra will remain (Metzner and Meffert 1997). A local politician is cited saying that when the forest dies his community will die with it (Bibelriether 2017, p. 115). The National Parks affirmative reaction to the natural disturbance event was hence rejected as an intentional destruction of "Heimat" (Müller 2011). The quarrel with local communities and local politicians included largescale protests, threats to National Park staff and Lawsuits.

Local antagonists of the National Park connected their appreciation of the landscape mainly to green and living trees. When trees died, they experienced this as a substantial negative change for the forest which is important for their understanding of "Heimat" and they kept urging the National Park management to take action against the bark beetles. Local's assessment of areas affected by bark beetles in North America show similar results and underline the importance of green trees in the public perception of a forest in a desirable condition (Flint 2006; McFarlane et al. 2006). This means that the value of cultural ecosystem services—in this case the forests meaning for regional identity—is connected to a certain aesthetic appearance. In a social constructivist terminology, one can say that mainly trees are integrated in the construction of the forest landscape by the locals. If trees are dying, the forest landscape is also perceived to be dying but as the forest landscape is so important for local self-description and for what local people call "Heimat", they felt their whole identity is in danger.

At this stage it can be concluded that the National Parks management strategy was insensitive to the cultural heritage value that the forest in its specific form had for the locals. The National Park authority brought an innovative approach of forest management but apparently at that time did not put much effort into understanding how this new strategy threatened the value of cultural ecosystem services for many locals. Cultural heritage values were not sufficiently considered and their potential for conflict were severely underestimated.

3.1.2 The Resurrection and Its Consequences

Light reached the forest floor when the mature spruce trees had died and a new generation of young and fast-growing trees conquered the post disturbance landscape. This rejuvenation sparked a lot of interest and gave rise to the first positive interpretation of the natural disturbance event that reached broad acceptance: the idea of a natural forest, whose recovery could be idealized as a symbol for the powerful force of nature. The National Park management picked up that interest in the development of the post-disturbance areas and enthusiastically celebrated the rejuvenation of the forest. One of the National Park administrations most successful Facebook-Posts (with more than 2000 reactions and close to 800 shares) shows a comparison of two pictures of one of the Parks most popular hiking trails: The picture on the left was taken in the year 2006 and shows a mostly brown area with dead silver-grey tree trunks. The other picture was taken ten years later and presents mainly lush green vegetation. The comparison intends to demonstrate the dynamic rejuvenation in the post-disturbance areas (see Fig. 2).

In the comments section, tourists and locals alike share their happiness about the rejuvenating green forest as the following comments exemplify:

- "Just fantastic. We often go hiking up there. Yesterday it was a ghost-forest, today it's a green oasis"
- "Great… Conservation has proven that it can really achieve something after all"
- "Nature thinks in longer time spans than we humans"
- "looks good again. We still know how it was in 2006"
- "Nature doesn't need humans…"
- "Thank God a new forest has grown"
- "back then nobody would believe that the forest can recover on its own"
- "This is the proof, that `let nature be nature' works"
- "Let nature be nature… the right way!"

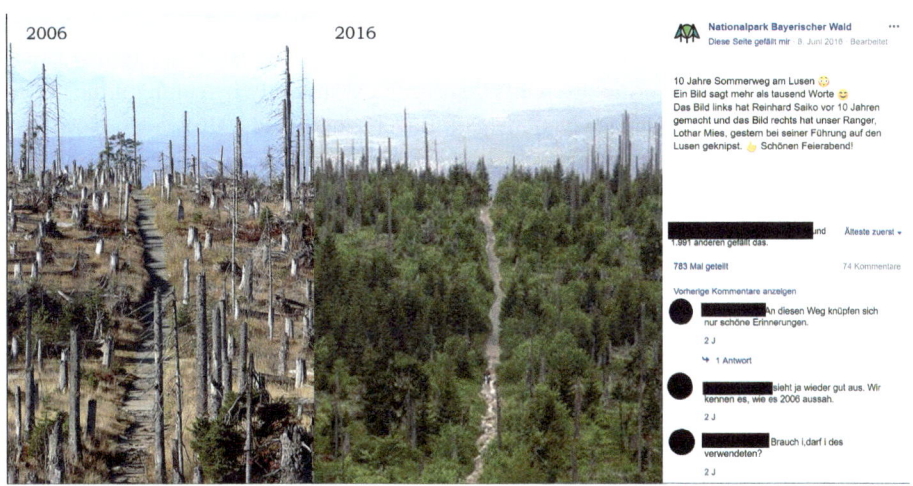

Fig. 2 Bavarian Forest National Park on Facebook. *Source* Facebook Fanpage Nationalpark Bayerischer Wald (2016)

In the year 2020 the comments section counts 97 comments that were all written in the year 2016 when the pictures where posted. From these comments 63 are positive affirmations of the National Parks management strategy. 14 comments are negative and criticize the National Park and 20 comments cannot be identified as negative or positive.

3.1.3 Stereotyping Nature

For many commentators the comparison of the two pictures contains more meaning than just a regrowing forest. For them the pictures symbolize the supremacy of nature over humans. These commentators prefer a naturally occurring change of landscape over one that they trace back to human action as one commentator put it: "What is growing there is much more beautiful and resistant and fascinating and authentic than what has been growing there before" (Facebook Fanpage Nationalpark Bayerischer Wald 2016). But the focus on the sheer re-growing and regreening of the forest provokes some opposition as well. The political head of the district authority comments the Facebook post by criticizing the National Park administration for showing off with a re-growing forest after deliberately destroying large areas in the Bavarian Forest while everyone knew that at some stage the forest will regrow and finishes with the question: "was all this necessary"? (Facebook Fanpage Nationalpark Bayerischer Wald 2016). Most negative comments were answered and challenged directly by other commentators with a positive attitude towards the National Park. Even the difference between touristic and (critical) local perspective remains not just implicit (e.g. by writing comments in local dialect) but is being discussed when one commentator criticizes the National Park for almost taking away the locals' identity. This local commentator is being challenged by various others, one of which states that the locals were only interested in timber prices and have put into place a monoculture forest that was now replaced by something better: a "healthy natural mixed forest". The local critic discredits this argument instantly by calling it the "blabla of a tourist" (Facebook Fanpage Nationalpark Bayerischer Wald 2016).

As the numbers suggest, far more positive comments have been written under the picture. Many locals as well as tourists celebrate the resurrection of the forest claiming a supremacy of the "rejuvenated" forest over the former forest. For this group the value of cultural ecosystem services has increased. They get inspiration out of the idealization of natural processes based on their opinion to value nature higher than human shaped environments. Others bemoan the changes that the bark beetle brought to the landscape and hold the National Park accountable for what they call the destruction of their "Heimat". For this group the value of cultural ecosystem services has decreased and seems to be only slowly re-growing with the growth of the young trees. This group also welcomes the regrowth of the forest but does not want to join in the celebration as they refuse to see a difference between the new and old forest and therefore moan the quarrels that they call unnecessary.

3.2   Travel Magazine-Landscapes—Tourists Like Stereotypes

Of high importance for the discussion on the natural disturbance event is the question if changes in the landscape will have negative effects on tourism, as tourism is one of the main arguments to demonstrate the national parks contribution to regional development. In the course of the bark beetle epidemic, opponents of the National Parks management strategy strongly questioned that the National Park was beneficial to regional development. The National Park reacted with scientific surveys. With reference to visitor numbers (Allex et al. 2016) and regional-economic effects (Mayer and Job 2014) the Parks contribution to regional economic development was clarified. Surveys among tourists couldn't find any proof that the post-disturbance landscape or the National Parks management strategy had any negative effect on tourism (Müller and Job 2009).

3.2.1 Landscape in Tourism Marketing

Besides surveys and their empirical results also the mechanisms of tourism marketing point to the direction that changes in the landscape in any given protected area tend to be overweighed from a local perspective. The tourism marketing relies heavily on landscape-stereotypes (Aschenbrand 2017) as they allow to compact complex geographical features like regions, cities or whole nations to one single statement and thereby positioning the given thing as a buyable product. Such landscape stereotypes are for example the "paradise beach" or the "sublime mountain scenery". These words bring pictures to the mind: maybe of white sand, palms, warm air and clear water or in the other case of mountain peaks, cold lakes, waterfalls and traditional mountain villages, herders and alpine meadows. Stereotypes like these are used constantly in tourism marketing and tourism related media like travel magazines (Aschenbrand 2017).

Stereotypical imaginations also exist also for the forest (Jenal 2019). These stereotypes differ between cultures (Bruns et al. 2015), as it is true for any geographical feature. Hence the cultural ecosystem services that a forest provides will also differ between cultures. In Germany, the term fairytale forest (Märchenwald) is common to describe a forest that is perceived to be of high aesthetic value. An ideal fairytale forest is dense and green with rather small and winding trails. Most important would be that it consists of very old giant trees that are often interpreted as symbols of wisdom and represent the strength of nature itself. A fairytale forest represents nature as an opposite pole to culture and society and was constructed prominently in the brother Grimms fairytales (Jenal 2019) as the antipode to civilization: a place that is fascinating and beautiful and that sets the stage for an initiation ceremony. Bestselling books and current trends like forest bathing show that this idealization of nature is increasingly popular as it is connectable to the rapidly growing climate change and environmental crisis narratives. As the industrialization with its large-scale pollution has sparked romanticism in Western Europe, the medial presence of the climate- and environmental crisis today might foster romantic and idealizing interpretations of nature.

3.2.2 Two Reasons Why the Post Disturbance Landscape was No Problem for Tourism

As we have described above, a forest is mainly thought of as consisting of green and healthy trees and also the common forest stereotype of the fairytale forest appears to be quite different from the post disturbance landscape. Two mechanisms can explain why the post disturbance landscape didn't cause problems for tourism.

Firstly, the National Parks claim "let nature be nature" (Natur Natur sein lassen) provided a framing that allowed the National Park management to tell a consistent story about the post-disturbance landscape and to integrate this rather small area in the Bavarian Forest into the large discourse of a society overusing and maltreating the planet. In a society that thoughtlessly consumes and destroys nature, here would be a place where nature comes to its right. As the comments in social media show, this perspective was inspiring for many visitors. Suddenly the post-disturbance landscape was a place where a widespread critique of civilization could materialize, where a feeling of individual guilt for being part of a society that is destroying nature could be released, a place where the resurrection of nature could take place. For visitors that have shared this perspective, visiting the post disturbance areas provided a meaningful experience.

Secondly, tourists simply are used to dealing with stereotypes. Neither the paradise beach, nor the alpine scenery or the fairytale forest from the advertisement is firmly connected to existing places. Tourism marketing uses these stereotypes to label the destinations that they want to sell. Stereotypes provide the tourist with a familiar image of an otherwise unknown place (Aschenbrand 2017). These stereotypes are important to shape imaginations and to spark motivation for travelling but once at their destination, tourists are used to the experience that not all expectations can be met at all stages of a journey (Aschenbrand 2017). For the consumer-satisfaction it is important that some expectations are met at some moments of the journey. Changes in the local landscape are mostly unproblematic for tourists as they normally don't follow the events in their holiday destinations. Therefore, they do not compare before and after and just take the given situation as normality. Hence tourists are much less sensible then locals to a negative interpretation of changes in local conditions of any kind as they just don't know about them. From this perspective regional tourism marketing organizations in the Bavarian Forest were doing right as they have largely ignored the landscape changes in their marketing and have continued to present the Bavarian Forest in a way that is consistent with existing stereotypes showing green rolling hills and romantic places at the shores of hidden lakes in the mountain forest as is illustrated by Fig. 3.

When assessing the impact that changes in the physical basis of landscape have on tourism, one cannot rely on the local's assessment of these changes. Locals will most likely overestimate negative impacts if their own attitude towards the given changes are negative (Aschenbrand and Grebe 2018). The case of the bark beetle crisis in the Bavarian Forest National Park shows that a compelling story can provide a positive framing of landscape-changes that appeared negative at first. National Parks and other protected areas have a strong position to create this kind of narratives because usually they are associated with competence concerning conservation issues and they are used in tourism marketing as indicators of fascinating landscape. There seems to be a wide range of landscape changes that National Parks can 'sell' their visitors without provoking their protest.

As the reactions to the Facebook post have indicated, for many visitors the National Park has increased cultural ecosystem services, especially because the National Park has created a compelling story that connected to the belief of a good nature and sinful humans. Based on this story, the post-disturbance landscape could be rated positively.

Fig. 3 The Bavarian Forest in Tourism Marketing. *Source* Tourismusverband Ostbayern e. V. (2020)

For visitors that are inspired by the claim "let nature be nature", the national park has created new value. It has in their perspective increased the value cultural ecosystem services.

3.3   Red List Landscapes—Conservationists Like Rare Species

While the National Park has adopted a clever and certainly effective communication strategy with a strong focus on the regrowth of the forest, the National Parks conservation scientists take a different perspective: In the conservationists expert discourse, biodiversity is the measure of value for geographic features (Beudert et al. 2015; Lehnert et al. 2013). Value is assessed based on biodiversity in general and the appearance of rare species in particular. The rarer the species are and the more indicative they are for the given ecosystem, the higher the value of the protected area is estimated.

3.3.1 Rare Species Benefit from Bark Beetle Crisis

Many rare species reacted positively to the bark beetle epidemic as conservation scientists soon realized. The dense forest canopy was lightened up and a lot of dead wood was available. Not only the landscape but also the living conditions changed. With this knowledge, the bark beetle could be reframed: in the researchers perspective it turned from pest to keystone species for the National Park (Müller et al. 2008). Conservation scientists intentionally drew a line between their biodiversity-based valuation of the post-disturbance landscape and the common valuation based on regrowth of the forest and on the idealization of natural processes. Two of the National Parks press releases exemplify this valuation of landscape based on biodiversity and rare species. One is reporting on a "sensational fungus-discovery in the Bavarian Forest National Park: Mycologists have detected fruiting bodys of Pseudorhizina sphaerospora.[…] This fungus is on the red list of endangered species and is classified as being extremely rare in Germany underlining the high value of Bavarian Forest National Park as a refuge for endangered species" (Bavarian Forest National Park 2009). The second press release reports on the discovery of a rare coarse woody-habitat beetle. This beetle is portrayed as looking "rather unimpressive to a non-professional but for a forest-ecologist it allows the conclusion that the Bavarian Forest National Park is now ranked number 1 in Bavaria's most valuable forest ecosystems" (Bavarian Forest National Park 2015).

3.3.2 Biodiversity as Measure of Value

Assessing the value of a given geographical feature on the basis of its biodiversity and the appearance of rare species is a common practice in nature conservation. Species are valued as they are seen as indicators of ecosystems (in this case as indicators for primary forests) and to some extend because of their sheer rareness. When classifying this approach in the scheme of ecosystem services we notice that neither the rare fungus nor the beetle is acknowledged explicitly for providing regulating, provisioning, or supporting ecosystem-services. We can classify the valuation of rare species and intact primary forest ecosystems as a cultural ecosystem service whose value has increased due to the natural disturbance event. However, the conservation scientists' valuation of landscape on the basis of rare species has not been uncontested. Local National Park-critics have picked up this valuation and ironized it in notes to the editor that were published in local newspapers saying: "Isn't it fantastic that the Bavarian Forest National Park preserves vitally important rarities for our region? [...] Only close to 10,000 hectares of forest had to be turned into dead-wood-zones to reach number 1 in Bavaria's most valuable forest ecosystems" (Letter to the editor Bayerwaldbote 2015). Another commentator gets more aggressive: "Discover your fungi and scan your biodiversity where you like but not here in our Bavarian Forest anymore!" (Letter to the editor Bayerwaldbote 2009). These and other commentators do not contest the appearance of rare species but they refuse to acknowledge them as a desirable result of good forest management decisions (see Fig. 4).

Like any valuation, the conservationists valuation of rare species is based on normative premises (Hupke 2015) that may be consistently considered as valid among nature conservation specialists and that are codified in national law as well as in international regulations. However, this does not mean that these premises are shared by other social groups. Here we want to point briefly at the fact that currently the ecosystem services framework still has difficulties to sharpen arguments for the conservation of rare species.

Fig. 4 Saproxylic beetle Synchita separanda: Evidence of an outstanding value of a forest? Synchita separanda— "number 1 in Bavaria's most valuable forest ecosystems". *Source* Bavarian Forest National Park (2008)

Ridder (2008) has argued more than a decade ago, that the contribution of many rare species to ecosystem services is often very difficult to detect. In order to promote species conservation, it is not sufficient to point at laws and legal regulations that codify protection status. For example, ethical and eudaimonistic values need to be discussed (Jax et al. 2013) and integrated in cultural ecosystem services assessments because otherwise the ecosystem services framework will rather obscure than highlight the value of many rare species.

For the conservation-perspective it can be concluded, that the National Park administration offered two different interpretations of the post disturbance landscape. One focused on the regrowth of a new forest and was attractive to many visitors and locals alike because it fit into large-scale narratives of an idealized nature endangered by reckless humans and offered a comprehensible solution: "let nature be nature"! The second interpretation is based on the valuation of rare species that are mostly unknown to the general public and has so far not received a wide public attention.

4   Conclusion

Protected area management can increase and decrease cultural ecosystem services at the same time for different stakeholder groups. At least three different forms of valuation can be observed in the discussion around the bark beetle epidemic and its effects on the landscape in the Bavarian Forest National Park. Many locals call the forest landscape their "Heimat". The forest is part of their self-description as "forest people" (Waidler) and therefore part of what they describe as their identity. They experienced undesired changes of the landscape as a threat to their identity. For this group the forests ability to provide cultural ecosystem services has decreased due to the natural disturbance events.

National Parks are often popular tourist attractions and National Park-centered tourism plays an important role for the host regions. This is why it is important to understand how tourism reacts to changes in the landscape. For tourists it's usually most important that a destination meets their expectations which are often based on landscape stereotypes that are spread in tour operators' advertisements, travel magazines and other media. But unlike locals, tourists are rather insensitive to changes in the landscape as they experience their destination at one given time and are most likely unaware of earlier changes that led to the current status. Even if they learn about the processes that created the prevailing state of things, this is experienced rather like a history lesson and received with more emotional distance compared to the locals' perspective. This is how we explain the results of existing research which concluded that the forests ability to be an attraction for tourism has remained largely unchanged. But we also found that the National Park was able to create a powerful narrative that framed the post disturbance landscape positively and that many visitors and locals could connect to. For the group of persons that were inspired by the idea of *letting nature be nature* the forest now provided a more meaningful experience which means that for them the cultural ecosystem services have increased.

For conservation scientists the value of the forest has increased due to the appearance of rare species, which they value as indicators of desired ecosystem conditions (Table 1).

Our results highlight the social dimension of cultural ecosystem services. For the analysis of diverging perspectives on landscape, social constructivist landscape theory (Kühne 2019) offers helpful tools. Linking the social constructivist perspective on landscape with cultural ecosystem services research appears fruitful to us even if there are epistemological concerns that have to be addressed. The social constructivist approach suggests the following perspective on landscapes and cultural ecosystem services: There is no firm and everlasting connection between a landscape's physical objects and the value of a landscape's cultural ecosystem services. Accordingly, the value of a landscape's cultural ecosystem services cannot be deduced from objects (or the modification of objects). However, estimating the (cultural) value of objects is exactly the intention of the (cultural) ecosystem services approach. Nevertheless, we argue that the integration of both perspectives can be useful, namely when attitudes towards nature are more or less stable in a given social context and/or when this approach helps to learn about the processes that lead to changing attitudes (which will consequently change the value of cultural ecosystem services). It is important to be aware that the value of cultural ecosystem services can quickly change not only with ecological changes but also when there is new information available or when a new cultural framing of natural becomes popular. From our perspective it would be helpful to imagine cultural ecosystem services as being co-created by social-ecological systems (Folke et al. 2005) as this allows to integrate socio-scientific perspectives. From an empirical perspective, we think that it is fruitful and feasible to analyze the value of cultural ecosystem services as this concept raises awareness for human dependence on ecosystems (Jax et al. 2013) and offers great

Table 1 Diverging perspectives on cultural ecosystem services in the Bavarian Forest National Park (source: own presentation)

Nature conservation perspective	Tourist perspective	Local perspective
Ecological value that is based on biodiversity and rareness of species/habitats defines valuation for landscape	Connectivity to touristic landscape stereotypes defines valuation of landscape	Familiarity defines valuation of landscape. Changes of landscape are therefore perceived as problems
↓	↓	↓
Value of landscapes CES increased because conditions for rare species improved	Value of landscapes CES unchanged, or positive because of positive framing: more meaningful experience for those who are inspired by the slogan „let nature be nature"	Value of landscapes CES decreased because of changes

potential for communication between different stakeholder groups involved in managing protected areas. It is also important to be aware of the limits of this approach, one of which is that any result concerning values is only valid for the given social context and time. Accordingly, we suggest that any value of cultural ecosystem services always has to be related to a social context. This social context needs to be specified when cultural ecosystem services are assessed.

We think that the analysis of conflicts about protected area management can be especially instructive for cultural ecosystem services research for a number of reasons:

- conflicts make the existence of diverging perspectives explicit
- analyzing diverging perspectives highlights the social aspect in the provision of cultural ecosystem services (e.g. Ecosystem services can be enhanced by innovative interpretations of natural features)
- acknowledging the social aspect in the provision of cultural ecosystem services helps to avoid ecological reductionism and contributes to the understanding of cultural ecosystem services as being co-created by social-ecological systems.

From a nature conservation perspective an important question is whether the ecosystem services approach obscures or highlights the value of rare species. As critics of the (cultural) ecosystem services approach (Ehrenfeld 1988; Ridder 2008; Trepl 2014) have argued, many rare species are probably unimportant for the provision of ecosystem services and for good reason Kirchhoff (2018) argues that it is inappropriate to characterize nature generally as an ecosystem. Other distinguished scholars claim that these values can be integrated in the ecosystem services approach as is the intention by Jax et al. (2013) who argue that the approach can be reclaimed and improved. In the early days of ecosystem services research David Ehrenfeld (1988) warned that the aim to attribute economic value to biodiversity would ultimately weaken arguments for conservation of rare species. Ehrenfelds assessment of the ecosystem services approach still waits to be finally disproved.

References

Allex, B., Preisel, H., Eder, R., Hußlein, M., & Arnberger, A. (2016). Touristen im Nationalpark Bayerischer Wald: Die Rolle des Nationalparks für den Besuch, die Einstellung zum Schutzgebiet und ihr raumzeitliches Verhalten. In M. Mayer & H. Job (Eds.), *Studien zur Freizeit- und Tourismusforschung: Band 12. Naturtourismus – Chancen und Herausforderungen. Mit 48 Abbildungen und 16 Tabellen.* Mannheim: Verlag MetaGIS-Systems.

Al-Khanbashi, M. (2020). Using matrix as a qualitative data display for landscape research and a reflection based on the social constructivist perspective. In D. Edler, C. Jenal, & O. Kühne (Eds.), *Modern approaches to the visualization of landscapes* (pp. 103–118). Wiesbaden: Springer VS.

Aschenbrand, E. (2017). *Die Landschaft des Tourismus: Wie Landschaft von Reiseveranstaltern inszeniert und von Touristen konsumiert wird* (1. Auflage 2017). *RaumFragen: Stadt – Region – Landschaft.* Wiesbaden: Springer Fachmedien Wiesbaden GmbH; Springer VS.

Aschenbrand, E., & Grebe, C. (2018). Erneuerbare Energie und ‚ intakte' Landschaft: Wie Naturtourismus und Energiewende zusammenpassen. In O. Kühne & F. Weber (Eds.), *Bausteine der Energiewende* (pp. 523–538). Wiesbaden: Springer Fachmedien Wiesbaden.

Bavarian Forest National Park (2009). *Sensationeller Pilzfund im Nationalpark Bayerischer Wald.* Grafenau, from Bavarian Forest National Park: Grafenau.

Bavarian Forest National Park (2015). *Käfer-Rarität im Nationalpark Bayerischer Wald entdeckt.* Grafenau, from Bavarian Forest National Park: Grafenau.

Bavarian Forest National Park/Frank Bielau (2020). *Der Nationalpark Bayerischer Wald im Porträt.* Retrieved March 19, 2020, from https://www.nationalpark-bayerischer-wald.bayern.de/ueber_uns/steckbrief/index.htm.

Bellini, A., & Leonardi, L. (2020). Prato: The social construction of an industrial city facing processes of cultural hybridization. In D. Edler, C. Jenal, & O. Kühne (Eds.), *Modern approaches to the visualization of landscapes* (pp. 549–572). Wiesbaden: Springer VS.

Berger, P. L., & Luckmann, T. (1966). *The social construction of reality: A treatise in the sociology of knowledge.* New York: Penguin Books.

Berr, K. (2020). Visuality, Aesthetics and Landscape. For the enlightenment and self-enlightenment of constructivist landscape research. In D. Edler, C. Jenal, & O. Kühne (Eds.), *Modern approaches to the visualization of landscapes* (pp. 189–215). Wiesbaden: Springer VS.

Beudert, B., Bässler, C., Thorn, S., Noss, R., Schröder, B., Dieffenbach-Fries, H., et al. (2015). Bark beetles increase biodiversity while maintaining drinking water quality. *Conservation Letters, 8*(4), 272–281.

Bibelriether, H. (2017). *Natur Natur sein lassen: Die Entstehung des ersten Nationalparks Deutschlands – der Nationalpark Bayerischer Wald* (1. Auflage 2017). Freyung: Edition Lichtland.

Binder, C. (2000). Waidlerklischees und Nazionalsozialistische Propaganda. *Ostbayerisches Magazin Lichtung, 1,* 10–14.

Bruns, D., Kühne, O., Schönwald, A., & Theile, S. (2015). *Landscape culture – Culturing landscapes.* Wiesbaden: Springer Fachmedien Wiesbaden.

Bürgerbewegung zum Schutz des Bayerischen Waldes (2018). *Unsere Ziele und Aufgaben.* Retrieved March 19, 2020, from https://www.bayerwald-schutzverein.de/.

Cernea, M. M., & Schmidt-Soltau, K. (2006). Poverty risks and national parks: Policy issues in conservation and resettlement. *World Development, 34*(10), 1808–1830.

Chan, K. M. A., Anderson, E., Chapman, M., Jespersen, K., & Olmsted, P. (2017). Payments for ecosystem services: Rife with problems and potential—For transformation towards sustainability. *Ecological Economics, 140,* 110–122.

Cosgrove, D. E. (Ed.) (1988). *Cambridge studies in historical geography: Vol. 9. The iconography of landscape: Essays on the symbolic representation, design and use of past environments.* Cambridge: Cambridge Univ. Press.

Cronon, W. (1996). The trouble with wilderness: Or, getting back to the wrong nature. *Environmental History, 1*(1), 7–28.

Dahrendorf, R. (1958). Toward a theory of social conflict. *Journal of Conflict Resolution, 2*(2), 170–183.

Daniel, T. C., Muhar, A., Arnberger, A., Aznar, O., Boyd, J. W., Chan, K. M. A., et al. (2012). Contributions of cultural services to the ecosystem services agenda. *Proceedings of the National Academy of Sciences of the United States of America, 109*(23), 8812–8819.

Edler, D., Jenal, C., & Kühne, O. (2020). Modern approaches to the visualization of landscapes—An introduction. In D. Edler, C. Jenal, & O. Kühne (Eds.), *Modern approaches to the visualization of landscapes* (pp. 3–15). Wiesbaden: Springer VS.

Ehrenfeld, D. (1988). Why put a value on biodiversity? In E. O. Wilson & F. M. Peter (Eds.), *Biodiversity* (pp. 212–216). Washington, DC: National Academy Press.

Facebook Fanpage Nationalpark Bayerischer Wald (2016). *10 Jahre Sommerweg am Lusen.* Retrieved September 20, 2019, from https://www.facebook.com/nationalpark.bayerischer.wald/photos/a.10150147542737901.282745.323649842900/10153796525532901/?type=3&theater.

Flint, C. G. (2006). Community perspectives on spruce beetle impacts on the Kenai Peninsula, Alaska. *Forest Ecology and Management, 227*(3), 207–218.

Folke, C., Hahn, T., Olsson, P., & Norberg, J. (2005). Adaptive governance of social-ecological systems. *Annual Review of Environment and Resources, 30*(1), 441–473.

Fontaine, D. (2020a). Landscape in computer games—The examples of GTA V and Watch Dogs 2. In D. Edler, C. Jenal, & O. Kühne (Eds.), *Modern approaches to the visualization of landscapes* (pp. 293–306). Wiesbaden: Springer VS.

Fontaine, D. (2020b). Virtuality and Landscape. In D. Edler, C. Jenal, & O. Kühne (Eds.), *Modern approaches to the visualization of landscapes* (pp. 267–276). Wiesbaden: Springer VS.

Hunziker, M., Felber, P., Gehring, K., Buchecker, M., Bauer, N., & Kienast, F. (2008). Evaluation of landscape change by different social groups. *Mountain Research and Development, 28*(2), 140–147.

Hupke, K.-D. (2015). *Naturschutz: Ein kritischer Ansatz.* Berlin: Springer Spektrum.

Jax, K., Barton, D. N., Chan, K. M. A., de Groot, R., Doyle, U., Eser, U., et al. (2013). Ecosystem services and ethics. *Ecological Economics, 93,* 260–268.

Jenal, C. (2019). *Das ist kein Wald, Ihr Pappnasen!" – Zur sozialen Konstruktion von Wald: Perspektiven von Landschaftstheorie und Landschaftspraxis.* Wiesbaden: Springer VS.

Jenal, C. (2020). Visualizations of 'landscape' in protest movements: On exclusive, inclusive patterns of perception, interpretation using the example of resistance to the expansion of the electricity grid in Germany. In D. Edler, C. Jenal, & O. Kühne (Eds.), *Modern approaches to the visualization of landscapes* (pp. 427–445). Wiesbaden: Springer VS.

Kahneman, D. (2012). *Thinking, fast and slow.* London: Penguin Books.

Kirchhoff, T. (2018). *Kulturelle Ökosystemdienstleistungen: Eine begriffliche und methodische Kritik = "Cultural ecosystems services"; A conceptual and methodological critique* (Originalausgabe). *Physis: Band 4.* Freiburg, München: Verlag Karl Alber.

Kirchhoff, T. (2019). Ökosystemdienstleistungen. In O. Kühne, F. Weber, K. Berr, & C. Jenal (Eds.), *Handbuch Landschaft* (pp. 807–822). Wiesbaden: Springer Fachmedien Wiesbaden.

Kühne, O. (2014). Das Konzept der Ökosystemdienstleistungen als Ausdruck ökologischer Kommunikation. Betrachtungen aus der Perspektive Luhmannscher Systemtheorie. *Naturschutz und Landschaftsplanung, 46*(1), 17–22.

Kühne, O. (2017a). Der intergenerationelle Wandel landschaftsästhetischer Vorstellungen. In O. Kühne, H. Megerle, & F. Weber (Eds.), *RaumFragen. Landschaftsästhetik und Landschaftswandel* (pp. 53–67). Wiesbaden: Springer VS.

Kühne, O. (2017). *Zur Aktualität von Ralf Dahrendorf: Einführung in sein Werk. Aktuelle und klassische Sozial- und Kulturwissenschaftler innen.* Wiesbaden: Springer Fachmedien Wiesbaden.

Kühne, O. (2018). *Landschaftstheorie und Landschaftspraxis.* Wiesbaden: Springer Fachmedien Wiesbaden.

Kühne, O. (2019). *Landscape theories: A brief introduction. RaumFragen : Stadt – Region – Landschaft.* Wiesbaden, Germany: Springer VS.

Kühne, O., & Duttmann, R. (2020). Recent Challenges of the ecosystems services approach from an interdisciplinary point of view. *Raumforschung Und Raumordnung Spatial Research and Planning, 78*(2), 1–14.

Kühne, O., & Jenal, C. (2020). The threefold landscape dynamics—Basic considerations, conflicts and potentials of virtual landscape research. In D. Edler, C. Jenal, & O. Kühne (Eds.), *Modern approaches to the visualization of landscapes* (pp. 389–402). Wiesbaden: Springer VS.

Lacy, T. de, & Whitmore, M. (2006). Tourism and recreation. In M. Lockwood, G. L. Worboys, & A. Kothari (Eds.), *Managing protected areas. A global guide* (pp. 497–527). London: Earthscan.

Lehnert, L. W., Bässler, C., Brandl, R., Burton, P. J., & Müller, J. (2013). Conservation value of forests attacked by bark beetles: Highest number of indicator species is found in early successional stages. *Journal for Nature Conservation, 21*(2), 97–104.

Linke, S. (2020). Landscape in internet pictures. In D. Edler, C. Jenal, & O. Kühne (Eds.), *Modern approaches to the visualization of landscapes* (pp. 139–156). Wiesbaden: Springer VS.

Letter to the editor Bayerwaldbote (2009, July 02). Herrschaftszeiten! *Bayerwald Bote.*

Letter to the editor Bayerwaldbote (2015, January 21). Forscherherz, was brauchst Du mehr?

Liebecke, R., Wagner, K., & Suda, M. (2011). *Die Akzeptanz des Nationalparks bei der lokalen Bevölkerung.* Grafenau: Nationalpark Bayerischer Wald.

Loda, M., Kühne, O., & Puttilli, M. (2020). The social construction of Tuscany in the German and English speaking world—Presented by the analysis of internet images. In D. Edler, C. Jenal, & O. Kühne (Eds.), *Modern approaches to the visualization of landscapes* (pp. 157–171). Wiesbaden: Springer VS.

Mayer, M., & Job, H. (2014). The economics of protected areas – A European perspective. *Zeitschrift für Wirtschaftsgeographie, 58*(1).

McFarlane, B. L., Stumpf-Allen, R. C. G., & Watson, D. O. (2006). Public perceptions of natural disturbance in Canada's national parks: The case of the mountain pine beetle (Dendroctonus ponderosae Hopkins). *Biological Conservation, 130*(3), 340–348.

McLean, K. (2020). Temporalities of the smellscape: Creative mapping as visual representation. In D. Edler, C. Jenal, & O. Kühne (Eds.), *Modern approaches to the visualization of landscapes* (pp. 217–246). Wiesbaden: Springer VS.

Metzner, W., & Meffert, C. (1997). Nationalpark Bayerischer Wald: Kaputtgeschützt. *Stern*, 20–25.

Milcu, A. I., Hanspach, J., Abson, D., & Fischer, J. (2013). Cultural ecosystem services: A literature review and prospects for future research. *Ecology and Society, 18*(3).

Millennium Ecosystem Assessment (2005). *Ecosystems and human well-being: synthesis.* Washington, DC.

Moltke, J. von (2005). *No place like home: Locations of Heimat in German cinema. Weimar and now: Vol. 36.* Berkeley: University of California Press.

Müller, J., Bußler, H., Goßner, M., Rettelbach, T., & Duelli, P. (2008). The European spruce bark beetle Ips typographus in a national park: From pest to keystone species. *Biodiversity and Conservation, 17*(12), 2979–3001.

Müller, M. (2011). How natural disturbance triggers political conflict: Bark beetles and the meaning of landscape in the Bavarian Forest. *Global Environmental Change, 21*(3), 935–946.

Müller, M., & Job, H. (2009). Managing natural disturbance in protected areas: Tourists' attitude towards the bark beetle in a German national park. *Biological Conservation, 142*(2), 375–383.

Oteros-Rozas, E., Martín-López, B., Fagerholm, N., Bieling, C., & Plieninger, T. (2018). Using social media photos to explore the relation between cultural ecosystem services and landscape features across five European sites. *Ecological Indicators, 94*, 74–86.

Plieninger, T., Dijks, S., Oteros-Rozas, E., & Bieling, C. (2013). Assessing, mapping, and quantifying cultural ecosystem services at community level. *Land Use Policy, 33*, 118–129.

Ridder, B. (2008). Questioning the ecosystem services argument for biodiversity conservation. *Biodiversity and Conservation, 17*(4), 781–790.

Roßmeier, A. (2020). Urban/Rural hybridity in pictures. The creation of neighborhood images using the example of San Diego's urbanizing inner-ring suburbs East Village and Barrio Logan. In D. Edler, C. Jenal, & O. Kühne (Eds.), *Modern approaches to the visualization of landscapes* (pp. 479–498). Wiesbaden: Springer VS.

Schaich, H., Bieling, C., & Plieninger, T. (2010). Linking ecosystem services with cultural landscape research. *GAIA – Ecological Perspectives for Science and Society, 19*(4), 269–277.

Scharnowski, S. (2019). *Heimat: Geschichte eines Missverständnisses.* Darmstadt: wbg Academic.

Siepmann, N., Edler, D., & Kühne, O. (2020). Soundscapes in cartographic media. In D. Edler, C. Jenal, & O. Kühne (Eds.), *Modern approaches to the visualization of landscapes* (pp. 247–263). Wiesbaden: Springer VS.

Sonter, L. J., Watson, K. B., Wood, S. A., & Ricketts, T. H. (2016). Spatial and temporal dynamics and value of nature-based recreation, estimated via social media . *PLoS ONE, 11*(9), e0162372.

Tourismusverband Ostbayern e. V. (2020). *Wandern im Bayerischen Wald.* Retrieved March 19, 2020, from https://www.bayerischer-wald.de/Urlaubsthemen/Wandern.

Trepl, L. (2014). Kulturelle Ökosystemdienstleistungen gibt es nicht. *Ökologisches Wirtschaften – Fachzeitschrift, 29*(2), 16.

Weber, F. (2020). Blurring the boundaries of landscape visualization: Welcome to Fabulous Las Vegas. In D. Edler, C. Jenal, & O. Kühne (Eds.), *Modern approaches to the visualization of landscapes* (pp. 461–478). Wiesbaden: Springer VS.

Weber, F., Kühne, O., & Hülz, M. (2019). Zur Aktualität von 'Heimat' als polyvalentem Konstrukt – eine Einführung. In M. Hülz, O. Kühne, & F. Weber (Eds.), *Heimat* (pp. 3–23). Wiesbaden: Springer Fachmedien Wiesbaden.

Erik Aschenbrand works in the UNESCO-Biosphere Reserve Middle Elbe as head of the Biosphere Reserves northern regional department. From 2015-2016 he worked in the Bavarian Forest National Park. He holds a Ph.D in Geography from University of Tübingen.

Thomas Michler works in Bavarian Forest National Park since 2007 and is responsible for the envi-ronmental education. He holds a Diploma in Social Work from Koblenz University of Applied Scienc-es.

Visualizations of 'landscape' in Protest Movements: On Exclusive, Inclusive Patterns of Perception, Interpretation Using the Example of Resistance to the Expansion of the Electricity Grid in Germany

Corinna Jenal

Abstract

While the energy system transformation being sought in Germany continues to enjoy great popular support, the physical manifestos of its actual on-site implementation, such as storage facilities, wind turbines, solar parks, and new or upgraded power lines, continue to meet with strong local protests. The focus of the opponents' argumentation is oriented towards the criticism of the landscape change associated with the measures, which for many critics entails an 'irretrievable destruction' of the landscape and the protest predominantly finds expression on a visual level through the images distributed by the opponents. For in the course of organizing resistance against measures to implement the energy revolution, protest movements are developing a set of visual patterns and codes that, especially on an emotional and, thus largely subconscious level, intend to regulate aesthetic criteria of social perception—demonstrated here using the example of the physical representations of the power grid expansion. The contribution focuses on the corresponding processes of landscape-related design formation, in which selected physical objects are explicitly included or excluded in the display and interpretation of landscape, in conjunction with the attempt to hegemonically embed one's own patterns of interpretation of landscape.

C. Jenal (✉)
Eberhard Karls Universität Tübingen, Tübingen, Germany
e-mail: corinna.jenal@uni-tuebingen.de

© Springer Fachmedien Wiesbaden GmbH, part of Springer Nature 2020
D. Edler et al. (eds.), *Modern Approaches to the Visualization of Landscapes*, RaumFragen: Stadt – Region – Landschaft,
https://doi.org/10.1007/978-3-658-30956-5_24

Keywords

Landscape · Visualization · Energy   transition · Resistance · Communication · Power
grid expansion · Citizens' initiatives

1   Introduction[1]

The final phase-out of nuclear power and the initiated energy transition still meet with
great approval and high acceptance levels among the German population (see among
others Agentur für Erneuerbare Energien 2015, 2019; BMUB and UBA 2017). However,
the actual physical implementation—for example in the form of wind turbines, solar
parks, and the expansion of the electricity grid—is generating growing opposition at
regional and local level, which is manifested in numerous citizens' initiatives (CI) and
networks all of which are increasingly opposing, delaying or even causing the implementation of planned projects to fail on the ground level (see in more detail Bauer 2015;
Hildebrand and Rau 2012; Riegel and Brandt 2015; Stegert and Klagge 2015; Weber
et al. 2016b; Weber and Kühne 2016).

Against this background, a large amount of research literature has been developed
that deals with a number of other aspects related to resistance towards large-scale infrastructure projects, such as the prevailing motivations of opponents, the conditions under
which initiatives and protest movements occur, their forms of protest, lines of argumentation, ethical aspects, and participation procedures (cf. Berr 2018; Hoeft et al. 2017;
Hübner and Hahn 2013; Kamlage et al. 2014; Kühne et al. 2016; Kühne 2018a; Kühne
and Weber 2018 [online first 2017]; Langer 2018; Marg et al. 2013; Weber et al. 2016a,
2017). Systematic surveys of their visual forms of communication, visual communication patterns, and the codes on which they are based, however, have so far been largely
ignored. In this context, visual forms of communication are central elements of communication for local protest movements against physical manifestations of the energy
system transformation with regard to the dissemination of aesthetic, emotional, and normative attributions of the planned projects: These range from the creation of their own
CI logos to detailed photo-documentation of the events and actions carried out to photomontages of the 'native normal landscape' (Kühne 2008, 2018c, 2019b) which stage,
(re)produce, and distribute via their own platforms the consequences of the physical
changes caused by the planning projects with the help of different presentation techniques according to their own interpretation and attributions (on aesthetics and landscape

---

[1]The present article is a revised and adapted version of the anthology article published in German
in 2018 Jenal, C. (2018). Ikonologie des Protests—Der Stromnetzausbau im Darstellungsmodus
seiner Kritiker(innen). In O. Kühne, & F. Weber (Eds.), *Bausteine der Energiewende* (pp. 469–
487). Wiesbaden: Springer VS.

see also Berr 2020a, b in this volume; see on conflict also Kühne and Jenal 2020b as well as Thomas 2020 in this volume).

Accordingly, the present contribution pursues the question of whether and how protest movements—relevant to the expansion and reconstruction of the transmission networks, in this instance—develop a set of visual patterns and codes in the course of their formation, using the images generated by them and distributed across the websites of the citizens' initiatives (CI). These not only depict an expressive form of self-portrayal by the actors, but at the same time also intend to regulate aesthetic criteria of social perception—exemplifying here the physical representatives of the expansion of the electricity grid and of 'landscape'—on an emotional and thus largely subconscious level. For, due to the usually unconscious processing of visual information in social perception and its more emotional evocation than cognitive control, "[v]isual codes […] are thus particularly effective in terms of behavior" (Fahlenbrach 2002, p. 43) and thus represent an important communication element of the protest movements, which needs to be examined more closely.

Before a systematic examination of the visual presentation modes of the electricity grid expansion on the home pages of the websites and *Facebook profiles* of citizens' initiatives, the theoretical framework of the analysis is briefly outlined (Chap. 2) and the methodological procedure described (Chap. 3). Chapter 4 then presents the results of the evaluation and, subsequently, a summary of the results and a brief outlook is given in the conclusion (Chap. 5).

2   General Preliminary Remarks

The starting point of the study is a social constructivist perspective (on the social constructivist perspective see also Fontaine 2020; Kühne 2020; Kühne and Jenal 2020b; Linke 2020; Loda et al. 2020; Roßmeier 2020; Siepmann et al. 2020, Weber 2020 in this volume), in the context of which—not only linguistic but also visual—signs and symbolic systems have a central function in the construction of social reality, and whose central statements are briefly presented (Sect. 2.1). Subsequently, the question of what protest actually is and what distinguishes it from other forms of social interaction (Sect. 2.2) is addressed, additionally, images and their 'dual nature' are outlined (Sect. 2.3).

2.1   (Visual) Signs and Symbolic Systems as Media and Mediation of Reality Constructions

From a social constructivist perspective, the perception of the everyday world takes place on the basis of a complex process of typification, which draws on socially mediated stocks of values, rules, and maxims of action, in order, for example, to assign interpretations and attributions to certain social situations, modes of action, and even physical manifestations. Against this background, the reality of the everyday world is experienced as an order of reality whose phenomena are pre-arranged according to patterns, which seem to be independent of how individuals experience them, and which, to a certain extent, overlap the experience of them (Berger and Luckmann 1966). Even if man is aware of the world as a variety of realities (Berger and Luckmann 1966) there is, among the many realities, one that presents itself as *the* reality and thus takes precedence as the 'supreme reality' (Berger and Luckmann 1966). With reference to this 'supreme reality', which has become decisive, it becomes possible for the individual to construct everyday life in a routine way and to regulate his behavior in it. As a rule, the reality of the everyday world is accepted as 'reality' and requires beyond its simple presence no additional verification" (Berger and Luckmann 1966). At the same time, it contains both problematic and unproblematic aspects of reality in equal measure: Problematic aspects—such as landscape change through physical manifestations of the energy system transformation—only become a problem when they 'destroy' the routine reality of the everyday world, thereby going beyond the boundaries of everyday reality and referring to completely different realities (see Berger and Luckmann 1966).

The reality character of the everyday world results primarily from objectivations coagulated in symbols and systems of symbols, whereby objectivations represent the objectification of subjective attributions and interpretations as something objectively given and are the results of social and societal interaction processes. The most important category of such objectivations are linguistic but also involve visual signs and symbolic systems, (vgl. dazu Loenhoff 2015). For signs and symbolic systems, which are used in everyday life, provide the individual with the sequencing in which these objectivations have meaning and in which the everyday world seems to make sense to the individual (Berger and Luckmann 1966). Starting points for the formation of meaning, i.e. whether something appears meaningful or nonsensical to individuals, are the formation of (system-internal) differences in social systems (Luhmann 1984) according to which "each processing of information is based on (to be understood exclusively as internal to the system) a differentiation (distinction) and a designation (indication)" (Staubmann 1997, p. 227). This is accompanied by an assignment in each case: For example, the distinction between beautiful and ugly is "not a distinction of the world per se [...], but one of an observing system". (Staubmann 1997, p. 227)—e.g. that of an observer of a wind turbine. On the basis of system-internal criteria, the viewer structures the world by assigning the wind turbine to the binary codes beautiful/ugly, superfluous/necessary, etc. according to the possible differences—or in other words—and thereby filling them with meaning. Accordingly, with Staubmann (1997, p. 227) that "sense enables the processing of information according to differences, differences that are not predetermined in the world, but are autopoietically produced from sense itself.

Transferred to landscape, this has the consequence that landscape "[...] is not an object that has certain characteristics, but rather represents a form of order and demarcation that cannot be founded in the essence of things, but rather refers to the processes of ordering and those ordering (observers)". (Miggelbrink 2002, p. 338). The 'lifeworld landscape' of lay people can be "divided into the constructs of the normal native landscape and stereotypical landscape" (Kühne 2013, p. 206). While the normal landscape of the native homeland emerges "in childhood and adolescence, in direct confrontation with the physical objects constructed as the normal native landscape with the mediation of parents, teachers, peers, etc." (Kühne 2013, p. 206), this is supplemented in the further course of life by "the socialization of stereotypical landscapes through secondary information" (Kühne 2013, p. 206). Even though both forms may experience changes in the further course of life through their intensification, modification, and questioning, they differ in terms of the (system-internal) expectations directed at them: While normal native landscapes must be 'familiar' and less aesthetic qualities are in focus, stereotypical landscapes are subject to this "social landscape assessment" (Kühne 2013, p. 207) and are assigned according to the respective system-internal differentiation such as 'beautiful'/'ugly'.

2.2   Protest and the Functions of Visual Protest Communication

If we turn to the social phenomenon of protest, then in a first approximation—even if there is no generally valid definition of what protest actually is and what distinguishes it from other social forms of interaction—protest can initially be regarded as a method of communication that can be characterized by certain features (Gherairi 2015, pp. 66ff.): a) articulation of a political and social concern, b) dissent on a decision/opinion that has been generally accepted up to this point, c) "performance of a communicative-persuasive display (protest technique) in public space" (Gherairi 2015, p. 67), as well as, d) the reference to the addressee, with the aim of "influencing public opinion in order to convince the authority with the power to decide and/or act of the need to change its decision and/or action in relation to the expressed concern" (Gherairi 2015, p. 66). According to Rucht (2001) protest always contains a 'twofold connotation': as a protest against one thing and for something else (Rucht 2001, p. 9).

   Another characteristic of protest is that the central concern of the protest is often "not the implementation or intended enforcement of a direct personal advantage or (exclusive) self-interest. On the contrary, protest is characterized precisely by the fact that the protesters, from their perspective as advocates, are working for a higher good or the common good (Gherairi 2015, p. 66). This means that in the subjective perception of the protesters, a social interest of their engagement is often constructed. The communication of the protest itself takes place "*within* society, […] but as *if it were from outside*. It expresses itself out of responsibility *for* society, but *against* it" (Luhmann 1996, p. 204; highlighting in the original).

   Central to this is the structural linking of protest movements to media communication forms and formats such as websites and *Facebook profiles*, because "[…] the final form of public opinion now seems to be the representation of conflicts […]. The planning of the protests also takes this into account. The protest stages 'pseudo-events' […], i.e. events that are staged for reporting from the outset and would not even take place if it were not for the mass media" (Luhmann 1996, p. 212). The function of visual protest communication becomes central not only in the articulation of the specific concern, but also regarding the formation of collective identities and value systems in relation to aesthetic patterns and the associated emotional and normative attributions and differentiations.

2.3   Images and Their 'Double Nature'

In attempting to define what a picture actually is, it quickly becomes clear that the field of perception behind the concept of picture or illustration in the sense of a sign that communicates specific content is much larger and more extensive than it may appear to the viewer at first glance. The practice of imagery is one of the oldest cultural techniques of mankind, and the identification and attempts to determine the concept of the image are still today the subject of an ever-widening range of research disciplines (see also among others Belting 2011 [1990]; Lobinger 2012; Müller and Geise 2015 [2003]; Sachs-Hombach 2002; see in this volume Edler et al. 2020, Schenk 2020).

Even though a number of other image theories have gained influence in recent decades (see also Doelker 1997; Mitchell 1990), these can be considered insufficient for practical application in visual communication research, as they are too fuzzy resulting from the integration of non-materially substantive pictorial concepts or, conversely, they completely exclude immaterial components of the image definition (see further details Müller and Geise 2015 [2003]).

With regard to the practical application towards the representation of images, a semiotic approach is followed in the present study, in which the image itself can be regarded as a symbol, or even as a combination of symbols (Lobinger 2012). These can form certain visual patterns—i.e. regularly repeating structures—and codes can be formed and communicated (Sachs-Hombach 2001). The specific quality of pictorial or perceptual signs (Sachs-Hombach 2002)—especially of photographs—is of a particular dual nature, which is expressed in a "simultaneousness of vividness and vagueness" (Lobinger 2012, p. 55). On a *pictorial* level, representational images are given a clear evidence with which they refer to the scene depicted. On a *symbolic* level, they exhibit a high degree of semantic imprecision and ambiguity […]" (Michel 2006, p. 46; emphasis in original).

Through the vividness of the pictures, especially photographs, on the pictorial level, the symbolic nature of pictures is often overlooked when viewed (Lobinger 2012) since, although they appear to be a faithful reproduction of 'excerpts of reality', they are not identical to them (Michel 2006). So, it can be illustrated with Drechsel (2005) that images are *"visual symbols whose meanings result from the interplay of glances and bearers"* (Drechsel 2005, p. 63; emphasis in original). A further special feature in the context of perception and interpretation of visual symbols is that the processing process is usually non-linear, since—in contrast to linguistic signs such as words or texts—there is no defined starting or end point, and thus that which is accentuated or ignored during the perception process is subjected even more to the individual cascades of perception.

In contrast, the approach of visual semiotics, which, following the semiotic concept of the image, also forms the basis of the present analysis, not only investigates the question of what or how the images may depict in a regularly repeating structure (pattern), but also whether they implicitly transmit certain ideas, values, or codes. So visual semiotics is asking according to van Leeuwen (2013)—building on the work of Roland Barthes (1915–1980)—two fundamental aspects of inquiry: "the question of representation (what do images represent and how?) and the question of the 'hidden meanings' of images (what ideas and values do the people, places, and things represented in the images stand for?)" (van Leeuwen 2013, p. 92; see also Breckner 2010). Which, accordingly, proposes the dual concept of levels of meaning: "The first level, also called the level of *denotation*, consists of what is depicted (who or what is shown in the picture?). The second level is the level of *connotations*, i.e. the ideas and values that are communicated (which ideas and values are expressed by what is shown and how it is represented? " (Lobinger 2012, p. 247; emphasis in original).

3   Methodical Approach

On the basis of a keyword-based Internet search,[2] One hundred twenty-three[3] online active citizens' initiatives saved the images on the home pages, the headers, and profile images of their *Facebook sites* in April 2017, so that a source database of n = 374 images could be compiled[4] on this basis. These were first divided quantitatively into different categories (logos, actions, posters/displays in the town, stereotypical 'beautiful' landscapes (siehe dazu Kühne 2008, 2012, 2013; Kühne and Jenal 2020a). The following information is classified as follows—e.g. power lines/poles, demonstrations, CI members, information events/stands, photomontages, underground cable/converter, graphics, and others.

---

[2]Keywords of the Internet research: Bürgerinitiative (Citizens' Initiative)*/Interessengemeinschaft (Interest Group)* (*in connection with): Südlink, 380 kV, power line, direct current line, monster line, monster power line, power monster line, extra high voltage, extra high voltage lines, giant masts, mega masts, megalomania, line madness, line opponents, action, action alliance, protest, overhead line, underground cable, Amprion, Tennet, 50 Hz, TransNetBW.

[3]In March 2017, a total of 131 websites of citizens' initiatives were initially identified as being opposed to the expansion of the electricity grid, but at the time of the image data backup, eight of these websites were no longer accessible.

[4]The citizens' initiatives and corresponding web addresses were compiled in a table and systematized with the numbers CI 1 - CI 131; the corresponding illustrations of the respective start page or *Facebook profile* were numbered in order of appearance and assigned to the respective CI by underscores. The underlying systematization can be made available on request.

In a next step, the categories formed were then assigned on the basis of their communicative functions, i.e. whether the images shown are *primarily* the depiction of persuasion, i.e. persuasive actions in public space, or whether the focus is on the explicit structuring of landscape-related aesthetic patterns.

On the basis of these classifications, a third step of analysis will now examine qualitatively in detail whether and to what extent sets of visual patterns and codes can be worked out in the individual categories, which—as already mentioned at the beginning—also represent an expressive form of self-representation of the actors, but which at the same time also shape aesthetic criteria of social perception on an emotional and thus largely unconscious level.

4   Evaluation Results

4.1   General Overview

In a first quantitative classification of the data, logos occupy a prominent position in the context of visual communication of citizens' initiatives with almost a quarter of the total (24.3%; see Fig. 1). This document is the clear endeavor of the CIs to generate visual elements that serve both to create a collective identity and to classify and structure the represented concern externally. For this purpose—as will be explained in more detail in Sect. 4.4—the graphic aim is to unite these aspects in a pictogram and to transfer them from the visual discourse to the linguistic one and to anchor them centrally through perpetuated use in public space. Furthermore, as expected, citizens' initiatives use their own websites and *Facebook profiles* for the documentation of CI activities, such as choreographed actions (route runs, action days, etc.; 11.8%), local posters or displays in the respective communities on site (11.0%) and demonstrations (10.2%) in roughly equal shares. CI members (7.0%) or information events/points (5.1%; see Fig. 1) are presented with somewhat less intensity. Additionally, the citation of stereotypically 'beautiful' landscapes (10.7%) and the focus on the presentation of selected physical objects such as power lines and power poles (10.2%)—also in the form of edited image files (4.8%)—are included as components of visual communication on the part of citizens' initiatives.

If these categories are assigned according to their respective *primary* communicative function, a clear focus on displaying actions of persuasion, i.e. protest stagings based on a conviction or 'persuasion' with regard to the positions represented in the public space, can be identified (44.9%; see Fig. 2). Followed by this, about a quarter (25.7%) of the illustrations aim at the explicit structuring and arrangement of landscape-related aesthetic patterns and classifications according to their own internal system codes, the aim being to structure which physical objects 'belong' to a landscape in the overall view and which are to be excluded from this overall view and not (or may not) be construed as a 'part' of the landscape. Furthermore, logos can be interpreted as an intersection of several communicative aspects. As such, the logos developed by the citizens' initiatives

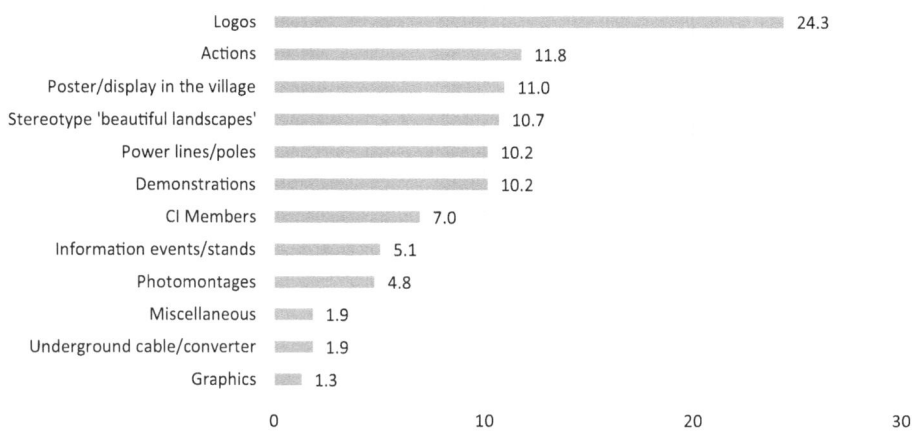

Fig. 1 Quantitative evaluation of the image data sets on the CI start pages or Facebook profiles and headers (n = 374). (*Source* Own survey and presentation; data in percent)

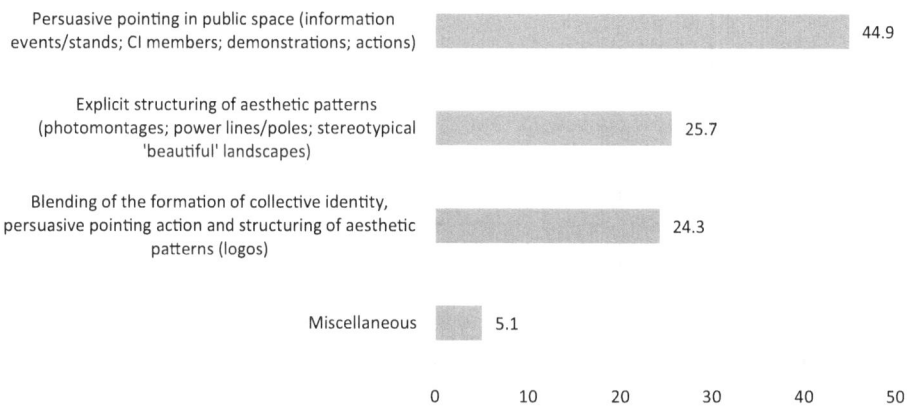

Fig. 2 Assignment of categories according to their communicative function (n = 374). (*Source* Own survey and presentation; data in percent)

can not only be regarded as visual communication signs that create identity and have a high recognition value in public space ('protest symbol'), but also as an intersection of the aspects of persuasive showmanship and structuring of aesthetic patterns—and thus as a category of their own—by graphically picking up core elements of resistance, stabilizing them through their recurring use in public space, and thus trying to convince and win over the public and the corresponding decision-makers concerning their protest issues. At the same time, the analyzed logos also structure aesthetic patterns related to the landscape and assign them according to variations that are applicable within the system, by integrating the interpretation of those physical objects that, from the perspective of the

Fig. 3 Examples of efforts demonstrating persuasive display in public space. Sources: above (from left to right): CI 65_01, CI 11_05, CI 01_08; below (from left to right): CI 36_16, CI 36_04, CI 110_02

opponents, are marked as excluded in the context of a landscape perceived as 'beautiful' (usually by warning signs, big red 'X' and circle with a slash mark, etc.) or are explicitly cited as an 'inventory' of a landscape perceived as 'beautiful' (e.g. trees, fields, hills, etc.). This is accompanied by the attempt to embed subjective attributions and interpretations as objectively given and to embed and (pre-)establish the underlying systemic codes in the social as well as individual constructions of landscape.

In the following, selected examples from the respective categories will be examined and presented in more detail in a qualitative analysis according to their communicative function. The central question is how citizens' initiatives structure, stage, and (re)produce the electricity grid expansion and its physical manifestations by means of different presentation techniques and sets of visual patterns or codes according to their own interpretations and attributions, thus shaping and regulating aesthetic criteria of social perception.

4.2 Persuasive Presenting in Public Space

A central pattern in the context of persuasive demonstrations by citizens' initiatives against the expansion of the power grid is the publicity-effective protest articulation and staging in public space at high-profile locations such as town centers and physical structures with a highly-visible profile (e.g. bridges, towers; Fig. 3—top/bottom center/right), often using bright symbolic colors that clearly stand out from the surrounding palette of

Fig. 4 Structuring of landscape-related aesthetic patterns. Sources: top (from left to right): CI 15_04 (‚our nightmare'), CI 42_01 (‚no to the monster line '), CI 59_01 (Aulataler, fight back!); middle: CI 90_01; bottom (from left to right): CI 01_05, CI 95_01 (For underground cable Neuss), CI 65_03

color (Fig. 3—top/bottom center, top right). While in the case of personal protest campaigns such as demonstrations, actions, and information events, the focus of external perspectives is on the number of committed protesters and interested citizens (Fig. 3—top center; bottom left), the focus of internal perspectives or protest forms not requiring human presence such as displays/posters is primarily on the messages being communicated (Fig. 3—top right, bottom center and right).

The determining codes of the recurring patterns, however, aim at a contextualization of the protest topic in the creation of distinctions as advocating or 'fighting' (on the part of the CI) for a social interest or common good as legitimate representatives and 'defenders'—the opponents from this perspective act as opposed to not advocating and not fighting (on the part of the decision-makers such as the state, energy suppliers, network operators). Furthermore, the assignment of the 'protectors' as antipode to the 'deliverers' can also be identified: 'objects' of protection are often aspects conceived as of 'greater good' such as landscape, home, nature, children, etc. (see example, Fig. 4—bottom middle, below), which cannot stand up for themselves, but require the protection of third parties—in this case the citizens' initiatives as 'protectors'—and which otherwise run the risk of being expropriated, for example by the state, energy suppliers, and network operators (on morality and ethics in landscape-related conflicts, see also Berr and

Kühne 2019a, b; Kühne 2018b, 2019c). At the center of the demonstrations is also the correlation of the perception of social responsibility through action and information (on the part of the CI; Fig. 3 top/bottom center/left) to an infringement (on the part of the state, energy suppliers, network operators). The internal system classification of true/ false, which is subject to the citizens' initiatives, also applies: By labeling opposing positions as a 'lie', one's own position is structured as the only 'true' one, which ethically and morally places one's own position above other existing assessments of the expansion of the electricity grid (see example Fig. 3—below right; poster with the inscription *"EVERY LIE/COMES TO/LIGHT/ BAVARIA/DOES NOT NEED THE/LINE"*).

In doing so, by singling out and confronting individual aspects in the protest communication, a certain simplification is brought about which, in view of the very demanding and complex interrelationships within the context of the initiated energy system transformation, offers a certain orientation for action amongst those locally affected, since breaking through the routine reality of their everyday world by changing the physical underpinnings of their landscape is often perceived as problematic and definitely requires further clarification. Critical reflection or even embedding of one's own position in the face of further existing perspectives, however, cannot be proven, at least in the realm of visual protest communication.

In connection with persuasive show acts in public space, it also becomes clear—as will be further elaborated in Sects. 4.3 and 4.4—that other physical manifestations of social needs, which overshadow the space just as concisely, may well be socially accepted or at least not be construed as 'problematic' in the landscape, rather—just the opposite—may even be used as 'advertising pillars' for spatial protest (Fig. 3—above/ below right: a former water tower and a road bridge).

4.3   Explicit Structuring of Landscape-Related Aesthetic Patterns

A further central function of visual protest communication involves the explicit structuring of landscape-related aesthetic patterns according to its own system logic. On the one hand, stereotypically 'beautiful' landscapes of the region will be cited and used as a benchmark for evaluating other possible, alternative landscape arrangements (Fig. 4— below; see also the text line in the Fig. 4—below left: *"We are committed to a high-voltage-free virgin forest"*). On the other hand, electricity pylons and power lines are used to stage the respective 'counterpart' as the difference to a landscape that is perceived as 'beautiful' (middle Fig. 4), namely a 'nightmare' induced by 'monster lines' (see Fig. 4—above, left and center)—sometimes also exaggeratedly through the creation of photomontages, which superimpose oversize power lines and place them disproportionately above or on existing residential buildings (Fig. 4—below).

Another perpetuated pattern of visual communication is that electricity pylons or power lines are construed as not belonging to a landscape, but rather as a landscape and habitat of 'destructive nightmares'. The problem is probably not so much the consideration of power lines or power poles as part of a normal *social* landscape (a land-scape arrangement as shown in the Fig. 4 above-right is quite common for large parts of the nation and is considered completely unproblematic in many places), but rather the change in the routinely constructed normal *domestic* landscape, which is caused by the 'breaking-in' of other realities—such as, for example, the 'destruction of the land-scape' or the 'creation of a nightmare'. Rather, it is the change of the routinely perceived domestic normal landscape that is 'threatened' by the 'collapse' of other realities—such as aspects in the context of energy system transformation for energy production and transport, which generally play a subordinate or no role at all in the composition of the everyday world.

The codes underlying the illustrations distinguish the landscape-related structuring of 'beautiful'/'ugly' (Berr 2020a, b; Kühne 2008, 2013; Linke 2017) and what is desirable in terms of the landscape, plus what is to be rejected, also often in a direct comparison (see example Fig. 4—middle row). In this process, a landscape 'ideal' is socially constructed, which is concentrated on a certain set of physical objects (meadow, forest, trees, paths, lake, village settlement, etc.) and in which certain objects—such as power supply lines, although they have been established in large parts of the country for several decades and form an elementary component of industrialized civilizations—are not integrated, but excluded. Instead, outright aesthetic lines of demarcation are drawn, which classify and structure cer-tain landscape arrangements in accordance with the logics internal to the system of aesthetic differentiations (middle Fig. 4). Alternative interpretations that integrate physical objects that deviate from the 'norm' thus placed into the landscape are subject to social exclusion. The self-generated landscape ideal and understanding is rendered absolutely, delimited against alternative interpretations and thus the social enclosure of a socially accepted land-scape image according to one's own interpretation is pursued (cf. Kühne 2018b, 2019c).

4.4 Logos as an Intersection of Identity, Demonstrations, and Aesthetic Structuring

As already mentioned at the beginning, logos can be understood as a combination of several communicative aspects, whereby a clear assignment of primary communication functions is difficult to carry out: On the one hand, they function as pictograms, which assign an identity-creating icon to the movement and protest in the course of their for-mation, and on the other hand, they aim both at a persuasion related to the situation and at the structuring of landscape-related perception through the graphic processing of the rejection of power lines/pylons by the citizens' initiatives.

Accordingly, certain patterns and differentiations from the two previous categories can also be found here: For example, the classification and explicit labelling of which physical objects are to be excluded from the viewpoint of those protesting in the con-text of landscape (electricity pylons/power lines usually marked with warning signs, big

Fig. 5 Citizens' initiative logos. Sources: above (from left to right): CI 60_01, CI 33_01, CI 94_01 (Energy transition Yes! Monster line No!); below (from left to right): CI 39_01 (No direct current line through the Upper Palatinate), CI 128_02 (For underground cable—arable), CI 129_01

red 'X' and circle with a slash mark, and similar; see example Fig. 5) and which, on the other hand, are accepted—i.e. do not want to be labelled accordingly, such as mountains or monoculturally cultivated landscapes (see example Fig. 5—bottom left, center). Through the use of ISO-standardized warning signs, however, the physical manifestations of the energy system transformation are provided with pictogram elements of the circular regulatory 'prohibited' sign already socialized in society, with red crossbars on a white background, which signalize what should be avoided for one's own safety. The dimensions of the power lines and pylons clearly exceed all other pictorial elements and extend beyond the frame (prohibited sign) and are thus intended to illustrate their relevant physical extension (see Fig. 5—top right and below right).

In parts, the towers are alienated in the depictions for demonic exaggeration by modifying them, for example—as can be seen in the Fig. 5—top right—to a one-eyed creature with tentacles and mischievous sinister grin with spiky teeth, which maltreats the tortured and vomiting earth.

With regard to the codes applied, the codes already mentioned in the previous Sects. 4.2 and 4.3 can also be identified here, such as the assignment of physical elements to the landscape as belonging to or even excluding it; the protection of the current physical foundations conveying what is to be striven for vs. the surrender to or handing over of such elements as the dangerous, the maliciously reprehensible.

5   Conclusion and Outlook

The analysis has shown that citizens' initiatives develop a set of visual patterns and differentiations according to their own logics via their own websites and *Facebook profiles*, which not only serve the self-representation of the actors, but at the same time also develop and intend to regulate aesthetic, emotional, and cognitive schemata for the social perception of both energy supply systems and 'landscape'. These are presented and (re) produced in accordance with their own—i.e. from the perspective of the citizens' initiatives—interpretations and attributions, thus making the understanding of landscape represented by the initiatives absolute, while rejecting and excluding alternative interpretations, thus creating a social enclosure of their own conception of landscape.

Despite the ambivalent and complex interrelationships in the context of the initiated energy system transformation and its spatially significant consequences, in the presentation of the electricity grid expansion by the citizens' initiatives, the differences between the two systems are still prevalent. These are—at a minimum, not discernable in the visual form of presentation—not subjected to any critical reflection or even opened up to other system logics in relation to the expansion of the electricity grid. Even if this simplification of the world means that the set of criteria presented for social perception in the abundance of information and aspects to be processed in connection with the expansion of the electricity grid, while at the same time 'threatening' the routinized construction of our own everyday world, may well represent an orientation for action. It should also be recalled in this context, that landscapes, especially in their physical foundations, are the result of constant anthropogenic adaptation and are also subject to constant transformation in their interpretations and attributions. Subsequentially, that landscape change resultant of changed or increasingly complex social needs as well as changes in their interpretations and attributions still represent the 'normal case' rather than the exceptional case. Furthermore, the opportunity for the productivity of landscape conflicts should be safeguarded, especially for social (further) development (Kühne 2018d, 2019a; Kühne et al. 2019).

References

Agentur für Erneuerbare Energien. (2015). Die deutsche Bevölkerung will mehr Erneuerbare Energien: Repräsentative Akzeptanzumfrage zeigt hohe Zustimmung für weiteren Ausbau. Retrieved 9 March 2016 from http://www.unendlich-viel-energie.de/die-deutsche-bevoelkerung-will-mehr-erneuerbare-energien.

Agentur für Erneuerbare Energien. (2019). Wichtig für den Kampf gegen den Klimawandel: Bürger*innen wollen mehr Erneuerbare Energien. Retrieved 18 March 2020 from https://www.unendlich-viel-energie.de/themen/akzeptanz-erneuerbarer/akzeptanz-umfrage/akzeptanzumfrage-2019.

Bauer, C. (2015). Stiftung von Legitimation oder Partizipationsverflechtungsfalle. Welche Folgen hat die Öffentlichkeitsbeteiligung beim Stromnetzausbau? *der moderne Staat – dms: Zeitschrift für Public Policy, Recht und Management, 8*(2), 273–293.

Belting, H. (2011 [1990]). *Bild und Kult. Eine Geschichte des Bildes vor dem Zeitalter der Kunst.* München: Beck.

Berger, P. L., & Luckmann, T. (1966). *The social construction of reality. A treatise in the sociology of knowledge.* New York: Anchor Books.

Berr, K. (2018). Ethische Aspekte der Energiewende. In O. Kühne & F. Weber (Eds.), *Bausteine der Energiewende* (pp. 57–74). Wiesbaden: Springer VS.

Berr, K. (2020a). Vom Wahren, Schönen und Guten. Philosophische Zugänge zu Landschaftsprozessen. In R. Duttmann, O. Kühne, & F. Weber (Eds.), *Landschaft als Prozess* (forthcoming). Wiesbaden: Springer VS.

Berr, K. (2020). Visuality, Aesthetics and Landscape. For the enlightenment and self-enlightenment of constructivist landscape research. In D. Edler, C. Jenal, & O. Kühne (Eds.), *Modern approaches to the visualization of landscapes* (pp. 189–215). Wiesbaden: Springer VS.

Berr, K., & Kühne, O. (2019a). Moral und Ethik von Landschaft. In O. Kühne, F. Weber, K. Berr, & C. Jenal (Eds.), *Handbuch Landschaft* (pp. 351–365). Wiesbaden: Springer VS.

Berr, K., & Kühne, O. (2019b). Werte und Werthaltungen in Landschaftskonflikten. In K. Berr & C. Jenal (Eds.), *Landschaftskonflikte* (pp. 65–88). Wiesbaden: Springer VS.

BMUB, & UBA, (Ed.). (2017). *Umweltbewusstsein in Deutschland 2016. Ergebnisse einer repräsentativen Bevölkerungsumfrage.* Berlin: Selbstverlag.

Breckner, R. (2010). *Sozialtheorie des Bildes. Zur interpretativen Analyse von Bildern und Fotografien.* Bielefeld: transcript.

Doelker, C. (1997). *Ein Bild ist mehr als ein Bild. Visuelle Kompetenz in der Multimedia-Gesellschaft.* Stuttgart: Klett-Cotta.

Drechsel, B. (2005). *Politik im Bild. Wie politische Bilder entstehen und wie digitale Bildarchive arbeiten.* Frankfurt (Main): Campus.

Edler, D., Jenal, C., & Kühne, O. (2020). Modern approaches to the visualization of landscapes—An introduction. In D. Edler, C. Jenal, & O. Kühne (Eds.), *Modern approaches to the visualization of landscapes* (pp. 3–15). Wiesbaden: Springer VS.

Fahlenbrach, K. (2002). *Protest-Inszenierungen. Visuelle Kommunikation und kollektive Identitäten in Protestbewegungen.* Wiesbaden: Westdeutscher Verlag.

Fontaine, D. (2020). Virtuality and Landscape. In D. Edler, C. Jenal, & O. Kühne (Eds.), *Modern approaches to the visualization of landscapes* (pp. 267–276). Wiesbaden: Springer VS.

Gherairi, J. (2015). *Persuasion durch Protest. Protest als Form erfolgsorientierter, strategischer Kommunikation.* Wiesbaden: Springer VS.

Hildebrand, J., & Rau, I. (2012). Die Akzeptanz des Netzausbaus. Ergebnisse einer umweltpsychologischen Studie. *EMF-Spektrum* (2), 4–7. Accessed: 12 February 2015.

Hoeft, C., Messinger-Zimmer, S., & Zilles, J. (Eds.). (2017). *Bürgerproteste in Zeiten der Energiewende. Lokale Konflikte um Windkraft, Stromtrassen und Fracking.* Bielefeld: transcript.

Hübner, G., & Hahn, C. (2013). *Akzeptanz des Stromnetzausbaus in Schleswig-Holstein.* Halle: Abschlussbericht zum Forschungsprojekt.

Jenal, C. (2018). Ikonologie des Protests – Der Stromnetzausbau im Darstellungsmodus seiner Kritiker(innen). In O. Kühne & F. Weber (Eds.), *Bausteine der Energiewende* (pp. 469–487). Wiesbaden: Springer VS.

Kamlage, J.-H., Nanz, P., & Fleischer, B., et al. (2014). Dialogorientierte Bürgerbeteiligung im Netzausbau. In H. Rogall, H.-C. Binswanger, F. Ekardt, A. Grothe, W.-D. Hasenclever, & I. Hauchler (Eds.), *Im Brennpunkt: Die Energiewende als gesellschaftlicher*

Transformationsprozess (Vol. Jahrbuch Nachhaltige Ökonomie 4, pp. 195–216). Marburg: Metropolis-Verlag.

Kühne, O. (2008). *Distinktion – Macht – Landschaft. Zur sozialen Definition von Landschaft.* Wiesbaden: VS Verlag für Sozialwissenschaften.

Kühne, O. (2012). *Stadt – Landschaft – Hybridität. Ästhetische Bezüge im postmodernen Los Angeles mit seinen modernen Persistenzen.* Wiesbaden: Springer VS.

Kühne, O. (2013). *Landschaftstheorie und Landschaftspraxis. Eine Einführung aus sozialkonstruktivistischer Perspektive.* Wiesbaden: Springer VS.

Kühne, O. (2018a). Neue Landschaftskonflikte' – Überlegungen zu den physischen Manifestationen der Energiewende auf der Grundlage der Konflikttheorie Ralf Dahrendorfs. In O. Kühne & F. Weber (Eds.), *Bausteine der Energiewende* (pp. 163–186). Wiesbaden: Springer VS.

Kühne, O. (2018b). Die Moralisierung von Landschaft – Überlegungen zu einer problematischen Kommunikation aus Sicht der Luhmannschen Systemtheorie. In S. Hennecke, H. Kegler, K. Klaczynski, & D. Münderlein (Eds.), *Diedrich Bruns wird gelehrt haben. Eine Festschrift* (pp. 115–121). Kassel: Kassel University Press.

Kühne, O. (2018c). *Landscape and power in geographical space as a social-aesthetic construct.* Dordrecht: Springer International Publishing.

Kühne, O. (2018d). Macht, Herrschaft und Landschaft: Landschaftskonflikte zwischen Dysfunktionalität und Potenzial. Eine Betrachtung aus Perspektive der Konflikttheorie Ralf Dahrendorfs. In K. Berr (Ed.), *Transdisziplinäre Landschaftsforschung. Grundlagen und Perspektiven* (pp. 155–170). Wiesbaden: Springer VS.

Kühne, O. (2019a). Die Produktivität von Landschaftskonflikten – Möglichkeiten und Grenzen auf Grundlage der Konflikttheorie Ralf Dahrendorfs. In K. Berr & C. Jenal (Eds.), *Landschaftskonflikte* (pp. 37–49). Wiesbaden: Springer VS.

Kühne, O. (2019b). *Landscape theories. A brief introduction.* Wiesbaden: Springer VS.

Kühne, O. (2019c). Vom 'Bösen' und 'Guten' in der Landschaft – das Problem moralischer Kommunikation im Umgang mit Landschaft und ihren Konflikten. In K. Berr & C. Jenal (Eds.), *Landschaftskonflikte* (pp. 131–142). Wiesbaden: Springer VS.

Kühne, O., & Jenal, C. (2020a). *Baton Rouge – The multivillage metropolis. A neopragmatic landscape biographical approach on spatial pastiches, hybridization, and differentiation.* Wiesbaden: Springer VS, in print.

Kühne, O., & Jenal, C. (2020). The threefold landscape dynamics—Basic considerations, conflicts and potentials of virtual landscape research. In D. Edler, C. Jenal, & O. Kühne (Eds.), *Modern approaches to the visualization of landscapes* (pp. 389–402). Wiesbaden: Springer VS.

Kühne, O., & Weber, F. (2018 [online first 2017]. Conflicts and negotiation processes in the course of power grid extension in Germany. *Landscape Research, 43*(4), 529–541. https://doi.org/10.1080/01426397.2017.1300639

Kühne, O., Weber, F., & Jenal, C. (2016). Der Stromnetzausbau in Deutschland: Formen und Argumente des Widerstands. *Geographie aktuell und Schule, 38* (222), 4–14.

Kühne, O., Weber, F., & Berr, K. (2019). The productive potential and limits of landscape conflicts in light of Ralf Dahrendorf's conflict theory. *Società Mutamento Politica, 10*(19), 77–90.

Kühne, O. (2020). The social construction of space and landscape in internet videos. In D. Edler, C. Jenal, & O. Kühne (Eds.), *Modern approaches to the visualization of landscapes* (pp. 121–137). Wiesbaden: Springer VS.

Langer, K. (2018). Frühzeitige Planungskommunikation – ein Schlüssel zur Konfliktbewältigung bei der Energiewende? In O. Kühne & F. Weber (Eds.), *Bausteine der Energiewende* (pp. 539–556). Wiesbaden: Springer VS.

Linke, S. (2017). Neue Landschaften und ästhetische Akzeptanzprobleme. In O. Kühne, H. Megerle, & F. Weber (Eds.), *Landschaftsästhetik und Landschaftswandel* (pp. 87–104). Wiesbaden: Springer VS.

Linke, S. (2020). Landscape in internet pictures. In D. Edler, C. Jenal, & O. Kühne (Eds.), *Modern approaches to the visualization of landscapes* (pp. 139–156). Wiesbaden: Springer VS.

Loda, M., Kühne, O., & Puttilli, M. (2020). The social construction of Tuscany in the German and English speaking world—Presented by the analysis of internet images. In D. Edler, C. Jenal, & O. Kühne (Eds.), *Modern approaches to the visualization of landscapes* (pp. 157–171). Wiesbaden: Springer VS.

Lobinger, K. (2012). *Visuelle Kommunikationsforschung. Medienbilder als Herausforderung für die Kommunikations- und Medienwissenschaft.* Wiesbaden: VS Verlag für Sozialwissenschaften.

Loenhoff, J. (2015). Die Objektivität des Sozialen. In B. Pörksen (Ed.), *Schlüsselwerke des Konstruktivismus* (pp. 131–147). Wiesbaden: VS Verlag für Sozialwissenschaften.

Luhmann, N. (1984). *Soziale Systeme. Grundriß einer allgemeinen Theorie.* Frankfurt (Main): Suhrkamp.

Luhmann, N. (1996). Protestbewegungen (1995). In K.-U. Hellmann (Ed.), *Protest. Systemtheorie und soziale Bewegungen* (pp. 201–215). Frankfurt (Main): Suhrkamp.

Marg, S., Hermann, C., Hambauer, V., & Becké, A. B. (2013). „Wenn man was für die Natur machen will, stellt man da keine Masten hin". Bürgerproteste gegen Bauprojekte im Zuge der Energiewende. In F. Walter, S. Marg, L. Geiges, & F. Butzlaff (Eds.), *Die neue Macht der Bürger. Was motiviert die Protestbewegungen? BP-Gesellschaftsstudie* (pp. 94–138). Reinbek bei Hamburg: Rowohlt.

Michel, B. (2006). *Bild und Habitus. Sinnbildungsprozesse bei der Rezeption von Fotografien.* Wiesbaden: VS Verlag für Sozialwissenschaften.

Miggelbrink, J. (2002). Konstruktivismus? ‚Use with caution'. Zum Raum als Medium der Konstruktion gesellschaftlicher Wirklichkeit. *Erdkunde, 56*(4), 337–350.

Mitchell, W. J. T. (1990). Was ist ein Bild? In V. Bohn (Ed.), *Bildlichkeit. Internationale Beiträge zur Poetik* (pp. 17–68). Frankfurt (Main): Suhrkamp.

Müller, M. G., & Geise, S. (2015 [2003]). *Grundlagen der visuellen Kommunikation* (UTB Medien- und Kommunikationswissenschaft, vol. 2414, 2., völlig überarb. Aufl.). Konstanz: UVK Verlagsgesellschaft.

Riegel, C., & Brandt, T. (2015). Eile mit Weile – Aktuelle Entwicklungen beim Netzausbau. *Nachrichten der ARL, 45,* (2, 10–16).

Roßmeier, A. (2020). Urban/Rural hybridity in pictures. The creation of neighborhood images using the example of San Diego's urbanizing inner-ring suburbs East Village and Barrio Logan. In D. Edler, C. Jenal, & O. Kühne (Eds.), *Modern approaches to the visualization of landscapes* (pp. 479–498). Wiesbaden: Springer VS.

Rucht, D. (2001). Protest und Protestereignisanalyse: Einleitende Bemerkungen. In D. Rucht (Ed.), *Protest in der Bundesrepublik. Strukturen und Entwicklungen* (pp. 7–26). Frankfurt (Main): Campus.

Sachs-Hombach, K. (2001). Kann die semiotische Bildtheorie Grundlage einer allgemeinen Bildwissenschaft sein? In K. Sachs-Hombach (Ed.), *Bildhandeln. Interdisziplinäre Forschungen zur Pragmatik bildhafter Darstellungsformen* (pp. 9–28). Magdeburg: Scriptum-Verlag.

Sachs-Hombach, K. (2002). Bildbegriff und Bildwissenschaft. *kunst – Gestaltung – Design* (8), 3–26.

Schenk, W. (2020). Visualization of the fundamental dimensions of "landscape" in landscape paintings around 1500 A.D. In D. Edler, C. Jenal, & O. Kühne (Eds.), *Modern approaches to the visualization of landscapes* (pp. 19–32). Wiesbaden: Springer VS.

Siepmann, N., Edler, D., & Kühne, O. (2020). Soundscapes in cartographic media. In D. Edler, C. Jenal, & O. Kühne (Eds.), *Modern approaches to the visualization of landscapes* (pp. 247–263). Wiesbaden: Springer VS.

Staubmann, H. (1997). Kapitel 10: Sozialsysteme als selbstreferentielle Systeme: Niklas Luhmann. In J. Morel, E. Bauer, T. Meleghy, H.-J. Niedenzu, M. Preglau, & H. Staubmann (Eds.), *Soziologische Theorie. Abriß der Ansätze ihrer Hauptvertreter* (5th ed., pp. 218–239). München: Oldenbourg.

Stegert, P., & Klagge, B. (2015). Akzeptanzsteigerung durch Bürgerbeteiligung beim Übertragungsnetzausbau? Theoretische Überlegungen und empirische Befunde. *Geographische Zeitschrift, 103*(3), 171–190.

van Leeuwen, T. (2013). Semiotics and iconography. In T. van Leeuwen (Ed.), *Handbook of visual analysis* (pp. 92–118). London: SAGE.

Weber, F. (2020). Blurring the boundaries of landscape visualization: Welcome to Fabulous Las Vegas. In D. Edler, C. Jenal, & O. Kühne (Eds.), *Modern approaches to the visualization of landscapes* (pp. 461–478). Wiesbaden: Springer VS.

Weber, F., & Kühne, O. (2016). Räume unter Strom. Eine diskurstheoretische Analyse zu Aushandlungsprozessen im Zuge des Stromnetzausbaus. *Raumforschung und Raumordnung, 74*(4), 323–338. https://doi.org/10.1007/s13147-016-0417-4

Weber, F., Kühne, O., Jenal, C., Sanio, T., Langer, K., & Igel, M. (2016a). Analyse des öffentlichen Diskurses zu gesundheitlichen Auswirkungen von Hochspannungsleitungen – Handlungsempfehlungen für die strahlenschutzbezogene Kommunikation beim Stromnetzausbau. Ressortforschungsbericht. Retrieved 17 Oct 2018 from https://doris.bfs.de/jspui/bitstream/urn:nbn:de:0221-2016050414038/3/BfS_2016_3614S80008.pdf.

Weber, F., Jenal, C., & Kühne, O. (2016b). Der Stromnetzausbau als konfliktträchtiges Terrain. The German power grid extension as a terrain of conflict. *UMID – Umwelt und Mensch-Informationsdienst* (1), 50–56. Retrieved 30 Aug 2017 from http://www.umweltbundesamt.de/sites/default/files/medien/378/publikationen/umid_01_2016_internet.pdf.

Weber, F., Jenal, C., Roßmeier, A., & Kühne, O. (2017). Conflicts around Germany's *Energiewende*: Discourse patterns of citizens' initiatives. *Quaestiones Geographicae, 36*(4), 117–130. https://doi.org/10.1515/quageo-2017-0040.

The Digitalizing Societys— Transformations and Challenges

Peter Martin Thomas

Abstract

The long-term consequences of the digital transformation associated with the further development of digital technologies and business models are difficult to predict. The developments are fast, diverse, complex and often contradictory. What is certain, however, is that we are in the midst of a fundamental process of change that is affecting the economy, politics, science, society and individuals, and that goes far beyond the technological developments initially perceived. The significance of the process of change for coexistence can be compared with the invention of letterpress printing. People are affected by the digital transformation in different ways and develop different attitudes and behaviour. A great variety of 'digital living worlds' is created. To the extent that, on the one hand, all areas of life are increasingly affected by digitization, on the other hand, efforts to preserve or regain one's own autonomy in the digitized world are growing. However, since it will not be possible to escape the digital transformation, the best option is to actively shape the digitisation or digital transformation of one's own living environment.

Keywords

Acceleration · Digital lifeworlds · Digital transformation · Digitalization · Media revolution · Singularities · VUCA paradigm

P. M. Thomas (✉)
Stuttgart, Germany
e-mail: petermartin@petermartinthomas.de

© Springer Fachmedien Wiesbaden GmbH, part of Springer Nature 2020   447
D. Edler et al. (eds.), *Modern Approaches to the Visualization of Landscapes*, RaumFragen: Stadt – Region – Landschaft,
https://doi.org/10.1007/978-3-658-30956-5_25

1   On Unknown Paths into A Digitalised Future

At re:publica 2019—one of the most influential conferences in the German-speaking world on current developments in the digital world—Mikael Colville-Andersen gave a lecture entitled "Back to the Future in Urban Design". He impressively illustrated to the audience the role of cities in combating climate change and the importance of the bicycle in this context (re:publica 2019). A topic that only at first glance has little to do with digitization, but where at second glance it becomes clear that digitization is also generating many new ideas and opportunities for urban planning. The fact that such a lecture on urban planning received any attention at all at the re:publica shows how closely interwoven topics such as digitization, urban planning, lifestyle, climate change, landscape, etc. are now.

The following article gives an overview and discusses how digital technologies affect all areas of life, the consequences in different social fields and how people deal with these changes.

We are in the middle of the digital transformation, sometimes called digitalization, digital revolution or fourth media revolution. The terms "digitization" and "digital revolution" (in reference to the industrial revolution) tend to refer to the technical and economic aspects of the changes. The term "fourth media revolution" sees the availability of digital media in a historical series with the development of language, the emergence of writing and the invention of letterpress. In this perspective, the fundamental change in media use that has now begun will lead to another, the so-called "next society" (Baecker 2007). The term "digital transformation" does not include a historical perspective in the same way, but it also describes the comprehensive consequences for all areas of life through the emergence of digital media. This transformation is a fundamental and in many respects open or unpredictable process of transformation. The concept of transformation also underlines the fact that it is not a sudden or one-off change, but rather a continuous one, the consequences of which are felt in many areas only slowly and with a time lag. This also applies to the topic of landscape: this is increasingly perceived on the basis of Internet information, which creates a further element in the synthesis of experience and experiencing as well as the cognitive processing of spatial and aesthetic impressions (Kühne 2017, 2019; in this volume: Kühne and Jenal 2020).

The "VUCA paradigm"which was developed in the late 1990s at the United States Army War College to describe the complex international relations after the Cold War—can be applied very well to the digital transformation as well. The four letters of the acronym VUCA stand for the volatility, uncertainty, complexity and ambiguity of future developments. Volatility describes the constant change of almost all systems and social areas, driven by an increasing speed of innovation. Uncertainty refers to the experience that forecasts of future developments appear (almost) impossible and that long-term plans are therefore not very meaningful. The complexity of developments and systems is shown, among other things, by the fact that simple cause-and-effect explanation models often no longer work and it is often not possible to overlook and consider all the interrelationships for a decision. Ambiguity ultimately stands for the ambiguity of information and decisions. This ambiguity is not necessarily new but is perceived much more intensively through the digital media. People looking for an answer to any question on the Internet—e.g. whether to vaccinate their child or which pension plan is the best option—will always receive diverse and contradictory answers.

If we make assumptions about the future digital transformation based on current perceptions, we are thus moving in a broad field of possible developments, in which we can neither fully grasp the connections nor the consequences. Even if we succeed in restricting and describing the field of future developments in one area or another in a reasonably manageable way, unforeseen events can still trigger completely different developments. For we tend to overestimate our ability to understand current events, overestimate factual information and distort historical events in retrospect, just as N. N. Taleb described in his book "Black Swan" on the importance of unforeseen, profound events (Taleb 2018).

Nevertheless, the following pages will attempt to describe some selected lines of development of the digital transformation in more detail. First of all, it will be shown how the digital transformation affects various areas of social and economic life and what effects this has on individuals. It will be worked out that different digital life worlds are emerging, which deal with the changes in their own way. This leads to the concluding thought that although different attitudes and behaviors can be developed with regard to the digital transformation, it will not be possible to escape it.

## 2	The Digital Transformation of All Areas of Society

The starting point of the digital transformation is new digital technologies and the associated business models, without which the respective technologies would hardly be able to reach the broad masses in their everyday lives. Consumers are familiar with MP3 files because they can be used to transfer and sell music. The possibilities of GPS positioning have largely replaced the car atlas and the map, as they have brought great progress for navigation in the car or with the smartphone (but see in this volume: Kleber et al. 2020). And even the three letters WWW (for "World Wide Web") would probably be meaningless to most people if the Internet were not primarily a huge business platform.

Digital technologies have an impact on companies and economic life, government and politics, science, research and teaching as well as on individuals, their social environment and society as a whole. Not only the so-called platform economy, which includes Uber, AirBnB and Spotify, for example, is highly dependent on digital technologies. All communication between companies and between companies, customers and clients is now dependent on the Internet. Even for the classic example of German industry—the automobile— "connectivity", i.e. the digital networking of the car with the smartphone, other road users and the Internet, has become a decisive factor for success and the future.

The surprising electoral success of a Donald Trump, the Brexit vote and the perhaps imminent abolition of summertime are not only but also consequences of the digitalization of political life. The debates in the so-called digital "social media" have a considerable influence on the formation of opinion. For many people, the point of reference for their information has long since ceased to be the former leading media of the major daily newspaper and public broadcasting (a contribution-financed, economically and politically independent radio station with a legally defined program mandate), but rather highly individualized sources on the Internet, above all various social media platforms. Increasingly, the question arises of a general political debate that transcends the individual's own (digital) opinion environment and is not damaged by hate and hatred in the digital space.

With ever-increasing billing services, innovative digital imaging processes, and the global networking of scientists, new dimensions have emerged for research that give rise to hopes of solving some as yet unsolved problems, especially in the areas of health and nutrition. Serious calculations on the future development of the climate or the world population are not possible without digital technologies. While 20 years ago a student's scientific cosmos was mainly available from the local university library and the limited possibilities of inter-library loan, today every student has—at least theoretically—access to all sources on this globe. However, this has multiplied not only knowledge, but also the challenge of being able to assess the relevance and accuracy of sources. At the same time, the best teachers—and also the others who spread untruths and myths—can be experienced by learners in livestreams, web-based trainings or on YouTube and often even be accessed worldwide via digital channels.

After all, the digital transformation has almost completely penetrated through the private and working lives of most people. Communication within the family and circle of friends has often shifted from physical to virtual space. With smartphones, whose great triumphant advance began with the first iPhone in 2007, the Internet became mobile and the distinction between "online" and "offline" became in some ways impossible. Anyone who wants can be physically present at almost any place at the same time and be connected to other people and places via the Internet. In a mobile life and a global migration society, this is perceived by many as a great freedom or even a necessity, because it allows one to stay in contact with people at a distance. Others experience the state of "always on" as an uninterrupted disturbance of an undivided attention on the spot. Consumers have become accustomed to the fact that they can obtain almost any information, product or service over the Internet.

Instead of linear television, series are streamed, and news are consumed on the tablet. So-called "wearables" measure one's own body, audio assistants such as Alexa, Siri and Cortana become relevant conversation partners and the home—even if it is still desired and used by relatively few people—is increasingly permeated by digital technologies as the "smart home".

What some experience as an increase in convenience and participation (for example, people with impaired vision or mobility) is feared by others as the entry point to total surveillance. The so-called "Internet of Things" is growing. More and more often, communication takes place not between people, but between people on the one hand and computers, devices and digital assistants on the other. The predictions of how robots (for example as care assistants), 3-D printers, more or less complex forms of so-called "artificial intelligence" and other technological developments will change our private everyday life are—as outlined in the VUCA paradigm described above—diverse, contradictory and rapidly changing.

3   People in the Digital Transformation

Even if many people are openly facing the changes described above or are using the new digital opportunities largely unconcernedly, uncertainties and unease about the developments can be observed at the same time, which are reflected in the public debate and the media. The book market—which despite digitization can still be a mirror of relevant social developments and debates—is full of books that directly or indirectly deal with the consequences of digitization or digital transformation and often achieve high rankings in the bestseller lists.

In 2017, for example, Yuval Noah Harari's "Homo Deus. A Story of Tomorrow" was one of the most widely read non-fiction books in Germany. In his book, he describes a possible future in which Homo Sapiens becomes the technologically enhanced Homo Deus, the "divine man" who constructs machines that can ultimately do everything better than man. Harari predicts that—when the fight against hunger, disease and war is finally won with these developments—the greed for immortality, happiness and divinity will grow and man will move further and further away from the "human" (Harari 2017). A different future is also possible but must be actively shaped. The digital transformation— this is how the book by Harari can be interpreted—does not necessarily or automatically lead to the "good".

Several years earlier, Hartmut Rosa had already described three dimensions of acceleration in his small and equally highly regarded volume "Beschleunigung und Entfremdung" (Rosa 2013), which also has to be seen in connection with digital transformation: An acceleration of social change can be observed. A great deal happens in a short time. The past can be inferred more and more briefly from the future. A (deliberate) acceleration of technical change is taking place—largely driven by digital technologies—with which communication, trade, production and much more is changing at a breathtaking pace. There is a discernible parallel to the volatility described above. Finally, many people are more or less voluntarily caught up in an acceleration of the pace of life. More and more experience units are to be accommodated in a shorter time. This is made possible by social media, the digitalized experience industry and comprehensive mobility reaching more and more people.

For many people, this acceleration not only results in more opportunities and freedom, but also in alienation. The body becomes a constantly and quickly changing styling object. Fast moving, often exclusively virtual relationships become collectible trophies and gainful employment is hardly a place of recognition but merely a source of money for further consumption. What is missing is a successful, resonant self- and world relationship, as Rosa explains in great detail in his book "Resonanz. A Sociology of the World Relationship" (Rosa 2019). The fact that it is not only possible to get in touch with other people and the world via digital media is an obvious thought.

As a third example of a publication that discusses the digital transformation in a comprehensive context, the title "Gesellschaft der Singularitäten" by Andreas Reckwitz can be mentioned (Reckwitz 2017). Just as with Hartmut Rosa, the publication is not essentially about digitalization. However, it does have a considerable influence on the developments diagnosed by Reckwitz. Both the "crisis of the general" he describes and the "explosion of the particular" are enabled and driven by digital media. Reckwitz sketches the picture of a singularistic way of life. In simple terms, this means that everyone wants to be something special, own special things and have a special job. The "general", the everyday seems—in the individual and public perception—to be of little value. As a result, three more crises in the "crisis of the general" have to be overcome: The crisis of recognition for people who are on the defensive in social and cultural terms. The crisis of self-realization for those who continuously struggle to "successfully realize themselves". And the crisis of the political: Politics is losing control; the political debate is shifting to autonomous partial public spheres (Reckwitz 2017). All three crises can only be overcome if a debate is also held on the role and function of digital media for identity formation, the world of work, social debates and political decision-making processes.

The greed for happiness and immortality, the acceleration and alienation and also the crises in general are not linear or exclusive consequences of the emergence of new digital technologies and business models. But neither can they be seen independently. The description of these developments confirms that with the digital transformation a fundamental and comprehensive process of upheaval is taking place, which in the end will have a direct impact on all areas of life.

Digitalization and digital transformation have the potential to improve the coexistence and everyday life of many people, science, the economy and political discourse. At present, however, the negative aspects often seem to predominate. Accordingly, criticism of current technological, economic, social and political developments is growing. People are increasingly looking for successful, alternative strategies for a (better) life and living together in the digitalised world. This is also reflected in book titles such as Jarrett Kobek's "I hate this Internet" (Kobek 2016), Jaron Lanier's "Reasons why you need to delete your social media accounts immediately" (Lanier 2018) or Cal Newport's "Digital Minimalism" (Newport 2019). The titles refer to the desire of many people for non-digitized or post-digital spaces in their own everyday lives, which can also include the conscious departure from social media or the trend towards haptics (the vinyl record, the Leica camera and the "do it yourself").

4 The Diversity of Digital Living Environments

In summary, these reactions to the digital transformation can be described as a mixture of uncertainty, fascination and excessive demands. Uncertainty due to the globalized, digitalized and accelerated world, which many people find uncertain and unpredictable. Fascination by the great possibilities that digital transformation and many other changes have brought, at least for a privileged part of humanity. And overwhelmed by the already long unmanageable and, with digitalization, ever-increasing variety of options in all areas of life.

People each show their own weighting of these reactions and derive from them a variety of behaviours with regard to digital media. Descriptions such as "Generation Y", "Generation Z" or, more recently, "Generation Alpha" are not very helpful in describing this diversity, because the attitude towards digital transformation cannot be determined exclusively by birth cohorts. Terms such as "digital immigrants" or "digital natives" are also too imprecise and are limited to a biographical aspect of access to the digital world. More helpful, however, are differentiating models such as the DIVSI Internet Milieus. Although they do not describe the overall attitude towards digital transformation, they are nevertheless able to describe diversity in terms of several aspects.

The German Institute for Trust and Security on the Internet (Deutsches Institut für Vertrauen und Sicherheit im Internet DIVSI)—which was dissolved at the end of 2018—describes seven Internet milieus from the Internet-distant insecure and the cautious skeptics to the unconcerned hedonists and net enthusiasts to the responsible elite, the efficiency-oriented performers and the sovereign realists. The chart below shows in each case the dominant attitude towards the Internet and the focus of the social situation of the individual Internet milieus (see Fig. 1).

Three very different Internet milieus will be highlighted as examples.

The so-called "Internet distant insecure" are strongly overstrained few-users or off-liners. They are rather helpless in the face of the net and therefore avoid it frequently. They virtually never participate actively in the Internet. They delegate the responsibility for security on the net to the state and to companies. Shopping and social contacts take place offline; documents are read in paper form. In addition, even navigation is often still done analogously with travel guides, maps and road atlases.

The "carefree hedonists", on the other hand, participate cluelessly in the many online opportunities. They tend to handle data protection issues rather carelessly, even out of uncertainty and fatalism. They operate only very limited security measures and tend to see other users in the obligation to be liable for possible damages. Shopping and social contacts are largely online. Instead of analogue text, content is preferably consumed via audio or video. Whether and who collects and stores their location or other data is of little concern to them.

The "Sovereign Realists" are unexcited intensive users with an enthusiasm for the Internet, but a distance to social networks. They are very confident in dealing with

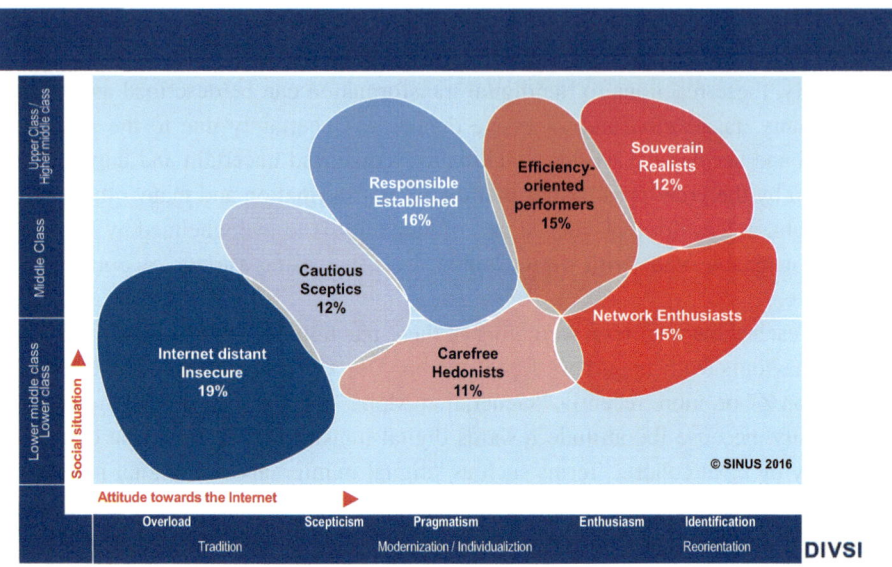

Fig. 1 Internet Milieus 2016. Source: DIVSI Internet Milieus © SINUS 2016

security issues. And they see the responsibility for security least of all with the state. Consumption and networking take place in an individual mix of online and offline activities. Content is searched for in a targeted manner and recorded in different forms. Open Street Map is more suited to one's own usage behavior than Google Maps (for other articles including and supporting open geodata in this volume, see: Edler et al. 2020a, b, Kleber et al. 2020; Meyer-Heß 2020; Siepmann et al. 2020; Stratmann et al. 2020; Vetter 2020), but one can also (re)get excited about the haptic experience of navigating with compass and map.

The four other Internet milieus also have their own independent profile. It is becoming apparent that among younger users the proportion of critical or at least less enthusiastic Internet users is increasing, as shown in the latest DIVSI study entitled "Euphoria was yesterday. The 'Internet generation' between happiness and dependence" (DIVSI 2018a, b).

The variety of changes and challenges resulting from the digital transformation is thus also confronted by a great variety of individual and/or life-world-influenced strategies for dealing with the Internet and the other aspects of digitization. Despite the economic supremacy of a few international corporations in the digital world, this means that the coming developments can to a certain extent be shaped socially and politically. However, this will only be the case if a greater awareness of the potential changes associated with the digital transformation prevails in broad sections of the population and internationally.

5   (Almost) No Place Outside the Digital Transformation

Since January 2019, the gadget blog Gizmodo features an article titled "How Cartographers for the U.S. Military Inadvertently Created a House of Horrors in South Africa" (Gizmodo 2019). The article describes in detail how the geographic mapping of innumerable IP addresses by the US National Geospatial-Intelligence Agency was applied to the backyard of John and Ann's house in Pretoria—not in the US, but in South Africa!x—led to them being accused of various crimes, such as the theft of iPads, possession of weapons or child abduction, several times a month. The IP address is the address of a device in computer networks and makes devices addressable or accessible. Every device that we use to access the Internet has such an IP address.

The example of John and Ann mentioned here only briefly at the end of the article is certainly an unusual but particularly impressive example of how the digital transformation can hit people unexpectedly and (at first) powerlessly. Ann and John have found a way to free their house from the assignment of IP addresses. Nevertheless, this required the support of a few experienced experts—and their networking via the Internet. There may be selected places without internet. A nice example is described in the article "The Land Where The Internet Ends" (New York Times 2019). However, even in the lives of the people in this place, the digital transformation is not meaningless. Because the digital transformation encompasses business, science, politics and society, it encompasses the lives of all people. The shaping of the digital transformation will therefore not succeed by frantically searching for the last "non-digital places" on this globe. Rather, it will be necessary to use one's own creative scope within the digital VUCA world with visions, understanding, clarity and agility. These are at least the four principles of a possible strategy to live with the VUACA paradigm.

References

Baecker, D. (2007). *Studien zur nächsten Gesellschaft*. Berlin: Suhrkamp.

Deutsches Institut für Vertrauen und Sicherheit im Internet DIVSI (2018a): *DIVSI Internet-Milieus 2016. Die digitalisierte Gesellschaft in Bewegung*. Hamburg: DIVSI.

Deutsches Institut für Vertrauen und Sicherheit im Internet DIVSI (2018b): *DIVSI-U25-Studie. Euphorie war gestern. Die „Generation Internet" zwischen Glück und Abhängigkeit*. Hamburg: DIVSI.

Edler, D., Jenal, C., & Kühne, O. (2020a). Modern approaches to the visualization of landscapes—An introduction. In D. Edler, C. Jenal, & O. Kühne (Eds.), *Modern approaches to the visualization of landscapes* (pp. 3–15). Wiesbaden: Springer VS.

Edler, D., Keil, J., & Dickmann, F. (2020b). From Na Pali to Earth—An 'Unreal' engine for modern geodata? In D. Edler, C. Jenal, & O. Kühne (Eds.), *Modern approaches to the visualization of landscapes* (pp. 279–291). Wiesbaden: Springer VS.

Gizmodo (2019). *Kashmir Hill: How cartographers for the U.S. military inadvertently created a house of horrors in South Africa*. Retrieved on 23 Dec 2019 from https://gizmodo.com/how-cartographers-for-the-u-s-military-inadvertently-c-1830758394.

Harari, Y. N. (2017). *Homo Deus. Eine Geschichte von morgen*. München: C. H. Beck.

Kleber, A., Edler, D., & Dickmann, F. (2020). Cartography and the sea: A javascript-based web mapping application for managing maritime shipping. In D. Edler, C. Jenal, & O. Kühne (Eds.), *Modern approaches to the visualization of landscapes* (pp. 173–186). Wiesbaden: Springer VS.

Kobek, J. (2016). *Ich hasse dieses Internet. Ein nützlicher Roman*. Berlin: S. Fischer.

Kühne, O. (2017). *Landschaft und Wandel: Zur Veränderlichkeit von Wahrnehmungen*. Wiesbaden: Springer. VS.

Kühne, O. (2019). *Landscape theories: A brief introduction*. Wiesbaden: Springer VS.

Kühne, O., & Jenal, C. (2020). The threefold landscape dynamics—Basic considerations, conflicts and potentials of virtual landscape research. In D. Edler, C. Jenal, & O. Kühne (Eds.), *Modern approaches to the visualization of landscapes* (pp. 389–402). Wiesbaden: Springer VS.

Lanier, J. (2018). *10 Gründe, warum du deine Social Media Accounts sofort löschen musst*. Hamburg: Hoffmann und Campe.

Meyer-Heß, F. (2020). Discovering forgotten landscapes. In D. Edler, C. Jenal, & O. Kühne (Eds.), *Modern approaches to the visualization of landscapes* (pp. 33–45). Wiesbaden: Springer VS.

New York Times (2019). Pagan Kennedy: The land where the internet ends. Retrieved on 23 Dec 2019 from https://www.nytimes.com/2019/06/21/opinion/sunday/wifi-wilderness-privacy-reserves.html?utm_source=pocket&utm_medium=email&utm_campaign=pockethits.

Newport, C. (2019). *Digitaler Minimalismus*. München: Redline Verlag.

re:publica (2019). *Mikael Colville-Andersen: Back to the future in urban design*. Retrieved on 22 Dec 2019 from https://www.youtube.com/watch?v=luzA74TgJI8.

Reckwitz, A. (2017). *Die Gesellschaft der Singularitäten: Zum Strukturwandel der Moderne*. Berlin: Suhrkamp.

Rosa, H. (2013). *Beschleunigung und Entfremdung*. Berlin: Suhrkamp.

Rosa, H. (2019). *Resonanz. Eine Soziologie der Weltbeziehung*. Berlin: Suhrkamp.

Siepmann, N., Edler, D., & Kühne, O. (2020). Soundscapes in cartographic media. In D. Edler, C. Jenal, & O. Kühne (Eds.), *Modern approaches to the visualization of landscapes* (pp. 247–263). Wiesbaden: Springer VS.

Stratmann, J., Ristea, A., Leitner, M., & Paulus, G. (2020). Exploring urban "Blightscapes" applying spatial video technology and geographic information system: A case study from Baton Rouge, USA. In D. Edler, C. Jenal, & O. Kühne (Eds.), *Modern approaches to the visualization of landscapes* (pp. 499–517). Wiesbaden: Springer VS.

Taleb, N. N. (2018). *Der Schwarze Schwan: Die Macht höchst unwahrscheinlicher Ereignisse*. München: Pantheon.

Vetter, M. (2020). Technical potentials for the visualization in virtual reality. In D. Edler, C. Jenal, & O. Kühne (Eds.), *Modern approaches to the visualization of landscapes* (p. 307–317). Wiesbaden: Springer VS.

Peter Martin Thomas is a certified and experiential educator, teacher for systemic consulting, coaching, supervision and organizational development (DGSF) as well as professional speaker GSA (SHB). He is Consultancy Partner of the SINUS Institute in Heidelberg and partner of the praxis institut für systemische beratung süd in Hanau.Peter Martin Thomas has been advising and accompanying organizations and companies on young people and other target groups of the future for around 30 years. For some years now, his work has focused on digital transformation and its effects on people's everyday lives, among other things.

He tries to make his own digital life more and more independent from the big companies and trusts in open source software. At the same time, as part of a bi-cultural partnership, he appreciates the benefits of global digitization for a successful social and family life across the globe.

Blurring the Boundaries of Landscape Visualization: Welcome to Fabulous Las Vegas

Florian Weber

Abstract

In recent years 'landscape' has received increasing attention as a processual construct, and one whose basis can be sought not only in the physical-material dimension of reality, but also in the virtual. The postmodern world in any case sees a fluid interchange between these domains. The article is concerned with visualizations of landscape as exemplified *par excellence* in Las Vegas—an urban rural hybrid in Nevada's Mojave Desert that has established itself as the Mecca of the tourist experience, with all the symbols and attributes imagination can bestow. The article shows how specific landscapes are simulated here both outwardly and inwardly, melting utopian and real space to enact artificial worlds. Las Vegas is, in this light, a simulated borderland, back-translating its virtual images into a dimension in which the individual can immerse, swimming across the blurred boundaries of our realities. To analyze landscape in this sense is to embark on a course that leaves traditional categories and apparent certainties increasingly behind.

Keywords

Landscape visualization · Urban rural hybrids · Boundaries · Social construction of landscape · Las Vegas

F. Weber (✉)
Universität des Saarlandes, Saarbrücken, Germany
e-mail: florian.weber@uni-saarland.de

© Springer Fachmedien Wiesbaden GmbH, part of Springer Nature 2020   459
D. Edler et al. (eds.), *Modern Approaches to the Visualization of Landscapes*, RaumFragen: Stadt – Region – Landschaft,
https://doi.org/10.1007/978-3-658-30956-5_26

1   Introduction: Simulated Landscapes

The approach to landscape as a social construct originated in the English-speaking world (e.g. Cosgrove 1984; Duncan 1990), but since the early 2000s it has also steadily been gaining ground in German-speaking academia (Kühne 2005, 2006; for an overview see also Kühne et al. 2018; Kühne 2019). Landscape is in this perspective no longer viewed as a natural construct, a given environmental factor, but as a complex, socially gener-ated category charged with specific cultural connotations (Bruns et al. 2015; Kühne et al. 2019a; see in this volume: Al-Khanbashi 2020; Aschenbrand and Michler 2020; Bellini and Leonardi 2020; Berr 2020; Edler et al. 2020a; Fontaine 2020a, b; Jenal 2020; Kühne 2020; Kühne and Jenal 2020; Linke 2020; Roßmeier 2020; Siepmann et al. 2020). It is hardly surprising, then, that the interpretation and use of landscape is often controver-sial—as visibly demonstrated in the acrimonious disputes about the energy transition and the extraction of raw materials—as well as general infrastructural projects—that have occupied German society in recent years (see e.g. Langer 2018; Leibenath and Otto 2014; Pasqualetti et al. 2002; Weber 2018; Weber et al. 2018). A striking aspect of these conflicts is that, even in basically constructivist perspectives, where landscape is not thought of as simply given, the argumentation generally starts from a material concep-tion of landscape as physically shaped space. And although the increasingly available virtual worlds of the Internet and video games have also been addressed in academic writing (see e.g. Kühne 2019, p. 9; Loda et al. 2020), they have rarely been analyzed—a prominent exception being Fontaine (2017)—despite the fact that precisely these con-structs have a discernible impact on our conception of 'real' landscape.

The researcher embarking on an analysis of virtual reality will inevitably—as evi-denced in the present collection of articles—confront the issue of the real versus the virtual (for an introduction see Edler et al. 2018; see in this volume: Edler et al. 2020b, Fontaine 2020b, Vetter 2020), a specific aspect being the many and various ways in which natural and artificial elements superimpose, blend and interact in the material dimensions of our world. A striking example of this hybrid postmodern transition is the Las Vegas agglomeration, whose "hyperreality" (Baudrillard 1994, p. 22) culminates in Las Vegas Boulevard—the famous 'Strip'. Here, over the years, prototypical visionary landscapes—in many cases imitative of real models—have been erected in bricks and mortar to create what Fontaine (2017) has aptly called 'simulated landscapes'.

The origins of this boom-town are somewhat unusual, lying as it does in "a flat desert landscape with sparse vegetation 600 m above sea level" (Grundmann and Synnatschke 2013, p. 420). But apart from its desert location, it is other qualities that stand out in most descriptions: "To think of Las Vegas is to be inundated with a veritable flood of images: images of a glittering neon city, of grandiose sparkling advertisements, of casino signs gesturing tirelessly in every color of the rainbow, […] and of extrava-gantly designed hotels" (Bieger 2007, p. 37). Las Vegas has become a city of the (illu-minated) signs that define it: If these were removed, Venturi et al. already suggested in

"Learning from Las Vegas" (1972), the city would in its de facto form cease to exist. For Las Vegas is constituted as pure experience, a chain of artificial urban landscapes staged and themed for consumption (Schmid 2006), an 'economy of fascination' (Firat and Dholakia 2003 [1998]) in which "the boundaries between imagined and real space" blur, producing "dimensions in which world and image fade over and into each other" (Bieger 2007, p. 9).

Against the background of these introductory remarks, the article will consider Las Vegas as an example of what might be called the 'floating boundaries' of landscape visualizations. My starting point is twofold: that such visualizations are not confined to virtual reality but also embrace physical space; and—the corollary of this—that the virtual imaginary can beget thoroughly material offspring. This will involve a brief preliminary treatment of some central theoretical aspects of landscape and its conceptions, before moving on to the origins and development of Las Vegas, followed by an analysis of specific materialized simulations of landscapes on the Las Vegas Strip. The article will end with an overview of the argument and of the conclusions that can be drawn from it.

2 Theoretical Background: Landscape as a Construct and Its Simulations

The traditional view of landscape as a given entity (see e.g. Paffen 1973), today widely rejected as "object fetishism" (Duncan 1990, p. 11; Gailing and Leibenath 2012, p. 97), has been generally replaced with acceptance (see Kühne et al. 2018, 2019b) of the insight that landscape is a social construct: "[I]n constructivist approaches the constitutive level of landscape is found in the individual or social construction" (Kühne 2019, p. 16). The cardinal insight of constructivism is that knowledge of the world cannot—in the premodern sense of the word—be objective: its basis does not lie in any 'natural' order (Berger and Luckmann 1966; Gergen 1999). It follows that interpretations of reality are never absolutely fixed and firm (Blumer 1973, p. 81); they develop in "processes of social interchange" (Flick 2007, p. 154). And this is the case, too, with constructs of landscape: they are negotiated and temporarily determined (see Weber 2016). In the individual instance, physical space plays a fundamental role in this process; and with Kühne we can define physical space as "the ordering of material objects in a spatial relationship irrespective of their social or individual designation as landscape" (Kühne 2018, p. 22). On both the individual and social—and hence, too societal—levels concepts of landscape are in this sense rooted in the material dimension (Kühne 2019, p. 9). Another element here is the growth of the Internet, which—along with the boom in video games and, more fundamentally, the digitization of our entire world—has profoundly subjected our constructs of landscape to the impact of virtual reality: "In the context of the development of modern communications media it [external space] increasingly also has virtual components" (Edler et al. 2018, p. 247). The relation between the material and the virtual is, therefore, constitutive of 'landscape' (Kühne 2019, p. 9).

The development in the conception of landscape must be seen in close connection with postmodernity (see in general e.g. Behrens 2008; Dear 2000; Lyotard 1979; see in this volume: Fontaine 2020a, b; Kühne 2020; Papadimitriou 2020), where plurality, deconstruction and hybridization have ousted the benchmarks of univocality and simple dichotomy as constitutive principles of thought (see Welsch 1987). That there should be one (and only one) truth is, after all, diametrically opposed to the basic insight of constructivism—an insight prototypically enacted in Disneyland and Las Vegas, with their radical break from traditional categories of description. Here fragmentation, overlaying, and referential proliferation have become the norm for urban structures: "Postmodernism cultivates […] a conception of the urban fabric as necessarily fragmented, a 'palimpsest' of past forms superimposed upon each other, and a 'collage' of current uses, many of which may be ephemeral" (Harvey 1989, p. 66). In this context hybridity—a decisive analytic concept for postmodern developments—must be understood as a "cultural strategy for mixing and negotiating differences" (Hein 2006, p. 55; see also Weber and Kühne 2017). Applied to landscape, this entails multiple transitions and a wide variety of—especially aesthetic—ascriptions whose increasing plurality is no longer seen as contradictory. Today, it is argued, "[a]lmost all aspects of social life have been aestheticized. This means that visual consumption can occur in many different contexts: shopping, eating and drinking, sport, leisure, education, culture and so on" (Lash and Urry 1994, p. 271; more generally Ipsen 2006). Paradigmatically in Las Vegas, the beautiful and the ugly, the sublime and the quaint, the comical and the kitschy form a continuum of fluid transitions (Kühne 2019, p. 48; see also Kühne and Weber 2019a).

Kitsch might seem a particularly appropriate description for Las Vegas, but to use the term with its received connotations of divergence from the tastes of high culture (Gelfert 2000; Illing 2006) would be erroneous. Postmodernism has relativized kitsch (Kühne 2008), making of it no longer "the false expression of false needs, nor [even] the false expression of true needs; kitsch, in our aesthetically tolerant world, is the true expression of true needs" (Liessmann 2002, p. 26f.). To describe as kitsch the active hyperbole with which Las Vegas meets and fulfills the wishes of its myriad customers (about this more later) would be to fall far short of a satisfactory analysis. What one finds in Las Vegas is something more: to use Fontaine's term, it is the simulated landscapes of postmodernity: "Committed to accept and re-legitimize mixed forms and replicas, the postmodern era offers a favorable climate for simulation. Where landscapes are imitated, specific atmospheres are created, always with an eye to aesthetic norms. Various aspects play a key role in this process: responding to social requirements, sustaining an ideal image, satisfying popular yearnings—such factors are of immense importance in this context" (Fontaine 2017, p. 17).

Here, the construction of landscapes has become a conscious activity, and one—following Kloock and Spahr (2007 [1986], p. 56)—that can be seen as the production of a hybrid, material-virtual-real-surreal hyperreality. For, as Brook puts it: "The 'hyperreality of the simulation' is how Jean Baudrillard characterizes a postmodern world of 'simulacra', in which reality has been reduced to representation" (Brook 2013, p. 96).

The consequence I draw from this is that visualizations of landscape extend not only—as observed in the Introduction—to virtual worlds, but also to material constructs, which can be the very embodiments of virtuality and simulation. Las Vegas, it will be seen, is a pre-eminent example. But, before we get there, a glance at the origins and history of the desert city will set the scene.

3   Historical Overview: Las Vegas—An Urban Rural Hybrid

The area of present-day Las Vegas has been permanently settled only from around the middle of the nineteenth century, when water and good grazing grounds were found there (Grundmann and Synnatschke 2013, p. 420; Hess 1999, p. 1980; Quack 1998, p. 513; Steffens 2014, p. 19). The original settlement grew into a railroad town when a central stop was set up on the San Pedro, Los Angeles and Salt Lake City line (City of Las Vegas 2019, n.p.), and the town of Las Vegas was officially founded in 1905 (Gottdiener et al. 1999, Chap. 1; Hess 1993, Chap. 1). The legalization of gambling by the State of Nevada in 1931 gave the local economy a boost, and in the following years the population grew with the many workers hired for the construction of the Hoover Dam (Lang and Nicholas 2012, p. 498). When completed, that project provided cheap electricity to feed the nascent city's insatiable appetite for air-conditioning and lighting (Grundmann and Synnatschke 2013, p. 420; Steffens 2014, p. 38). But it was not until the 1940s that things really began to take off (Bieger 2007, p. 33f.; City of Las Vegas 2019, n.p.), with a casino and hotel culture developing on a three-and-a-half mile-long stretch of the highway toward Los Angeles, the so-called 'Strip'. Powerfully aided by the surging U.S. automobile culture (Bieger 2007, p. 24; Böker 2013, p. II), this soon crystallized into the center of the present-day urban agglomeration.

El Rancho Vegas, the "first themed resort hotel on the Strip" (City of Las Vegas 2019, n.p.), opened in 1941, bringing "an extravagant blend of Western rusticality and ambient luxury that if only for reasons of space would have been unaffordable downtown" (Bieger 2007, p. 36). In hindsight, this was the starting point for a radical change in the city's profile: "Before the El Rancho the highway was dotted with a handful of scattered casinos, billboards, and gas stations […]. The hotel established a pattern of roadside landmarks, vistas, and signs that broke with the traditions of downtown Las Vegas hotels and realized a vision that would mould the city's current form" (Hess 1993, p. 26).

The Flamingo Hotel followed in 1946, bringing the architecture of Miami and above all Los Angeles—"the glamour of other places" (Bieger 2007, p. 43)—to Las Vegas (Hess 1993, pp. 40ff.), and in the process transforming what had once been a sleepy Western town into a Mecca of gambling, driven by the competition between the casinos to ever new superlatives. These superlatives were expressed in ever bigger and ever brighter neon signs, until "the buildings and the entire urban space were defined after nightfall by light" (Hess 1999, p. 1980) and the city itself became a sign (Venturi et al. 1972). In 1966 Caesars Palace opened, with an actively staged landscape of Roman references (Bieger 2007, p. 114) that stood in marked contrast to the surrounding casinos. This constituted an important new milestone on the way to Circus Circus, whose opening in 1968 became the catalyst for the many and various theme hotels that have subsequently set their mark on the city (Steffens 2014, p. 102).

Moreover, if until the early 1970s Las Vegas was wholly dedicated to gambling (Lang and Nicholas 2012, p. 497), the later decades of the twentieth century saw a broadening of the city's image as a center for trade fairs and congresses and a leisure paradise for families (Martin 1999, p. 1978). The controlled demolition (with high explosives) of the Dunes in 1993 and the Sands in 1996 gained wide media coverage; in their place the Bellagio and the Venetian added "Italian feeling" to the already existent attractions of the Nevada desert (Davis 1999, p. 1990; Quack 1998, p. 516). Today, more than 150,000 hotel rooms on the Strip bear witness to the gigantic dimensions of the Las Vegas tourist industry (Lang and Nicholas 2012, p. 500). According to Hess, this is "a post-industrial city established on the basis of services, tourism, and entertainment for a mass public" (Hess 1999, p. 1980), built from the bare desert within a century. In 1910 the area had a mere 1500 inhabitants (Quack 1998, p. 515); by 2018 Las Vegas City counted almost 650,000 (United States Census Bureau 2018b, n.p.) and the surrounding suburbs account for some 700,000 more (Grundmann and Synnatschke 2013, p. 420). Already in 1995 Clark County, which includes Las Vegas, passed the million mark (Quack 1998, p. 513), and by 2018 its population amounted to 2.2 million (United States Census Bureau 2018a, n.p.). This vast growth rate has given rise to what in urban geographical terms can be called an urban–rural hybrid (Kühne et al. 2016; Kühne and Weber 2019b; Weber and Kühne 2017), a settlement form in which simple differentiations blur (see Fig. 1). And the heart of this development is the Strip.

On the one hand the settlement complex extends ever further out into the desert: "Seemingly endless estates of detached family homes stretch to the horizon in a modular grid of square mile units" (Martin 1999, p. 1978f.; see also Davis 1999, p. 1992); and these flourishing automobile-based suburbs form "sealed-off worlds of their own" (Davis 1999, p. 1995). On the other hand there is a counter-movement: "The realities of aridity, steep mountainsides and large federal land holdings have served to constrain outward growth and produce more densely settled metropolitan areas than in the East" (Lang and Nicholas 2012, p. 505). Scott Brown and Venturi speak here of a turn "from rampant urban expansion to concentration" (1999, p. 1977). Environmental problems of water and energy consumption are particularly evident (Davis 1999), adding to the problematic social developments common to other large cities (Gottdiener et al. 1999). Against this background the Strip with its abundant greenery and artificial lakes stands in surreal contrast—an anthropogenic visualization of landscape as the expression of what human ingenuity can achieve. We can now look at this in greater detail.

Fig. 1 Las Vegas, an urban–rural hybrid (Photos: Florian Weber 2018)

4   A Landscape of Fluid Visualizations

Despite the central importance of its entertainment industry, Las Vegas cannot, from a socio-economic, political, or ecological point of view, be reduced to the densely crowded casinos and hotels of the Strip. These form the nucleus of the city's staging and simulation of landscape, but the Strip itself can only be understood in terms of its interplay with the wider environment. The key factor in the settlement of the south-western USA was the ability to engineer access to water in quantities that, at least for several decades, allowed unrestricted consumption (see especially Bierling 2006; also Kühne and Weber 2019a; Starr 2007). Las Vegas, too, lives from a supply of water abundant enough to create an oasis in the desert—an oasis exponentially transcended on the Strip. This is all many visitors experience of the city: "The desert as a living, inhabited landscape is actually not there; it is no more than the dark, brooding background for [...] a man-made neon Babel" (Davis 1999, p. 1992).

In the middle of Highway 95 stands a colossal sign bearing the slogan 'Welcome to Fabulous Las Vegas'. The message is programmatic of the city's invitation to "enter a different reality" (Bieger 2007, p. 122)—a reality of landscape constructs where boundaries blur, where the outer draws the visitor into an inner, virtual world, or where, as in the 'Venetian' or 'Paris Las Vegas', the two flow seamlessly into each other. That the Strip is experienced not on foot but from the automobile is an aspect of U.S. normality: hence the glittering, oversized billboards crying the city's wares (Scott Brown and Venturi 1999, p. 1974f.; Venturi et al. 1972, p. 9). Urban space is assumed here into the expression of architectonic communication (Venturi et al. 1972, p. 6), or as Bieger puts it: "The increasing compression of the urban landscape of the Strip [...] has engendered an ever greater intensity of visual stimuli by means of which competing hotels seek, in increasingly crowded space, to gain an advantage over each other" (Bieger 2007, p. 123). In the second decade of the twenty-first century this has peaked in façades that are themselves gigantic screens for the projection of a virtual reality (see Fig. 2). Schmid speaks in this context of an "artificial and themed urbanity" (Schmid 2006, p. 348; see also Light 1999; Zukin 1993). Here urban space has itself become virtual.

Again Bieger emphasizes the visual quality of the Las Vegas experience: this is "an urban space constituted by—and primarily interpretable through—its images" (2007, p. 42), a city of "hyperreality" (2007, p. 207). A glance even from the outside at the themed visualizations of landscape created by the Treasure Island, Mirage, Venetian, Bellagio, New York-New York, or Paris hotels and casinos is undoubtedly impressive. Pirate battles and erupting volcanoes are among the spectacular stagings of a parallel world that brings Italy, Paris, and New York to the Nevada desert. Along the Strip "hotel investors have created an artificial world for which, in European entertainment parks, for example, one would have to pay a high entrance fee" (Quack 1998, p. 523).

Themed on Lake Como the Bellagio features a "12-hectare [i.e. almost 30 acre] artificial lake [...] with grandiose computerized water shows" (Braunger 2016, p. 247). Landscape is clearly the central point of reference here, not only in the form of a "generous 4500 m^2 of water in front of the hotel", but also in "the gentle incline leading up to the main entrance, and the luxuriant Mediterranean vegetation of pine trees, oleanders and cypresses, that frames the ensemble. The lake at the foot of the hotel is bordered by charming Tuscan buildings with colorful coach houses, balconies and forecourts, while scattered on the lake are a few well-placed boats. Lake, boats, typical plants, colors, and façades—that is the semiotic material transported and translated here. The image thus created is reminiscent of a well-composed landscape painting, albeit one at whose center,

Fig. 2 Urban landscape as a surface for VR projections (Photos: Florian Weber 2018)

instead of a mountain peak, there towers the main building of a hotel" (Bieger 2007, p. 172). From the point of view of landscape aesthetics, the associations evoked here are of the beautiful and picturesque—by no means only of kitsch.

The dense symbolic language of the New York-New York hotel and casino complex (opened in 1997, see Fig. 3) exemplifies Kühne's concept of "stereotypical landscapes" (2019, p. 136), with a panoramic collage (Fontaine 2017, p. 37) featuring key elements of the East Coast metropolis set around the façade of the hotel like an oversized

Fig. 3 The Bellagio and New York-New York as simulated landscapes (Photos: Florian Weber 2018)

glass-ball souvenir rather than a simple postcard motif (Bieger 2007, p. 167). The "atmosphere of some Manhattan districts", Braunger observes (2016, p. 250), has been recreated with some success, and for Quack (1998, p. 516) a "real miniature city" has been transplanted. From the point of view of landscape atmosphere (see in general Kazig 2007, 2019), it is again the density of the experience that is dominant. Bieger describes the overall scene as "an ever more concentrated vision of urban hyperreality […] in which Paris rubs shoulders with New York, imperial Rome shines in pristine glory, and

Venice is saved from the never-ending threat of inundation" (Bieger 2007, p. 10). Visitors are invited to immerse themselves in a "synthetic parallel world [...], the computer-simulated image of a super-city whose referential reality is manipulated through data processing technologies (cut and paste, cleansing, smoothing etc.) to achieve an immense spatial and temporal density" (Bieger 2007, p. 208). The transition from the material to the virtual is fluid and ongoing; all the more so as the city's visualizations of landscape have found a place in many well-known movies and television series, among them *Casino*, *Leaving Las Vegas*, and *CSI: Vegas*.

Depending on the level of idealization, outer simulation passes seamlessly into inner, where it plays on in the imagination. If, for instance, Caesars Palace could in its early years be understood as history intensified into kitsch, the extensions added to the complex from the late 1980s—of which the Colosseum theater is a prime example—bring a touch of "authentic reproduction" (Bieger 2007, p. 116) that takes the visitor on an imaginary trip to Rome. Themed architecture has here become "architainment" (Klein 2004, p. 11).

An even more striking example of the themed indoor visualization of landscape is in my opinion the Venetian, a hotel-casino resort featuring canals on which singing gondoliers glide beneath blue skies dotted with clouds as if after a cooling rain shower (Fig. 4). Depending on the visitor, in this "American subsidiary of Italy's lagoon city" (Braunger 2016, p. 245) the question of authenticity fades into the background: "Italian *gondolieri* steer [...] Oklahoman lovebirds along an artificial *Canale Grande* between smart boutiques and gourmet restaurants, and their passengers no longer feel the need to embark on a journey to distant Venice" (ibid., p. 240f.). The Venice of the mind in all its variants finds material form in the Venetian, where—no matter whether day or night—indirect lighting and perfectly blended colors create unchanging conditions (Bieger 2007, p. 200). The out-of-doors is brought indoors: boundaries blur and become irrelevant. Easily recognizable elements of the Venetian landscape are recreated in a model landscape that allows of multiple interpretations: for some it may be kitsch, the exaggerated stereotypes of the North American imagination, for others a cool, clean-smelling miniature of the real thing, and for yet others a life-size video game with a VR headset, like the GTA V simulation of Los Angeles.

The hotel landscape has here become a theater (Cosgrove 1984, xxvi; Duncan 1995, p. 415; Goffman 2011 [1959]) in which seemingly familiar motifs are juxtaposed, staged, and interpreted anew (Bieger 2007, p. 159). The result is a hybrid visualization of landscape that evades any attempt at clear and unambiguous reading. Our common cultural system still dictates that only a visit to the "real" Venice or the "real" Paris actually counts; but in a postmodern perspective the densely compacted simulations of Las Vegas are also valid realizations of mental-emotional constructs which are themselves the product of—or at least conditioned by—omnipresent reproductive media. In a constructivist view, taken to its logical conclusion, this parallel landscape is, in its own way, no less "authentic" than the original.

Fig. 4 The illusion of Venice: the Venetian Resort Las Vegas (Photos: Florian Weber 2018)

5   Conclusion: Simulated Landscapes, Blurred Boundaries

What conclusions may, then, be drawn from this article? My basic perspective throughout has been that landscape is a social construct, and I have presented an exemplary instance in which the borderline between the material and virtual dimensions of that construct blur. Our standard view of landscape assumes that it is composed of physical elements in certain spatial relationships. But the impact of television and Internet images and videos, and the concomitant growth in influence of virtual reality has given a slant to this perspective that must increasingly be taken into account in the context of landscape construction processes.

Las Vegas is a striking example of a composite melding of real and utopian space, enhanced by the associations of the surrounding desert, whose harsh climate and sublime austerity meet in that 'neon Babel' with a picturesque and, at its center, actively exaggerated urban simulation. At night the darkness of the desert stands in stark contrast with the city lights, culminating in the floodlit glamour of the Strip. The gondoliers of the Grand Canal, the fountains of Rome, the pirates battling on the oceans, the erupting volcanoes are the "material translation of a virtual computer image" (Bieger 2007, p. 208), a virtual reality (or real virtuality) of simulated landscapes into whose atmosphere the visitor is invited to plunge with the promise of enhanced experience. Here a 'culture of simulation' (Opaschowksi 2000) is the leitmotif of a theatrically staged vision whose material-mental interface is as hybrid as the indoor-outdoor world of the casino-hotels. To borrow the terminology of border studies, the artificial landscapes of Las Vegas are a 'simulated hybrid borderland' where boundaries of the real, unreal, surreal, physical, and virtual interweave (see in general Banerjee and Chen 2013; Blake 2000; Newman 2003, p. 18, 2011, pp. 37ff.; Pavlakovich-Kochi et al. 2004; Zorko 2015).

The postmodern perspective on visualizations of landscape presented here entails an understanding of the analog and digital, the material and virtual, as linked constructs in a single continuum—an understanding readily derivable from the opening slogan 'Welcome to Fabulous Las Vegas', but by no means confined to that example. The argument can certainly be taken further. Any future gain in the outreach of virtual reality and the scope of what Fontaine (2017, p. 117) calls "lived hyperreality" will predictably extend and widen scientific as well as popular constructs and imaginaries of landscape and call for more systematic research. The present article represents a postmodern slant on already existent landscape visualizations and a plea to think of landscape in a more complex way.

References

Aschenbrand, E., & Michler, T. (2020). Linking socio-scientific landscape research with the ecosystem service approach to analyze conflicts about protected area management—The case of the Bavarian Forest National Park. In D. Edler, C. Jenal, & O. Kühne (Eds.), *Modern approaches to the visualization of landscapes* (p. xx). Wiesbaden: Springer VS.

Banerjee, P., & Chen, X. (2013). Living in in-between spaces: A structure-agency analysis of the India–China and India–Bangladesh borderlands. *Cities, 34,* (18–29). https://doi.org/10.1016/j.cities.2012.06.011.

Baudrillard, J. (1994). *Simulacra and simulation.* Ann Arbor: University of Michigan Press.

Behrens, R. (2008). *Postmoderne* (2nd ed.). Hamburg: Europäische Verlagsanstalt.

Bellini, A., & Leonardi, L. (2020). Prato: The social construction of an industrial city facing processes of cultural hybridization. In D. Edler, C. Jenal, & O. Kühne (Eds.), *Modern approaches to the visualization of landscapes* (pp. 549–572). Wiesbaden: Springer VS.

Berger, P. L., & Luckmann, T. (1966). *The social construction of reality. A treatise in the sociology of knowledge.* New York: Anchor books.

Berr, K. (2020). Visuality, Aesthetics and Landscape. For the enlightenment and self-enlightenment of constructivist landscape research. In D. Edler, C. Jenal, & O. Kühne (Eds.), *Modern approaches to the visualization of landscapes* (pp. 189–215). Wiesbaden: Springer VS.

Bieger, L. (2007). *Ästhetik der Immersion. Raum-Erleben zwischen Welt und Bild. Las Vegas, Washington und die White City.* Bielefeld: transcript Verlag.

Bierling, S. (2006). *Kleine Geschichte Kaliforniens* (Beck'sche Reihe, vol. 1702). München: C. H. Beck.

Blake, G. (2000). Borderlands under stress: Some global perspectives. International boundary studies series. In M. Pratt & J. A. Brown (Eds.), *Borderlands under stress* (Vol. 4, pp. 1–16). London: Kluwer Law International.

Blumer, H. (1973). Der methodologische Standort des symbolischen Interaktionismus. In Arbeitsgruppe Bielefelder Soziologen (Ed.), *Alltagswissen, Interaktion und gesellschaftliche Wirklichkeit* (Band 1, pp. 80–146). Reinbek bei Hamburg: Rowohlt.

Braunger, M. (2016). *USA. Der Südwesten* (4th ed.). Ostfildern: DuMont Reiseverlag.

Brook, V. (2013). *Land of smoke and mirrors. A cultural history of Los Angeles.* New Brunswick: Rutgers University Press.

Bruns, D., Kühne, O., Schönwald, A., & Theile, S. (Eds.). (2015). *Landscape culture – Culturing landscapes. The differentiated construction of landscapes.* Wiesbaden: Springer VS.

Böker, P. (2013). Niete oder Jackpot? Auswirkungen des Tourismus auf Las Vegas. *Geographie und Schule, 35* (204), I–IV. Kreative Denk- und Lernaufgaben.

City of Las Vegas. (2019). Las Vegas Then & Now. Retrieved 27 June 2019 from https://www.lasvegasnevada.gov/Visitors/History.

Cosgrove, D. E. (1984). *Social formation and symbolic landscape.* London: University of Wisconsin Press.

Davis, M. (1999). Las Vegas Versus Nature. *StadtBauwelt* (143), 1990–1997.

Dear, M. J. (2000). *The postmodern urban condition.* Malden: Wiley-Blackwell.

Duncan, J. S. (1990). *The city as text: The politics of landscape interpretation in the Kandyan Kingdom.* Cambridge: Cambridge University Press.

Duncan, J. (1995). Landscape geography, 1993–94. *Progress in Human Geography, 19*(3), 414–422. https://doi.org/10.1177/030913259501900308.

Edler, D., Jenal, C., & Kühne, O. (2020a). Modern approaches to the visualization of landscapes—An introduction. In D. Edler, C. Jenal, & O. Kühne (Eds.), *Modern approaches to the visualization of landscapes* (pp. 3–15). Wiesbaden: Springer VS.

Edler, D., Keil, J., & Dickmann, F. (2020b). From Na Pali to Earth—An 'Unreal' engine for modern geodata? In D. Edler, C. Jenal, & O. Kühne (Eds.), *Modern approaches to the visualization of landscapes* (pp. 279–291). Wiesbaden: Springer VS.

Edler, D., Kühne, O., Jenal, C., Vetter, M., & Dickmann, F. (2018). Potenziale der Raumvisualisierung in Virtual Reality (VR) für die sozialkonstruktivistische Landschaftsforschung. *KN – Journal of Cartography and Geographic Information, 68*(5), 245–254.

Firat, A. F., & Dholakia, N. (2003 [1998]). *Consuming people. From political economy to theaters of consumption.* London: Routledge.

Flick, U. (2007). Konstruktivismus. In U. Flick, E. v. Kardorff, & I. Steinke (Eds.), *Qualitative Forschung. Ein Handbuch* (pp. 150–164). Reinbek bei Hamburg: Rowohlt.

Fontaine, D. (2020b). Virtuality and Landscape. In D. Edler, C. Jenal, & O. Kühne (Eds.), *Modern approaches to the visualization of landscapes* (pp. 267–276). Wiesbaden: Springer VS.

Fontaine, D. (2017). *Simulierte Landschaften in der Postmoderne. Reflexionen und Befunde zu Disneyland, Wolfersheim und GTA V*. Wiesbaden: Springer VS.

Fontaine, D. (2020a). Landscape in computer games—The examples of GTA V and Watch Dogs 2. In D. Edler, C. Jenal, & O. Kühne (Eds.), *Modern approaches to the visualization of landscapes* (pp. 293–306). Wiesbaden: Springer VS.

Gailing, L., & Leibenath, M. (2012). Von der Schwierigkeit, „Landschaft" oder „Kulturlandschaft" allgemeingültig zu definieren. *Raumforschung und Raumordnung, 70*(2) 95–106. https://doi.org/10.1007/s13147-011-0129-8.

Gelfert, H.-D. (2000). *Was ist Kitsch?* Göttingen: Vandenhoeck und Ruprecht.

Gergen, K. J. (1999). *An invitation to social construction*. London: Sage.

Goffman, E. (2011 [1959]). *Wir alle spielen Theater. Die Selbstdarstellung im Alltag*. München: Piper.

Gottdiener, M., Collins, C. C., & Dickens, D. R. (1999). *Las Vegas. The social production of an All-American city*. Malden: Blackwell.

Grundmann, H.-R., & Synnatschke, I. (2013). *USA. Der ganze Westen. Das Handbuch für individuelles Entdecken* (19., komplett überarbeitete und erweiterte Auflage). Westerstede: Reise Know-How-Verlag

Harvey, D. (1989). *The condition of postmodernity: An enquiry into the origins of cultural change*. Oxford: Blackwell.

Hein, K. (2006). *Hybride Identitäten. Bastelbiografien im Spannungsverhältnis zwischen Lateinamerika und Europa*. Bielefeld: transcript.

Hess, A. (1993). *Viva Las Vegas. After-hours architecture*. San Francisco: Chronicle Books.

Hess, A. (1999). Eine kurze Geschichte von Las Vegas. *StadtBauwelt* (143), 1980–1987.

Illing, F. (2006). *Kitsch, Kommerz und Kult. Soziologie des schlechten Geschmacks*. Konstanz: UVK Verlagsgesellschaft.

Ipsen, D. (2006). *Ort und Landschaft*. Wiesbaden: VS Verlag für Sozialwissenschaften.

Jenal, C. (2020). Visualizations of 'landscape' in protest movements: On exclusive, inclusive patterns of perception, interpretation using the example of resistance to the expansion of the electricity grid in Germany. In D. Edler, C. Jenal, & O. Kühne (Eds.), *Modern approaches to the visualization of landscapes* (pp. 427–445). Wiesbaden: Springer VS.

Kazig, R. (2019). Atmosphären und Landschaft. In O. Kühne, F. Weber, K. Berr, & C. Jenal (Eds.), *Handbuch Landschaft* (pp. 453–460). Wiesbaden: Springer VS.

Kazig, R. (2007). Atmosphären – Konzept für einen nicht repräsentationellen Zugang zum Raum. In C. Berndt & R. Pütz (Eds.), *Kulturelle Geographien. Zur Beschäftigung mit Raum und Ort nach dem Cultural Turn* (pp. 167–187). Bielefeld: transcript.

Al-Khanbashi, M. (2020). Using matrix as a qualitative data display for landscape research and a reflection based on the social constructivist perspective. In D. Edler, C. Jenal, & O. Kühne (Eds.), *Modern approaches to the visualization of landscapes* (pp. 103–118). Wiesbaden: Springer VS.

Klein, N. M. (2004). *The Vatican to Vegas. A history of special effects*. New York: New Press.

Kloock, D., & Spahr, A. (2007 [1986]). *Medientheorien. Eine Einführung* (UTB). München: Fink.

Kühne, O. (2006). *Landschaft in der Postmoderne. Das Beispiel des Saarlandes*. Wiesbaden: DUV.

Kühne, O. (2018). *Landscape and power in geographical space as a social-aesthetic construct*. Dordrecht: Springer International Publishing.

Kühne, O. (2019). *Landscape theories. A Brief introduction*. Wiesbaden: Springer VS.

Kühne, O. (2020). The social construction of space and landscape in internet videos. In D. Edler, C. Jenal, & O. Kühne (Eds.), *Modern approaches to the visualization of landscapes* (pp. 121–137). Wiesbaden: Springer VS.

Kühne, O., & Jenal, C. (2020). The threefold landscape dynamics—Basic considerations, conflicts and potentials of virtual landscape research. In D. Edler, C. Jenal, & O. Kühne (Eds.), *Modern approaches to the visualization of landscapes* (pp. 389–402). Wiesbaden: Springer VS.

Kühne, O., Weber, F., Berr, K., & Jenal, C. (Eds.). (2019a). *Handbuch Landschaft.* Wiesbaden: Springer VS.

Kühne, O., Weber, F., & Jenal, C. (2019b). Neue Landschaftsgeographie. In O. Kühne, F. Weber, K. Berr, & C. Jenal (Eds.), *Handbuch Landschaft* (pp. 119–134). Wiesbaden: Springer VS.

Kühne, O., & Weber, F. (2019b). Postmoderne Zugriffe und Differenzierungen von Stadt und Land(schaft): Stadtlandhybride, räumliche Pastiches und URFSURBS. In O. Kühne, F. Weber, K. Berr, & C. Jenal (Eds.), *Handbuch Landschaft* (pp. 755–770). Wiesbaden: Springer VS.

Kühne, O. (2005). *Landschaft als Konstrukt und die Fragwürdigkeit der Grundlagen der konservierenden Landschaftserhaltung – Eine konstruktivistisch-systemtheoretische Betrachtung. 2005.* Wien: Selbstverlag.

Kühne, O. (2008). Landschaft und Kitsch – Anmerkungen zu impliziten und expliziten Landschaftsvorstellungen. *Naturschutz und Landschaftsplanung, 44,* (12), 403–408.

Kühne, O., & Weber, F. (2019a). *Hybrid California. Annäherungen an den Golden State, seine Entwicklungen, Ästhetisierungen und Inszenierungen.* Wiesbaden: Springer VS.

Kühne, O., Schönwald, A., & Weber, F. (2016). Urban/Rural hybrids: The urbanisation of former suburbs (URFSURBS). *Quaestiones Geographicae, 35*(4), 23–34. https://doi.org/10.1515/quageo-2016-0032.

Kühne, O., Weber, F., & Jenal, C. (2018). *Neue Landschaftsgeographie. Ein Überblick* (Essentials). Wiesbaden: Springer VS.

Lang, R. E., & Nicholas, C. (2012). Las Vegas: More than a one-dimensional world city? In B. Derudder, M. Hoyler, P. J. Taylor, & F. Witlox (Eds.), *International handbook of globalization and world cities* (pp. 497–507). Cheltenham: Edward Elgar Publishing.

Langer, K. (2018). Frühzeitige Planungskommunikation – ein Schlüssel zur Konfliktbewältigung bei der Energiewende? In O. Kühne & F. Weber (Eds.), *Bausteine der Energiewende* (pp. 539–556). Wiesbaden: Springer VS.

Lash, S., & Urry, J. (1994). *Economies of signs and space.* London: Sage.

Leibenath, M., & Otto, A. (2014). Competing wind energy discourses, contested landscapes. *Landscape Online* (38), 1–18. https://doi.org/10.3097/LO.201438.

Liessmann, K. P. (2002). *Kitsch! oder Warum der schlechte Geschmack der eigentlich gute ist.* Wien: Brandstätter.

Light, J. S. (1999). From city space to cyberspace. In M. Crang, P. Crang, & J. May (Eds.), *Virtual geographies. Bodies, space and relations* (pp. 109–130). London: Routledge.

Linke, S. (2020). Landscape in internet pictures. In D. Edler, C. Jenal, & O. Kühne (Eds.), *Modern approaches to the visualization of landscapes* (pp. 139–156). Wiesbaden: Springer VS.

Loda, M., Kühne, O., & Puttilli, M. (2020). The social construction of Tuscany in the German and English speaking world—Presented by the analysis of internet images. In D. Edler, C. Jenal, & O. Kühne (Eds.), *Modern approaches to the visualization of landscapes* (pp. 157–171). Wiesbaden: Springer VS.

Lyotard, J.-F. (1979). *La condition postmoderne. Rapport sur le savoir.* Paris: Les Éditions de Minuit.

Martin, V. (1999). Städtebauliche Daten zu Las Vegas. *StadtBauwelt, (143,* 1978–1979).

Newman, D. (2011). Contemporary research agendas in border studies: An overview. In D. WastlWalter (Ed.), *The Ashgate research companion to border studies* (pp. 33–47). Farnham: Ashgate.

Newman, D. (2003). On borders and power: A theoretical framework. *Journal of Borderlands Studies, 18*(1), 13–25. https://doi.org/10.1080/08865655.2003.9695598.

Opaschowksi, H. (2000). Kathedralen und Ikonen des 21. Jahrhunderts: zur Faszination von Erlebniswelten. In A. Steinecke (Ed.), *Erlebnis- und Konsumwelten* (pp. 44–54). München: Oldenbourg.

Paffen, K. (Ed.). (1973). *Das Wesen der Landschaft*. Darmstadt: WBG.

Papadimitriou, F. (2020). Visualization of future landscapes, postmodern cinema and geographical education. In D. Edler, C. Jenal, & O. Kühne (Eds.), *Modern approaches to the visualization of landscapes* (pp. 351–369). Wiesbaden: Springer VS.

Pasqualetti, M. J., Gipe, P., & Righter, R. W. (Eds.). (2002). *Wind power in view: Energy landscapes in a crowded world*. San Diego: Academic Press.

Pavlakovichx, V., Morehouse, B. J., & Wastl-Walter, D. (Eds.). (2004). *Challenged borderlands. Transcending political and cultural boundaries*. Aldershot: Ashgate.

Quack, U. (1998). *Kalifornien. Reiseführer*. Dormagen: Iwanoski GmbH.

Roßmeier, A. (2020). Urban/Rural hybridity in pictures. The creation of neighborhood images using the example of San Diego's urbanizing inner-ring suburbs East Village and Barrio Logan. In D. Edler, C. Jenal, & O. Kühne (Eds.), *Modern approaches to the visualization of landscapes* (pp. 479–498). Wiesbaden: Springer VS.

Schmid, H. (2006). Economy of fascination: Dubai and Las Vegas as examples of themed urban landscapes. *Erdkunde, 60*, (4, 346–361).

Scott Brown, D., & Venturi, R. (1999). Las Vegas heute. *StadtBauwelt* (143), 1974–1977.

Siepmann, N., Edler, D., & Kühne, O. (2020). Soundscapes in cartographic media. In D. Edler, C. Jenal, & O. Kühne (Eds.), *Modern approaches to the visualization of landscapes* (pp. 247–263). Wiesbaden: Springer VS.

Starr, K. (2007). *California. A history* (A Modern Library chronicles book, vol. 23). New York: Modern Library.

Steffens, D. (2014). *Nichts bleibt für die Ewigkeit. Wie sich die amerikanische Entertainment-City Las Vegas immer wieder neu erfindet*. Hamburg: disserta Verlag.

United States Census Bureau. (2018a). Quick Facts. Clark County, Nevada. Retrieved 30 June 2019 from http://www.census.gov/quickfacts/clarkcountynevada.

United States Census Bureau. (2018b). QuickFacts. Las Vegas city, Nevada. Retrieved 27 June 2019 from https://www.census.gov/quickfacts/lasvegascitynevada.

Venturi, R., Scott Brown, D., & Izenour, S. (1972). *Learning from Las Vegas*. Cambridge: MIT Press.

Vetter, M. (2020). Technical potentials for the visualization in virtual reality. In D. Edler, C. Jenal, & O. Kühne (Eds.), *Modern approaches to the visualization of landscapes* (p. 307–317). Wiesbaden: Springer VS.

Weber, F. (2018). *Konflikte um die Energiewende. Vom Diskurs zur Praxis*. Wiesbaden: Springer VS.

Weber, F., Kühne, O., Jenal, C., Aschenbrand, E., & Artuković, A. (2018). *Sand im Getriebe. Aushandlungsprozesse um die Gewinnung mineralischer Rohstoffe aus konflikttheoretischer Perspektive nach Ralf Dahrendorf*. Wiesbaden: Springer VS.

Weber, F. (2016). The Potential of Discourse Theory for Landscape Research. *Dissertations of Cultural Landscape Commission, (31*, 87–102). Retrieved 30 Aug 2017 from http://www.krajo-braz.kulturowy.us.edu.pl/publikacje.artykuly/31/6.weber.pdf.

Weber, F., & Kühne, O. (2017). Hybrid suburbia: New research perspectives in France and Southern California. *Quaestiones Geographicae, 36*(4), 17–28. https://doi.org/10.1515/quageo-2017-0033.

Welsch, W. (1987). *Unsere postmoderne Moderne*. Weinheim: VCH Acta Humaniora.

Zorko, M. (2015). The construction of socio-spatial identities alongside the Schengen border: Bordering and border-crossing processes in the Croatian-Slovenian Borderlands. Border regions series. In C. Brambilla, J. Laine, J. W. Scott, & G. Bocchi (Eds.), *Borderscaping: Imaginations and practices of border making* (pp. 97–107). Burlington: Ashgate.

Zukin, S. (1993). *Landscapes of pssower: From Detroit to Disney World.* Berkeley: University of California Press.

Florian Weber studied geography, business administration, sociology and journalism at the University of Mainz and gained his doctorate at the University of Erlangen-Nuremberg with a thesis comparing German and French area-based politics in light of discourse theory. After working from 2012-2013 as a project manager in Würzburg, he took up an appointment as researcher and project coordinator in the transregional university cooperation UniGR at Kaiserslautern University of Technology and after that Weihenstephan-Triesdorf University of Applied Sciences. Since October 2016 he has been Associate Professor (Akademischer Rat) at the University of Tübingen, where he completed his post-doctoral degree (Habilitation) in 2018. In April 2019 he was appointed Junior Professor of European Studies at Saarland University, with special reference to Western Europe and border regions.

Urban/Rural Hybridity in Pictures. The Creation of Neighborhood Images Using the Example of San Diego's Urbanizing Inner-Ring Suburbs East Village and Barrio Logan

Albert Roßmeier

Abstract

As extensive privately driven development efforts in Downtown San Diego started to cross its southeastern boundaries in the early 2000s, the former industrial sites and inner-ring suburbs East Village and Barrio Logan were facing fundamental urbanization and infill processes. Within these developments functional restructuring, new uses and reutilizations, symbolic charges and staging took place. In this context the present article uses a poststructuralist, discourse theoretical-oriented *Google* image analysis to examine the question to which extent the medial, pictorial representations of the neighborhoods East Village and Barrio Logan are characterized by similar motifs and thus lead to discursive determinations of meaning and the creation of neighborhood images. The analysis revealed clearly different medial representations of the two neighborhoods which testify to regularities, recurring argumentations, and certain breaks, as well as heterogeneities: While the pictorial representations of East Village strengthen a rather clear image of an upmarket, urban environment for young professionals, Barrio Logan obtains a plural and diverse image of drastic upheavals in a pastiche-like, hybrid manner.

Keywords

Urban/rural hybridity · Inner-ring suburbs · Urban infill · Neighborhood · Image · Picture analysis · Discourse theory · San Diego

A. Roßmeier (✉)
Bad Füssing, Germany
e-mail: albert.rossmeier@uni-tuebingen.de

© Springer Fachmedien Wiesbaden GmbH, part of Springer Nature 2020   477
D. Edler et al. (eds.), *Modern Approaches to the Visualization of Landscapes*, RaumFragen: Stadt – Region – Landschaft,
https://doi.org/10.1007/978-3-658-30956-5_27

1   Introduction: A Discourse Theoretical-Oriented Picture Analysis of the Inner-Ring Neighborhoods East Village and Barrio Logan in San Diego

Downtown San Diego and adjacent neighborhoods have been affected by extensive reurbanization and restructuring measures for several years (Appleyard and Stepner 2018; Kayzar 2006; Kühne and Schönwald 2015a; Kühne et al. 2016; Roßmeier 2019; Weber and Kühne 2017). Starting with the rebuilding measures in the historic Gaslamp Quarter in the 1970s (Comer-Schultz 2011; Costello et al. 2003; Eddy 1995; Ervin 2007), the development efforts have increasingly moved east to East Village and south east to Barrio Logan since the early 2000s (Delgado and Swanson 2019; Kühne and Schönwald 2015b; Roßmeier 2019; Rumpf 2016). In particular, the construction of Petco Park in East Village – Downtown's baseball stadium opened in 2004 – acted as a catalyst project for downtown redevelopment, leading to ancillary development, increased property values, and the influx of higher income households in the urban core (Erie et al. 2010). In this course, San Diego, which has been fiscally strapped throughout the late 20th and early 21st centuries due to its limiting tax policy (Erie et al. 2011; Kühne and Schönwald 2015b), turned to a public-private partnership for its inner-city redevelopment plans, which has been "one of the largest redevelopment projects in North America [in the early 2000s and thus] has encouraged more than $1 billion in private investment" (Erie et al. 2010, p. 645). Fundamental restructuring and extensive densification took place in the former industrial, rebranded East Village, largely without urban planning control as a result of reduced administrative expenses (Kühne and Schönwald 2015b). Accordingly, the developments have been spilling over the borders of downtown San Diego towards other inner-ring neighborhoods, especially to the Mexican-American community neighborhood Barrio Logan, which "has recently been re-codified as an 'up-and-coming' artistic enclave offering 'authentic' cultural experiences for adventurous urbanites" (Delgado and Swanson 2019, p. 13). Thus, the downtown redevelopment entails urban infill projects, attracts new uses and user groups and therefore initiates the urbanization of the former suburban inner-ring (urfsurbs) as it is being traced for other US and European metropolises (Charles 2013; Kühne 2016; Kühne and Schönwald 2015a; Kühne et al. 2016; Kühne and Weber 2019; Markley 2018; Roßmeier 2019; Sweeney and Hanlon 2017; Weber 2019b; Weber and Kühne 2017).

In this process, attempts are being made to create an image and to reshape or reorient the inner-city neighborhoods of San Diego and their visions (Gaslamp Quarter Association 2019; Kayzar 2006; San Diego Tourism Authority 2019a, b, c). In addition, recent community plan updates serve to adapt land use and assign new functions creating designated urban neighborhoods out of industrial sites and port structures (CCDC 2015; City of San Diego 2013, 2015, 2019; Rumpf 2016) and thus turning East Village into "Downtown San Diego's eclectic hipster neighborhood and another hotspot for tech and innovation startups" (DSDP 2016, p. 9). These landscape or urban/rural hybrid[1] upheavals need to be accompanied scientifically in order to be able to work out how the social perception of the different neighborhoods and individual associations change and what role medial representations can play here (cf. Weber 2019a, p. 111).

Accordingly, this article presents the results of a qualitative poststructuralist, discourse theoretical-oriented *Google* picture analysis which is dedicated to the medial representation of urban/rural hybrid spaces specifically the neighborhoods East Village and Barrio Logan, and investigates the extent to which similar motifs and regularities in the pictorial elements can be identified. Following a constructivist, discourse theoretical premise in the tradition of Laclau and Mouffe (1985) no supposedly objective realities will be worked out but rather it will be traced which moments are regularly linked to the respective nodal points East Village and Barrio Logan and which elements are not (re) produced in each case. For it is assumed that in the course of recurring combinations pictorial representations can break up social realities, temporarily anchor them, and push alternative interpretations into the background (cf. Glasze 2013, p. 119; Jørgensen and Phillips 2002, p. 33; Mattissek 2010; Mattissek and Reuber 2004, p. 229; Weber 2018, pp. 39, 132).

In the following, the research perspective (Sect. 2.1) and the methodological approach (Sect. 2.2) of the picture analysis will be explained further. The next section is introduced with comments on the revitalization efforts in downtown San Diego starting in the 1970s (Sect. 3.1). Then, the more recent lines of development of the neighborhoods East Village and Barrio Logan are traced, which can be read as the urbanization of the formerly industrial and suburban inner-ring. Within the framework of the picture analysis special attention is paid in accordance with the underlying poststructuralist perspective to *what extent* regularities and similar motifs can be identified in the medial representation of the neighborhoods East Village (Sect. 3.2) and Barrio Logan (Sect. 3.3) and "*how* meaning and thus social reality is constituted in pictures"[2] (Glasze 2013, p. 119). For it can be assumed that the social perception and thus the images of the neighborhoods are adapting as a result of similar and regularly occurring arguments and combinations of pictorial elements. To conclude, a short summary of the analysis and an outlook on possible starting points for further research is given (Sect. 4).

---

[1]In the present article it is rather spoken of *urban/rural hybrids* than of *landscape* in general in order to take into account the differentiation, fragmentation and complexity of today's urban-rural transitions (see fundamentally Kühne 2012, 2016; Kühne et al. 2016, 2017).

[2]In this article, German and non-English citations will be translated without the original text parts being given.

2   Theoretical Perspective and Methodological Approach

In order to illuminate processes of how certain patterns (re)produce themselves, a discourse-theoretical research perspective (Glasze 2013; Laclau and Mouffe 1985) as a variant of constructivist-oriented science is taken as a basis (Berger and Luckmann 1966; Kühne et al. 2019, pp. 21–24; Weber 2015, 2018), as it will be explained in the following.

2.1   Discourse Theoretical Approach

Laclau and Mouffe conceived a concept of discourse that oscillates in the field of tension between, on the one hand, structures that are never finally completed and, on the other, relationships that seem to be firmly anchored in everyday life (Laclau 1993, p. 435). For them "any discourse is constituted as an attempt to dominate the field of discursivity, to arrest the flow of differences, to construct a centre" (Laclau and Mouffe 1985, p. 112) and to reduce possibilities by excluding other meanings. Accordingly, within discourses meaning is repeatedly renegotiated and produced anew (Torfing 1999, p. 40) and in this way temporarily provided with specific interpretations (Kühne et al. 2019, pp. 21–24; Laclau 2007, p. 69). Therefore, discourses can be understood as *temporary* fixations of meaning, whereby elements are placed in relation to and delimited from each other and thus their identity is changed and stabilized (Glasze 2015, p. 25; Laclau and Mouffe 1985, p. 112). This assumption is based on the rejection of the structuralist understanding of "language as a stable, unchangeable and totalising structure" (Jørgensen and Phillips 2002, p. 10; Torfing 1999, p. 4) and is referred to as poststructuralist approach (Gailing and Leibenath 2015, p. 126; Weber 2019a, p. 107).

Central work in this field comes from Barthes (2007 [1970]) and Derrida (1999 [1972]). Barthes particularly emphasizes the ambiguity of texts through the plurality of signifiers (2007 [1970]). According to Derrida (1999 [1972]), meaning can never be definitively determined, since "each signifier always refers to different preceding or subsequent signifiers" (Glasze 2013, p. 68). This *mutability* of meaning entails the impossibility of a "comprehensive, fixed social structure" (Glasze 2013, p. 74) which also has consequences for space and thus for urban/rural hybrids and their perception. Building on the premises of social constructivism, Laclau and Mouffe understand "human reality as socially constructed and articulated in discourse" (Glynos and Stavrakakis 2004, p. 203; see in this volume: Al-Khanbashi 2020, Aschenbrand and Michler 2020, Berr 2020, Edler et al. 2020, Fontaine 2020a, b, Kühne 2020, Kühne and Jenal 2020, Linke 2020, Loda et al. 2020, Siepmann et al. 2020, Weber 2020). They do not limit the concept of discourse to language or text, but conceive the entire social reality as discursive (Glasze 2013, p. 67; Laclau 2007, p. 68), thus also visual elements as pictures, videos and films (Weber 2015, pp. 107–108).

In contrast to Foucault's concept of discourse, Laclau and Mouffe make no distinction between discursive and non-discursive objects, thus ultimately emphasizing the interweaving of language, practices, material and social (Glaze and Mattissek 2009, p. 12; Jørgensen and Phillips 2002, p. 19; Laclau 2007, p. 249; Weber 2018, p. 22), because for them "any social reality is always a discursive reality" (Gailing and Leibenath 2015, p. 131). This opens up the possibility of developing discourse analytical methods and approaches with a spatial reference, since the present discourse concept – in an anti-essentialist way (Gailing and Leibenath 2015; cf. Weber 2019a, p. 107) – also conceives space, and therefore urban/rural hybrids, as linguistically conveyed and therefore individually constructed (Leibenath and Otto 2012, p. 121), as it will be further illustrated below.

2.2 Qualitative Discourse Theoretical-Oriented Picture Analysis

Analytically, in the present example, regularities and heterogeneities in medial, pictorial representations come into focus. It is fundamental to note that "photographs, pictures and films (…) are omnipresent today and (…) impressions and attitudes have a decisive influence. They play a central role in the constitution of social-space relations" (Weber 2015, p. 107) because visual units are verbalized in the process of perception and thus ultimately can be analyzed (cf. Miggelbrink 2009, p. 180; Weber 2018, p. 132). In the following, with the help of the "meta-theoretical approach" (Kühne 2014, p. 79) of the discourse concept of Laclau and Mouffe (1985), which is simultaneously applied to individual linguistic signs, it will be worked out how specific patterns of interpretation develop. In the poststructuralist tradition, spaces, urban/rural hybrids, or neighborhoods are not understood as "physical conditions or sensory impressions, but [as] signifiers that are connected to other signifiers by relations of difference and thus acquire meaning. (…) These meanings are not stable, however, but always fragile and unstable" (Leibenath and Otto 2012, p. 121; cf. Weber 2018, p. 68).

Accordingly, the present analysis of the representations of neighborhoods and quarters assumes the changeability of meaning and focuses on regularities and certain breaks, so-called "dislocations" (Weber 2019a, p. 107). For it is assumed that neighborhood images do not prevail as fixed meanings in society, but they are socially constructed. Neighborly interpretations are thus subject to constant processes of change, the individual attributions may break up and get reassembled in the wake of signifiers that are strung together in different ways (Gailing and Leibenath 2015; Torfing 1999). Time and again, temporary fixations of meaning emerge, "which *seem* ‚irrevocable'–to be understood as hegemonic discourses" (Weber et al. 2017, p. 216)–and push alternative interpretations into the background. Thereby discourses, here neighborly images, gain power, in particular also by demarcation from a discursive outside, that is, from what they are *not*. But this outside may also strengthen, "whereby it becomes a constitutive outside: It is on the one hand identity-forming, but at the same time also identity threatening (…).

This interplay comes into focus in discourse-theoretical analyses, or becomes a guiding principle of analysis" (Weber 2019a, p. 107), as it is in the following.

Thus, within the present picture analysis, it is necessary to examine which moments (e.g. physical-spatial elements such as new apartment buildings) connect to the nodal points East Village and Barrio Logan and which are not or no longer linked with them, in order to draw conclusions about current hegemonies and thus about the constitution of neighborly images. It is not a matter of tracing the representation of (individual) pictures, but rather of working out the "recurring argumentation logics" (Stakelbeck and Weber 2013, p. 239) in a multitude of pictures from the pictorial elements they contain, i.e. which moments are comprehensively recognizable and which elements are not (re) produced or even excluded (cf. Kühne et al. 2013, p. 42). It is of central interest what is regularly represented in a comprehensive way, what is put into relation with each other and thus constitutes meaning.

For this purpose, the first 200 pictures of a *Google* picture search with the keywords "East Village" and "Barrio Logan" have been saved and combined into two corpora with 100 pictures each. The pictures were individually systematized according to regularly occurring, recognizable moments such as new apartment buildings in postmodern, urban cubature or industrial structures or land uses. An inductive approach was used, the element categories were developed and conceived from the visual material in order to be able to draw conclusions about current hegemonies in the discourse around the neighbourhoods. Subsequently, two visualizations of the nodal points East Village and Barrio Logan and the respective recurring moments were created.

3   San Diego's Urbanizing Inner-Ring – A Postmodern Pastiche Between Differentiation, Polarization and Economization

Following the theoretical orientation as well as the explanations on the methodology, the empirical results of the study are now presented. With the help of the *Google* picture analysis, it will be traced to what extent certain regularities, recurring argumentation logics, and similar motifs occur in the medial, pictorial representations of the neighborhoods East Village and Barrio Logan. For it can be assumed that in addition to the social, functional, and structural transformation processes in the neighborhoods, the social perception and interpretations of the neighborhoods, and thus ultimately their images are constantly changing and adapting with the media also playing a major role.

3.1   Downtown San Diego's Spillover Into the Next Ring: The Urbanization of East Village and Barrio Logan

Since its founding as 'New Town' in 1867, downtown San Diego went through several boom and bust phases, from times of prospering real estate industry and market collapses in its early years, to the Panama-California Exposition upswing after 1915, to a declining place of urban problems and poverty–especially in the post-war period, to a segregative attraction stage for more affluent parts of the population. "What began as a city improvement project to refurbish San Diego's historic town plaza, mushroomed into one of the major downtown redevelopment plans of the 1970s and 1980s" (Eddy 1995, n.pag.). Under the slogan "*America's Finest City*," the 1970s renovations of the historic Gaslamp Quarter and the construction of the downtown shopping center Horton Plaza marked the 'revitalization' of the downtown area by parts of the City Council, then Republican Mayor Pete Wilson and developer Ernest W. Hahn, through which the city entered a new threshold of self-staging and location marketing (Comer-Schultz 2011; Costello et al. 2003; Eddy 1995; Ervin 2007). "More than most cities, San Diego has embraced redevelopment P3s [public-private partnerships] as an integral downtown revitalization strategy. (…) Horton Plaza, an outdoor mall based on an Italian hill design, was completed in 1985, at a cost of $140 million, including $40 million in public funds and a timely loan from the state. In addition to designing the shopping center, Hahn pushed local officials to build new housing and a convention center nearby and improve downtown public transit" (Erie et al. 2010, p. 655). In this process, particularly in the 21st century with the major project Petco Park, also built as a public-private partnership (Erie et al. 2010, 2011), the building trend gradually encompassed adjacent neighbourhoods such as East Village and Barrio Logan, where property values are increasing and higher income households are moving in (Kühne and Schönwald 2015a, 2015b; Roßmeier 2019; Rumpf 2016).

It is evident, that the redevelopment efforts and associated gentrification processes in the urban core "have begun to cross the central-city boundary" (Charles 2013, p. 1505) causing functional, structural, and social upheavals in the neighborhoods of the inner-ring (Kühne et al. 2017). These developments are not only occurring as a result of the spatial limitation of San Diego's downtown area, but are promoted by social trends, economic, and fiscal interests as well as changing local planning policies dedicated to the urbanization of the former suburban inner-ring, the emerging urfsurbs (City of San Diego 2019; Gallagher 2014; Kühne and Schönwald 2015b; Kühne et al. 2016; Markley 2018, p. 606; Markley and Sharma 2016; Sweeney and Hanlon 2017, pp. 255–256). As a result, marginalized sections of the population are facing gentrification processes and residential segregation (Ervin 2007, p. 188; Roßmeier 2019, p. 603; Weber and Kühne 2017, p. 26). The previously occurring "spill over effects of blighted areas from the inner cities to the inner-ring suburbs" (Lee and Leigh 2007, p. 148) are now taking place in the first ring, causing the "displacement of working-class residents" (Markley 2018, p. 607) from the first-tier into the neighborhoods of the next ring (cf. Kühne et al. 2017, p. 183), from East Village and Barrio Logan to more affordable but peripheral places (Delgado and Swanson 2019).

In contrast, members of the middle class and the creative and cultural economy in San Diego (Erie et al. 2010) are lured for new consumption, work, and especially living opportunities in the prestigious loft and shopping complexes in converted industrial structures meeting the "stereotypical visual expectations of the urban populace in a distinctly postmodern 'architecture of experience'" (Weber and Kühne 2017, p. 25). Initiated by the inner-city restructuring on the one hand, and the postmodern increase in the attractiveness of urban life on the other, gentrification processes are beginning to take place in the peripheral areas of downtown, increasing pressure on the real estate and housing market as well as certain spatial conflicts (see Figs. 1 and 2). With the influx of the urbanophile population there are also changes in the infrastructure, new patterns of consumption promote the settlement of breweries and art galleries, cafés and other stereotypical urban uses in the former warehouse district East Village and the Mexican-American community neighborhood Barrio Logan (Roßmeier 2019; Rumpf 2016). "Scattered throughout the East Village are studios, artist lofts, galleries, restaurants

Fig. 1 Physical representations of the urfsurbanization in San Diego: A hybrid mixture of small and large, old and new structures and the affluent and underprivileged society at the eastern edge of East Village. [*Source* Albert Roßmeier (2019)]

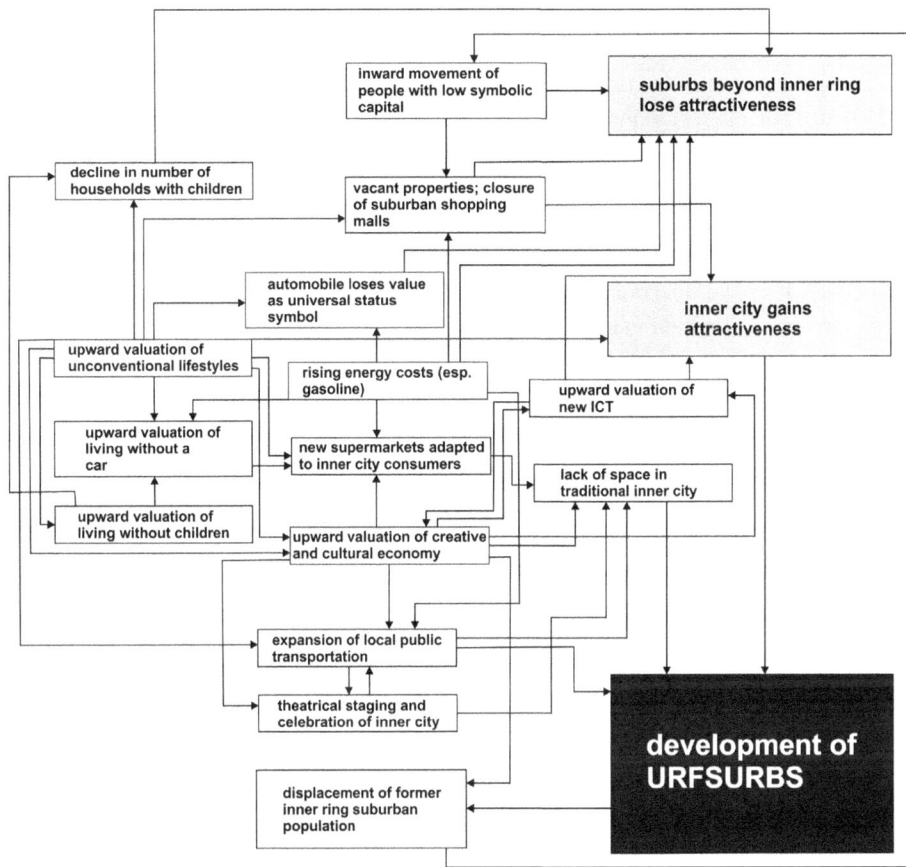

Fig. 2 The urbanization process of the former suburban inner-ring and its interrelations. (*Source* Kühne 2016; Kühne et al. 2017)

and shops" (DSDP 2016, p. 9). The San Diego Tourism Authority (2019aa) describes Barrio Logan as follows: "For years, the Barrio Logan neighborhood has been one of San Diego's best kept secrets. An epicenter of the city's Mexican-American culture, this neighborhood is quickly becoming a hotspot for cutting-edge art and authentic local culture. (…) In recent years, formerly vacant warehouses have turned over into funky and creative spaces for art exhibitions, music performances, and much more." Thus, the urfsurbanization process in San Diego is not to be seen as "further expansion into socially as well as architecturally already urbanized areas, but a complete (often also structural) transformation from suburban to urban" (Weber and Kühne 2017, p. 26) with effects and interrelations on the physical, social, as well as the infrastructural level. These developments are schematically visualized in Fig. 2.

In the following, the results of the *Google* picture analysis are presented. The aim is to workout the recurring moments, which are connected to the nodal points East Village and Barrio Logan and constantly put into relation with each other (cf. Weber et al. 2017, p. 217). It is of interest which moments occur regularly within one corpus that are only interlinked marginally with or even excluded within the other. What aspects of the urbanization processes and upheavals in the inner-ring of San Diego are represented in the *Google* search results? And to what extent are heterogeneous and contradictory moments interwoven in each corpus which reveal certain breaks and changes of interpretation? Ultimately, it is the question of the extent to which the two corpora testify to hybrid characteristics and which meanings and neighborhood images are promoted by the regularities and heterogeneities in the gathered pictures. To address these questions in the following, the nodal points East Village and Barrio Logan and the strikingly combined moments have been visualized in two illustrations (Fig. 3 and 4). It became clear that the two corpora, each consisting of 100 *Google* search results, contain clearly different moments, which in turn are also linked in different ways and thus underline the diversity of today's inner-ring suburbs (cf. Puentes and Orfield 2002, p. 2; Sierra 2019, p. 4).

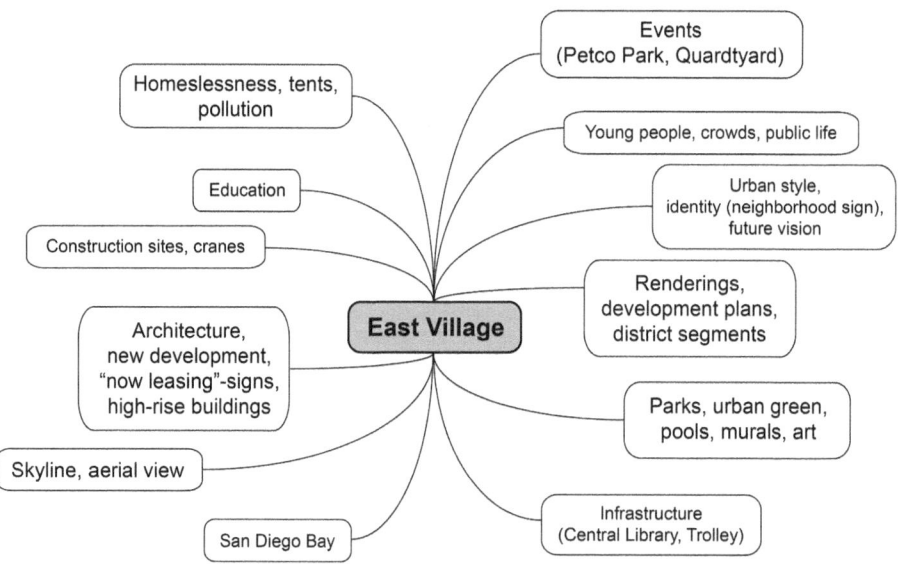

Fig. 3 Recurring and regularly interlinked picture elements in the *Google* search results for "East Village." The font size indicates a quantitative frequency of the different elements. (*Source* own survey and illustration)

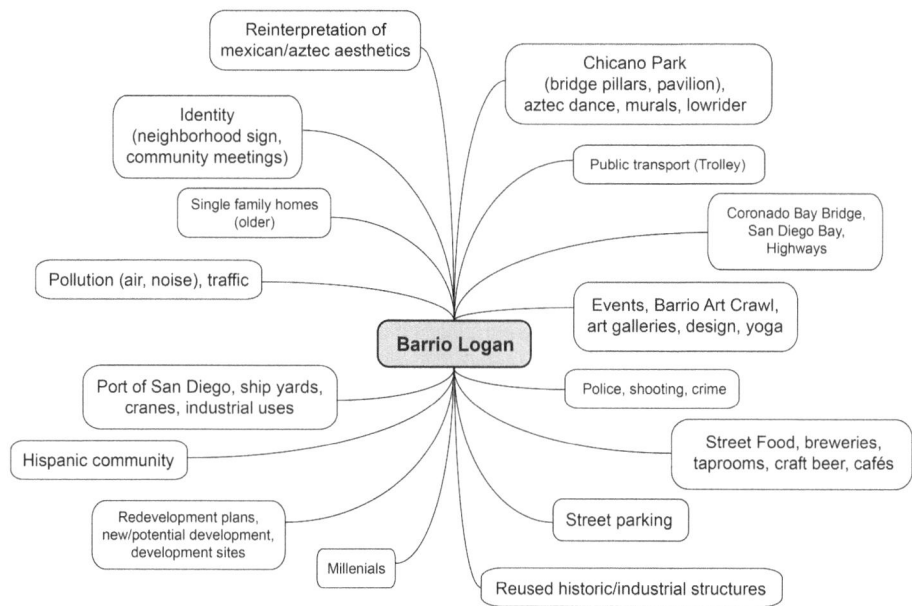

Fig. 4 Recurring and regularly interlinked picture elements in the pictorial representation of "Barrio Logan." The font size indicates a quantitative frequency of the different elements. (*Source* own survey and illustration)

3.2 The Medial, Pictorial Representations of East Village: A Stereotypical Upmarket, Urban Neighborhood

In the example of the former industrial quarter East Village, the overall picture the research corpus shows is rather consistent, as will be explained in the following (see Fig. 3). The district is regularly represented by aerial photographs of new high-rise buildings, apartment complexes or projects under construction. Advertisements for new and vacant apartments (in the form of "now leasing"-sings) can be seen across the pictures, which are presented in combination with architectural renderings and neighborhood features such as planned, urban green spaces. East Village is visualized as a logical extension of the downtown area, as a neighborhood that, together with other high-rise projects, fits into the San Diego skyline both visually and functionally. Further elements are, for example, construction cranes which underline the status of a developing, prospering quarter.

Corresponding to the visualized development processes, the corpus also contains development plans for individual building projects, but also larger-scale subdivisions of the quarter into four neighbourhood segments called "Post Office" in the northwest of East Village, "I.D.E.A. District" in the northeast, "Ballpark" in the southwest around Petco Park, and "East Village South" in the southeast on the border with Barrio Logan.

These illustrations can be interpreted as attempts to establish identity within individual downtown sections for (real estate) advertising purposes, which are further underscored by occasionally depicted designs of a neighborhood sign for East Village. Ultimately, this also highlights the dominance of private interests within San Diego's urban development (cf. Erie et al. 2010, 2011; Kühne and Schönwald 2015b). In addition, the regular appearance of elements such as wall murals and other art projects, green spaces, and in particular the depiction of young people or larger groups at local events traces the neighborhood as an interesting, young, and urban new district of downtown San Diego. Large event locations such as Petco Park, but also alternative interim use projects such as "Quartyard" underline the 'exciting experience character' of the quarter which is represented in the corpus. Only sporadically can moments such as the San Diego Bay or infrastructure elements such as the Central Library or the San Diego Trolley be found in the collected pictures (see Fig. 3).

What is striking, however, is what is *not* integrated into the corpus: the industrial past of the neighborhood, the partially still existing warehouses and fallow land are not reproduced in the pictures. Thus, meaning is not only constituted through the regular depiction of stereotypical urban elements, but is especially generalized, synthesized, and temporarily anchored through the omission of others. The industrial, physical rudiments of the quarter drift into the outside of the discourse around the now urban downtown district East Village, and become identity-forming through their demarcation, i.e., a constitutive outside (cf. Weber et al. 2017, pp. 216–217). Also missing are moments around the settlement of art galleries, which are regularly communicated for image purposes, especially by the San Diego Tourism Authority (2019bc) and other central nonprofit organizations dedicated to the further development of East Village such as the Downtown San Diego Partnership (2016). In summary, the medial, pictorial representations of East Village are ultimately focused on real estate and development aspects.

The East Village quarter is thus depicted in a largely stringent manner across the gathered pictures, the related signifiers draw a clear, unambiguous image. Urban stereotypes such as high-rise and apartment buildings are reproduced, which constitute and temporarily fix meaning (cf. Weber 2015, pp. 107–108). However, to certain extent, contrasts and breaks become apparent: in addition to the physical and functional representations of prosperous urban development, social problems such as the increasing homelessness in the East Village are also reflected in the corpus (see Figs. 1 and 3). As a consequence of rising real estate prices, the resulting displacement in the urbanization process of the inner-ring (Weber and Kühne 2017, see also Fig. 2) and the cumulation of homeless services in the eastern downtown area, homelessness in East Village is coming to a head, which–also (re)produced in the corpus–is increasingly manifested by the presence of homeless people, their tent settlements, and accumulations of garbage in public spaces (see Fig. 1).

Thus, the corpus of the medial representations of East Village reveals predominantly complementary moments; heterogeneities are not reproduced except for the aspects related to homelessness. East Village is ultimately represented as an urfsurb, a place in the former suburban ring which is characterized by the "expansion of 'urban' lifestyles" (Kühne and Weber 2019, p. 760), functions, and structures which transports and strengthens a corresponding image. The medial perception of the quarter is not (any longer) characterized by a hybrid mixture of industrial structures and new apartment complexes, which indicates that the quarter has already made significant progress in the process of urfsurbanization (Kühne and Weber 2019).

3.3   Barrio Logan in the Media: A Spatial Pastiche of Opposites and Contradictory Attributions

In contrast, the corpus of the Mexican-American neighborhood Barrio Logan is more plural and thus more hybrid (see Fig. 4). On the one hand, there are regular moments around the Hispanic community of the neighborhood, image elements show traditional (neighborhood) events and their visitors, aztec dance and lowrider events held in Chicano Park. The neighborhood is comprehensively represented by photographs of the park and the pavilion as the central place of meetings, especially the murals on the bridge pillars of the Coronado Bay Bridge are quantitatively often represented in the examined pictures. These ultimately reproduce a 'protest character' of the neighborhood, which is pictorially strongly linked to Chicano Park and visualized within the murals in a kind of postmodern reinterpretation of Mexican or Aztec aesthetics (e.g., the slogan "Varrio Si! Yonkes No!" which can be interpreted as a protest call against the settlement of junk yards and other industrial uses in the neighborhood). The pictorial elements regularly contained in the analyzed corpus, such as the neighborhood sign in the Cesar E. Chavez Parkway or photographs of events such as community meetings, promote the creation of identity, which is also emphasized by the comprehensive illustration of Spanish language elements and thus temporarily fixing meaning.

On the other hand, the pictures representing Barrio Logan also reproduce moments that visualize a different society and its uses, leisure activities and consumer habits. In addition to the Hispanic community, the pictures regularly show white, young residents indulging in trendy sports activities such as yoga or spending their free time in the neighborhood's cafés and art galleries (Kühne and Schönwald 2015b, pp. 53–54). Aesthetic aspects are in the foreground, pastiche-like combinations of stylistic elements, and references to Mexican or Aztec aesthetics which adorn the new consumer offerings such as breweries and fast food shops are regularly found in the pictures of the corpus. In accordance with a postmodern aesthetic (Kühne and Weber 2019, p. 759; Schönwald 2017), the pictures also include representations of converted, historical or industrial structures, in particular the artistic elements depicted are linked to the aestheticized, industrial structures: for example, the "Bread & Salt" gallery shown here is located in a former food factory in Barrio Logan. According to other elements regularly reproduced in the pictures, events such as the regular "Barrio Art Crawl" strengthen the perception of the neighbourhood as a young and creative neighbourhood, which is 'still undiscovered' and therefore exciting and new (cf. Delgado and Swanson 2019). Barrio Logan is communicated as "a hidden hub for art" (San Diego Tourism Authority 2019a), and thus as a leisure and entertainment district, the neighborhood is given meaning. In general, the plurality and heterogeneity of possible interpretations and, in particular, the hybridity of the neighborhood become apparent.

In addition, the analysis also reveals pictorial elements of the physical-constructional level to be clearly heterogeneous. On the one hand, stereotypical suburban aspects are regularly reproduced in the images, (older) single-family homes with front gardens and their access roads with parked cars are shown in a comprehensive manner. On the other hand, the pictorial elements of the corpus also bear witness to developmental aspects: Representations and graphics of new or localizations of potential building projects and redevelopment plans reveal urfsurbanization processes that are moving from downtown via East Village to Barrio Logan (Roßmeier 2019). A hybrid and ambiguous picture emerges from rather older, smaller and suburban structures, which are replaced and supplemented by new, larger and thus rather urban structures. The regularly reproduced elements draw a plural and contradictory picture of the neighbourhood, a quarter in transition that cannot be clearly classified (cf. Kühne and Weber 2019; see Fig. 4).

In addition to the pictorial elements that can be ascribed to residential and leisure use, Barrio Logan's corpus also contains moments around existing industrial use. The Port of San Diego, the ship yards, and the cranes clearly visible in the background of photos are regularly reproduced elements in the *Google* search results. However, this is also accompanied by aspects such as noise, air, and also traffic pollution, which are linked together as a disturbing effect of the industrial uses in Barrio Logan. Also woven into the overall picture of the neighborhood are elements such as police operations, crime or gun related violence, but rather marginalized. This includes moments such as the Coronado Bay Bridge and San Diego Bay, the highways surrounding the neighborhood and the San Diego Trolley, which are only occasionally depicted in the pictures.

Unlike the corpus of the East Village quarter, in the case of Barrio Logan no moments such as high-rise buildings or the skyline of San Diego are (re)produced. New real estate projects or structural developments, which dominate the visual representations of the East Village quarter in quantitative terms, are less conspicuous in the Barrio Logan example and are mixed with moments of older, rather suburban building stock. What is included and what is excluded in the discourse around Barrio Logan, what meanings are produced and temporarily fixed, is not exactly clear and remains open. Many different and at first glance incompatible moments are linked together in the example. The constitutive elements of Barrio Logan are the breaks and the heterogeneities, the hybrid, with which the neighbourhood stands in contrast to the rather stereotypically represented downtown quarter East Village. The neighborhood Barrio Logan becomes a spatial pastiche (Kühne 2012), it is characterized by different compartments of varying degrees of hybridity (see Hofmeister and Kühne 2016; Kühne and Weber 2019, p. 758). In the present example, hybridity can be located on different levels, for example on a physical or physical-constructional, social, aesthetic and functional level, which is strengthened rather than cancelled out in the incipient process of urfsurbanization (Weber and Kühne 2017, p. 18). Here, rather urban and rather suburban structures and lifestyles mix, the analysis shows Barrio Logan as a spatial pastiche which can also be described as an urban/rural hybrid due to its "(increasing) differentiation, fragmentation, complexity" (Kühne and Weber 2019, p. 757; Schönwald 2017).

4   Conclusion

The goal of the present article was to examine the question to which extent the medial, pictorial representations of the neighborhoods East Village and Barrio Logan are characterized by similar motifs and thus lead to discursive determinations of meaning and the creation of neighborhood images. In both neighborhoods, processes of urfsurbanization (Kühne et al. 2016; Kühne and Weber 2019; Weber and Kühne 2017) and associated spatial and social upheavals are currently taking place, which are leading to changing uses and user groups and thus also to reinterpretations. The underlying discourse and hegemony theory of Laclau and Mouffe (1985) was operationalized within a picture analysis that traces regularly occurring pictorial elements and their interweaving (cf. Miggelbrink 2009, p. 180; Weber 2015, p. 107; 2018, p. 132). Of importance in this context was what is comprehensively connected with each other in the pictures, but also what does *not* appear in the formed corpora or whether certain breaks and heterogeneities can be traced. Within the analysis it was specifically asked what was regularly connected to the nodal points East Village and Barrio Logan, how the individual moments were related to each other, and what was represented only marginally or not at all. In addition, in the context of the urfsurbanization of San Diego's inner-ring, it was possible to ask about the stringency and homogeneity of the argumentation logics, to what extent the reproduced moments are correspondingly related to each other or whether there are rather plural, ambiguous and thus hybrid combinations of contradictory pictorial elements. In this way, conclusions could be drawn about current hegemonies in the interpretation of neighborhoods and thus certain differences between the two corpora could be worked out. In the tradition of Laclau and Mouffe (1985, p. 112), it is assumed that meaning is constituted and *temporarily* fixed as a result of regularly occurring elements, and thus neighborhood images are created, strengthened, and become hegemonic (Glaze 2015, p. 25).

The *Google* picture analysis of the first 200 search results for the keywords "East Village" and "Barrio Logan" revealed clearly different medial images and representations of the two neighborhoods: While the pictorial representations of East Village produce a rather clear and unambiguous image of an upmarket, urban environment for young professionals (see Fig. 3), Barrio Logan is represented by a plural and diverse image of drastic upheavals in a pastiche-like, hybrid manner (see Fig. 4). The results can be classified with the upheavals in the context of an urban/rural hybridization and an incipient urfsurbanization: the urbanization processes in East Village have already progressed significantly since the 2000s and the construction of the Ballpark. Ancillary development, increased property values, and the influx and presence of higher income households characterizes not only the urban core of San Diego but also its eastern extension, East Village. In contrast, in the adjacent neighborhood Barrio Logan, there are currently only isolated representations of the incipient urbanization of the inner-ring. Existing structures are being replaced, extended and supplemented by newer and larger ones (Roßmeier 2019; Rumpf 2016). In comparison to East Village, stock structures are still largely in place and currently continue to exist in a hybrid mixture. In a kind of spatial pastiche, different compartments of varying degrees of hybridity are united here, which become even more pluralized, further differentiated, and thus more complex with the onset of the urfsurbanization process. Accordingly, the term urban/rural hybrid is used to cover the clearly variable and ambiguous developments in the former suburban inner-ring of San Diego, which take on clearly different forms in the course of an uniform development direction.

In the future, the question arises how the developments of the urfsurbanization will continue to inscribe themselves in the neighborhood of Barrio Logan. It will have to be examined whether, in accordance with the results for East Village, hybrid structures will also fade in Barrio Logan in favor of homogeneous, stereotypical urban elements. For currently there are signs of corresponding developments on the physical, functional and social levels, which are progressing rapidly. There is a need for future research in this area in order to be able to accompany the upheavals adequately. In the sense of a 'new applied landscape geography' (Kühne et al. 2019), the potential to be explored lies in a neo-pragmatic approach (Chilla et al. 2015; Kühne 2019), which, through a wise combination of theoretical approaches and empirical methodology, is able to comprehensively address processes of change and their individual social classification, in order to be able to also investigate power-specific questions such as *how* and *by whom* landscape or urban/rural hybrids will be visualized in the future and *with what goal*.

References

Al-Khanbashi, M. (2020). Using matrix as a qualitative data display for landscape research and a reflection based on the social constructivist perspective. In D. Edler, C. Jenal, & O. Kühne (Eds.), *Modern approaches to the visualization of landscapes* (pp. 103–118). Wiesbaden: Springer VS.

Appleyard, B., & Stepner, M. (2018). Toward the dreams and realities of temporary paradise? Lynch and Appleyard's look at the special landscape of San Diego/Tijuana. *Journal of the American Planning Association, 84*(3–4), 230–236.

Aschenbrand, E., & Michler, T. (2020). Linking socio-scientific landscape research with the ecosystem service approach to analyze conflicts about protected area management – The case of the Bavarian Forest National Park. In D. Edler, C. Jenal, & O. Kühne (Eds.), *Modern approaches to the visualization of landscapes* (p. xx). Wiesbaden: Springer VS.

Barthes, R. (2007 [1970]). *S/Z*. Frankfurt am Main: Suhrkamp.

Berger, P. L., & Luckmann, T. (1966). *The social construction of reality*. New York: Anchor.

Berr, K. (2020). Visuality, Aesthetics and Landscape. For the enlightenment and self-enlightenment of constructivist landscape research. In D. Edler, C. Jenal, & O. Kühne (Eds.), *Modern approaches to the visualization of landscapes* (pp. 189–215). Wiesbaden: Springer VS.

CCDC. (2015). *Downtown Community Plan.* Centre City Development Corporation. Retrieved Nov 27, 2019 from http://civicsd.com/wp-content/uploads/2015/02/Downtown-Comunity-Plan-All-1.pdf.

Charles, S. L. (2013). Understanding the determinants of single-family residential redevelopment in the inner-ring suburbs of Chicago. *Urban Studies, 50*(8), 1505–1522.

Chilla, T., Kühne, O., Weber, F., & Weber, F. (2015). Neopragmatische' Argumente zur Vereinbarkeit von konzeptioneller Diskussion und Praxis der Regionalentwicklung. In O. Kühne & F. Weber (Eds.), *Bausteine der Regionalentwicklung* (pp. 13–24). Wiesbaden: Springer VS.

City of San Diego, P. D. (2013). *Barrio logan community plan and local coastal program draft.* Retrieved Nov 5, 2019, from https://www.sandiego.gov/sites/default/files/legacy/planning/community/cpu/barriologan/pdf/bl_cpu_full_w_historic_res_091913.pdf.

City of San Diego, P. D. (2015). *Southeastern San Diego Community Plan October 2015.* Retrieved Nov 5, 2019 from https://www.sandiego.gov/sites/default/files/legacy/planning/community/cpu/southeastern/pdf/sesd_community_plan_reduced.pdf.

City of San Diego, P. D. (2019). *Why pilot villages are important.* City of San Diego Planning Department. Retrieved Nov 5, 2019 from https://www.sandiego.gov/planning/genplan/pilotvillage/important.

Comer-Schultz, J. (2011). *History and historic preservation in San Diego Since 1945: Civic Identity in America's Finest City.* (PhD), Arizona State University.

Costello, D., Schemass, L., Mendelsohn, R., Canby, A., & Bender, J. (2003). *The returning city: Historic preservation and transit in the age of civic revival.* Retrieved Nov 5, 2019 from https://www.planning.dot.gov/Documents/CaseStudy/Cities/returning_city.htm#sd.

Delgado, E., & Swanson, K. (2019). *Gentefication* in the barrio: Displacement and urban change in Southern California. *Journal of urban affairs,* 1–16.

Derrida, J. (1999 [1972]). *Randgänge der Philosophie.* Wien: Passagen-Verlag.

DSDP. (2016). *Downtown San Diego: The innovation economy's next frontier. A Data driven exploration of San Diego's urban renaissance.* Downtown San Diego Partnership & UC San Diego Extension Center For Research On The Regional Economy. Retrieved Nov 5, 2019 from https://downtownsandiego.org/wp-content/uploads/2016/05/DSDP-Demographic-Study-2016.pdf.

Eddy, L. (1995). Visions of paradise. *The Journal of San Diego History 41* (Summer), n.pag.

Edler, D., Jenal, C., & Kühne, O. (2020). Modern approaches to the visualization of landscapes—An introduction. In D. Edler, C. Jenal, & O. Kühne (Eds.), *Modern approaches to the visualization of landscapes* (pp. 3–15). Wiesbaden: Springer VS.

Erie, S. P., Kogan, V., & MacKenzie, S. A. (2010). Redevelopment, San Diego Style: The limits of public—private partnerships. *Urban Affairs Review, 45*(5), 644–678.

Erie, S. P., Kogan, V., & MacKenzie, S. A. (2011). *Paradise plundered: Fiscal Crisis and Governance Failures in San Diego.* Stanford: Stanford University Press.

Ervin, J. C. (2007). Reinventing downtown San Diego: A spatial and cultural analysis of the Gaslamp Quarter. *The Journal of San Diego History, 53*(4), 188–217.

Fontaine, D. (2020a). Landscape in computer games—The examples of GTA V and Watch Dogs 2. In D. Edler, C. Jenal, & O. Kühne (Eds.), *Modern approaches to the visualization of landscapes* (pp. 293–306). Wiesbaden: Springer VS.

Fontaine, D. (2020b). Virtuality and Landscape. In D. Edler, C. Jenal, & O. Kühne (Eds.), *Modern approaches to the visualization of landscapes* (pp. 267–276). Wiesbaden: Springer VS.

Gailing, L., & Leibenath, M. (2015). The social construction of landscapes: Two theoretical lenses and their empirical applications. *Landscape Research, 40*(2), 123–138.

Gallagher, L. (2014). *The end of the suburbs: Where the American dream is moving*. New York: Penguin.

Gaslamp Quarter Association. (2019). *Gaslamp's History. Gaslamp Quarter Association*. Retrieved Nov 19, 2019 from https://www.gaslamp.org/history/.

Glasze, G. (2013). *Politische Räume. Die diskursive Konstitution eines „geokulturellen Raums" – die Frankophonie*. Bielefeld: transcript.

Glasze, G. (2015). Identitäten und Räume als politisch: die Perspektive der Diskurs-und Hegemonietheorie. *Europa Regional, 21*(1–2), 23–34.

Glasze, G., & Mattissek, A. (2009). Diskursforschung in der Humangeographie: Konzeptionelle Grundlagen und empirische Operationalisierung. In G. Glasze & A. Mattissek (Eds.), *Handbuch Diskurs und Raum. Theorien und Methoden für die Humangeographie sowie die sozial- und kulturwissenschaftliche Raumforschung* (pp. 11–59). Bielefeld: transcript.

Glynos, J., & Stavrakakis, Y. (2004). Encounters of the real kind. Sussing out the limits of Laclau's embrace of Lacan. In S. Critchley & O. Marchart (Eds.), *Laclau: A critical reader* (pp. 201–216). London: Routledge.

Hofmeister, S., & Kühne, O. (Eds.). (2016). *StadtLandschaften. Die neue Hybridität von Stadt und Land*. Wiesbaden: Springer VS.

Jørgensen, M. W., & Phillips, L. J. (2002). *Discourse analysis as theory and method*. London: SAGE Publications.

Kayzar, B. A. (2006). *Analyzing revitalization outcomes in downtown San Diego*. (PhD), University of California Santa Barbara and San Diego State University.

Kühne, O. (2012). *Stadt – Landschaft – Hybridität: Ästhetische Bezüge im postmodernen Los Angeles mit seinen modernen Persistenzen*. Wiesbaden: Springer VS.

Kühne, O. (2014). Wie kommt die Landschaft zurück in die Humangeographie? Plädoyer für eine, konstruktivistische Landschaftsgeographie'. *Geographische Zeitschrift, 102*(2), 68–85.

Kühne, O. (2016). Transformation, Hybridisierung, Streben nach Eindeutigkeit und Urbanizing former Suburbs (URFSURBS): Entwicklungen postmoderner Stadtlandhybride in Südkalifornien und in Altindustrieräumen Mitteleuropas – Beobachtungen aus der Perspektive sozialkonstruktivistischer Landschaftsforschung. In S. Hofmeister & O. Kühne (Eds.), *StadtLandschaften. Die neue Hybridität von Stadt und Land* (pp. 13–36). Wiesbaden: Springer VS.

Kühne, O. (2019). Sich abzeichnende theoretische Perspektiven für die Landschaftsforschung: Neopragmatismus, Akteur-Netzwerk-Theorie und Assemblage-Theorie. In O. Kühne, F. Weber, K. Berr, & C. Jenal (Eds.), *Handbuch Landschaft* (pp. 153–162). Wiesbaden: Springer VS.

Kühne, O. (2020). The social construction of space and landscape in internet videos. In D. Edler, C. Jenal, & O. Kühne (Eds.), *Modern approaches to the visualization of landscapes* (pp. 121–137). Wiesbaden: Springer VS.

Kühne, O., & Jenal, C. (2020). The threefold landscape dynamics—Basic considerations, conflicts and potentials of virtual landscape research. In D. Edler, C. Jenal, & O. Kühne (Eds.), *Modern approaches to the visualization of landscapes* (pp. 389–402). Wiesbaden: Springer VS.

Kühne, O., & Schönwald, A. (2015a). *San Diego: Eigenlogiken, Widersprüche und Hybriditäten in und von, America's finest city'*. Wiesbaden: Springer VS.

Kühne, O., & Schönwald, A. (2015b). San Diego: Trouble in Paradise?Zwischen Stadterneuerung, Reurbanisierung und restriktiver Steuerpolitik. *Geographische Rundschau, 67*(5), 49–54.

Kühne, O., Schönwald, A., & Weber, F. (2016). Urban/rural hybrids: The urbanisation of former suburbs (URFSURBS). *Quaestiones Geographicae, 35*(4), 23–34.

Kühne, O., Schönwald, A., & Weber, F. (2017). Die Ästhetik von Stadtlandhybriden. In O. Kühne, H. Megerle, & F. Weber (Eds.), *Landschaftsästhetik und Landschaftswandel* (pp. 177–197). Wiesbaden: Springer VS.

Kühne, O., & Weber, F. (2019). Postmoderne Zugriffe und Differenzierungen von Stadt und Land(schaft): Stadtlandhybride, räumliche Pastiches und URFSURBS. In O. Kühne, F. Weber, K. Berr, & C. Jenal (Eds.), *Handbuch Landschaft* (pp. 755–770). Wiesbaden: Springer VS.

Kühne, O., Weber, F., & Jenal, C. (2019). Neue Landschaftsgeographie. In O. Kühne, F. Weber, K. Berr, & C. Jenal (Eds.), *Handbuch Landschaft* (pp. 119–134). Wiesbaden: Springer VS.

Kühne, O., Weber, F., & Weber, F. (2013). Wiesen, Berge, blauer Himmel. Aktuelle Landschaftskonstruktionen am Beispiel des Tourismusmarketings des Salzburger Landes aus diskurstheoretischer Perspektive. *Geographische Zeitschrift 101* (1), 36-54.

Laclau, E. (1993). Discourse. In R. E. Goodin & P. Pettit (Eds.), *A Companion to contemporary political philosophy* (pp. 431–437). Oxford: Blackwell.

Laclau, E. (2007). *On Populist Reason*. London: Verso.

Laclau, E., & Mouffe, C. (1985). *Hegemony and socialist strategy. Towards a radical democratic politics*. London: Verso.

Lee, S., & Leigh, N. G. (2007). Intrametropolitan spatial differentiation and decline of inner-ring suburbs: A comparison of four US Metropolitan Areas. *Journal of Planning Education and Research, 27*(2), 146–164.

Leibenath, M., & Otto, A. (2012). Diskursive Konstituierung von Kulturlandschaft am Beispiel politischer Windenergiediskurse in Deutschland. *Raumforschung und Raumordnung, 70*(2), 119–131.

Linke, S. (2020). Landscape in internet pictures. In D. Edler, C. Jenal, & O. Kühne (Eds.), *Modern approaches to the visualization of landscapes* (pp. 139–156). Wiesbaden: Springer VS.

Loda, M., Kühne, O., & Puttilli, M. (2020). The social construction of Tuscany in the German and English speaking world—Presented by the analysis of internet images. In D. Edler, C. Jenal, & O. Kühne (Eds.), *Modern approaches to the visualization of landscapes* (pp. 157–171). Wiesbaden: Springer VS.

Markley, S. (2018). Suburban gentrification? Examining the geographies of New Urbanism in Atlanta's inner suburbs. *Urban Geography, 39*(4), 606–630.

Markley, S., & Sharma, M. (2016). Gentrification in the Revanchist Suburb: The Politics of Removal in Roswell. *Georgia. southeastern geographer, 56*(1), 57–80.

Mattissek, A. (2010). Analyzing city images. Potentials of the „French School of Discourse Analysis". *Erdkunde 64*(4), 315–326.

Mattissek, A., & Reuber, P. (2004). Die Diskursanalyse als Methode in der Geographie – Ansätze und Potentiale. *Geographische Zeitschrift, 92*(4), 227–242.

Miggelbrink, J. (2009). Verortung im Bild. Überlegungen zu ‚visuellen Geographien'. In J. Döring & T. Thielmann (Eds.), *Mediengeographie: Theorie – Analyse – Diskussion* (pp. 179-202). Bielefeld: transcript.

Puentes, R., & Orfield, M. (2002). *Valuing America's First Suburbs: A Policy Agenda for Older Suburbs in the Midwest*: The Brookings Institution Center on Urban and Metropolitan Policy.

Roßmeier, A. (2019). Quo vadis, Temporary Paradise'?Urfsurbanisierung und räumliche Konflikte im StadtLandHybriden San Diego. In K. Berr & C. Jenal (Eds.), *Landschaftskonflikte* (pp. 591–616). Wiesbaden: Springer VS.

Rumpf, L. E. (2016). *The synthesis of social reality and the evolution of Barrio Logan: A critical view of the framing and representation of urban redevelopment*. Fielding Graduate University.

San Diego Tourism Authority. (2019a). *Barrio Logan: A Hidden Hub for Art*. San Diego Tourism Marketing District Corporation. Retrieved Oct 31, 2019 from https://www.sandiego.org/articles/downtown/barrio-logan.aspx.

San Diego Tourism Authority. (2019b). *The Craft Beer Capital of America: San Diego, CA*. San Diego Tourism Marketing District Corporation. Retrieved Dec 8, 2019 from https://www.sandiego.org/campaigns/good-stuff/craft-beer.aspx.

San Diego Tourism Authority. (2019c). *Downtown & Gaslamp Quarter: The Hottest Nightlife on the West Coast.* San Diego Tourism Marketing District Corporation. Retrieved Oct 31, 2019 from https://www.sandiego.org/explore/downtown-urban/downtown.aspx.

Schönwald, A. (2017). Ästhetik des Hybriden. Mehr Bedeutungsoffenheit für Landschaften durch Hybridisierungen. In O. Kühne, H. Megerle, & F. Weber (Eds.), *Landschaftsästhetik und Landschaftswandel* (pp. 161–175). Wiesbaden: Springer VS.

Sierra, A. C. (2019). Inner Suburbs. In A. M. Orum (Ed.), *The Wiley Blackwell encyclopedia of urban and regional studies* (pp. 1–5). Chichester: Wiley Blackwell.

Siepmann, N., Edler, D., & Kühne, O. (2020). Soundscapes in cartographic media. In D. Edler, C. Jenal, & O. Kühne (Eds.), *Modern approaches to the visualization of landscapes* (pp. 247–263). Wiesbaden: Springer VS.

Stakelbeck, F., & Weber, F. (2013). Almen als alpine Sehnsuchtslandschaften: Aktuelle Landschaftskonstruktionen im Tourismusmarketing am Beispiel des Salzburger Landes. In D. Bruns & O. Kühne (Eds.), *Landschaften: Theorie, Praxis und internationale Bezüge* (pp. 235–252). Schwerin: Oceano Verlag.

Sweeney, G., & Hanlon, B. (2017). From old suburb to post-suburb: The politics of retrofit in the inner suburb of Upper Arlington, Ohio. *Journal of urban affairs, 39*(2), 241–259.

Torfing, J. (1999). *New theories of discourse: Laclau, Mouffe and Žižek.* Oxford: Wiley.

Weber, F. (2015). Diskurs – Macht – Landschaft. Potenziale der Diskurs- und Hegemonietheorie von Ernesto Laclau und Chantal Mouffe für die Landschaftsforschung. In S. Kost & A. Schönwald (Eds.), *Landschaftswandel – Wandel von Machtstrukturen* (pp. 97–112). Wiesbaden: Springer VS.

Weber, F. (2018). *Konflikte um die Energiewende. Vom Diskurs zur Praxis.* Wiesbaden: Springer VS.

Weber, F. (2019a). Diskurstheoretische Landschaftsforschung. In O. Kühne, F. Weber, K. Berr, & C. Jenal (Eds.), *Handbuch Landschaft* (pp. 105–117). Wiesbaden: Springer VS.

Weber, F. (2019b). Von divergierenden Grenzziehungen und Konflikten im StadtLandHybriden des Grand Paris. In K. Berr & C. Jenal (Eds.), *Landschaftskonflikte* (pp. 637–663). Wiesbaden: Springer VS.

Weber, F. (2020). Blurring the boundaries of landscape visualization: Welcome to Fabulous Las Vegas. In D. Edler, C. Jenal, & O. Kühne (Eds.), *Modern approaches to the visualization of landscapes* (pp. 461–478). Wiesbaden: Springer VS.

Weber, F., & Kühne, O. (2017). Hybrid suburbia: New research perspectives in France and Southern California. *Quaestiones Geographicae, 36*(4), 17–28.

Weber, F., Roßmeier, A., Jenal, C., & Kühne, O. (2017). Landschaftswandel als Konflikt. In O. Kühne, H. Megerle, & F. Weber (Eds.), *Landschaftsästhetik und Landschaftswandel* (pp. 215–244). Wiesbaden: Springer VS.

Albert Roßmeier studied landscape architecture with a focus on urban planning at the University of Applied Sciences Weihenstephan-Triesdorf, followed by human geography at the University of Tuebingen. At the University of Applied Sciences Weihenstephan-Triesdorf he was employed from winter 2015 to winter 2016 as a research assistant in a project on landscape change in the wake of the energy revolution funded by the Federal Agency for Nature Conservation. From winter 2016 to spring 2019, he conducted research at the University of Tuebingen in the EU project 'LIFE living Natura 2000,' among others. He is currently working on his doctoral project on the transformation of inner-ring suburbs and urban bordering processes in San Diego. His work focuses on energy system transformation, landscape change and urban development in the European as well as the US context.

Exploring Urban 'Blightscapes' Applying Spatial Video Technology and Geographic Information System

Judith Stratmann, Alina Ristea, Michael Leitner and Gernot Paulus

Abstract

The term urban 'blightscape' refers to cartographic visualizations of urban blight locations. Such locations describe disordered neighborhoods in urban areas characterized by the deterioration of properties and the environment. The creation of urban 'blightscapes' is explored from data collection to final mapping in five selected neighborhoods of Baton Rouge, US. The spatial video is applied to gather urban blight incidents. This technology consists of a video stream with an embedded location and timestamp included in each video frame. It allows the digitizing and storage of identified urban blight locations in a Geographic Information System (GIS) for subsequent analysis and visualization. Different spatial analysis methods are applied to create urban 'blightscapes'. Results show 'blightscape' hot spots exclusively found in the two northern neighborhoods, where crime is (very) high. In contrast, 'blightscape'

J. Stratmann (✉) · G. Paulus
Carinthia University of Applied Sciences, Villach, Austria

G. Paulus
e-mail: G.Paulus@fh-kaernten.at

A. Ristea
Northeastern University, Boston, USA
e-mail: a.ristea@northeastern.edu

M. Leitner
Louisiana State University, Baton Rouge, USA
e-mail: mleitne@lsu.edu

© Springer Fachmedien Wiesbaden GmbH, part of Springer Nature 2020   497
D. Edler et al. (eds.), *Modern Approaches to the Visualization of Landscapes*, RaumFragen: Stadt – Region – Landschaft,
https://doi.org/10.1007/978-3-658-30956-5_28

cold spots are solely identified across most of the southern three neighborhoods, where crime is moderate to low. Results also indicate that spatial video and GIS are appropriate technologies to establish urban 'blightscapes' in a relatively quickly, complete, and standard way.

Keywords

Urban 'blightscape' · Spatial video technology · Geographic information system · Local moran's I spatial autocorrelation · Kernel density estimation · Baton Rouge

1   Introduction

Urban blight describes disordered neighborhoods in urban areas characterized by the deterioration of properties and the environment (e.g., abandoned cars, barred windows, dumping, graffiti, infrastructural degradation, overgrowth, etc.) and by social and economic problems (Maghelal et al. 2013; Sampson and Raudenbush 1999). A standardization of urban blight indicators is required in order to quantify the physical quality and the level of blight at an appropriate spatial scale. No standardized measure of blight indicators currently exists to operationalize and visualize urban blight.

This research introduces a novel geospatial technology, termed Spatial Video Acquisition System (SVAS), or spatial video for short, to collect urban blight occurrences. A geographic information system (GIS) is used to digitize these occurrences into a GIS layer, where they can be stored, analyzed, and mapped (for other GIS-based studies in this volume, see: Edler et al. 2020a, Kleber et al. 2020; Poplin et al. 2020). A series of visualization methods are applied to derive landscapes of urban blight, including the Local Moran's I Spatial Autocorrelation (LISA) and the kernel density estimation (KDE). These urban blight landscapes are referred to as urban 'blightscapes' in the remainder of this book chapter.

1.1   Developing a Catalogue for Physical Urban Blight Indicators

In general, a distinction is made between social and physical disorder/blight. Social urban blight refers to the anti-social behavior of mostly unpredictable people, and can include verbal harassment in public places, open solicitation for prostitution, school truancy, public urination, people sleeping in public, nuisance neighbors, etc. (Sampson and Raudenbush 1999; Skogan 1990; Kelling and Wilson 1982). This research focuses on physical urban blight, only. For this reason, a criteria catalogue for mapping physical urban blight is derived from the existing literature (Gau and Pratt 2010; Hinkle and Weisburd 2008; Maghelal et al. 2013; Ross and Mirowsky 2001; Skogan 1990; Weisburd et al. 2010) to ensure consistency in the analysis. Each urban blight indicator is listed and described in more detail in Table 1. For illustration purposes, a sample image is

Table 1 Criteria catalogue for physical urban blight indicators (Gau and Pratt 2010; Hinkle and Weisburd 2008; Maghelal et al. 2013; Ross and Mirowsky 2001; Skogan 1990; Weisburd et al. 2010, all pictures own representation)

Physical urban blight indicator	Description	Image
1. Property		
Abandoned property	Properties that show signs of decay or abandonment (e.g. doors are boarded up); nobody lives in the house	
Broken window/door	Broken windows/doors are left unrepaired	
Boarded window/door	Windows are boarded up with plywood or other material or covered with plastic, foil, etc.	
No glass in window/door	Windows/doors without glass and are left unrepaired	
Building graffiti	Graffiti on buildings *Note: Graffiti that is not situated on buildings is considered as infrastructural graffiti in environmental/infrastructural blight*	
Structural integrity	Severe cracks in the foundation of the building's structure; holes in plaster, wood, masonry, rooftop; unstable foundations	
Building overgrowth	Unsafe amount of vegetation touching the building; constitutes fire, health, or safety hazard *Note: Overgrown vegetation that does not touch a building is considered as overgrown vegetation in environmental/infrastructural blight*	

(continued)

Table 1 (continued)

Physical urban blight indicator	Description	Image
2. Environment/Infrastructure		
Overgrown vegetation	Any vegetation that appears to be overgrown in garden/vacant lots etc.; may block sidewalks *Note: In contrast to building overgrowth, this indicator does not appear on buildings, but exclusively in empty spaces or on infrastructure*	
Litter	Single pieces of trash that are not within a garbage disposal bag; trash left on the ground in a public place	
Illegal dumping	Trash piles: unlawful deposit of any type of waste material (e.g. construction materials, tires, asbestos, etc.)	
Unkempt areas	Empty areas (e.g. vacant lots) that appear unkempt (e.g. overgrown grass, trash, etc.)	
Illegal parking	Car parked illegally on sidewalk; should be reported to city agency for removal	
Abandoned vehicle	Number plates missing, flat tires, missing wheels, broken windows, etc.; any vehicle that has been parked illegally for > 72 h	
Infrastructural graffiti	Graffiti on walls or other surfaces *Note: Graffiti on buildings is considered as building graffiti in property blight*	

also included in the same table. The selected indicators deem to be relevant for studying urban blight in the study area (Baton Rouge, US), but may need to be adapted when applied to another study area.

The physical urban blight catalogue distinguishes between indicators that lead to the destruction of buildings and indicators that lead to environmental or infrastructural decay (Table 1). The latter set of indicators would normally also include damaged sidewalks and damaged roads. However, both indicators were omitted from this research, since few sidewalks were found in the study area and when they existed, their surface was heavily damaged. Similarly, the pavement of many street segments showed structural damages with many cracks and potholes and their collection would have been too time-consuming.

1.2 Methods to Collect Urban Blight Indicators

Sampson and Raudenbush (1999) utilized the Systematic Social Observation (SSO) method to assess disordered public spaces by applying analogue videos, with permanent visual records being collected while driving through neighborhoods. Physical disorder indicators and social disorder indicators (e.g. loitering, public consumption of alcohol, presumed drug sales, etc.) were extracted from video recordings to describe the quantity of urban blight.

Google Street View (GSV) has been used in recent studies as an alternative to the SSO method. By utilizing GSV, the systematic observation can be conducted remotely. This method is beneficial, because it is cost-effective and safe, and research can be expanded to larger areas where physical presence is not necessary. GSV is freely available and includes high-resolution 360° images of many areas in the world (related in this volume: Prisille and Ellerbrake 2020). Limitations of GSV are the flexibility and stability in the temporal component of images (Curtis et al. 2013a; Marco et al. 2017).

Spatial video is a newly developed powerful technology that allows a collection of contextual field data for a fine-scale research. This technology can be applied to a wide range of disciplines, particularly in a geographic context. With the spatial video approach, spatio-temporal data can be collected to analyze geographic phenomena and environmental changes. Global Positioning System (GPS) sensors are linked to the spatial video, so that each recorded frame contains information about its geographical location, including a timestamp. Spatial video cameras can be mounted onto different surveying vehicles, such as cars, motorbikes, bicycles, or boats (Curtis et al. 2013a; Curtis et al. 2015).

Spatial video can be used as a new powerful tool to assess and evaluate objective measures of urban blight. The utilization of spatial video has considerable advantages over other methods. It enables a time-efficient, cost-effective, and easy-to-use data acquisition method that collects contextual data together with their associated attributes. The required time for fieldwork can be reduced and the processed data can be readily used for spatial analysis within a GIS. This method is similar to the SSO method, but its advantages are that images are linked to geographic coordinates and timestamps and that data can be integrated into a GIS for further analysis and visualization purposes. Unlike

GSV, the collection of spatial video data is under the control of the researcher and therefore more flexible to use (Curtis et al. 2015).

The City-Parish government of Baton Rouge publishes many datasets (e.g., housing and development, culture and recreation, public safety, etc.) on their Open Data BR portal (https://data.brla.gov/). Included in this portal are also all requests for service received from the City-Parish 311 Call Center through concerned citizens. Many of these requests identify urban blight indicators that can be mapped, since locations of indicators in the form of geographic coordinates are also stored inside the portal. While this is a great service for citizens, we believe that such collection efforts result in an urban blight dataset that does not follow a standardized collection procedure and is less comprehensive and complete as the dataset collected with the SVAS technology. In Sect. 3.2 below, kernel density estimation models of an urban 'blightscape' derived from 311 calls are compared with an urban 'blightscape' collected with the SVAS technology (for other articles supporting open data initiatives in this volume, see: Edler et al. 2020a, 2020b, Kleber et al. 2020; Meyer-Heß 2020; Siepmann et al. 2020; Thomas 2020; Vetter 2020).

2 Study Area, Data Collection, and Spatial Analysis Methods

2.1 Study Area

This study is implemented in the City of Baton Rouge, the capital city of the State of Louisiana, US. Baton Rouge is the major urban area in the East Baton Rouge Parish[1] (EBRP). It should be noted that any other city in the US that, similar to Baton Rouge, experiences widespread physical urban blight, could have been chosen as the study area. The main reason for choosing Baton Rouge was due to the extensive familiarity of the authors, especially the corresponding author, with this city. Baton Rouge consists of 58 neighborhoods with a total population of 229,422 according to the most recent census in 2010. Since then, the estimated population has decreased to 221,599 in July 2018 (www.census.gov). Five neighborhoods within the city limits of Baton Rouge are selected for the collection of physical urban blight data utilizing the spatial video technology.

Neighborhoods are chosen depending on their crime density measured as the total number of crimes per km^2. The quintile classification method resulted in the following five different density classes, each class being represented by one neighborhood: (1) very low crime, (2) low crime, (3) moderate crime, (4) high crime, (5) very high crime (Fig. 1). The rationale for using crime density for the selection of the five neighborhoods assumes that crime is highly correlated with urban blight and that different levels of crime densities would result in different counts of urban blight incidents across the study

---

[1]An administrative subdivision in Louisiana that corresponds to a county in other US states (https://www.thefreedictionary.com/parish).

Fig. 1 Study area with five neighborhoods showing varying crime density (Stratmann 2019)

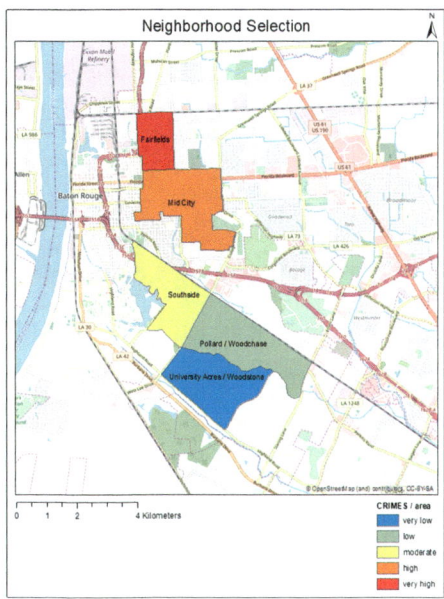

area. Additional criteria that were considered in the selection of the five neighborhoods were that they included no highways and no big lakes; being connected or located close to each other to minimize travel distances between them; had many buildings, easy to navigate through them by automobile, and possessed street networks with similar total street lengths.

The five selected neighborhoods exhibit crime densities decreasing from north to south (Fig. 1). The northernmost neighborhood (Fairfields) represents the neighborhood with a very high crime density. It is bordered in the south by the Mid City neighborhood that shows a high crime density, followed by the Southside neighborhood with a moderate crime density, and the Pollard/Woodchase neighborhood with a low crime density. The lowest crime density can be found in the southernmost selected neighborhood (University Acres/Woodstone). In total, the five selected neighborhoods consist of 22 census block groups.

The spatial video technology is utilized in this research to collect a complete as possible set of physical urban blight indicators in each neighborhood. This requires that all street segments inside each neighborhood are covered by a vehicle along an optimally designed route (considering, e.g., one-way streets) to minimize travel time. The 'Ride with GPS[2]' application is utilized to plan optimal routes based on both Open Street Map and Google Maps, since both applications sometimes differ in the map information provided. However, each designated optimal route include some street segments that have to be driven more than once.

---

[2]https://ridewithgps.com/

Fig. 2 Arrangement of spatial video equipment; a) camera on the middle inside windshield, b) two cameras on the right inside window of the backseat, c) two cameras on the left inside window of the backseat, d) charger equipment (Stratmann 2019)

2.2 Data Collection with the Spatial Video Technology

The spatial video technology allows the acquisition of video data that include informa-tion about the geographical location (with GPS) and a timestamp of each recorded video frame. Five extreme sport cameras from the brand Contour+2[3] are utilized to record the video data, while the vehicle is moving along the optimally designed route. Using suc-tion window clamps, all video cameras are mounted to the inside windows of the vehicle. One camera is mounted to the middle of the front windshield (Fig. 2a). The other four cameras are attached to the backseat windows of the car, two to the right and two to the left (Fig. 2b, c). The data collection could proceed with only one camera mounted to each of the backseat windows and one mounted to the front windshield (three cam-eras in total), however, one extra camera on each backseat window is used to guaran-tee a backup in case one camera fails to properly record the spatial video or loses the

---

[3]https://contour.com/

GPS connection. Since the data collection is partly implemented in environments with a high level of criminal activities, the cameras are attached to the inside of the vehicle for unobtrusive data collection. In power mode, the battery lasts for approximately two hours, but can be extended when connected to the vehicle's charger (Fig. 2d). A four-hour HD video recording can be saved on a 32 GB memory card and is saved in mp4 format with embedded GPS tracks (Curtis et al. 2013a; Curtis et al. 2015; Mills et al. 2010; Strelnikova et al. 2018). The placement of the cameras enables a wide viewshed for the next step in the data collection process, the digitization.

In total, 383.84 km were driven in 14 h and 36 min split up over eight days during spring 2019. All trips were conducted during daytime in good weather condition without precipitation. During the drives, every street segment, except highways, were covered inside the five selected neighborhoods. The average speed was 26.29 km/h. High speed should be avoided due to negative impacts on the quality of video recordings. Some cameras, for unknown reasons, stopped recording after about 30 min. Therefore, before 30 min of video recording were completed, cameras were turned off and on again in order to secure an appropriate data collection/storage. This was done every 30 min during each ride.

After field data collection, all video recordings with integrated GPS tracks and timestamps were viewed in a video player software. Every time a physical blight indicator, based on the developed criteria catalogue, was detected, the video was stopped and the blight indicator digitized (recorded) as a point symbol into a GIS layer at its location using ArcMap 10.6 (ESRI 2018).

2.3   Spatial Analysis Methods

After completing the digitizing process, all physical urban blight locations are spatially analyzed, either individually or aggregated at the level of census block groups. The main focus is to visualize the overall spatial distribution and to identify statistically significantly high and low spatial concentrations/clusters of urban blight indicators inside the study area. For the aggregated case, Local Moran's I Spatial Autocorrelation (LISA) statistics are calculated with GeoDA 1.12 (Anselin 2019). Kernel Density Estimations (KDE) statistics are derived from the individual urban blight locations. KDEs are calculated with CrimeStat 4.02 (Levine 2015) and their resulting density surfaces ('blightscapes') are visualized in ArcMap 10.6 (ArcMap 2018).

The LISA statistic derives local spatial hot- and cold spots, and spatial outliers of physical urban blight indicators aggregated to census block groups. Local spatial hot spots consist of census block groups with a high number of blight incidents that are surrounded by other block groups with a similarly high number of blight incidents. The relationship of the high blight counts between the central block group and all its neighboring block groups can be statistically described as local positive spatial autocorrelation. Likewise, local spatial cold spots are defined as block groups with low numbers of blight incidents surrounded by block groups with similarly low numbers of blight incidents. This spatial arrangement is also referred to as local positive spatial autocorrelation. In contrast, local spatial outliers, are either high incident urban blight block groups surrounded by low incident urban blight block groups, or vice versa.

LISA is the (spatial) extension of (global) spatial autocorrelation, which can be measured with the Moran's I statistic (Moran 1950). This statistic ranges between -1 and $+1$. A negative value can be interpreted as a spatial arrangement in which nearby block groups have dissimilar numbers of blight incidents, whereas a positive value is defined as a spatial arrangement in which nearby block groups have similar numbers of blight incidents. This latter arrangement became to be known as Tobler's First Law (TFL) of Geography, which states that "everything is related to everything else, but near things are more related than distant things" (Tobler, 1970). Finally, a Moran's I value around 0 points to a random distribution of varying numbers of blight incidents contained in each census block group.

Kernel Density Estimation (KDE) is both a spatial interpolation and hot/cold spot method for spatially discrete point data, such as physical urban blight incidents. A regular grid is placed over the study area and a density value calculated for each grid cell. The more urban blight locations are found inside and in the immediate vicinity of one cell, the higher the density of the cell is. KDE results are sensitive to the kernel function, a three-dimensional function placed over each urban blight location, and the function's bandwidth. Normal and quartic are the most prominent kernel functions applied. The shorter the bandwidth, the more peaked the resulting density distribution is, the wider the bandwidth, the more generalized the density estimation appears. The cell size does not impact the density estimation results, but influences how estimates are visualized (Eck et al. 2005; Smith and Bruce 2008).

3   Spatial Distribution of Physical Urban Blight

In total, 1,717 urban blight locations are identified and collected in the five neighborhoods of the selected study area (Fig. 3). Environmental/infrastructural blight accounts for about two-thirds of all locations (68.84% or 1,182 points), which is more than double the amount of property blight locations with 535 points (31.16%). It should be noted that any one location can possess more than one blight indicator. For example, a building can have a broken window, a blocked window, and graffiti. When considering blight indicators individually, 880 property blight indicators and 1,498 environmental/infrastructural blight indicators for a total of 2,378 blight indicators are identified in the selected study area. Overall, with 798 incidents, 'litter' is by far the most frequently occurring physical urban blight indicator. Among property blight indicators, 'structural integrity' and 'blocked windows/doors' are the most prevalent. In contrast, 'broken windows/ doors', 'building graffiti', and 'building overgrowth' occur least frequently. 'Overgrown vegetation', 'dumping', and 'unkempt areas' are behind 'litter' the most common

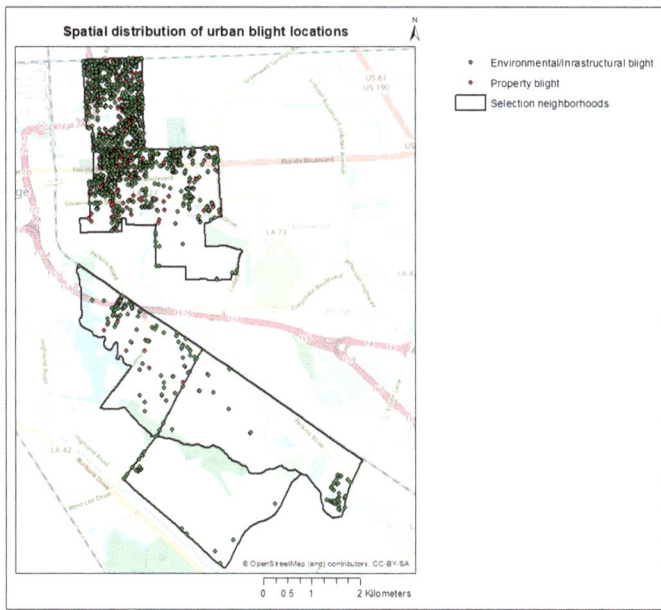

Fig. 3 Spatial distribution of individual property and environmental/infrastructural blight locations (Stratmann 2019)

environmental/infrastructural blight indicators. The vast majority of physical urban blight locations are spatially clustered in the two northern neighborhoods 'Fairfields' and 'Mid City', especially north of Clay Cut Rd., which is located south of and runs parallel to Government St., a major thoroughfare in Baton Rouge from downtown to the west. Due to the high number of points in these two neighborhoods, it is difficult to interpret a general trend in the spatial distribution of physical urban blight. This is the main drawback of the chosen visualization type. The few physical urban blight locations in the other three neighborhoods are mostly concentrated in the northwest and southeast.

3.1 Physical Urban 'Blightscapes' Applying the Local Moran's I Spatial Autocorrelation

Choropleth mapping is an alternative visualization method to the one in the previous subsection (i.e., common dot map), with points being aggregated to spatial units. In this research, physical urban blight locations are aggregated to 22 census block groups and the number of blight locations for each block group recorded. This allows the Moran's I statistic, measuring the spatial autocorrelation, to be calculated. The scatterplot in Fig. 4 shows the number of urban blight locations in each census block group (calculated as a z-score) on the x-axis and the number of spatially-lagged urban blight locations for

Fig. 4 Moran's I value and scatter plot for urban blight aggregated to census block groups (Stratmann 2019)

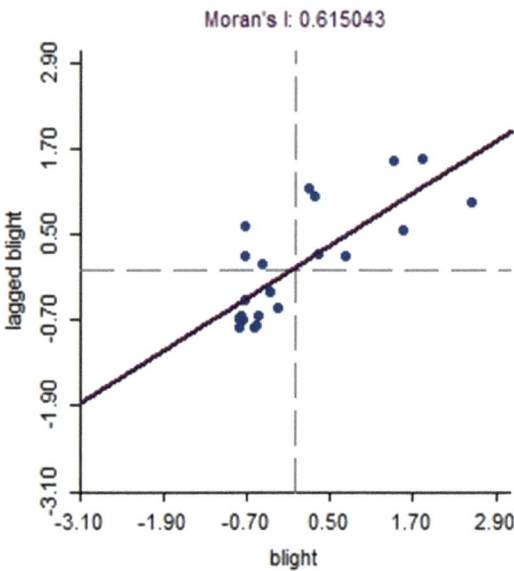

census block groups (calculated as a z-score, as well) on the y-axis. A spatially-lagged blight location is calculated as the average count of blight locations in all census block groups surrounding a (central) census block group. In this research, neighbors are defined with the first order Queen contiguity, meaning that two census block groups are considered neighbors, if they share a common edge or a common vertex. The slope of the univariate linear regression line that is fit into the point pattern of the scatterplot is equal to the Moran's I value (Anselin 1995). This value (0.615043) is displayed on top of the scatterplot in Fig. 4 and can be interpreted as a high positive spatial autocorrelation. This points to a spatial arrangement in which nearby block groups have similar numbers of blight incidents and thus confirms TFL of Geography.

The Moran's I scatter plot can be divided into four quadrants showing spatial associations of physical urban blight counts between a census block group and its first order Queen contiguity neighbors (Fig. 4). The portion of the point pattern falling into the lower left quadrant includes census block groups with below average urban blight counts surrounded by block groups that, on average, also exhibit similar below average urban blight counts. The spatial association of six of these census block groups surrounded by their Queen contiguity neighbors are statistically significant different from a chance association (based on 999 random permutations of the original counts of urban blight locations aggregated to census block groups) and are referred to as significant cold spots in the LISA cluster map (Fig. 5, left and designated as 'Low-Low (6)' in the map legend). The associated level of significance is displayed in the LISA significance map (Fig. 5, right) and ranges from 0.05 (four census block groups) to 0.01 (two census block groups).

In contrast, the upper right quadrant in the scatterplot is defined by block groups with above average urban blight counts surrounded by (Queen contiguity) block groups that,

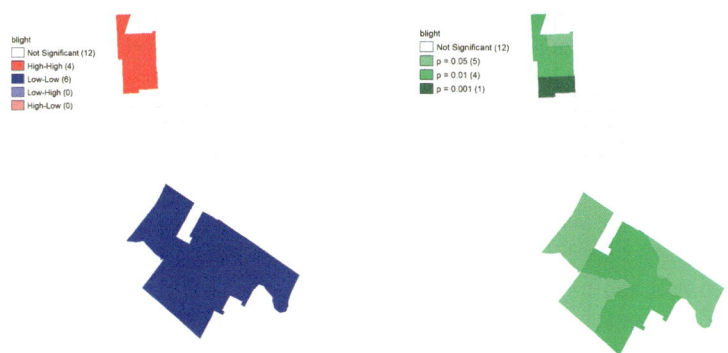

Fig. 5 Cluster map (left) and significance map (right) of physical urban blight aggregated to census block groups (Stratmann 2019)

on average, also exhibit similar above average urban blight counts. Of all points falling into this quadrant, four have a spatial association between the central block group and their neighbors that can be interpreted as statistically significantly different from chance. These four points are referred to as significant hot spots in the LISA cluster map (Fig. 5, left and designated as 'High-High (4)' in the map legend). The same four points have a statistical significance of 0.05 (one census block), 0.01 (two census blocks), and 0.001 (one census block).

Three of the 22 census block groups fall into the upper left quadrant of the scatter plot. They are defined by block groups with below average urban blight counts surrounded by (Queen contiguity) block groups that, on average, exhibit above average urban blight counts. None of these three block groups exhibit a spatial association with their adjacent neighboring block groups that is statistically significant different from a chance association. These three points are referred to as non-significant spatial outliers in the LISA cluster map (Fig. 5, left and designated as 'Low–High (0)' in the map legend) and are filled with a gray color in the LISA significance map (Fig. 5, right). Finally, no census block groups fall into the lower right quadrant.

As expected, significant urban 'blightscape' hot spots are exclusively found in the northern two neighborhoods of the study area. In contrast, all but three census block groups in the southern three neighborhoods are defined as statistically significant urban 'blightscape' cold spots. No significant spatial outliers are found in the urban 'blightscape'.

3.2 Physical Urban 'Blighscapes' Applying the Kernel Density Estimation

As mentioned in Subsection 1.2, the East Baton Rouge Parish and its largest urban area, the city of Baton Rouge, have developed an Open Data Portal to provide citizens with information on housing and development, culture and recreation, government, business and financial, public safety, and transportation and infrastructure (https://data.brla.gov/). Concerned citizens can also make requests for service submitted through the City-Parish 311 Call Center. Among other things, citizens can self-report urban blight indicators identified in their neighborhood. All requests for services are listed in the Open Data Portal and can be visualized through standard mapping functions. Indicators in the Open Data Portal consist of broader categories than the one collected with the spatial video technology. Examples for such categories are recycling, sewer/wastewater, street/traffic issue, environmental issues, blighted property, garbage, etc. From these broad categories, environmental issues and blighted property, are selected to closely resemble the physical urban blight indicators collected with the spatial video technology. For the same five neighborhoods, a total of 708 self-reported blight locations are selected from the Open Data Portal.

Two KDE 'blightscape' surfaces, comparing the spatial distribution of urban blight indicators derived from the spatial video with the self-reported blight locations from the Open Data Portal, are illustrated in Fig. 6. Both calculations are based on the same parameter settings, with final kernel density estimates normalized from 0–100 to better account for the difference in the total amounts of blight locations (1,717 and 708, respectively). The 'blightscape' surface derived from the spatial video shows a higher number and larger hot spot areas in Fairfields (northernmost neighborhood) and the northern part of Mid City (neighborhood immediately adjacent to Fairfields in the south), as compared

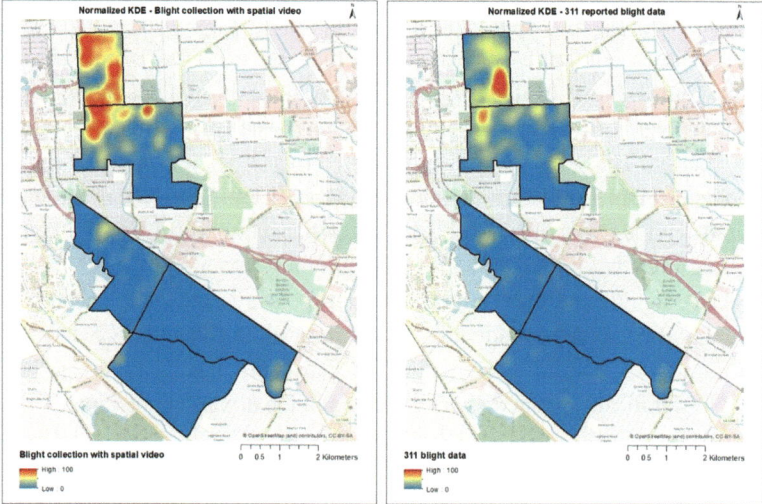

Fig. 6 Comparison between KDE-derived 'blightscape' surfaces of physical urban blight locations collected with the spatial video (left) and self-reported blight locations from the Open Data Portal (right) (Stratmann 2019)

to the self-reported 'blightscape' surface. The two 'blightscape' surfaces for the other three neighborhoods are very similar between the two sets of urban blight data.

The final step in the analysis is to construct KDE-derived 'blightscape' surfaces for the six most prevalent urban blight indicators collected with the spatial video technology. The top three of these 'blightscape' surfaces in Fig. 7 show property blight indicators that occur most frequently in the study area. They include abandoned property, blocked window/door, and structural integrity. The bottom three 'blightscape' surface maps include the most occurring environmental/infrastructural blight indicators. These are dumping, litter, and overgrown vegetation. All 'blightscape' surfaces apply the same parameter settings, with a normal kernel function and a grid cell size of 100*100 m. Final density values are classified into five ranges, according to the quintile classification method (Fig. 7).

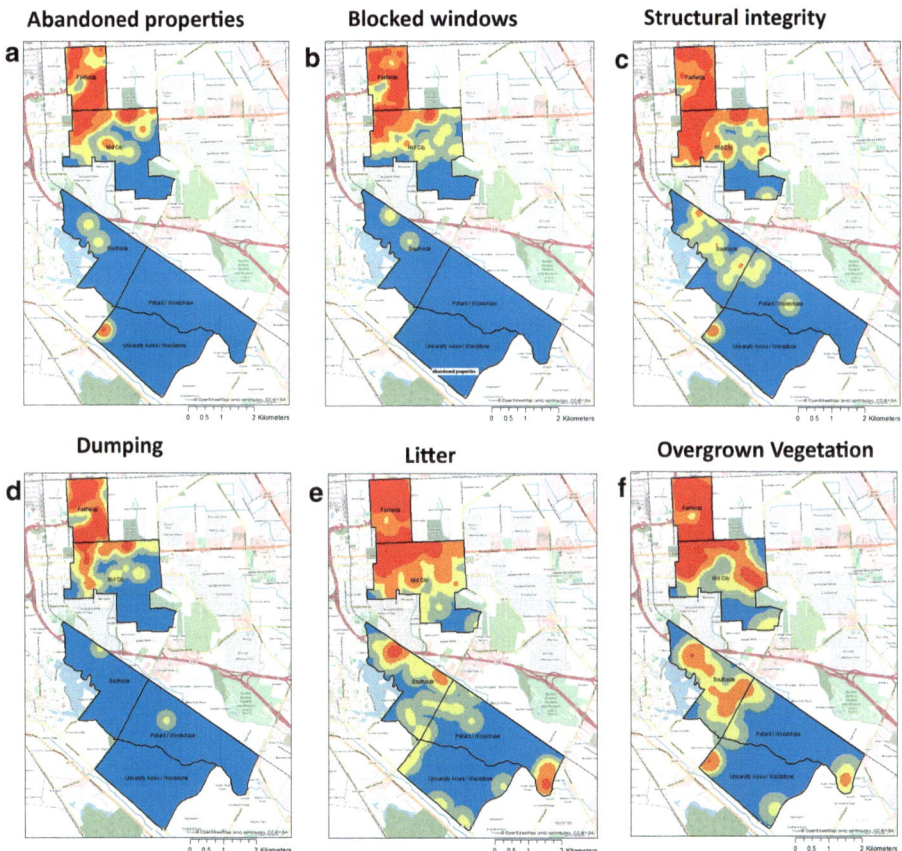

Fig. 7 KDE-derived 'blightscape' surfaces of the most prevalent urban blight indicators collected with the spatial video technology: (a) abandoned properties, (b) blocked windows, (c) structural integrity, (d) dumping, (e) litter, (f) overgrown vegetation (Stratmann 2019)

The three blight indicators, abandoned properties, blocked windows, and dumping show very similar 'blightscape' surfaces with high spatial concentrations mainly in Fairfields (northernmost neighborhood) and in the northern part of Mid City (neighborhood immediately adjacent to Fairfields in the south). The other three blight indicators, namely structural integrity of buildings, litter, and overgrown vegetation, also show (very) high 'blightscape' KDEs in Fairfields and Mid City. However, pockets of moderate to very high 'blightscape' values for these three indicators can also be found inside the other three adjacent neighborhoods in the south of the study area. These latter three blight indicators seem to be more common phenomena in Baton Rouge that are not necessarily and exclusively associated with higher crime neighborhoods. Validating this statement would obviously require the collection of the same urban blight indicators with the video technology across all city neighborhoods. It is also worth noting that out of all five neighborhoods, Mid City shows a very high diversity in each of the six most prevalent urban blight indicators, as all five (color) categories from each blight indicator of the quintile classification appear in this neighborhood. The main reason is that Mid City includes two socio-economically and ethnically diverse neighborhoods. The boundary between the two 'sub neighborhoods' runs from East to West through about the middle of Mid City somewhere between Government St. and Clay Cut Rd. (compare the discussion in Sect. 3 and Fig. 3).

4   Summary and Future Research

Urban blight continues to be a serious problem in any city's neighborhood. Fighting urban blight requires knowledge about the seriousness of this problem and its spatial distribution across neighborhoods. However, fine-scale data about urban blight are difficult to come by. One possibility is to establish an online and open data portal, where concerned citizens can self-report urban blight issues occurring inside their neighborhood. The City-Parish government of Baton Rouge has gone this route and provides an Open Data Portal (https://data.brla.gov/), where citizens can submit requests for service, including urban blight issues, to their local government. While this is a great service for the citizens of Baton Rouge, we believe that such collection efforts result in an urban blight dataset that does not follow a standardized collection procedure, introduces selection bias, and for this reason is less comprehensive and complete. A second possibility is the application of a fairly novel geospatial technology termed Spatial Video Acquisition System (SVAS) or spatial video, for short. This is the data collection procedure that we introduce in this research for the collection, analysis, and mapping of fourteen physical urban blight indicators in five neighborhoods in the city of Baton Rouge.

Results show that spatial video is a fairly reliable and low-budget technology to capture urban blight indicators in a comprehensive and standardized way. In total, 1,717 blight locations with 2,378 individual blight indicators – any one location can include more than a single blight indicator - are collected with the spatial video technology. Two different spatial analysis methods are applied to build 'blightscape' surfaces with the collected data across the five neighborhoods in Baton Rouge. In a first analysis step, the Local Moran's I Spatial Autocorrelation (LISA) statistic is applied to physical urban blight locations aggregated to 22 census block groups subdividing the five neighborhoods. This results in a 'blightscape' surface that allows statistically significant spatial hot-/ cold spots and spatial outliers to be identified. In a second step, KDE-derived 'blightscape' surfaces (KDE is an interpolation method for spatially discrete data, such as urban blight locations) are created for (1) all blight locations and (2) their individual indicators. Both LISA and KDE analyses show similar 'blightscape' surfaces with hot spots exclusively found in the northern two neighborhoods (Fairfields and Mid City) and cold spots solely identified across most of the southern three neighborhoods (Southside, Pollard/Woodchase, University Acres/Woodstone). A comparison of KDE-derived 'blightscape' surfaces between urban blight locations collected with the spatial video and self-reported blight locations from the Open Data Portal shows a higher number and a larger extent of hot spots for the spatial video data in the two northernmost neighborhoods. For the rest of the study area, the two urban 'blightscape' surfaces calculated from the two different collection methods show very similar results.

Future research can continue in many different directions. First, in order to find out whether results from this research can be generalized, the collection of physical urban blight indicators with the spatial video technology would need to be expanded to a larger number of neighborhoods in Baton Rouge and replicated in other urban areas. Second, the impact of future interventions by the city government to alleviate urban blight issues in Baton Rouge could be measured with the collection of urban blight data at regular time intervals, especially before and after interventions, in the target- and selected control areas. Third, the spatial video technology could be enhanced with a geo-narrative approach (audio data collected in the form of interviews with experts that can be linked to the location of the video with a time stamp) to gain further insights into the urban blight problem and to solicit solutions for its reduction.

Acknowledgements   This research was funded by the Austrian Science Fund (FWF) through the Doctoral College GIScience at the University of Salzburg (DK W 1237-N23). Alina Ristea and Judith Stratmann are very thankful to the Austrian Marshall Plan Foundation for receiving a Marshall Plan Scholarship to fund their research stay in Baton Rouge to cooperate in this project.

References

Anselin, L. (1995). Local Indicators of Spatial Association — LISA. *Geographical Analysis, 27,* 93–155.

Anselin, L., Syabri, I. & Kho, Y. (2019). *GeoDa (Version 1.12)* [Computer software]. Retrieved November 12, 2019, from https://geodacenter.github.io/download.html

Curtis, A., Blackburn, J. K., Widmer, J. M., & Morris, J. G. (2013a). A ubiquitous method for street scale spatial data collection and analysis in challenging urban environments: Mapping health risks using spatial video in Haiti. *International Journal of Health Geographics, 12,* 21. https://doi.org/10.1186/1476-072X-12-21

Curtis, J. W., Curtis, A., Mapes, J., Szell, A. B., & Cinderich, A. (2013b). Using Google Street View for systematic observation of the built environment: Analysis of spatio-temporal instability of imagery dates. *International Journal of Health Geographics, 12,* 53. https://doi.org/10.1186/1476-072X-12-53

Curtis, A., Curtis, J. W., Shook, E., Smith, S., Jefferis, E., & Porter, L. (2015). Spatial video geonarratives and health: Case studies in post-disaster recovery, crime, mosquito control and tuberculosis in the homeless. *International Journal of Health Geographics, 14,* 22. https://doi.org/10.1186/s12942-015-0014-8

EBRGIS Open Data (2019). *East Baton Rouge GIS Map Portal.* City of Baton Rouge and Parish of East Baton Rouge. Retrieved May 10, 2019 from https://data-ebrgis.opendata.arcgis.com/

Eck, J. E., Chainey, S., Cameron, J. G., Leitner, M., & Wilson, R. E. (2005). *Mapping Crime: Understanding Hot Spots.* Washington, DC: U.S. Department of Justice.

Edler, D., Keil, J., & Dickmann, F. (2020a). From Na Pali to Earth—An 'Unreal' engine for modern geodata? In D. Edler, C. Jenal, & O. Kühne (Eds.), *Modern approaches to the visualization of landscapes* (pp. 279–291). Wiesbaden: Springer VS.

Edler, D., Jenal, C., & Kühne, O. (2020b). Modern Approaches to the Visualization of Landscapes—An Introduction. In D. Edler, C. Jenal, & O. Kühne (Eds.), *Modern Approaches to the Visualization of Landscapes* (pp. 3–15). Wiesbaden: Springer VS.

Environmental Systems Research Institute (ESRI) (2018). *ArcGIS Desktop. Version 10.6.* Retrieved November 12, 2019, from https://desktop.arcgis.com/de/arcmap/

Gau, J. M., & Pratt, T. C. (2010). Revisiting Broken Windows Theory: Examining the Sources of the Discriminant Validity of Perceived Disorder and Crime. *Journal of Criminal Justice, 38*(4), 758–766.

Hinkle, J. C. & Weisburd, D. (2008). The irony of broken windows policing: A micro-place study of the relationship between disorder, focused police crackdowns and fear of crime. *Journal of Criminal Justice, 36*(6), 503–512.

Kelling, G. L. & Wilson, J. Q. (1982). Broken Windows. The police and neighborhood safety. *Atlantic Monthly 249, 3,* 29–38.

Kleber, A., Edler, D., & Dickmann, F. (2020). Cartography and the sea: A javascript-based web mapping application for managing maritime shipping. In D. Edler, C. Jenal, & O. Kühne (Eds.), *Modern approaches to the visualization of landscapes* (pp. 173–186). Wiesbaden: Springer VS.

Levine, N. (2015). *CrimeStat. Version v 4.02.* A Spatial Statistics Program for the Analysis of Crime Incident Locations. Houston: Ned Levine & Associates; National Institute of Justice.

Maghelal, P., Andrew, S. A.; Arlikatti, S. & Jang, H. S. (2013). From blight to light. Assessing blight in the city of Dallas. Final Report. *Department of Public Administration University of North Texas.* Retrieved November 12, 2019, from https://www.dallasareahabitat.org/wp-content/uploads/2015/09/Blight-Study.pdf

Marco, M., Gracia, E., Martín-Fernández, M., & López-Quílez, A. (2017). Validation of a Google Street View-Based Neighborhood Disorder Observational Scale. *Journal of Urban Health: Bulletin of the New York Academy of Medicine, 94*(2), 190–198.

Meyer-Heß, F. (2020). Discovering forgotten landscapes. In D. Edler, C. Jenal, & O. Kühne (Eds.), *Modern approaches to the visualization of landscapes* (pp. 33–45). Wiesbaden: Springer VS.

Moore, H. C. (2018). State of crime: 2018. A look at crime and response needs in East Baton Rouge Parish. *East Baton Rouge District Attorney's Office.* 6–75.

Moran, P. A. P. (1950). Notes on continuous stochastic phenomena. *Biometrika, 37,* 17–23.

Open Data BR (2019). Open Data Baton Rouge. *City of Baton Rouge.* Retrieved May 5, 2019, from https://data.brla.gov/.

Poplin, A., Andrade, B. de, & Mahmud, S. (2020). Exploring tangible and intangible landscapes of evocative places: Case study of the City of Vitória in Brazil. In D. Edler, C. Jenal, & O. Kühne (Eds.), *Modern approaches to the visualization of landscapes* (pp. 519–547). Wiesbaden: Springer VS.

Prisille, C., & Ellerbrake, M. (2020). Virtual Reality (VR) and geography education: Potentials of 360° 'Experiences' in secondary schools. In D. Edler, C. Jenal, & O. Kühne (Eds.), *Modern approaches to the visualization of landscapes* (pp. 321–332). Wiesbaden: Springer VS.

Ross, C. E., & Mirowsky, J. (2001). Neighborhood Disadvantage, Disorder, and Health. *Journal of Health and Social Behavior, 42,* 3. https://doi.org/10.2307/3090214

Sampson, R. J., & Raudenbush, S. W. (1999). Systematic Social Observation of Public Spaces: A New Look at Disorder in Urban Neighborhoods. *American Journal of Sociology, 105*(3), 603–651.

Siepmann, N., Edler, D., & Kühne, O. (2020). Soundscapes in cartographic media. In D. Edler, C. Jenal, & O. Kühne (Eds.), *Modern approaches to the visualization of landscapes* (pp. 247–263). Wiesbaden: Springer VS.

Skogan, W. G. (1990). Disorder and decline. Crime and the spiral of decay in American neighborhoods. *New York: Free Press, 52.*

Smith, S. C., & Bruce, C. W. (2008). *CrimeStat III.* Washington, DC: User Workbook. The National Institute of Justice.

Stratmann, J. (2019). *Fine-scale Analysis and Modeling of Urban Blight and Crime Applying Geospatial Technology: A Case Study in Baton Rouge, Louisiana.* (Unpublished Master Thesis). Spatial Information Management, Carinthia University of Applied Sciences, Villach, Austria.

Thomas, P. M. (2020). The digitalizing society—Transformations and challenges. In D. Edler, C. Jenal, & O. Kühne (Eds.), *Modern Approaches to the Visualization of Landscapes* (p. 447–457). Wiesbaden: Springer VS.

Tobler, W. R. (1970). A computer movie simulating urban growth in the Detroit region. *Economic Geography, 46,* 234–240.

United States Census Bureau (2018): QuickFacts. *City of Baton Rouge,* Louisiana. Retrieved April 17, 2019, from https://www.census.gov/quickfacts/batonrougecitylouisiana.

Vetter, M. (2020). Technical potentials for the visualization in virtual reality. In D. Edler, C. Jenal, & O. Kühne (Eds.), *Modern approaches to the visualization of landscapes* (p. 307–317). Wiesbaden: Springer VS.

Weisburd, D., Hinkle, J. C., Famega, C. & Ready, J. (2010). Legitimacy, Fear and Collective Efficacy in Crime Hot Spots: Assessing the Impacts of Broken Windows Policing Strategies on Citizen Attitudes. *U.S. Department of Justice.* Retrieved November 12, 2019, from https://www.ncjrs.gov/pdffiles1/nij/grants/239971.pdf

Ms. Judith Stratmann is a master student in Spatial Information Management at Carinthia University of Applied Sciences in Villach, Austria.

Dr. Alina Ristea is a postdoctoral research associate, Boston Area Research Initiative, School of Public Policy and Urban Affairs at Northeastern University in Boston, USA.

Dr. Michael Leitner is a professor of geography in the Department of Geography and Anthropology, Louisiana State University in Baton Rouge, USA.

Dr. Gernot Paulus is a professor of geoinformation at Carinthia University of Applied Sciences in Villach, Austria.

Exploring Tangible and Intangible Landscapes of Evocative Places: Case Study of the City of Vitória in Brazil

Alenka Poplin, Bruno de Andrade and Shoaib Mahmud

Abstract

This paper explores tangible and intangible characteristics of places. It concentrates on gathering characteristics, emotions, memories and stories related to self-selected evocative places in a city. Evocative places are defined as places that evoke images, memories or emotions. There are two goals identified for this article. The first goal is to study which words citizens use to describe the main characteristics of their self-selected evocative places. The second goal is to map emotions associated with the self-selected evocative places. The case study selected in this research is the city of Vitória in Brazil. We collected 192 evocative places and their characteristics with the help of an online mapping platform that links an online questionnaire with an interactive map. This paper summarizes the main results gathered empirically about evocative places in Vitória, their characteristics and the emotions felt at these places. These places are then mapped in a geographic information system (GIS) in order to understand their locations and concentrations. On the basis of this empirical work in Vitória, and the work accomplished in the cities of Hamburg (Germany), Vienna

A. Poplin (✉) · S. Mahmud
Iowa State University, Ames, USA
e-mail: apoplin@iastate.edu

S. Mahmud
e-mail: smahmud@iastate.edu

B. de Andrade
University College Dublin, Dublin, Ireland

Delft University of Technology, Delft, The Netherlands
e-mail: B.deAndrade@tudelft.nl

© Springer Fachmedien Wiesbaden GmbH, part of Springer Nature 2020   517
D. Edler et al. (eds.), *Modern Approaches to the Visualization
of Landscapes*, RaumFragen: Stadt – Region – Landschaft,
https://doi.org/10.1007/978-3-658-30956-5_29

(Austria), Ames and Grinnell (both Iowa, USA), we also designed and expanded the conceptual model of evocative places presented in this paper for the first time. The conceptual model includes four main categories with which an evocative place can be described including its physical characteristics, experiences, senses and values. We conclude the article with a discussion and further research directions.

Keywords

Place characteristics · Place experiences · Emotions felt at places · Conceptual model of evocative places · City of Vitória, Brazil

1   Introduction

People and places interact to form the experience. By better understanding components that contribute to positive place experiences, designers can create spaces that promote comfort, a sense of belonging, and a bond between people and places (Waxman 2006). What are the characteristics of places that may evoke positive emotions and contribute to increased happiness of the citizens? How do these characteristics affect people, the inhabitants of these cities and their affective states? This research aims to contribute to these discussions. The study concentrates on how people feel at certain places and focuses especially on public places that evoke images, memories and emotions (for additional studies addressing cartographic visualizations of emotions, see in this volume: McLean 2020; Siepmann et al. 2020). They are called evocative places. Research presented in this article builds on previous literature in this area (de Andrade and de Almeida 2016; de Andrade 2019; Poplin 2017, 2018, 2020).

Recent discussions on happiness and happy places/cities have intensified and resulted in the first World Happiness Report being published in 2012 in support of the UN High Level Meeting on happiness and well-being (Helliwell et al. 2019). Happiness has been considered as an important measure of social progress and one of the goals of public policy. The OECD is also committed to put people's well-being and happiness at the center of public policy and governments' efforts (OECD 2017). Paying more attention to emotions, affects, and to happiness should be part of our efforts to achieve both human and sustainable development (Helliwell et al. 2017). The way we design places and our communities plays an important role in how we experience our lives (Walljasper 2017).

This study concentrates on the city of Vitória in Brazil as one of the co-authors of this article studying the heritage and values (de Andrade and de Almeida 2016, de Andrade 2019) of this city initiated this additional research on evocative places for the same city. We explore the image of the place as remembered, memorized, stored by their citizens. What do they recollect and associate with their self-selected places? What are the main characteristics of these places and which emotions and stories can they share in relation to these places?

The research methodology uses an online interactive map-based platform for data collection called maptionnaire (Maptionnaire 2020). The platform combines an interactive map with a questionnaire and saves the data in the back-end of the platform. This data can then be downloaded and imported in a geographic information system (GIS) which enables to visualize the collected data on maps.

There are two main results of this paper. The first one is the conceptual model for evocative places, places that evoke emotions, images and memories. It is based on a set of experiments conducted by the main author of this article in different cities, countries and continents (Poplin 2017, 2018, 2020). The experiments conducted in Vitória add dimensions to the model from the perspective of South American continent. The second result is a collection of the evocative places in the city of Vitória and the intangible landscapes of emotions associated with these places and mapped for this city. Intangible landscapes in this context refer to landscapes to which certain meaning, memory, lived experience and attachment, in relation to people's connection to locality and landscape, can be traced, detect, and even map and/or analyse (Müller 2008). With the research on evocative places we aim to contribute to a better understanding of intangible dimensions of places including emotions and how are they related to the characteristics of places and activities people perform at these places. It also contributes to the discussion about places and the way we can structure and organize the knowledge and descriptions of places.

2   Place, Emotions and Values

2.1   Defining the Notion of a Place

Cresswell (2004) in his book *Place: a short introduction* reviews the concepts and definitions of place and makes a clear distinction to space, landscape and location. According to him, places are "spaces that people have made meaningful" (Cresswell 2004, p. 8); a place is therefore defined as a meaningful location. "As well as being located…places must have some relationship to humans and the human capacity to produce and consume meaning" (Cresswell 2004, p. 8). A place can be described as the intersection of a setting's physical characteristics, a person's individual perceptions, and the actions or uses that occur in a particular location (Bonnes and Secchiaroli 1995; Bott et al. 2003; Canter 1977; Pretty et al. 2003).

People establish relationships with places and we may talk about people-place bonding or an attachment to location (Altman and Low 1992). Place attachment can be considered as an interplay of emotions, knowledge, beliefs, and behaviors in reference to a place (Low and Altman 1992; Proshansky and Fabian 1983; Waxman 2006). The word *attachment* refers to affect while the word *place* refers to the "environmental settings to which people are emotionally and culturally attached" (Low and Altman 1992, p. 5). When relationships develop between people and places, the result is often a feeling of place attachment. "Places root us - to the earth, to our own history and memories, to our families and larger community" (Cooper-Marcus and Francis 1998, p. xi). Tuan (1980) suggested the existence of a state of rootedness in which one's personality merges with one's place.

2.2   Place and Restorative Experience

People and place interact together to form the experience. The experience of place is unique to each individual and is directly related to his or her lived experiences and personal disposition. Yi-Fu Tuan (1977) writes that people construct their reality through their experience. The experience can range from "more direct and passive senses of smell, taste, touch, to active visual perception and the indirect mode of symbolization" (Tuan 1977, p. 8), and is "compounded of feeling and thought". Steele (1981) distinguishes among immediate feelings and thoughts, views of the world, intimate knowledge of one spot or location, memories or fantasies, and personal identification.

Places can also denote emotional support, restorative experience or fulfill people's emotional needs (Kaiser and Fuhrer 1996). We may talk about restorative experiences found and felt in specific places when people seek places to recharge or recuperate which may sometimes happen after traumatic, difficult or negative experience in their lives. Restorative experiences of these places may involve positive mood changes, positive sensations, or feelings of being at peace or even experiencing the feeling of happiness.

2.3   Place and Emotions

In environmental psychology, research on emotions and affect are considered important topics (Kaplan and Kaplan 1989; Kaplan 1995; Korpela et al. 2001; Korpela 2012; Russell and James 2003; Russell and Pratt 1980). "*Affect is central to conscious experience and behavior in any environment, whether natural or built, crowded or unpopulated. Because virtually no meaningful thoughts, actions, or environmental encounters occur without affect*" (Ulrich 1983, p. 85). In addition, the cognitive component is of considerable value in experiencing the physical environment as well (Ittelson 1973, Russell and Pratt 1980). In recent studies by Shoval, Zeile and their collaborators (Shoval et al. 2018; Zeile et al. 2016), researchers study objective and subjective measurement methods for recording emotions in the cities, which is not an easy task. How can you separate body sensations (usually measured) from the emotions and how can you identify those?

Schlosberg's (1954) was the first one that proposed a model which illustrates emotions in a circular form which represent the range of facial expressions. Figure 1 demonstrates the model taken from his original publication. The emotions are indicated on

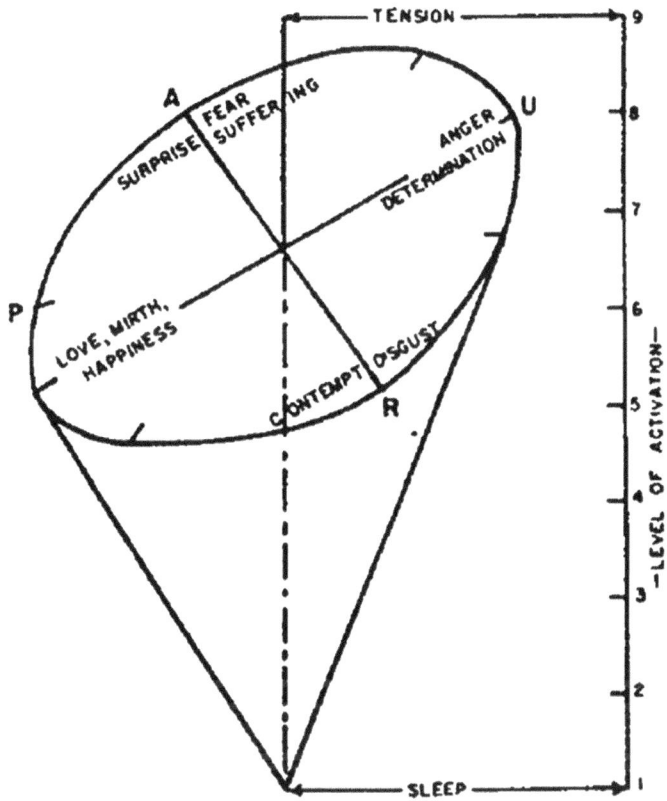

Fig. 1 Schlosberg's model of emotions from 1954 (Schlosber 1954)

the ordinate with respect to their maximum level of activation (sleep-tension). "The top surface is sloped to show that anger and fear can reach higher levels of activation than can contempt. The other two dimensions are pleasantness-unpleasantness and attention-rejection (Schlosberg 1954).

Inspired by Schlosberg's work Russell (1980) proposed *A Circumplex Model of Affect* that includes a two-dimensional space with eight variables arranged in a circle (Fig. 2). The horizontal, east-west dimension is the pleasure-displeasure dimension. The vertical, north-south dimension is arousal-sleep dimension. The diagram introduces four quadrants. On the north-east we find excitement and the polar opposite is depression on the south-west. Distress is located on the north-west with its polar opposite, contentment, on the southeast. This model is still used today in social psychology to assess affect (Russell 1980).

Fig. 2 Russell's (1980) Circumplex model of affect, representing eight affect concepts in a circular order

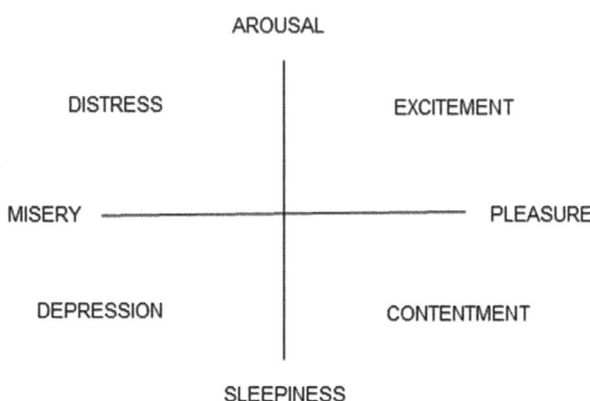

2.4   Place and Values

Place can also be looked from the perspective of values and memories. The division comes from the tangible perspective addressing the use of place and its material characteristics vs. its intangible nature of storing the memories, sense of belongings and collective values (Geddes 1949; Magnaghi 2005; Poli 2013, 2015; Rossi 1982). The place as a collective heritage has a value of existence of an intangible dimension that concerns the transmission of the legacy of past generations towards the future, and a value of use as a resource that must support local development and ensure the reproducibility of historical narrative and active memory (de Andrade and de Almeida 2016, de Andrade 2019; Ecléa 1994; Pierre 1996;). The collective values may encompass stories and memories attached to this place. We follow the distinction of the values of use from the values of existence. Values of use are material sediments, legacy objects of morphology, physicality of place, and landscape. Values of existence are cultural and identities sediments, legacies of collective memory, symbolic values and characters of belonging (Magnaghi 2005). The moment a place and its utility (of use and of exchange) is revealed socially, it enters the collective memory and acquires a value of existence (Poli 2013). The value of use can be altered over time, but the value of existence remains because they relate to the symbolic representations of the place that are stored as an active memory in the narratives of everyday life and keeping the collective imagination alive.

3   Research Focus and Methodology

3.1   Research Goals: Evocative Places and their Characteristics

This study concentrates on positive places, i.e. places that evoke images, memories and emotions, at which people can recharge, relax and recuperate. We call them *evocative places*. The definition of an evocative place is grounded in the conceptual literature on places (Tuan 1974, 1977, 1980, Cresswell 2004). An evocative place is defined as a meaningful location (Cresswell 2004, Authors' references), and by giving emotional/ affective meanings to this place, people form place attachments (Low 1992). They construct places – and their reality – through experience (Tuan 1977) and their affect (Russell 1980). Therefore, evocative places are meaningful locations with which people associate meaning and affect as a result of their experience of this place.

There are two main research goals set for this article: a. to better understand locations and characteristics of evocative places in a city and to construct a universal model of an evocative place that can be used across continents, countries, cultures, languages and cities; and b. gather and map the intangible landscapes of emotions associated with the self-selected evocative places. The result of the first goal is the conceptual model that aims to provide the so called "image of the place" a tool that provides categories with which an evocative place can be described. We call it The Conceptual Model of an Evocative Place. It not only includes the location of the place, but also its characteristics, experience, and values. The result of the second goal are maps of the selected city with locations of evocative places and analysis of the most commonly expressed emotions associated with these self-selected places.

3.2   Research Material and Methodology

The research methodology is concentrated around online data collection. The data collected are self-selected evocative places, their characteristics and emotions felt at these places. The main question asked for the self-selected evocative places was "Select a place in the city at which you can relax, recharge and at which you feel at peace". The survey participants were given the link to the online survey and asked to respond to the online questions related to their self-selected evocative places. The online map-based survey consisted of the following main parts:

1. Identify the location of self-selected evocative places and mark them on the online interactive map
2. Describe the characteristics of the self-selected evocative places with your own words; choose up to three words that best describe your self-selected evocative place
3. Choose the words that best describe the emotions felt at these places from the list of emotions; select up to three words for emotions from the list of emotions given to you in the survey
4. Communicate which transportation mean is used to access the self-selected evocative places; choose among feet/walking, bike, car, public transportation or other
5. Share personal stories, memories and values connected to the self-selected evocative place

The survey was implemented in Maptionnaire (Maptionnaire 2020) environment. The link to the online survey can be found at the following address https://app.maptionnaire.com/en/3690. This platform enabled to link a questionnaire with online interactive maps (related in this volume: Kleber et al. 2020). The data collected was then stored in the back-end of the platform and accessible to the researcher for download. After the completion of the survey, the data collected had to be exported into the form of shape files or/and Excel spreadsheets.

3.3   Study Example: The City of Vitória

The selected study case for this research was the city of Vitória in Brazil. There are several reasons for this choice. The main reason was that wanted expand research on evocative places to different continents and explore how people describe their places and emotions in different continents, states and cities. This selection contributes to the expansion of the conceptual model of an evocative place that embraces the notions coming from different cultural backgrounds. The other, minor, reason is that one of the co-authors of the paper, at the time of writing this paper, lived and worked in this city.

The city of Vitória is the capital of the State of Espírito Santo state in Brazil (Fig. 3). It is located on the eastern coast region of Brazil, with a 96,536 km^2 area, and 327.801 inhabitants according to the last census of the Brazilian Institute of Geography and Statistics in 2010 (IBGE 2010). The states of Rio de Janeiro borders on the south, Bahia borders on the north, and Minas Gerais borders on the west side of the city. The history of the city goes back to the 16th century, when the king of Portugal, D. João III, divided the Brazilian lands into hereditary captaincies, and designated the nobleman Vasco

Fig. 3   The location of the city of Vitória (left) and a closer look into the city (©google maps)

Fernandes Coutinho to the Espírito Santo captaincy. In 1551, the Portuguese won a fierce battle against the Goytacaz Indians at the island across the bay where they first settled, and began to call the place Victoria, to honor this achievement/conquest.

There are several additional reasons for studying Vitória, including its history, its unique geographic configuration as an island surrounded by a bay and the sea, the increased altimetry as one approaches the center of the island, the currently forming extensive areas for the preservation of vegetation cover, as well as the raise of the informal settlements in the city. The iron industry, the agricultural products of the state's interior, and two important ports have been supporting the local economy. The city expanded in the last decade, including the emerging new green and public spaces on the waterfront making it increasingly attractive as a tourist destination.

3.4   Survey Participants

Table 1 summarizes the characteristics of our participants. Altogether, 73 inhabitants of the city of Vitória responded to the online survey. The majority of the participants (52 all together) were of the age group between 19 and 25 years old. If we add 11 participants in the age between 26 and 35 we understand that the survey mostly captures the responses of a young population of this city. This is comparable with other studies by the authors of this articles (Authors' references) which concentrated on interviewing the

Table 1 The main characteristics of the survey participants (source: the author's own work)

Category	Count	Percentage
Age		
19–25	52	71.23%
26–35	11	15.07%
36–25	3	4.11%
46–55	2	2.74%
Not answered	5	6.85%
Total	73	100%
Gender		
Male	18	24.66%
Female	50	68.49%
Not answered	5	6.85%
Total	73	100%
Employment		
Student (Estudante)	47	64.38%
Full-time (Tempo Integral)	5	6.85%
Part-time (Tempo Parcial)	12	16.44%

(continued)

Table 1 (continued)

Category	Count	Percentage
Unemployed (Desempregado)	1	1.37%
Not Answered	8	10.96%
Total	73	100%
Years of living in the city		
Since Childhood (desde a infância)	34	46.58%
More than 5 years (mais de 5 anos)	12	16.44%
Between 2-5 years (entre 2-5 anos)	11	15.07%
Less than 2 years (menos de 2 anos)	7	9.59%
Not Answered	9	12.33%
Total	73	100%

students. The majority of the participants were students (47) and part-time employees (12). The majority of the participants lives in Vitória since their childhood (34) or more than 5 years (12). This means that they know the city quite well and are suitable to participate in our online survey. The distribution among the gender groups is not completely equal; there are more female participants (50 all together) than male participants (18 participants). In the next data collection for the same city we will particularly target more male participants to get approximately equal number of both genders. For the purpose of this study, this inequality does not represent a problem.

4   Characteristics of Places: Conceptual Model of an Evocative Place

4.1   Locations and Concentrations of Evocative Places

Each participant of the online survey on evocative places could mark up to three self-selected evocative places. Figure 4 shows the locations of 192 collected evocative places in and around the city of Vitória. Each of the indicated evocative place is represented by a black dot.

The majority of the evocative places in Vitória is located at the waterfronts, in the southwest around the old town, in the southeast where the bay meets the ocean, in the northeast in the seashore, and at the university. Some evocative places can be found in the hill's belvederes or homes with sea views. Some additional evocative places can be found on the south of the main city, in Vila Velha city. They mainly include the locations on the coast and a few of them in the nearby hills. These locations show a preference for the east region of the island close to the beaches, in contrast to the west region of the island. The eastern side is the richer part of the city, with well-functioning infrastructure and pleasantly arranged public spaces, in contrast to the poorer west part of the city, with precarious infrastructure and low-quality public spaces.

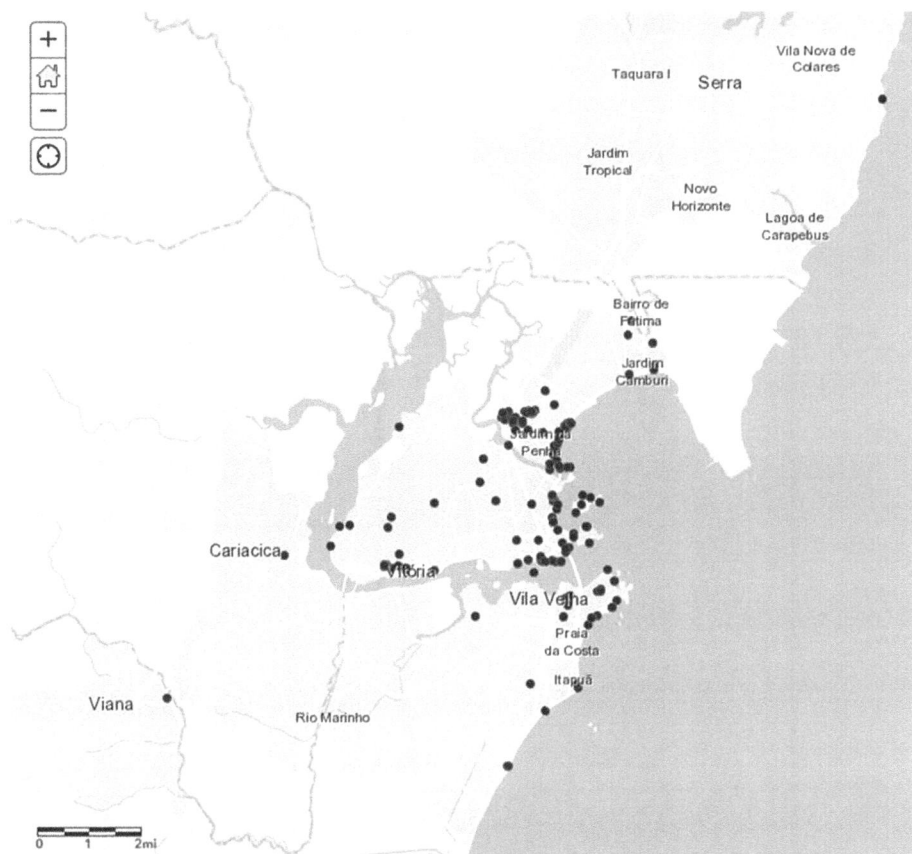

Fig. 4 Locations of evocative places in Vitória, Brazil. (*Source* data collected by the authors and visualized on google maps)

The density analysis shows high concentrations of evocative places in darker colors (Fig. 5) with a zoom-in into the most-dense areas. It suggests one very highly concentrated area and four emerging areas with high concentrations of evocative places. These five emerging high concentration areas include:

- Northeast – the highest concentration (the top of the map, Jardim da Penha). Jardim da Penha is a university neighborhood, predominantly residential. Many students that study and work at the Federal University of Espírito Santo, a big campus to the west, live in this area.
- West – the second highest concentration (next to the name Vitória). This is the old center/town of Vitória, where the city was first built. There is a "lower city" with reclaimed land and an "upper city" containing the eldest architecture ensemble. It is a predominantly commercial area.

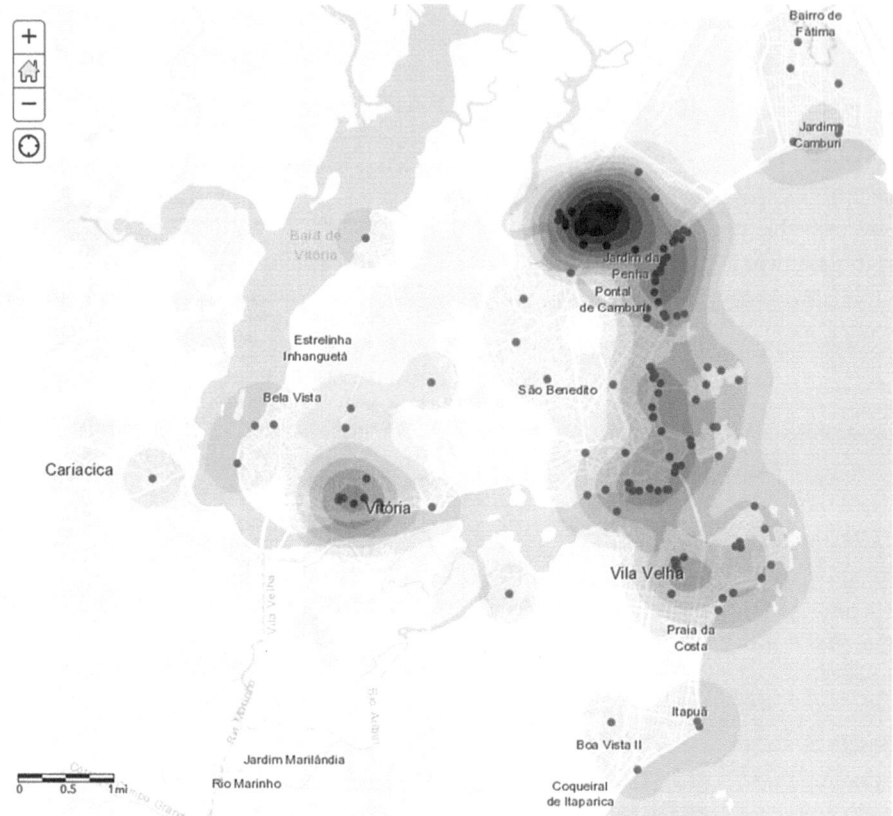

Fig. 5 Density analysis of the evocative places in Vitória (authors' analysis)

- East along the coast (central). From south to north: Curva da Jurema, Praia do Canto and Jardim da Penha. The two islands (predominantly residential) are Ilha do Boi (south) and Ilha do Frade (north). They represent the main beachside area full of parks and infrastructure for recreation.
- Southeast (Vila Velha). This neighborhood is called Praia da Costa and it is the most often visited place in Vila Velha city. It is also located near the Convent of Penha. Praia da Costa includes a hill called Morro do Moreno; a place for the citizens to hike, walk the trail, or climb. It has an amazing view of both cities Vitoria and Vila Velha.
- Northeast (Jardim Cambury). Jardim Camburi is predominantly residential area, with a lot of green areas between Jardim da Penha and the airport.

5 The Conceptual Model of an Evocative Place

5.1 Descriptions of Evocative Places in Vitória

The participants of our online survey were free in choosing three words that best describe their self-selected evocative place; there was no taxonomy for these descriptions available to the participants. They were given the freedom of inspiration to choose up to three words of their choice that best describe their self-selected evocative place in Vitória. We ended up with more than 800 words for characteristics of evocative places. Studying the collected words we searched for similarities and possibilities to form categories. The result of our study was a definition of four main groups of characteristics, which included Characteristics, Experiences, Senses and Values. The category Characteristics includes the physical characteristics of an evocative place. In the case of the City of Vitória, the different objects (benches, buildings, fountains, etc.) of places play an important role. The next categories are blue and green spaces in which the blue spaces represent the presence of the sea and the element of water, and the green spaces include the nature and parks.

The second category is Experience. Other mapping experiments executed in Hamburg, Germany and Ames, Iowa USA confirm that *experiences* are very important characteristic of an evocative place (Poplin 217, 208, 2020). They seem to be the most important across cultures and continents, which is a significant finding. It is not just the objects, or infrastructure that makes up an evocative place. It is the experience with its variety consisting of the activities that can be performed at an evocative place (paddling, yoga, meditation, jogging, socializing), experience of healing and restoration (…the square offers an opportunity to contemplate about life), emotions expressed and felt as part of the experience (a place without worry and for pure pleasure), and different stimulations experienced at a place.

The participants in Vitória used the most words (38.6%) to describe the experience of their self-selected evocative place. This additionally communicates how important experience of a place is. Tables 2, 3, 4 and 5 show some examples of such experiences

Table 2 Category Experience—Activity (source: the author's own work)

Original examples in Portuguese	Translation to English
Costumo fazer Stand-up na praia e vislumbrar a incrível paisagem de Vitória, a água, as ilhas, as áreas verdes, os edifícios e as pessoas relaxando, fazendo esportes ou socializando. Costumo correr na orla também, e encontrar os amigos.	I usually do stand up paddleboarding at the beach and have a glimpse of the incredible landscape of Vitória, the water, the islands, the green areas, the buildings and people relaxing, doing sports or socializing. I often go for a run at the waterfront, and meet my friends.
Costumava ir até o Parque da Pedra da Cebola para grupos de meditação, yoga, e de estudos. Ia também para namorar e encontrar amigos, por exemplo, para um pique-nique.	I used to go to Pedra da Cebola Park for meditation, yoga, and study groups. I also went there for dating and meeting friends, for example, for a picnic.

Table 3 Category Experience—Restoration/Healing (source: the author's own work)

Original examples in Portuguese	Translation to English
Minha mãe trabalhava no décimo segundo andar do edifício, de onde se avistava a praça. Cresci experimentando as sensações de vivenciá-la na rua e do alto. Hoje, está circunscrita no perímetro de minha pesquisa de mestrado. A praça traz oportunidade à contemplação da vida da cidade, concentrada em um espaço atrativo e de qualidade impar. Compartilho conversas com pessoas especiais e recarrego as energias quando vou sozinho.	My mother used to work on the twelfth-floor of the building, from where the square was visible. I grew up experiencing the sensations of the street and of their healing effects. Today, it became part of my master thesis research. The square offers an opportunity to contemplate the city's everyday life, concentrated in an attractive space and with a unique quality. I share conversations with special people and recharge my energies when I go there alone.

Table 4 Category Experience—Emotions felt at the place (source: the author's own work)

Original examples in Portuguese	Translation to English
O lugar não é importante para mim, mas as emoções proporcionadas por ele sim. A possibilidade de contemplar a paisagem de Vitória e Vila Velha, sem preocupação e por puro prazer, transformou o local numa espécie de "espaço seguro", um espaço onde em momentos em que muitos sentimentos vem à tona recorro à ele para possa refletir e buscar a tranquilidade para retomar os rumos e objetivos com clareza.	The place is not important to me, but the emotions provided by it are. The possibility of contemplating the landscape of Vitória and Vila Velha, without having to worry and for pure please, transformed the place into a kind of "safe space", a space where in moments when many feelings emerge I reach there so that I can reflect and seek tranquility to take back the wheel and have clarity on my objectives.

Table 5 Category Experience—Stimulations (source: the author's own work)

Original examples in Portuguese	Translation to English
Primeiro lugar que conheci no Centro Histórico de Vitória, fiquei impressionado com a variedade de épocas refletidas nos edifícios e em outros elementos do bairro num espaço relativamente pequeno. Gosto de caminhar no calçadão, sentar no banco e ficar contemplando com gratidão a Deus esse lugar lindo.	First place I went in the historic center of Vitória, I was impressed by the variety of times reflected on the buildings and other elementos of the neighborhood in a relatively small space. I like to go for a walk on the main boardwalk, sit on the bench and contemplate with gratitude to God this beautiful place.

combined with the information from the stories shared with the experimenters by the inhabitants of the city of Vitória.

The third category are Senses. *Senses* can be visual, they can include sounds, smells, or tastes. We added tastes after conducting experiments in Vienna (a study not yet published) where people would often mention food as something that is memorable and they can associate an evocative place with. Sometimes people sense and notice temperatures that affect them in certain ways. An example of the expressed senses:

- *"Gosto de lembrar do cheiro da maresia associado aos peixes dos pescadores, gosto de contemplar a bela vista, fotografar"* [I like to remember the smell of the sea associated with the fish of the fishermen, I like to contemplate the beautiful view, to photograph];
- *"É um local onde sento pra desacelerar e ver a vida passar"* [It's a place where I sit to slow down and watch life go by];
- *"Por ser próximo à praia, normalmente há vento, o que me ajuda ainda mais a respirar fundo e diminuir a velocidade da vida"* [Being close to the beach, usually there is wind, which helps me to breathe deeply and slow the pace of life].

The last category are Values. **Values** can be associated with attachment, memories and stories, an intangible dimension that makes a place. An example of values as expressed by the inhabitants of Vitória:

- *Levo minha filha para brincar. Meus pais me levavam quando criança. Um lugar lúdico e histórico, carregado de simbolismos e de memórias* [I take my daughter here to play. My parents took me when I was a child. A playful and historical place, loaded with symbolism and memories]

We summarize the number of responses for all 192 evocative places captured for the city of Vitória in Table 6. The numbers in the table indicate how often each of the categories were mentioned on the first, second or third level in the online survey and summarizes the percentage of each of the categories used in the inhabitants' descriptions.

Table 6 Descriptions of evocative places gathered for Vitória, Brazil (source: the author's own work)

	Char 1	Char 2	Char 3	Sum	Percentage	Sub-total
Characteristics						
Green Space	40	24	10	74	14.9%	
Blue Space	31	7	4	42	8.5%	
Object	16	6	2	24	4.8%	
Subject	0	5	1	6	1.2%	
Infrastructure	5	17	11	33	6.7%	
Openness	1	2	3	6	1.2%	
Accessibility	1	0	2	3	0.6%	
						38.0%
Experiences						
Activities	15	28	20	63	12.7%	
Healing/Restoration	1	0	1	2	0.4%	

(continued)

Table 6 (continued)

	Char 1	Char 2	Char 3	Sum	Percentage	Sub-total
Emotions	42	48	30	120	24.2%	
Stimulation	1	2	3	6	1.2%	
						38.6%
Senses						
Visuals	17	13	10	40	8.1%	
Sounds	1	2	4	7	1.4%	
Smells	2	1	1	4	0.8%	
Temperatures	3	15	11	29	5.9%	
						16.2%
Values						
Memories	0	1	1	2	0.4%	
Attachments	10	8	7	25	5.1%	
Stories	4	4	1	9	1.8%	
						7.3%
SUM	190	183	122	495		100%

The physical characteristics and experiences are almost equally important to the citizens of Vitória. The most important category is Experiences and in this category the emotions that the citizens can feel and experience at their self-selecting evocative places. Among Characteristics, green and blue spaces are the most valuable to the citizens of Vitória. Experiencing positive visual impressions is also very valuable to the citizens.

5.2 The Conceptual Model of an Evocative Place

The conceptual model as presented and discussed in this section is based on the experiments in the City of Vitória, Brazil presented in this article, and additional experimental work on evocative places in Hamburg, Germany (Poplin 2017, 2018), Ames, Iowa, USA (Poplin 2018, 2020), Grinnell, Iowa and Vienna, Austria (by Poplin, studies not yet published). It is designed by the main author of this article and may serve as a useful tool that can help researchers and practitioners describe evocative places. It also contributes to a better understanding of how evocative places can be described and what constitutes a place that may evoke positive emotions. What are the noticeable and felt elements of a place? How people describe evocative places. The conceptual model summarizes categories with which evocative places can be described on different continents taking into account different cultures and languages. The categories listed were added as a result of conducting experiments on evocative and power places in Europe, North America and South America. This is also the current limitation of the model as it has not yet been tested in other regions and/or continents.

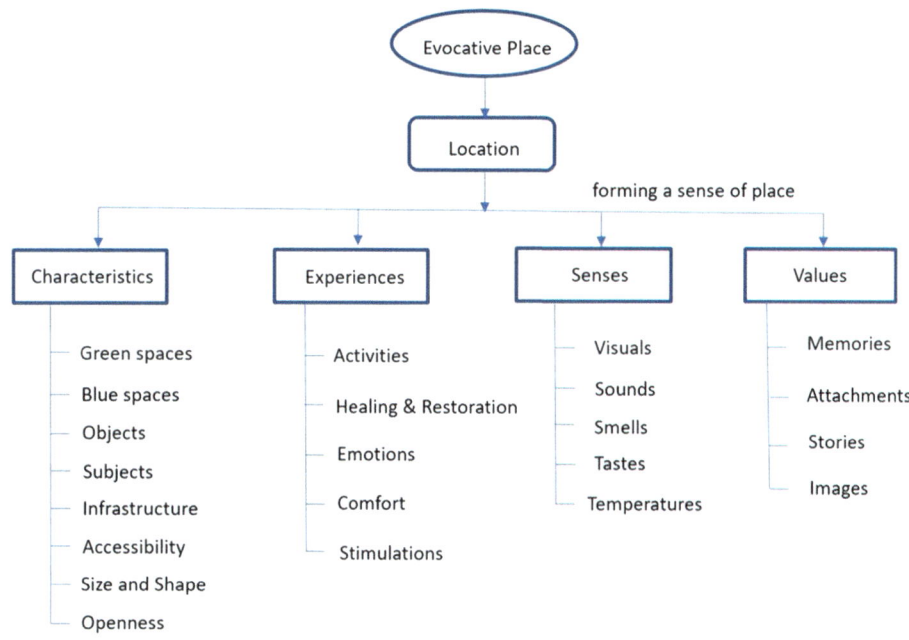

Fig. 6 The Conceptual Model of an Evocative Place. (*Source* the author's own work)

It is the result of our numerous experiments on evocative places and our attempt to provide a taxonomy for the descriptions of evocative places. A structure that can be used by other researchers and practitioners while referring to evocative places. It is also useful for researchers on places and place-making as the majority of such places can be categorized as evocative places. Figure 6 summarizes this effort. We entitled it *The Conceptual Model of an Evocative Place*.

The description starts with a geographic location. Every place has a location. The location is the only physical and rational dimension of a place that can be mapped and described with coordinates representing a location on Earth. All other elements of a place – including characteristics, experiences, senses and values – are subjective and depend on the human being observing, sensing and remembering.

Characteristics This category includes the physical characteristics of a place; that which is tangible and visible. These physical characteristics may include *green areas* such as parks, meadows, trees, or flowers. *Blue spaces* include water in many different forms. Water is a significant element of many places and may include fountains, rivers, lakes, canals, waterfronts, or oceans. Places usually consist of *Objects* such as benches, buildings, swimming pools, restaurants, coffee places, libraries, university buildings or other objects. People—*Subjects*—many times add to the value of place, and on the contrary, often the absence of people may have a special meaning contributing to the place to be more serene and tranquil. An important part of every place is its *Infrastructure* that can come in the form of paths, roads, streets, or trails. The *Accessibility* of a place indicates how can this place be physically accessible. Every place also has its *Size and Shape* that may be important to humans. How open is the place; *Openness* is a characteristic of a place important to some of their users.

Experiences of a place are related to some intangible characteristics of a place and may include activities, healing and restoration, emotions, comfort or stimulation. *Activities* describe the activities the (power) place enables and may include jogging, walking, swimming, reading, studying or other activities. Some places have *Restorative, Healing effects* on people in that they enable a transformation of negative, traumatic, painful feelings into positive, peaceful, or contented states. These may be places of meditation, inner peace, tai chi movements, or just walking and observing the healing power of a green color. *Emotions* are the emotions this place evokes. Experiences also includes the level of *comfort* at this place and the *stimulation* people experience by this place which can be motivating, fun, inspiring or invigorating.

Senses This category includes the methods of perceptions which can be visual observing, *Visuals* – visual impressions – through their sight. Places may be significant due to their *Sounds* or *Smells*. *Tastes* refers mostly to food found at evocative places that may be significant. *Temperatures* are felt by people as well and they may describe places as warm, or with nice and pleasant temperatures.

Values Places may have historical values, or personal values. These values can be tangible or intangible. They may include *Memories* of what a person experienced at this place that can also form an *Attachment* demonstrating its significant imprint. People may have *Stories* connected to this particular place they may want to share with others and are specific for this particular place.

6 Emotions of Evocative Places in the City of Vitória

6.1 Prevailing Emotions in the City and their Locations

The words for emotions were given to the survey participants as a list of words (Table 7). This taxonomy of emotions was taken from the book titled *Non-violent communication* written by Rosenberg (Rosenberg 1999). For the purpose of this study, the taxonomy including all the words for emotions were translated into Portuguese.

The participants were able to use this taxonomy as part of their online survey and choose the words that best describe their emotions felt at their self-selected evocative places. Table 8 summarizes the categories of the emotions and how often they were selected by the citizens. The majority of the participants feel peaceful (28.4%) at their evocative places. In contrast to that, others feel excited (20.7%) which is an indication

Table 7 The list of emotions according to Rosenberg (1999) translated into Portuguese (source: Translated to Portuguese from the original English, original by Rosenberg (1999))

AFETUOSO	EMPOLGADO	INEBRIADO
Compassivo	Impressionado	Maravilhado
Amigável	Animado	Extasiado
Amoroso	Acalorado	Eufórico
Aberto	Desperto	Encantado
Bondoso	Estupefato	Exuberante
Simpático	Deslumbrado	Radiante
Tenro	Ávido	Extático
Caloroso	Energizado	Emocionado
ENGAJADO	Entusiasmado	**TRANQUILO**
Assimilado	Contente	Calmo
Alerta	Revigorado	Lúcido
Curioso	Vivo	Concentrado
Absorto	Apaixonado	Confortável
Encantado	Surpreso	Centrado
Em transe	Vibrante	Contente
Fascinado	**AGRADECIDO**	Preenchido
Interessado	Apreciativo	Suave
Intrigado	Comovido	Quieto
Envolvido	Grato	Relaxado
Enfeitiçado	Tocado	Aliviado
Estimulado	**INSPIRADO**	Satisfeito
ESPERANÇOSO	Impressionado	Sereno
Expectante	Admirado	Silencioso
Encorajado	Maravilhado	Tranquilo
Otimista	**ALEGRE**	Confiante
CONFIANTE	Divertido	**REVIGORADO**
Empoderado	Encantado	Vivificado
Aberto	Contente	Rejuvenecido
Orgulhoso	Feliz	Renovado
À salvo	Jubiloso	Descansado
Seguro	Satisfeito	Restaurado
	Cativado	Revivido

Table 8 Emotions felt at evocative places in Vitória, Brazil (source: the author's own work)

	Emo_1_Crit	Emo_2_Crit	Emo_3_Crit	SUM	Percentage
Emotions					
Affectionate/Afetuoso	19	5	9	33	5.89%
Confident/Confiante	17	12	15	44	7.86%
Engaged/Engajado	13	22	15	50	8.93%
Inspired/Inspirado	45	41	31	117	20.89%
Excited/Empolgado	4	9	5	18	3.21%
Exhilarated/Inebriado	9	8	20	37	6.61%
Grateful/Agradecido	6	6	2	14	2.50%
Hopeful/Esperançoso	14	3	12	29	5.18%
Joyful/Alegre	8	19	16	43	7.68%
Peaceful/Tranquilo	43	57	49	149	26.61%
Refreshed/Revigorado	9	5	12	26	4.64%
SUM	187	187	186	560	100%

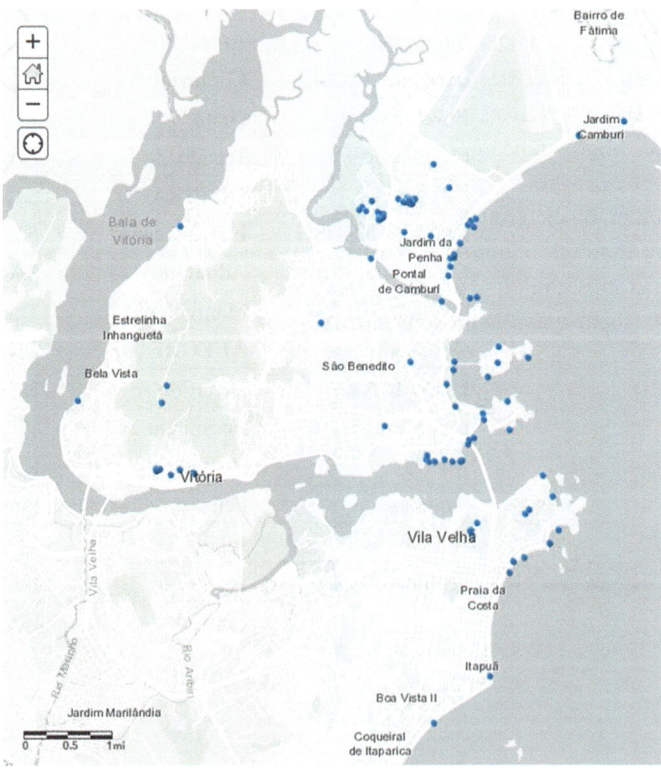

Fig. 7 Evocative places at which citizens feel 'PEACEFUL'. (*Source* data collected by the authors and visualized on google maps)

of energy and stimulation of evocative places for the students in this city. These were the prevailing categories of emotions selected by the participants. If one adds the percentage of the students that feel engaged (8.9%) we end up with 29.6% of the students being excited and engaged at their evocative places. This demonstrates a high level of engagement, stimulation, and excitement in the city. Energy is in the air in this city!

We mapped these categories on a map and visualized the top four categories of emotions. Figures 7, 8, 9 and 10 demonstrate the locations of the evocative places according to the most often selected categories of these emotions. The maps with data and visualizations can be found online in ArcGIS Online under the following link: https://arcg.is/eeCz5. The company ESRI provides the base maps for these maps.

The majority of the "peaceful" emotions can be found near the cost, close to the ocean, at the beaches. Additionally, the area Jardim da Penha on the north of the city, is the source of feeling peaceful as well. The citizens/students feel excited at a variety of locations across the city area. Engaged they feel mostly in two areas in the north, namely Jardim da Penha and Jardim Camburi. The locations for the feeling "confident" mainly concentrate in the area Jardim da Penha, at some beach locations and in Vila Velha.

Fig. 8 Evocative places at which citizens feel 'EXCITED'. (*Source* data collected by the authors and visualized on google maps)

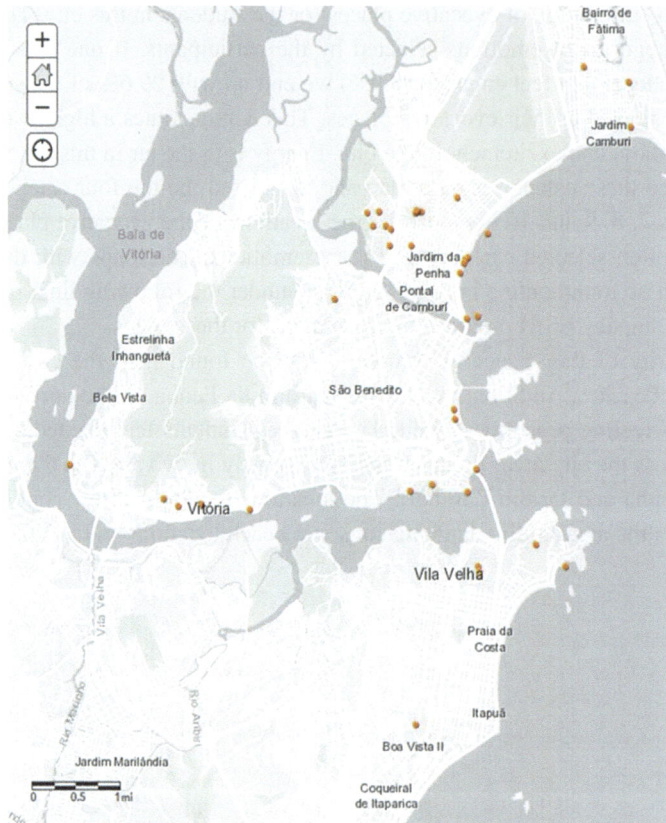

Fig. 9 Evocative places at which citizens feel 'ENGAGED'. (*Source* data collected by the authors and visualized on google maps)

The most often selected emotions were calm, tranquil, energetic, open, invigorated, relaxed, giddy, thankful, wonder, happy, comfortable, appreciative, and serene. We represent the emotions selected in the Russell's (1980) Circumplex model of affect. The model represents eight affect concepts in a circular order (Fig. 11). All emotions mentioned more than ten times are represented in the model with the number of mentions indicated in the brackets. All mapped emotions can be found on the positive axis. The majority of the emotions can be found in the quadrant pleased—aroused and pleased—sleepy.

6.2   How Planners Can Use this Information

What can landscape architects, urban planners, designers and architects learn about the most evocative place in the city of Vitoria? How can they use this analysis and the

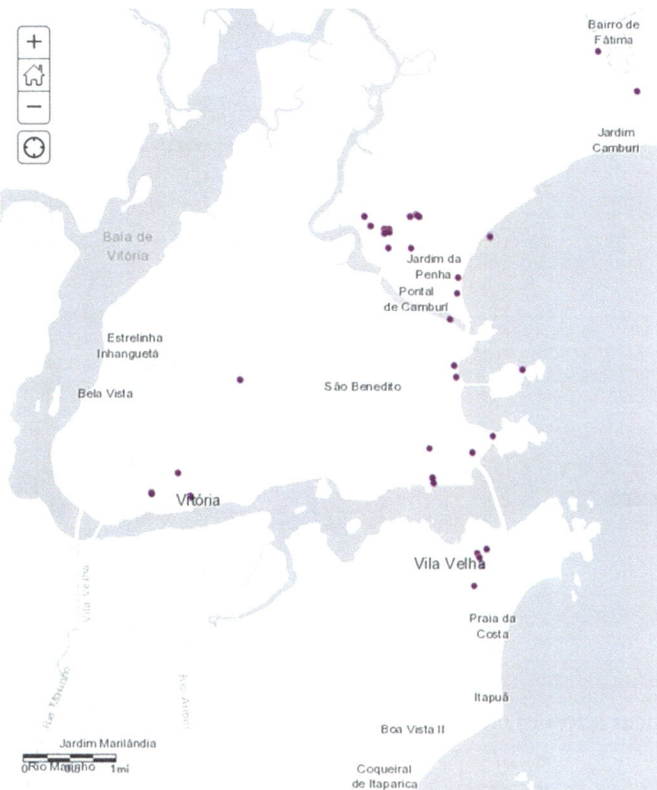

Fig. 10 Evocative places at which citizens feel 'CONFIDENT'. (*Source* data collected by the authors and visualized on google maps)

knowledge provided in this research? Knowing the evocative places and their high concentrations in the cities can provides a useful feedback to them learning about the places at which the citizens feel positive emotions. What can they learn from such places? As an example in this research, the city's neighborhood Jardim da Penha represents the highest concentration of evocative places. Figure 12 shows the location of the neighborhood.

Figure 13 also shows a very specific street design. It is a design that has been carefully planned and designed. Figure 8 shows a detail of this planning. The streets end in a concentric circle; in the middle of this circle is a park which is a public space available for the citizens to socialize, meet and enjoy the public space for their own use. It is an example that is successful and could be replicated in other neighborhoods or cities. This is also something that needs to be studied further. How do these publically available spaces contribute to a better quality for life for the citizens and their happiness in the city in which they live? The city is car-centered, but these parks in the middle represent the possibility for the citizens to walk and have an experience of a walkable part of the city (see Fig. 13). One can imagine moms and dads bringing their children to play there, or students bringing their books to study, elderly to sit and observe others, have a good chat or place to read and socialize.

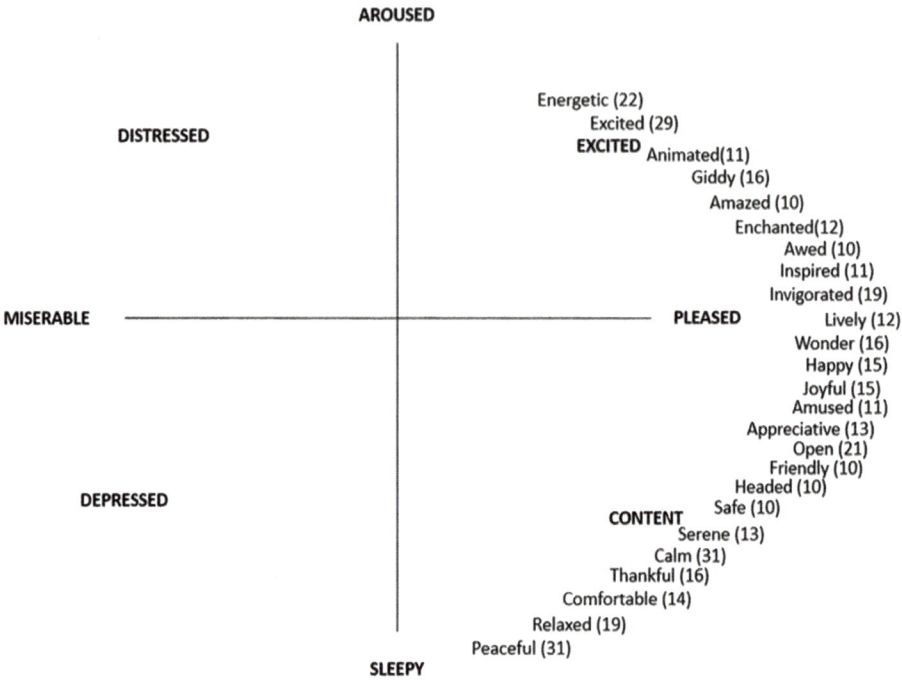

Fig. 11 Emotions expressed in Vitória, Brazil organized in Russell's (1980) Circumplex model of affect

Fig. 12 Jardim da Penha (©google maps)

Fig. 13 Street design in Jardim da Penha neighborhood (@riamscheidegger)

7   Discussion and Conclusions

The study presented focuses on evocative places, places that evoke positive affect and attempts to contribute to a better understanding of how people express emotions related to places and which emotions do they associate with their self-selected evocative places. This article summarizes characteristics and emotions felt at 192 evocative places in the City of Vitória in Brazil. One of the significant results of this article is *The Conceptual Model of an Evocative Place* which adds additional elements for the descriptions of evocative places. It aims to provide cross-cultural and multi-dimensional ways of describing evocative places. The conducted experiments in Europe (Poplin 2017, 2018), in North America (Poplin 2018, 2020) and South America (this article) show the importance of tangible/physical and intangible/non-physical characteristics of evocative places. Often, the intangible characteristics become the most important category as the majority of our participants/inhabitants of the city express appreciation of their experience at these places as being the most important category with which they describe their places. Additionally, important are also senses and values. People are receptive to things they see, smell, hear, taste and remember. This research aims to contribute to the discussion on intangible characteristics of places that are difficult to acquire and map. It also aims to contribute to the discussion in positive places, places that evoke positive emotions and to understand where such places are located in the cities and how are they perceived by their citizens. A better knowledge about emotional landscapes and attachments to places may help create happier cities where the inhabitants can have positive experiences.

We are aware of some limitations of this study. The participants of the survey were students and it therefore represents the view from the perspective of the students and cannot be generalized to all inhabitants of the city. However, it was also our intention to focus on this population in order to better compare the results gained in this city with previous experiments accomplished in Hamburg, Germany, Ames and Grinnell, both in Iowa, and in Vienna, Austria. It was also our intention to focus on the positive feelings with which we then excluded places that may evoke other emotions found on the opposite side of the Russell's model. Another study could look into the whole spectrum of emotions in order to understand the positive vs. negative places in the city. The sample size was big enough to be able to complete the conceptual model of evocative places and to continue the discussion about the concentration of places that evoke positive emotions as presented on the density maps. Alternative ways of representing emotions need to be additionally studied. Representing them with a dot attached to the selected location is possible in a geographic information system as they are attached to the locations of self-selected places (and represented as an attribute to the location). However, the current state of GIS development only enables to represent objects with well-defined boundaries. Emotions are not a typical GIS object and can only be represented with visualizations currently offered by the limited concept of representation in a GIS reduced to point, line and polygon object types.

In the next step we intend to concentrate on an in-depth study of public spaces in the city. The knowledge of public places, their characteristics and how people feel at these places may be a valuable input for the development of novel co-design approaches that can safeguard and intensify common goods and the well-being of the citizens. The evocative places regarding values in the City of Vitória possess a set of attributes of representations (legitimate history) and social practices (supporting collective memory) that goes back to the origins of the city and the way its citizen built attachment articulated with the element water. Moreover, to a city as an island, the evocative places experiment might have identified a pattern in this type of geographical configuration, as most of evocative places were located in the waterfronts.

In the next steps of this research the team intends to study the role of positive public places, their concentrations in the cites (forming emotional landscapes) and in particular and in more detail their relevance for urban planning. It is important to reflect upon the emotions and collect additional data to be able to better understand the correlation of the emotions and the characteristics of these places. The plan is to continue with data collection in Brazil and study cultural differences of evocative places, the characteristics of these public places and the ways people express their emotions related to places on different continents. This research aims to contribute to a better understanding of intangible landscapes of the cities and how cities can be planned and designed so that they evoke positive emotions and lead to happier and healthier cities of the future.

Acknowledgements We acknowledge the professors and staff at the Department of Architecture and Urbanism at the Federal University of Espírito Santo, Vitória city, for their support. Special thanks go to all the students that have undertaken the online surveys and shared their urban experiences in Vitória. Thank you to Stephen Poplin for the language improvements of this text.

References

Altman, I., & Low, S. M. (1992). *Place attachment*. New York: Plenum Press.

Bonnes, M., & Secchiaroli, G. (1995). *Environmental psychology: A psycho-social introduction*. London: Sage.

Bott, S., Cantrill, J. G., & Myers, O. E., Jr. (2003). Place and the promise of conservation psychology. *Human Ecology Review, 10*(2), 100–112.

Canter, D. V. (1977). *The psychology of place*. New York: Palgrave Macmillan.

Cooper-Marcus, C., & Francis, C. (1998). *People places*. New York: Wiley.

Cresswell, T. (2004). *Place a short introduction*. Oxford, UK: Blackwell Publishing.

Ecléa, B. (1994). *Memória e sociedade: Lembranças dos velhos (Memories and Sociaety: Memories of elderly people)*. Companhia das Letras: São Paulo.

Geddes, P. (1949). *Cities in evolution*. London: William and Norgate Limited.

Helliwell, J., Layard, R., & J. Sachs (2017). World Happiness Report. https://s3.amazonaws.com/happiness-report/2017/HR17.pdf. Accessed 4 November 2020.

Helliwell, J., Layard, R., & J. Sachs (2019). *World happiness report 2019*. New York: Sustainable Development Solutions Network. Retrieved 12 Dec 2019 from https://worldhappiness.report.

IBGE (2010). *Brazilian Institute of Geography and Statistics*. Retrieved 20 May 2018 from https://cidades.ibge.gov.br/brasil/es/vitoria/.

Ittelson, W. H. (1973). *Environment and cognition*. Oxford: Seminar Press.

Kaiser, F. G., & Fuhrer, U. (1996). Dwelling: Speaking of an unnoticed universal language. *New Ideas in Psychology, 14*, 225–236.

Kaplan, S. (1995). The restorative benefits of nature: toward an integrative framework. *Journal of Environmental Psychology, 15*, 169–182.

Kaplan, R., & Kaplan, S. (1989). *The experience of nature: A psychological perspective*. Cambridge: Cambridge University Press.

Kleber, A., Edler, D., & Dickmann, F. (2020). Cartography and the sea: A javascript-based web mapping application for managing maritime shipping. In D. Edler, C. Jenal, & O. Kühne (Eds.), *Modern approaches to the visualization of landscapes* (pp. 173–186). Wiesbaden: Springer VS.

Korpela, K. (2012). Place attachment. In S. Clayton (Ed.), *The Oxford handbook of environmental and conservation psychology* (pp. 148–163). New York: Oxford University Press.

Korpela, K. M., Hartig, T., & F. G. Kaiser (2001). Restorative experience and self-regulation in favorite places. *Environment and Behavior*, 33 (572).

Low, S. (1992). *Symbolic ties that bind*. Book chapter in the book: Place attachment. I. Altman & S. Low (eds). Plenum Press, New York.

Low, S., & L. Altman (1992). *Place attachment: A conceptual inquiry*. Book chapter in the book: Place attachment. I. Altman and S. Low (eds.). Plenum Press, New York.

Magnaghi, A. (2005). *The urban village: A charter for democracy and sustainable development in the city*. London: Zed books.

Maptionnaire, (2020), Online participatory software, Retrieved 2 March 2020 from www.maptionnaire.com.

McLean, K. (2020). Temporalities of the smellscape: Creative mapping as visual representation. In D. Edler, C. Jenal, & O. Kühne (Eds.), *Modern approaches to the visualization of landscapes* (pp. 217–246). Wiesbaden: Springer VS.

Müller, L. (2008). Intangible and tangible landscapes: an anthropological perspective based on two South African case studies. *SAJAH, 23*(1), 118–138. ISSN 0258-3542.

OECD. (2017), How's Life? 2017: Measuring Well-being. OECD Publishing, Paris. Retrieved 2 March 2020 from https://doi.org/10.1787/how_life-2017-e.

Pierre, N. (1996). *Realms of memory* (p. 98). New York: Columbia UP.

Poli, D. (2013). Democrazia e Pianificazione del Paesaggio: Governance, Saperi Contestuali e Partecipazione per Elevare la Coscienza di Luogo (Democracy and Landscape Planning: Governance, Contextual Knowledge and Participation to Elevate the Place Consciousness). *Rivista Geografica Italiana, 120*, 255–273.

Poli, D. (2015). Patrimonio Territoriale fra Capitale e Risorsa nei Processi di Patrimonializzazione Proattiva (Territorial Heritage between Capital and Resource in Proactive Patrimonialization Processes). In B. Meloni (Ed.), *Aree interne e progetti d'area* (pp. 123–140). Rosenberg & Sellier: Turim.

Poplin, A. (2018). Cartographies of fuzziness: Mapping places and emotions. *The Cartographic Journal, 54*(4), 291–300.

Poplin, A. (2017). Mapping expressed emotions: Empirical experiments on power places. *KN – Journal of Cartography and Geographic Information, Special issue: Methods and Applications of Empirical Cartography* 67/2, 83–91.

Poplin, A. (2020). Exploring evocative places and their characteristics. *The Cartographic Journal.* Published online on March 2, 2020. https://www.tandfonline.com/doi/full/10.1080/00087041.2019.1660502.

Pretty, G. H., Chipuer, H. M., & Bramston, P. (2003). Sense of place amongst adolescents and adults in two rural Australian towns: The discriminating features of place attachment, sense of community and place dependence in relation to place identity. *Journal of Environmental Psychology, 23*, 273–287.

Proshansky, H. M., & Fabian, A. K. (1983). Place-identity: Physical world socialization of the self. *Journal of Environmental Psychology, 3*, 57–83.

Rosenberg, M. B. (1999). *Nonviolent communication: A language of compassion.* Encinitas, CA: PuddleDancer Press.

Rossi, A. (1982). *The architecture of the city.* Cambridge, MA: MIT Press.

Russell, J. A. (1980). A circumplex model of affect. *Journal of Personality and Social Psychology, 39*(6), 1161–1178.

Russell, J. A., & James, A. (2003). Core affect and the psychological construction of emotion. *Psychological Review, 110*(1), 145–172.

Russell, J. A., & Pratt, G. A. (1980). A description of the affective quality attributed to environments. *Journal of Personality and Social Psychology, 38*, 311–322.

Schlosberg, H. (1954). Three dimensions of emotion. *Psychological Review, 61*(2), 81–88.

Shoval, N., Schvimer, Y., & Tamir, M. (2018). Tracking technologies and urban analysis: Adding the emotional dimension. *Cities, 72*, 34–42.

Siepmann, N., Edler, D., & Kühne, O. (2020). Soundscapes in cartographic media. In D. Edler, C. Jenal, & O. Kühne (Eds.), *Modern approaches to the visualization of landscapes* (pp. 247–263). Wiesbaden: Springer VS.

Steele, F. (1981). *The Sense of Place.* Boston: CBI Publishing.

Tuan, Y. F. (1974). *Topophilia: A study of environmental perception, attitudes, and values.* NJ, Prentice Hall: Englewood Cliffs.

Tuan, Y. F. (1977). *Space and place. The Perspective of Experience*. Minneapolis: University of Minnesota Press.

Tuan, Y. F. (1980). Rootedness versus sense of place. *Landscape, 24*, 3–8.

Ulrich, S. R. (1983). Aesthetic and affective response to natural environment, behavior and the natural environment. *Human Behavior and Environment, 6*, 85–125.

Walljasper, J. (2017). How to design our neighborhoods for happiness. YES! Magazine, YES!, published on 29 March 2017.

Waxman, L. (2006). The coffee shop: Social and physical factors influencing place attachment. *Journal of Interior Design, 31*(3), 35–53.

Zeile, P., Resch, B., Loidl. M., Petutschnig, A., & I. Doerrzapf (2016). Urban emotions and cycling experience – Enriching traffic planning for cyclists with human sensor data. *Proceedings of the GI_Forum*, 201–216.

de Andrade, B., & R. H. de Almeida (2016). I valori del patrimonio territoriale: un'analisi sui discendenti di immigrati germanici in un'area montana di Espírito Santo, Brasile (The values of the territorial heritage: an analysis of the descendants of Germanic immigrants in a mountain area of Espírito Santo, Brazil). *Rivista Scienze del Territorio*, vol. 4, pp. 206–215. https://dx.doi.org/10.13128/Scienze_Territorio-19407.

de Andrade, B. (2019). *Planning, children and games: Geodesign in the identification of quotidian and symbolic values in the territory*. PhD thesis, Postgraduate Program in Architecture and Urbanism, Federal University of Minas Gerais. Belo Horizonte, Brazil.

Alenka Poplin is Assistant Professor at the College of Design, Iowa State University, USA—email: apoplin@iastate.edu.

Bruno de Andrade is Assistant Professor at the Faculty of Architecture and the Built Environment, Delft University of Technology, NL—email: b.deandrade@tudelft.nl

Shoaib Mahmud is Ph.D. candidate at the Department of Civil, Construction and Environmental Engineering, Iowa State University, USA—email: smahmud@iastate.edu

Prato: The Social Construction of an Industrial City Facing Processes of Cultural Hybridization

Andrea Bellini and Laura Leonardi

Abstract

This chapter deals with a widely studied case, that is, Prato, a middle-sized city with rooted industrial traditions, in the Centre of Italy. Prato is a *textile industrial district* embedded in the so-called Third Italy—an area characterized by the presence of small firms spread throughout the territory, linked together in supply and subcontracting relationships—which, in the last twenty years, has undergone a profound transformation as a consequence of the crisis of textile and immigration, leading to the formation of a large Chinese community. The related changes brought with them problems of social cohesion and sustainable development. The authors address these issues by analyzing both academic and public discourses on Prato. Their basic idea is that common stereotypes act as drivers of a public discourse that prevents the city to re-negotiate its identity. The analysis concludes that different forms of hybridization—particularly cultural hybridization—are occurring, which would need further investigations.

Keywords

Industrial district · factory city · Chinese immigration · environment · landscape
stereotypes · Covid-19 · cultural hybridization

A. Bellini (✉) · L. Leonardi
Università degli studi Firenze, Firenze, Italy
e-mail: andrea.bellini@unifi.it
L. Leonardi e-mail: laura.leonardi@unifi.it

© Springer Fachmedien Wiesbaden GmbH, part of Springer Nature 2020   547
D. Edler et al. (eds.), *Modern Approaches to the Visualization
of Landscapes*, RaumFragen: Stadt – Region – Landschaft,
https://doi.org/10.1007/978-3-658-30956-5_30

1 The City of Prato in the 21st Century: A Short Introduction

Prato is a middle-sized city with rooted industrial traditions. It is located in the North of Tuscany, in the Centre of Italy. Despite its being embedded in a metropolitan area and its proximity to Florence, it distinguishes itself for a strong identity, built around the idea of *industrial district*. This term refers to a form of development characterized by the territorial concentration and sectoral specialization of a high number of small and medium-sized firms or, to put it with its earliest promoter, Alfred Marshall (1890, p. 198), a place with an *industrial atmosphere*, where "the mysteries of the trade become no mysteries, but are as it were in the air". It was, then, reprised and developed by the Florentine economist Giacomo Becattini (1962, 1979, 1987, 2004) and other eminent scholars (Bellandi and Russo 1994; Brusco 1989; Trigilia 1986; Viesti 2000). It is no coincidence that these authors, first and foremost Becattini (1997, 2000), devoted special attention to Prato, considered as the archetype of Italian industrial districts.

As such, Prato is part of the so-called *Third Italy* (Bagnasco 1977), referring to the particular trajectories of socioeconomic development of the regions of the Centre and North-East, distinct from the North-West and the South of the country. This type of development is characterized by the presence of small firms spread throughout the territory, linked together in supply and subcontracting relationships. This structure is highly flexible and, during the 1970s and 1980s, demonstrated a greater capacity to adapt to globalization's challenges. The changes that occurred over time in the distribution of economic activities and of the population in the metropolitan area had a significant impact on the city, modifying its functions and morphology.

In the last twenty years, in fact, Prato has undergone a profound transformation as a consequence of the enduring crisis of its productive specialization, that is, the textile industry, and of migration dynamics, which led to the formation of the second-largest Chinese community in the country after Milan. While it has not abandoned its industrial vocation, Prato deals with the tertiarization of its economy and its becoming more and more multicultural. These changes brought with them problems of social cohesion. Furthermore, they would imply re-negotiating the identity of the city.

This chapter addresses the above issues by analyzing the discourses on Prato, focusing on the stereotypes that make it difficult to confront the processes of change in a way that allows the renovation of the public image of the city. For this purpose, we proceed as follows. In Sect. 2, we draw a historical, geographical, and socioeconomic outline of Prato, to introduce the case. In Sect. 3, we present a literature review aimed at identifying thematic patterns in academic research. In Sect. 4, then, we investigate social representations through an analysis of Internet videos, as practical means for the social construction of places (see in this volume: Kühne 2020; for movies in general: Papadimitriou 2020; for internet images: Linke 2020; Loda et al. 2020; for introductory remarks on the current media availability: Edler et al. 2020). After a brief description of the research strategy, we examine the main patterns of representation and related *landscape stereotypes*. Drawing on Kühne's (2018, 2019) constructivist approach, we look at the *landscape*—specifically *urban* landscape—in its dual character of "individual construct based on social patterns of interpretation and evaluation" (Kühne et al. 2019, p. 78). In this sense, landscape is a 'mediating' concept, which allows us to understand how the reality 'as it is' is interpreted by individuals and translated into collective meanings, of which stereotypes are accessible codifications (Kühne and Bellini 2019; see in this volume: Kühne and Jenal 2020). In the concluding section, we compare academic and public discourses to get inputs to refine the research agenda. More generally, we reflect on how to redraw the picture.

2   Historical, Geographical, and Socioeconomic Sketches

Prato is a city with a long history. The first human settlement in the area where the city stands today was of the Etruscan era. The foundations of the city, however, date back to the Middle Age, and the name Prato came into use in the 11th century. At that time, the rivalries with the neighboring cities gave rise to frequent conflicts, through which the city gained the status of a free commune. In the subsequent centuries, however, Prato was subjected to foreign dominations. In 1351, it passed to the Florentine Republic, although preserving a certain autonomy and further developing its identity. In those years, Francesco di Marco Datini, also known as the Merchant of Prato, conducted his businesses, mostly in textile products' manufacturing and trade. Afterward, Prato continued to grow and maintain a balance between city and countryside, at least until the half of the 19th century, when the transition from an artisan to an industrial mode of production took place. Between the two world wars, then, Prato was an important industrial city, although it was in the second post-war period that it achieved its fame.

Prato differs from other areas of Tuscany because of the presence of widespread entrepreneurship in the industrial sector, whose origins, nevertheless, are in the type of organization of the agricultural sector that characterized it in the past. Small farms guided by sharecroppers (*mezzadri*), helped to form competencies and skills, but also a labor culture, which was functional to the transition to industrial entrepreneurship. Since the beginning, Prato is known as the 'city of rags' (*città degli stracci*) for the recycling of used fabrics, which became its distinguishing feature. Family-run micro-firms initially formed inside the houses, in the urban context. For this reason, a typical representation of Prato is that of a 'factory city' (*città fabbrica*), in which there is no solution of continuity between home and place of production.

From 1950 to 1980, the number of textile workers grew at a steady pace. The development of Prato was based on a widespread culture linked to a work ethic, the centrality of the family, the sharing of political values—the 'red' political subculture (Trigilia 1986), linked to resistance to fascism—and the presence of deep-rooted civil associationism, which allowed for administrative continuity. The city developed around its industrial core attracting workers from other regions, most of them from Southern Italy, guaranteeing them full citizenship, with integration into the production system and the local society. Inward migration flows reshaped the urban environment giving rise to new neighborhoods and industrial areas and laying the foundations for a polycentric development, which became more accentuated with the arrival of Chinese immigrants during the 1990s. That said, the decade between the 1980s and the 1990s was a period of decline, which led to a downsizing of the district and its repositioning in international markets, with a partial relocation of production abroad in Eastern Europe. A reversal of this trend coincided with the arrival of the Chinese, who brought some phases of production back in Prato, reducing transaction costs.

Geographically, Prato is located in the center of a metropolitan area that covers more than 1,200 km^2 from Pistoia to Florence and counts over 1 million inhabitants. As such, it is at the center of one of the most important commuting areas in Tuscany (see: Fig. 1). The city of Florence is the major pole of attraction. More than 10,000 people, indeed, commutes from Prato to Florence every day for business purposes. A more limited, bidirectional flow connects Prato with Pistoia (Iommi and Marinari 2020).

The proximity to Florence and the possibility to commute to work, associated with the rise in real estate market prices in the capital of Tuscany, partly explains the increase in the number of inhabitants of Prato, now close to 200,000. The rise in the number of foreign residents also contributed to this growth. Foreigners, indeed, are over 40,000 (more than 20 percent of the resident population); among them, almost 25,000

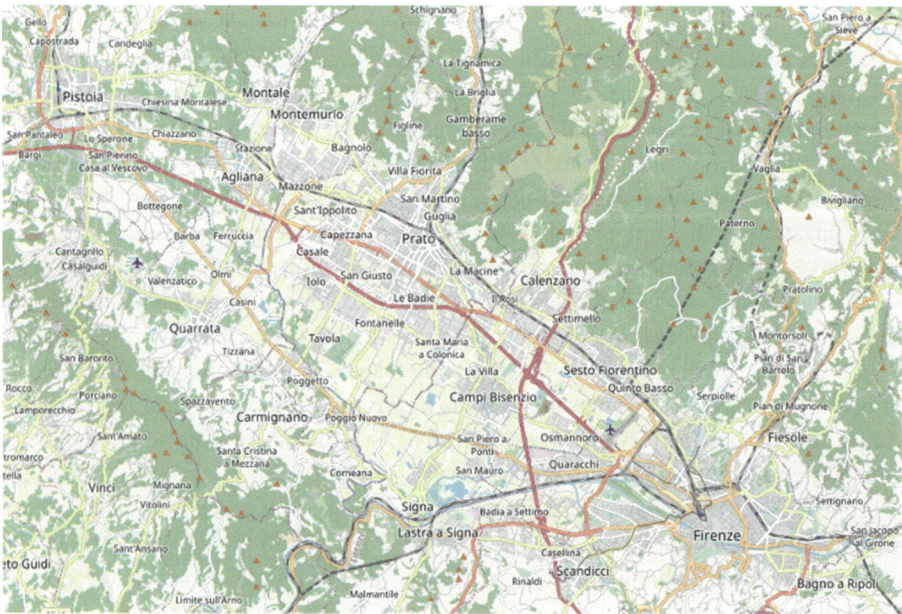

Fig. 1 Map of the metropolitan area of Pistoia-Prato-Florence. (*Source* OpenStreetMaps (https://www.openstreetmap.org/); image retrieved on May 18th, 2020.)

(12 percent of the total) are Chinese people (source of data: Municipality of Prato, at December 31st, 2019). It is worth noting that the increase of foreigners rather than offset the decrease of natives allows the population to continue to grow.

The configuration of the city also changed considerably. Due to the presence of mountains that draw a physical border at the North-East and tighten the city in a funnel that flows into the Bisenzio Valley in the North, Prato expanded from the North-West to the South-East, where the white roofs of the factories replaced green fields (see: Fig. 2). The creation of new industrial areas outside the urban perimeter aimed to overcome the model of factory city.

This process met increasing difficulties because of the concentration of Chinese immigration in the so-called 'Macrolotto 0', adjacent to the city center. This implied the reproduction of the old model of urban development. Figure 3 gives an idea of how the Chinese community occupied the territory. It is known, in fact, that there was a tacit division of labor between Italian and Chinese firms, with the former concentrated in textile, the latter in apparel. As Fig. 3 shows, apparel firms spread throughout the city occupying the same areas as textile firms, but with a higher concentration in the 'Macrolotto 0' (in the middle of the picture on the right) and the 'Macrolotto 1' (below the red line, that is, the artery that connects Pistoia, Prato, and Florence).

Figure 4 gives further information on the spatial distribution of the population, focusing on the distribution of socioeconomic disease. The map on the left—divided into basic statistical units, similar to neighborhoods—reveals that wealthy areas are concentrated

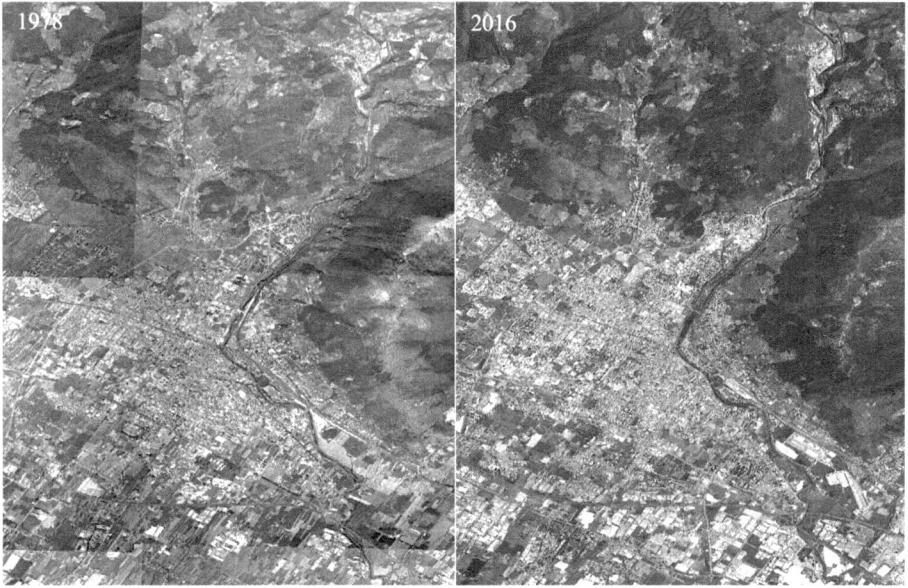

Fig. 2 Aerial photos of the city of Prato, in 1978 and 2016. (*Source* Municipality of Prato, GeoServer Web Map Service)

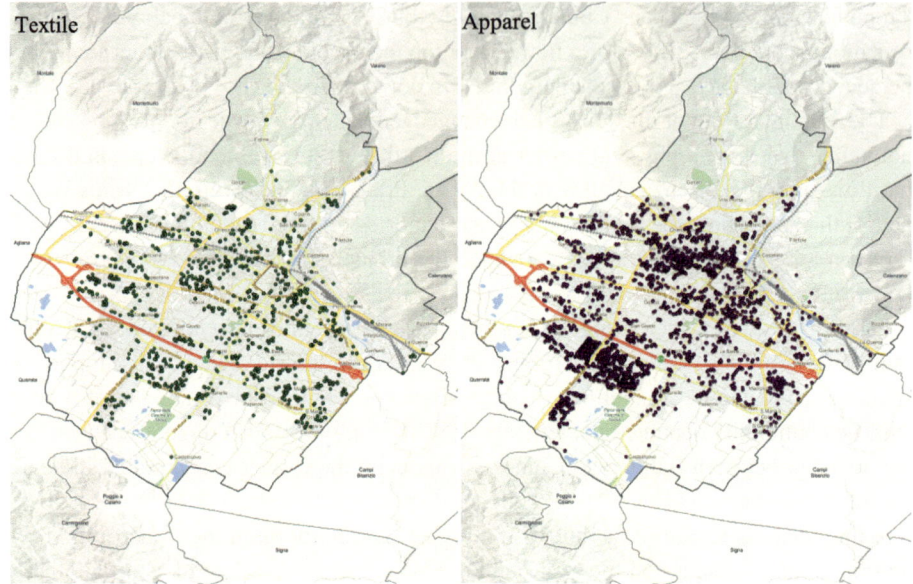

Fig. 3 Spatial distribution of textile and apparel firms. (*Source* Municipality of Prato, GeoServer Web Map Service)

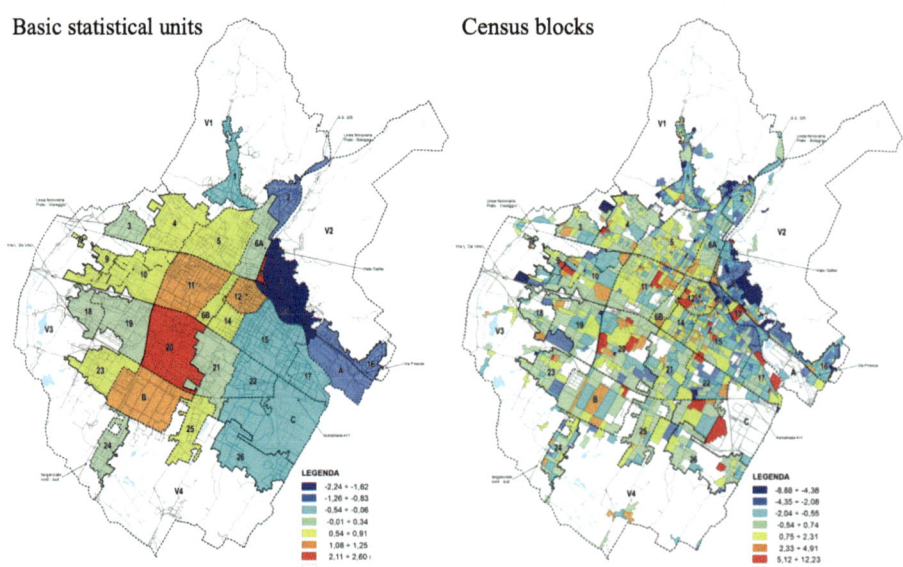

Fig. 4 Spatial distribution of socioeconomic disease (deprivation index, 2011). [*Notes* (1) the index used is that created by Caranci et al. (2010), applied to data from the 2011 Census of Population and Housing; (2) it is a sum of the frequencies of five variables, namely low level of education, unemployment, one-parent family, and home rental and home overcrowding; (3) a higher value indicates higher deprivation. *Source* Statistical Office of the Municipality of Prato (2015, pp. 8, 11)]

in the North-East (units 2, 8, 13, A, in dark tones of blue). The Bisenzio River takes on the role of a physical barrier that isolates 'bourgeois' residential areas. If we compare Figs. 3 and 4, it is evident that these are the only areas where living spaces are clearly separated from the places of production. On the other hand, the map on the right—based on smaller units, namely census blocks—depicts the rest of the city as a mosaic, where clear-cut patterns cannot be recognized except that at the level of small areas, such as that between Via Filzi and Via Pistoiese (unit 11, red and orange blocks), that is, Chinatown.

Since the 1990s, the city of Prato has faced increasing social cohesion problems, combined with the textile industry's crisis and the expansion of the Chinese community. The local production system has changed significantly, having undergone a process of *tertiarization*, only in part related to textile. The global crisis of 2008 intensified this trend. Now, two different paths of development can be identified based on manufacturing and services. Another constitutive element of the district identity is in question. The roots of the 'red' subculture, in fact, have become weaker. In this regard, clear signs came from the electorate, which reoriented its vote in a more volatile manner, resulting in an unprecedented alternation of right- and left-wing coalitions to the government of the city.

To sum up, the long history of Prato, its character as an industrial and, increasingly, multicultural city, together with the processes of change that have taken place, make it a case of great interest, today more than ever. The tension between continuity and change is evident in the tie the people have with the past, based on rooted stereotypes. Our idea is that common stereotypes act as drivers of a public discourse that prevents the city to renegotiate its identity. In the following sections, we investigate this issue by analyzing the discourses on Prato and the underlying stereotypes.

3   A Lively Academic Debate: A Multidisciplinary Literature Review

Preliminarily, we conducted a literature review to identify thematic patterns in academic research. We performed the literature search in February 2020 in Web of Science (WoS), a widely recognized and trusted multidisciplinary database, which gathers a large number of high-quality research products.

As a starting point, we entered the following criteria in the search engine: *search string* [topic] (Prato); *timespan* [custom year search] (2000–2019); *search language* (English). The choice of the reference period finds justification in that, since the early 2000s, the effects of the crisis of textile and of immigration have become evident. As noticed, these processes have transformed Prato profoundly. Instead, English was a forced decision since we aimed to make this work comprehensible for the broadest possible audience.

Subsequently, we refined the search by restricting it to journal articles and removing 'false positives'. Articles, in effect, are shorter and more focused than other academic products, such as monographs; this makes them more suitable for identifying thematic patterns. A further selection was needed, due to the presence of articles that reported 'Prato' as a keyword, although with different semantic values.

The application of these criteria yielded a total of 72 publications, classified in 75 research areas. Most of them were articles of 'Environmental Sciences Ecology' and 'Public Environmental Occupational Health', followed by 'Geography', 'Business Economics', 'Toxicology', 'Demography', and many others.

Three preliminary remarks must be made here. First, Prato is the subject of a lively academic debate. Second, this debate finds space in influential international journals, which indicates that Prato is a breeding ground for research in different disciplinary areas. Third, environmental sciences hold a primary position on this ground.

Once we completed the selection process, we used a computer-assisted qualitative data analysis software—i.e., QSR NVivo 12 Plus for Windows (see: Bazeley and Jackson 2013)—to store, manage, code, and explore the texts. As a first step, we coded the data sources and created two project items, which gathered together keywords and abstracts, respectively. After that, we ran a word frequency query in the 'keywords' item. The search identified 320 words that, for simplicity, were displayed in a word cloud (see: Fig. 5, below). Then, we read through the list and further selected 50 terms, grouping those with

Fig. 5 Word cloud of the most frequently used keywords in journal articles. *Note*: based on a word frequency query limited to words with a minimum length of 3 characters and stemmed words. (*Source* authors' processing of the texts of selected articles retrieved from WoS on February 23rd, 2020)

similar meanings. We ran text searches in the 'abstract' item on each of them and created results as nodes. Finally, we drew a map relating themes (nodes) to articles (cases) to identify thematic patterns (see: Fig. 6, below).

As Fig. 5 shows, the most frequently used keywords are 'industrial OR industry' and 'district OR districts', followed by 'Chinese' and 'textile OR textiles'. These terms refer to typical representations of Prato as a *textile industrial district* and a city that hosts a large *Chinese community*. Unexpected indications come from other keywords, such as 'cancer', 'chromium', 'environmental', 'health', 'noise', 'reuse', 'sewage', 'waste', 'wastewater', 'n-alkanes', 'asbestos', 'hexavalent', and 'pollution'. These terms relate to challenging discourses, worth exploring.

The analysis revealed the existence of six patterns built around six main themes, reported in Fig. 6, numbered from 1 to 6 from most to least significant: (1) *district*; (2) *production*, primarily referred to the textile industry, (3) *immigration*, related to the tale of the Chinese community; (4) *policy*; (5) *environment*; (6) *wastewater*.

The map shows, that these patterns are related to each other, but three of them—numbered from 1 to 3—are interconnected to such an extent that they may be considered as a unique, broader pattern. Most of the studies converged on the district dimension, either developing a meta-reflection on the inherent nature of an industrial district, analyzing its specific mode of production, or focusing on the migrant population, which works in these sectors, primarily composed of Chinese people. The district also lies behind other patterns. Local development policies, for instance, are a constitutive element of the model of governance of the district. Being an industrial city, then, implies dealing with

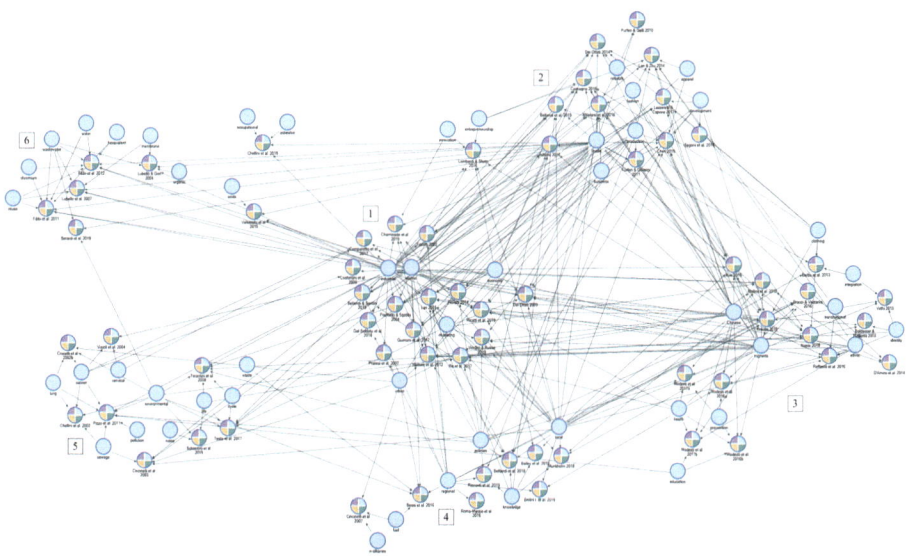

Fig. 6 Map of thematic patterns. (*Source*: authors' processing of the texts of selected articles retrieved from WoS on February 23rd, 2020)

environmental issues related to pollution and sewage disposal, which are, in turn, connected with severe health problems such as cancer.

In detail, the *district* pattern emphasizes the specificity of Prato as a Marshallian industrial district. Several studies reviewed in this chapter refer to this concept (Bellandi et al. 2019; Bellandi and Santini 2019; Chaminade et al. 2009; Dei Ottati 2009, 2014; Fioretti 2002; Lan 2015; Milanesi et al. 2016). Though, a dominant thematization is that looking at the *transformation* of the district, as a consequence of the *decline* of the textile industry (Adamo 2016; Bellandi and Santini 2019; Bellandi et al. 2018; Chaminade et al. 2009; Dei Ottati 2009) and the expansion of the *Chinese community* (Ceccagno 2015; Lan 2015; Ricatti et al. 2019; Verdini and Russo 2019).

The *production* pattern looks specifically at economic aspects (Bailey et al. 2010; Bellandi et al. 2019; Lazzeretti and Capone 2017; Milanesi et al. 2016) and technological issues (Furferi and Gelli 2010; Furferi and Governi 2011), concentrating on sectoral dynamics in the *textile, apparel, and fashion industries*. Furthermore, some studies analyze the role of *Chinese firm*s in reorganizing production and repositioning the district in the global value chain (Chen 2015; Dei Ottati 2014; Lan and Zhu 2014).

A definite *immigration* pattern, however, developed in parallel, focusing on the problems related to the integration of Chinese people (Baldassar and Raffaetà 2018; Bracci and Valzania 2016; Raffaetà et al. 2016) and their forms of entrepreneurship (Barbu et al. 2013; Krause 2015; Molina et al. 2018). A singular thematization has arisen in the field of medicine. Here, the interest is due to the empirical evidence that, in European countries, ethnic minorities tend to have higher health risks than autochthone populations (Modesti et al. 2016a, b, 2017a, b).

Then, the *policy* pattern is represented by contributions that address particular issues, shifting the attention from district to urban governance, dealing with the *unwanted consequences* of industrial production and immigration. Besides those focusing on local development, thus, we find studies referring to *urban regeneration policies* adopted to combat the social segregation of Chinese people (Verdini and Russo 2019), and *policies against illegality*, again connected with the presence of a large ethnic minority and the related cultural conflicts (Munkholm 2018; Riccardi et al. 2019). Others focus on the problem of *management and recovery of urban waste*, with specific attention to the waste produced in industrial areas (Bessi et al. 2016; Tarantini et al. 2009; Testa et al. 2017).

A substantial group of articles is framed in the *environment* pattern. Part of them addresses the problem of *air pollution*, connected with the concentration of atmospheric aerosols, a great matter of concern in urban and industrialized areas due to their toxic and carcinogenic properties (Cincinelli et al. 2003, 2004, 2007; Pratesi et al. 2007). Another part deals with the *risk of cancer*, indeed, related to environmental pollutants (Chellini et al. 2002, 2015; Crocetti et al. 2002a, b; Crocetti et al. 2001; Pizzo et al. 2011; Seniori Costantini 2008; Visioli et al. 2004). Among other relevant environmental issues is *noise pollution* (Schiavoni et al. 2015).

The *wastewater* pattern, finally, is worth a separate mention. The treatment of wastewaters, in fact, is strategic in the context of an industrialized area where a large part of wastewaters comes from the textile industry (Berardi et al. 2019; Fibbi et al. 2011, 2012; Lubello et al. 2007; Lubello and Gori. 2004; Valsecchi et al. 2015).

To conclude, the analysis returned an all-round image of Prato, as a complex city affected by multiple problems. On the other hand, this operation did not allow us to overcome the monolithic representation of Prato as a textile industrial district. Quite the opposite, the academic discourse and its manifold thematizations developed from this central theme.

4　Social Representations of a City in Transformation: An Analysis of INTERNET Videos

4.1　Brief Methodological Notes

In the following pages, we investigate how the city of Prato is represented in Internet videos, looking at how urban landscape is constructed and how stereotypes are produced (on the social construction of landscape see: Kühne 2019). Specifically, we searched and analyzed YouTube videos, as YouTube is one of the most important video-sharing platforms, offering a wide variety of user-generated and corporate videos. We performed the search in April 2020.

There follows a description of the search strategy and analytical methods. First, we installed a clean search browser to exclude progression-related pre-selections and avoid distortions due to personal and location settings. Second, we entered the following criteria in the search engine: *search string* (Prato); *type* (video); *sort by* (relevance). Third, we selected the fifty most relevant videos, being careful to rule out false positives. After a first summary analysis of the results, we decided to exclude 14 sports videos—these were brief journalistic reports on sporting events that made no significant references to the city and urban landscape—and we replaced them with the next 14 videos in order of relevance. Fourth, we captured the videos with QSR NCapture, a web-browser extension that enables us to gather web content to import into NVivo. We, then, imported the videos into NVivo and classified them on the basis of 8 variables, grouped as follows: *technical aspects* (the type of product and degree of professionalism); and *content-related aspects*, further divided into *general* (the type of contents, the intensity of specific spatial references, and the degree of stereotipicity) and *issue-specific* (the intensity of references to industrial identity, immigration, and the past). After that, we conducted a descriptive analysis of the database aiming to identify essential aspects related to the dominant themes and the associated spatial stereotypes. Fifth, we further selected six videos to analyze in depth, taking into account the variety of contents. In particular, we focused on recurrent subjects, themes, and spaces to understand how their symbolic values translate into stereotypes, which, in turn, contribute to constructing the public image of the city.

4.2   Patterns of Representation

The selected Internet videos were published from 2007 to 2020. Choosing videos based on relevance meant that the most recent videos were overrepresented; indeed, 22 out of 50 are published in 2019 and the first three months of 2020. The number of views varies considerably, between 4 and 1,274,003. Instead, the length is under half an hour, the types of products being constituted almost exclusively by relatively short videos and a small number of excerpts from documentary films.

Interesting indications come from the examination of the degree of professionalism in the implementation of the videos (see: Fig. 7). To categorize the videos, following Kühne in this book, we took as references—in terms of editing, image, and sound quality—the standard of a national news channel for a 'professional realization' and of an amateur video for an 'unprofessional realization'. Here, it is worth noting that 16 out of 50 videos are classified as professional realizations, mostly consisting of journalistic reportages and excerpts from daily news launched by national tv channels, which means that Prato finds space in the national media. Stereotypes play a primary part in constructing a public image that is closely connected with discourses developed at the national level—e.g. concerning immigration—and leaves it open to political exploitation. The comparison with the realizations approaching professionalism, such as those consisting of excerpts from the daily news of local tv channels, reveals that stereotypes are handled differently at the two levels: indeed, to reinforce a public or political discourse, at the national level; to deal with concrete problems, but neutrally, at the local level. Here, a counterintuitive finding is that the higher is the degree of professionalism, the more stereotypes have a constitutive effect.

In terms of content, 12 inductively obtained categories can be identified: advertising videos; amateur videos; daily news; documentaries; food-related videos; journalistic reportages; movie trailers; music videos; photo collections; satirical videos; videos for territorial marketing; travel videos. Among them, journalistic reportages prevail.

Different contents are related to different ways of approaching space. Space is represented in most videos, although with a different intensity, being frequently conducive to action or thematized (see: Fig. 8). This is truer for amateur videos and documentaries, while, in journalistic reportages and daily news, space tends to play a secondary part. Here, again, a difference can be observed, since journalistic reportages, especially when transmitted by national tv channels, more often thematize space. In this case, stereotypical references to Chinatown and the Chinese community prevail.

Immigration, indeed, is one of the most referenced themes (see: Fig. 9). Besides, references to places or issues related to the industrial identity remain in the background, while another constitutive theme is connected to the idea of a 'lost' past. In effect, these three themes are somewhat interrelated in a coherent field of meaning. The nostalgic references to the past, indeed, rely on spatial stereotypes that have to do with the industrial identity. In this case, space is conducive to action, and stereotypes have a constitutive

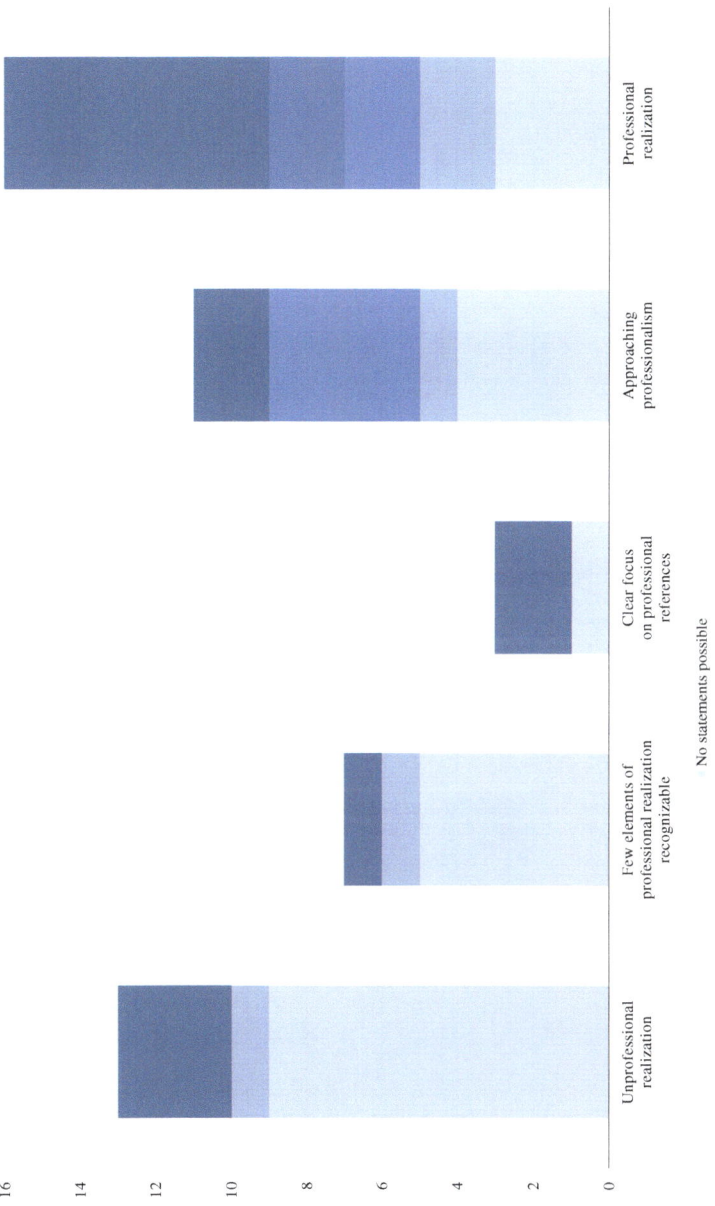

Fig. 7 Degree of professionalism and degree of stereotipicity (frequencies). (*Source* authors' processing of the videos retrieved from YouTube on April 10th, 2020)

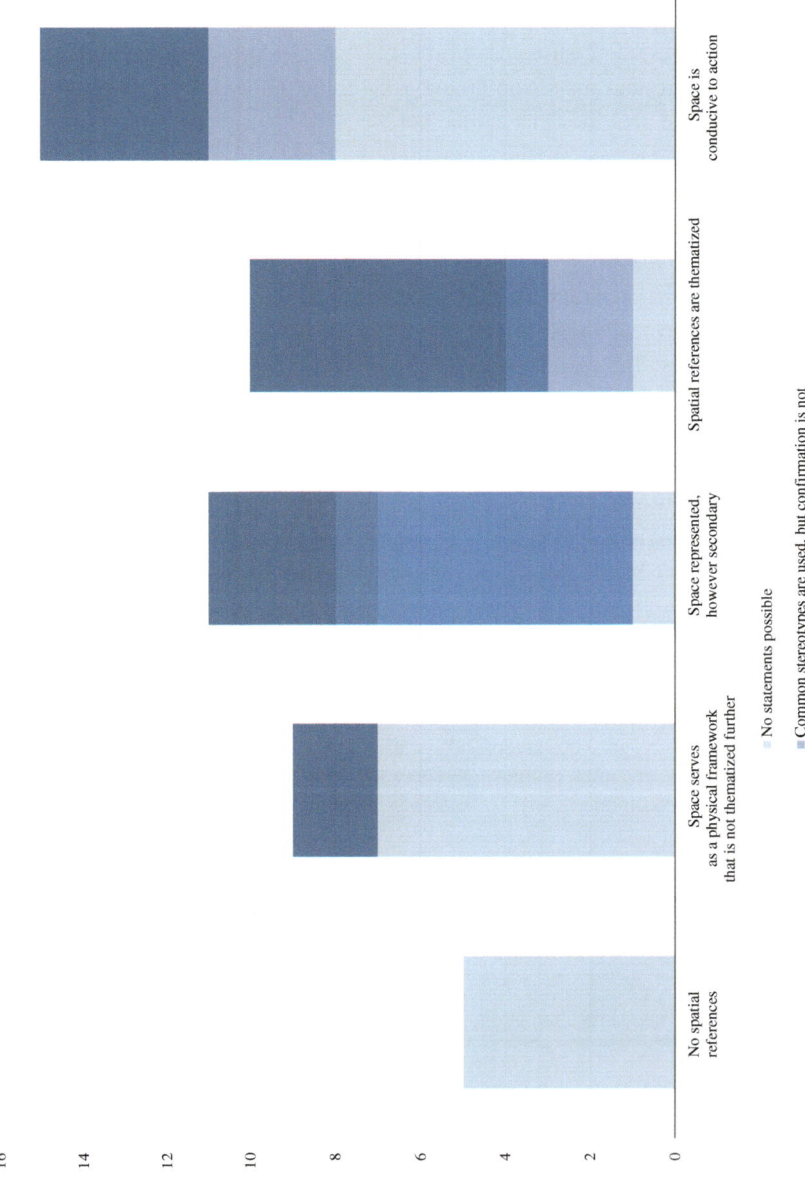

Fig. 8 The intensity of specific spatial references and degree of stereotipicity (frequencies). (*Source* authors' processing of the videos retrieved from YouTube on April 10th, 2020)

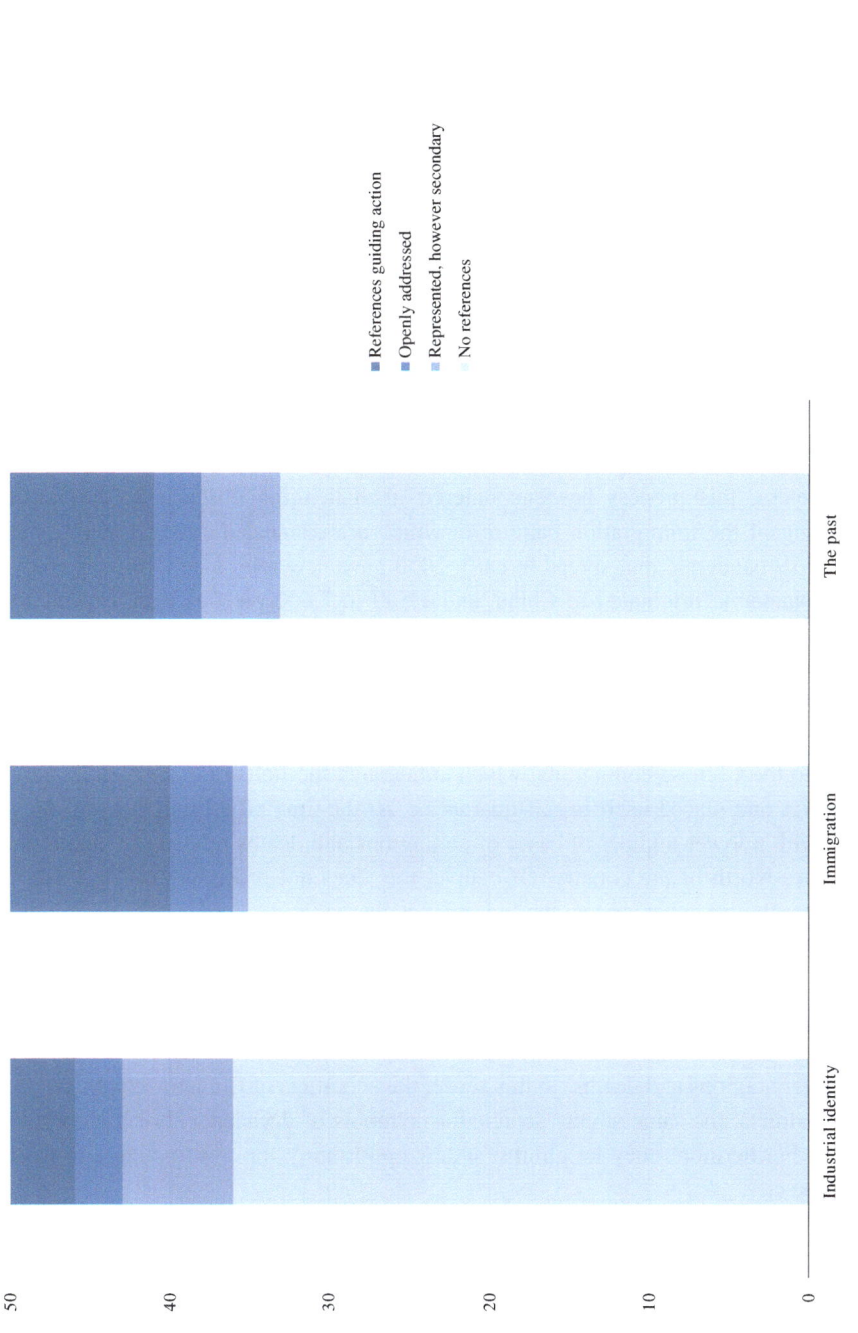

Fig. 9 The intensity of specific thematic references (frequencies). (*Source* authors' processing of the videos retrieved from YouTube on April 10th, 2020)

effect, although these are, in a sense, 'neutral' stereotypes, not implying the use of a discriminatory language. Instead, immigration is seen as a threat to the industrial identity, contributing to the city losing its past. This view, in truth, emerges as a minority, though amplified by the national media.

Generally speaking, stereotypes are very present. Moreover, they have more often a constitutive effect when they rely on connotative or thematized spatial references, and when immigration is the focal theme.

That said, in inductive research design, the analysis of the deviance from the norm is as significant as that of recurrences. From this perspective, some minor thematic patterns are of great significance. These patterns rely on 'new' aesthetic references, such as those to historical heritage, cultural sights, modern art, and good food, which are struggling to replace 'old' aesthetic stereotypes related to the symbolic signs of the industrial and multi-ethnic identity of Prato. These patterns are also put forward as a strategy of territorial marketing to change the public image of the city.

Nevertheless, this process has encountered a challenging obstacle in the further strengthening of the immigration pattern, to which are connected stronger stereotypes. The Covid-19 pandemic that spread in early 2020 gave new relevance to this discourse. Since the pandemic originated in China, and 1,500 to 2,500 persons were expected to come back after participating in the Chinese New Year festivities in the country of origin, Prato was under watch. The videos addressing this issue, also those handling common stereotypes neutrally, testify the concern of local authorities. However, Prato has proven to be able to respond better than the neighboring cities to this emergency, primarily thanks to the Chinese community, which anticipated the Italian government's restrictive measures and placed itself in self-quarantine. At the time of writing, Prato is one of the cities with a lower number of cases of contagions and deaths related to Coronavirus in the Centre-North of the country. Of course, this does not prove any causal relationship between the two phenomena, while it reveals that what we know about the Chinese community in Prato is mostly based on stereotypes. The Chinese, in Prato, are often stereotyped as people who do not respect the law. Yet, with the Covid-19 crisis, they are now being praised as models of responsible behavior. The Mayor of Prato has publicly thanked them for enabling Prato to be one of the Italian cities less vulnerable to the virus in terms of contagions and deaths. In this sense, the shootings of Chinatown with no people on the streets, the same streets depicted as symbols of decadence, break a common stereotype. Furthermore, they let glimpse a new opportunity for constructing a different image of the city.

4.3   A Critical Analysis of Landscape Stereotypes

While it allowed the identification of recurrent subjects, themes, and spaces, the in-depth analysis of the selected six videos (see: Table 1) also led to the isolation of different 'narrations' of the city of Prato and the basic elements of its 'multiple identities'. These

Table 1 List of the selected videos. *Source*: retrieved from YouTube on April 10th, 2020

No.	Title and URL	Year
1	*Blebla - Prato* https://youtu.be/lwsacE4JEQ4	2012
2	*Prato 1981* https://youtu.be/vr3ZRinWhQ8	2007
3	*La Chinatown di Prato - Piazza Pulita 11 ott 2013* https://youtu.be/pUa4GwOrW98	2013
4	*Noi non siamo un virus. Viaggio a Prato, nella Cina italiana discriminata per il Coronavirus* https://youtu.be/xnvUsQ3QT9k	2020
5	*Prato - Piccola Grande Italia* https://youtu.be/4SuAl7jAsiI	2016
6	*Coronavirus, Matteo Biffoni (Sindaco di Prato): "Siamo pronti a prendere ulteriori misure"* https://youtu.be/WSCn2xB_Te4	2020

narrations rely on rooted stereotypes partly linked to each other, such as those related to Prato's industrial identity, the threat of immigration, and the nostalgia for the past. Besides, aesthetic stereotypes are emerging that depict Prato as a beautiful city and a city of culture.

The first notable finding is that most of the spatial references in the six videos—with due differences depending on the focal theme—relate to spaces and symbols that are not directly connected with the city's industrial identity. Moreover, the majority of them are *cultural sights*, either related to Prato's pre-modern history (i.e., the Bacchino Fountain, the Cathedral of St. Stephen, the Church of St. Domenico, the City walls, the Emperor's Castle, the Monument to Francesco Datini, and the Praetorian Palace) or modern and contemporary art and culture (i.e., the Exegi Monumentum Aere Perennius, the Mazzocchio, the Metastasio Theatre, Moore's sculpture Square Form with Cut, the Pecci Amphitheatre, and the Pecci Centre for Contemporary Art). Among others, the Textile Museum, together with the Lazzerini Library, is part of the Campolmi Pole, a symbol of Prato's industrial origins and an exemplar of industrial archaeology. Besides, the Buddhist Temple—namely a sign of the presence of the Chinese community and social change towards multiculturalization—also finds space. On the other hand, Chinatown is the most referenced *city place*, while other symbolic places remain in the background (e.g., the City Hall, Duomo Square, the Lungobisenzio Stadium, Magnolfi Street, Mazzoni Street, Mercatale Square, Slaughterhouses Square, and Town Hall Square). *Industrial sites*, instead, are mostly referenced to trace back space to the discourse on Chinese immigration: in particular, Chinese factories in 'Macrolotto 0'—that is, Chinatown—and 'Macrolotto 1' are used as a background to characterize the way of working and living within the Chinese community.

The subjects in the videos can be easily clustered based on their ethnic origins, namely Italians and Chinese. Two remarks are worth making concerning the processes of cultural hybridization, which are revealed by the—either explicit or implicit—references to two common stereotypes. The term *cultural hybridization* refers to "external flows", related to globalization, which "interact with internal flows producing a unique cultural hybrid that combines their elements" (Ritzer 2010, p. 255; on this concept see: Appadurai 1996). On the one hand, we find Italian people, interviewed on the road, who speaks with a Southern dialectal inflection. This phenomenon is related to inward migrations, dating back to the years of economic boom, in the 1950s and the 1960s. Today, the nephews of the first generations of migrants are full citizens of Prato, even if they maintain a link with their origins, indeed, developing a hybrid identity—as the urban rapper Blebla says, in video 1, "half Tuscan, half Southerner" (*mezzo Toscano, mezzo terrone*). On the other hand, we have a sharp contrast between Chinese workers interviewed at the workplace, who do not speak Italian, and the Chinese young people speaking Italian fluently that appear in videos 1 and 4 to break a common stereotype (*no Italiano*) and testify an ongoing process of hybridization.

Figure 10 gives an idea of how the 'Chinese question' is thematized. The map relates themes (nodes) to videos (cases) to identify thematic patterns.

The comparison between the two central patterns is highly significant. The two related videos, in fact, are based on journalistic reportages launched within the same tv program, on a national tv channel, but thematize the question differently. Video 3 reproduces common stereotypes related to the idea of Chinese factories as places of *illegality*—i.e., firms working 24 hours a day, organized as factories-dormitories, making systematic use of undeclared work, not paying taxes, and transferring wealth abroad through money transfers. The correspondent defines them as "the first enemy"—of Italian entrepreneurs, in the grip of crisis. In this video, Chinatown is the place where "expensive cars" run, driven by Chinese people. On the other hand, Video 4, dealing with Coronavirus in the early days of contagion, attempts to break an emerging stereotype,

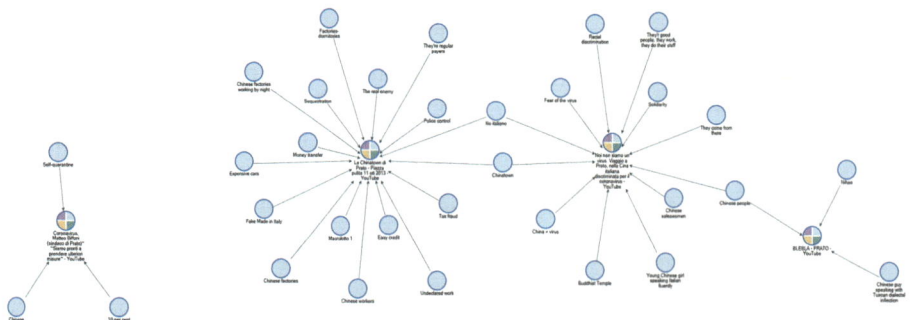

Fig. 10 Map of thematizations of the 'Chinese question' *Source* authors' processing of the texts of selected articles retrieved from YouTube on April 10th, 2020.

expressed by the equivalence "China = virus". To give strength to the main argument—that is, "We [Chinese people] are not a virus"—it reports the testimony of a Chinese girl speaking Italian fluently. Chinatown, with no people on the streets, because of their rigidly adhering to self-imposed quarantine, becomes the place of *virtuosity*.

These examples reveal that the definition of the identity of the city relies on how the local community deals with its multiple identity sources. Among them are not only *ethnic* sources but also *cultural* ones. The latter, in particular, play a crucial role in mediating between emotional ties to the past and a pragmatic look at the future.

5   Lines of Hybridization—Refining the Research Agenda

This chapter has dealt with a well-known and widely studied case, that is, the city of Prato. The great availability of studies on this subject, certainly, is an added value; on the other hand, it makes it difficult to deviate from pre-determined and well-established binaries. This seems to produce effects of path-dependence in academic research, which, as such, developed around a central theme, that is, Prato as a textile industrial district, with its economic, social, and—increasingly—environmental and health implications. The review of academic literature conducted in Sect. 3, however, revealed an unsuspected complexity of this city, which has undergone a profound transformation, bringing with it problems of social cohesion and sustainable development.

That said, the analysis of the public discourse, as conveyed by Internet videos, revealed that academic research, overall, has since now failed to grasp the inherent meanings of change just like the local community has not been able to deal with the mutations of its identity. The resilience of established stereotypes, associated with a nostalgic attachment to the past, is proof of that.

Nevertheless, the images disseminated through the Internet videos clearly show that this process is accompanied by a new geography of the places, landscapes, and spaces of the city, redrawn by the flows and settlements of people, especially new Chinese immigrants.

As highlighted, immigration is the most critical issue. It receives a high priority in social research and catches the attention of mass media, probably because it is perceived as a point of conjunction between the lost past and an uncertain future. Chinatown is the place where the changes related to immigration processes come to life. By the time, it is a settled and lively pole of attraction in Prato's polycentric development model. The experience of the Covid-19 pandemic further confirmed this fact.

On the other hand, a discourse that gives expression to Prato's artistic vocation is gaining space. As a result, aesthetic symbols related to arts and cultural heritage are gradually breaking the monopoly of material symbols related to industrial production.

This leaves room for the construction of a new imaginary, which, even in the references to places and landscapes, is projected towards the future, using cultural hybridization as a driver for change.

To conclude, what emerges is that different lines of hybridization are active that are reconfiguring structure and functions, as well as spaces and symbols, of the city of Prato. These are economic, social, and, indeed, cultural lines of hybridization. Along these lines, academic research should develop new paths of inquiry.

References

Adamo, S. (2016). The Crisis of the Prato industrial district in the works of Edoardo Nesi: A blend of nostalgia and self-complacency. *Modern Italy, 21*(3), 245–259.

Appadurai, A. (1996). *Modernity at large: Cultural dimensions of globalization.* Minneapolis, MN-London: University of Minnesota Press.

Bagnasco, A. (1977). *Tre Italie. La problematica territoriale dello sviluppo italiano.* Bologna: Il Mulino.

Bailey, D., Bellandi, M., Caloffi, A., & De Propris, L. (2010). Place-renewing leadership: Trajectories of change for mature manufacturing regions in Europe. *Policy Studies, 31*(4), 457–474.

Baldassar, L., & Raffaetà, R. (2018). It's complicated, Isn't It: Citizenship and ethnic identity in a mobile world. *Ethnicities, 18*(5), 735–760.

Barbu, M., Dunford, M., & Weidong, L. (2013). Employment, entrepreneurship, and citizenship in a globalised economy: The Chinese in Prato. *Environment and Planning A: Economy and Space, 45*(10), 2420–2441.

Bazeley, P., & Jackson, K. (2013). *Qualitative data analysis with NVivo* (2nd ed.). London: Sage.

Becattini, G. (1962). *Il concetto di industria e la teoria del valore.* Torino: Bollati Boringhieri.

Becattini, G. (1979). Dal 'settore' industriale al 'distretto' industriale. Alcune considerazioni sull'unità d'indagine dell'economia industriale. *Rivista di Economia e Politica Industriale, 1,* 7–21.

Becattini, G. (Ed.). (1987). *Mercato e forze locali: il distretto industriale.* Bologna: Il Mulino.

Becattini, G. (Ed.). (1997). *Prato. Storia di una città. 4. Il distretto industriale (1943–1993).* Firenze: Le Monnier.

Becattini, G. (2000). *Il bruco e la farfalla. Prato nel mondo che cambia (1954–1993).* Firenze: Le Monnier.

Becattini, G. (2004). *Industrial districts: A new approach to industrial change.* Cheltenham, UK: Edward Elgar.

Bellandi, M., De Propris, L., & Santini, E. (2019). An evolutionary analysis of industrial districts: The changing multiplicity of production know-how nuclei. *Cambridge Journal of Economics, 43*(1), 187–204.

Bellandi, M., & Russo, M. (Eds.). (1994). *Distretti industriali e cambiamento economico locale.* Torino: Rosenberg & Sellier.

Bellandi, M., & Santini, E. (2019). Territorial servitization and new local productive configurations: The case of the textile industrial district of Prato. *Regional Studies, 53*(3), 356–365.

Bellandi, M., Santini, E., & Vecciolini, C. (2018). Learning, unlearning and forgetting processes in industrial districts. *Cambridge Journal of Economics, 42*(6), 1671–1685.

Berardi, C., Fibbi, D., Coppini, E., Renai, L., Caprini, C., Agata Scordo, C. V., et al. (2019). Removal efficiency and mass balance of polycyclic aromatic hydrocarbons, phthalates, ethoxylated alkylphenols and alkylphenols in a mixed textile-domestic wastewater treatment plant. *Science of the Total Environment, 674,* 36–48.

Bessi, C., Lombardi, L., Meoni, R., Canovai, A., & Corti, A. (2016). Solid recovered fuel: An experiment on classification and potential applications. *Waste Management, 47,* 184–194.

Bracci, F., & Valzania, A. (2016). Hidden selectivity: Irregular migrants and access to socio-health services in a heated local context. *Cambio. Rivista sulle Trasformazioni Sociali, 5*(10), 141–148.

Brusco, S. (1989). *Piccole imprese e distretti industriali. Una raccolta di saggi.* Torino: Rosenberg & Sellier.

Caranci, N., Biggeri, A., Grisotto, L., Pacelli, B., Spadea, T., & Costa, G. (2010). L'indice di deprivazione italiano a livello di sezione di censimento: definizione, descrizione e associazione con la mortalità. *Epidemiologia e Prevenzione, 34*(4), 167–176.

Ceccagno, A. (2015). The mobile emplacement: Chinese migrants in italian industrial districts. *Journal of Ethnic and Migration Studies, 41*(7), 1111–1130.

Chaminade, C., Bellandi, M., Plechero, M. & Santini, E. (2019). Understanding processes of path renewal and creation in thick specialized regional innovation systems. Evidence from two textile districts in Italy and Sweden. *European Planning Studies, 27*(10), 1978–1994.

Chellini, E., Cherubini, M., Chetoni, L., Seniori Costantini, A., Biggeri, A., & Vannucchi, G. (2002). Risk of respiratory cancer around a sewage plant in Prato, Italy. *Archives of Environmental Health, 57*(6), 548–553.

Chellini, E., Martino, G., Grillo, A., Fedi, A., Martini, A., Indiani, L., et al. (2015). Malignant mesotheliomas in textile rag sorters. *Annals of Occupational Hygiene, 59*(5), 547–553.

Chen, C. (2015). Made in Italy (by the Chinese): Migration and the rebirth of textiles and apparel. *Journal of Modern Italian Studies, 20*(1), 111–126.

Cincinelli, A., Del Bubba, M., Martellini, T., Gambaro, A., & Lepri, L. (2007). Gas-particle concentration and distribution of n-alkanes and polycyclic aromatic hydrocarbons in the atmosphere of Prato (Italy). *Chemosphere, 68*(3), 472–478.

Cincinelli, A., Mandorlo, S., Dickhut, R. M., & Lepri, L. (2003). Particulate organic compounds in the atmosphere surrounding an industrialised area of Prato (Italy). *Atmospheric Environment, 37*(22), 3125–3133.

Cincinelli, A., Stefani, A., Seniori Costantini, S., & Lepri, L. (2004). Characterization of N-Alkanes and PAHS in PM10 Samples in Prato (Italy). *Annali di Chimica, 94*(4), 281–293.

Crocetti, E., Bernini, G., Tamburini, A., Miccinesi, G., & Paci, E. (2002a). Incidence and survival cancer trends in children and adolescents in the Provinces of Florence and Prato (Central Italy), 1985–1997. *Tumori Journal, 88*(6), 461–466.

Crocetti, E., Ciatto, S., & Zappa, M. (2001). Prostate cancer: Different incidence but not mortality trends within two areas of Tuscany, Italy [2]. *Journal of the National Cancer Institute, 93*(11), 876–877.

Crocetti, E., Miccinesi, G., Paci, E., & Cislaghi, C. (2002b). What Is hidden behind urban and semiurban cancer incidence and mortality differences in Central Italy? *Tumori Journal, 88*(4), 257–261.

Dei Ottati, G. (2009). An industrial district facing the challenges of globalization: Prato Today. *European Planning Studies, 17*(12), 1817–1835.

Dei Ottati, G. (2014). A transnational fast fashion industrial district: An analysis of the Chinese businesses in Prato. *Cambridge Journal of Economics, 38*(5), 1247–1274.

Edler, D., Jenal, C., & Kühne, O. (2020). Modern approaches to the visualization of landscapes—An introduction. In D. Edler, C. Jenal, & O. Kühne (Eds.), *Modern approaches to the visualization of landscapes* (pp. 3–15). Wiesbaden: Springer VS.

Fibbi, D., Doumett, S., Colzi, I., Coppini, E., Pucci, S., Gonnelli, C., et al. (2011). Total and hexavalent chromium removal in a subsurface horizontal flow (h-SSF) constructed wetland operating as post-treatment of textile wastewater for water reuse. *Water Science and Technology, 64*(4), 826–831.

Fibbi, D., Doumett, S., Lepri, L., Checchini, L., Gonnelli, C., Coppini, E., et al. (2012). Distribution and mass balance of hexavalent and trivalent chromium in a subsurface, horizontal flow (SF-h) constructed wetland operating as post-treatment of textile wastewater for water reuse. *Journal of Hazardous Materials, 199–200,* 209–216.

Fioretti, G. (2002). Information structure and behaviour of a textile industrial district. *Journal of Artificial Societies and Social Simulation, 4*(4), 123–135.

Furferi, R., & Gelli, M. (2010). Yarn strength prediction: A Practical model based on artificial neural networks. *Advances in Mechanical Engineering, 2010,* 1–11.

Furferi, R., & Governi, L. (2011). Prediction of the spectrophotometric response of a carded fiber composed by different kinds of coloured raw materials: An artificial neural network-based approach. *Color Research & Application, 36*(3), 179–191.

Iommi, S., & Marinari, D. (2000). *Mobilità residenziale e pendolarismo in Toscana.* Firenze: IRPET.

Krause, E. (2015). Fistful of tears: Encounters with transnational affect, Chinese Immigrants and Italian Fast Fashion. *Cambio. Rivista sulle Trasformazioni Sociali, 5*(10), 27–40.

Kühne, O. (2018). *Landscape and power in geographical space as a social-aesthetic construct.* Dordrecht: Springer International Publishing.

Kühne, O. (2019). *Landscape theories: A brief introduction.* Wiesbaden: Springer VS.

Kühne, O. (2020). The social construction of space and landscape in internet videos. In D. Edler, C. Jenal, & O. Kühne (Eds.), *Modern approaches to the visualization of landscapes* (pp. 121–137). Wiesbaden: Springer VS.

Kühne, O. & Bellini, A. (2019). Landscape conflicts and the making of contemporary European Societies: A dialogue with Olaf Kühne, *Cambio. Rivista sulle Trasformazioni Sociali, 9*(18), published online on March 20th, 2020.

Kühne, O., & Jenal, C. (2020). The threefold landscape dynamics—Basic considerations, conflicts and potentials of virtual landscape research. In D. Edler, C. Jenal, & O. Kühne (Eds.), *Modern approaches to the visualization of landscapes* (pp. 389–402). Wiesbaden: Springer VS.

Kühne, O., Weber, F., & Berr, K. (2019). The productive potential and limits of landscape conflicts in Light of Ralf Dahrendorf's Conflict Theory. *SocietàMutamentoPolitica, 10*(19), 77–90.

Lan, T. (2015). Industrial district and the multiplication of labour: The Chinese Apparel Industry in Prato. *Italy. Antipode, 47*(1), 158–178.

Lan, T., & Zhu, S. (2014). Chinese Apparel Value Chains in Europe: Low-end fast fashion, regionalization, and transnational entrepreneurship in Prato, Italy. *Eurasian Geography and Economics, 55*(2), 156–174.

Lazzeretti, L., & Capone, F. (2017). The transformation of the prato industrial district: An organisational ecology analysis of the co-evolution of Italian and Chinese Firms. *Annals of Regional Science, 58*(1), 135–158.

Linke, S. (2020). Landscape in internet pictures. In D. Edler, C. Jenal, & O. Kühne (Eds.), *Modern approaches to the visualization of landscapes* (pp. 139–156). Wiesbaden: Springer VS.

Loda, M., Kühne, O., & Puttilli, M. (2020). The social construction of Tuscany in the German and English speaking world—Presented by the analysis of internet images. In D. Edler, C. Jenal, & O. Kühne (Eds.), *Modern approaches to the visualization of landscapes* (pp. 157–171). Wiesbaden: Springer VS.

Lubello, C., Caffaz, S., Mangini, L., Santianni, D., & Caretti, C. (2007). MBR pilot plant for textile wastewater treatment and reuse. *Water Science and Technology, 55*(10), 115–124.

Lubello, C., & Gori, R. (2004). Membrane bio-reactor for advanced textile wastewater treatment and reuse. *Water Science and Technology, 50*(2), 113–119.

Marshall, A. (1890). *Principles of economics*. London: Macmillan.

Milanesi, M., Guercini, S., & Waluszewski, A. (2016). A Black Swan in the District? An IMP perspective on immigrant entrepreneurship and changes in industrial districts. *IMP Journal, 10*(2), 243–259.

Modesti, P. A., Calabrese, M., Malandrino, D., Colella, A., Galanti, G., & Zhao, D. (2017a). New findings on type 2 diabetes in first-generation chinese migrants settled in Italy: Chinese in Prato (CHIP) Cross-Sectional Survey. *Diabetes/Metabolism Research and Reviews, 33*(2), e2835.

Modesti, P.A., Calabrese, M., Marzotti, I., Bing, H., Malandrino, D., Boddi, B., Castellani, S., & Zhao, D. (2017b). Prevalence, awareness, treatment, and control of hypertension among chinese first-generation migrants and Italians in Prato, Italy: The CHIP Study. *International Journal of Hypertension, 2017*.

Modesti, P.A., Calabrese, M., Perruolo, E., Bussotti, A., Malandrino, D., Bamoshmoosh, M., Biggeri, A., & Zhao, D. (2016a). Sleep history and hypertension burden in first-generation Chinese Migrants Settled in Italy: The Chinese in Prato Cross-Sectional Survey. *Medicine (United States), 95*(14).

Modesti, P. A., Colella, A., & Zhao, D. (2016b). Atrial Fibrillation in first generation Chinese migrants living in Europe: A proof of concept study. *International Journal of Cardiology, 215*, 269–272.

Molina, J. L., Martínez-Cháfer, L., Molina-Morales, F. X., & Lubbers, M. J. (2018). Industrial districts and migrant enclaves: A model of interaction. *European Planning Studies, 26*(6), 1160–1180.

Munkholm, L. (2018). Creating a new type of labour law enforcer: The law technician in Prato. *Journal of Law and Society, 45*(4), 538–562.

Papadimitriou, F. (2020). Visualization of future landscapes, postmodern cinema and geographical education. In D. Edler, C. Jenal, & O. Kühne (Eds.), *Modern approaches to the visualization of landscapes* (pp. 351–369). Wiesbaden: Springer VS.

Pizzo, A. M., Chellini, E., & Seniori Costantini, A. (2011). Lung cancer risk and residence in the neighborhood of a sewage plant in Italy. A Case-Control Study. *Tumori Journal, 97*(1), 9–13.

Pratesi, G., Zoppi, M., Vaiani, T., & Calastrini, F. (2007). A morphometric and compositional approach to the study of ambient aerosol in a medium industrial town of Italy. *Water, Air, and Soil Pollution, 179*(1–4), 283–96.

Raffaetà, R., Baldassar, L., & Harris, A. (2016). Chinese immigrant youth identities and belonging in Prato, Italy: Exploring the intersections between migration and youth studies. *Identities, 23*(4), 422–437.

Ricatti, F., Dutto, M., & Wilson, R. (2019). Ethnic enclave or transcultural edge? Reassessing the Prato district through digital mapping. *Modern Italy, 24*(4), 369–381.

Riccardi, M., Milani, R., & Camerini, D. (2019). Assessing money laundering risk across regions. An application in Italy. *European Journal on Criminal Policy and Research, 25*(1), 21–43.

Ritzer, G. (2010). *Globalization: A basic text*. Malden, MA: Wiley-Blackwell.

Schiavoni, S., D'Alessandro, F., & Conte, A. (2015). The contribution of LIFE+ NADIA project on the implementation of the European Directive on Environmental Noise. *Noise Mapping, 2*(1), 13–30.

Seniori Costantini, A., Martini, A., Puliti, D., Ciatto, S., Castiglione, G., Grazzini, G., et al. (2008). Colorectal cancer mortality in two areas of Tuscany with different screening exposures. *Journal of the National Cancer Institute, 100*(24), 1818–1821.

Statistical Office of the Municipality of Prato. (2015). *Progetto URBES 2015. Le aree di disagio socio-economico analizzate attraverso un indice di deprivazione*. Retrieved 23rd February 2020 from http://statistica.comune.prato.it/.

Tarantini, M., Dominici Loprieno, A., Cucchi, E., & Frenquellucci, F. (2009). Life cycle assessment of waste management systems in Italian Industrial Areas: Case Study of 1st Macrolotto of Prato. *Energy, 34*(5), 613–622.

Testa, F., Nucci, B., Iraldo, F., Appolloni, A., & Daddi, T. (2017). Removing obstacles to the implementation of LCA among SMEs: A collective strategy for exploiting recycled wool. *Journal of Cleaner Production, 156,* 923–931.

Trigilia, C. (1986). *Grandi partiti e piccole imprese.* Bologna: Il Mulino.

Valsecchi, S., Rusconi, M., Mazzoni, M., Viviano, G., Pagnotta, R., Zaghi, C., et al. (2015). Occurrence and Sources of Perfluoroalkyl Acids in Italian River Basins. *Chemosphere, 129,* 126–134.

Verdini, G., & Russo, E. (2019). The Chinese in Southern Europe: Has urban regeneration addressed their new form of clustering? *Documents d'Anàlisi Geogràfica, 65*(1), 163.

Viesti, G. (2000). *Come nascono i distretti industriali.* Bari: Laterza.

Visioli, C. B., Zappa, M., Ciatto, S., Iossa, A., & Crocetti, E. (2004). Increasing trends of cervical adenocarcinoma incidence in Central Italy despite extensive screening programme, 1985–2000. *Cancer Detection and Prevention, 28*(6), 461–464.

Andrea Bellini is an Assistant Professor of Sociology at the University of Florence, Italy. He writes on middle classes, creative labor, professions, and, more generally, on issues of economic and social regulation. He has published two books in the Italian language: Il puzzle dei ceti medi (2014, FUP); Una professione plurale. Il caso dell'avvocatura fiorentina (2017, with F. Alacevich, A. Tonarelli, FUP). His recent works include: Small Firms and the External Context: Embeddedness Versus Dependency (2020, with V. Fortunato, Palgrave); Not Only Riders. The Uncertain Boundaries of Digital Creative Work as a Frontier for Emerging Actors in Interest Representation (2019, with S. Lucciarini, in PaCo); Professions Within, Between and Beyond. Varieties of Professionalism in a Globalising World (2018, with L. Maestripieri, in Cambio).

Laura Leonardi is a Professor of Sociology at the "Cesare Alfieri" School of Political Science at the University of Florence. Since 2013 to date, she is Director of CESVI (Centre for European Studies on Local and Regional Development) at the Department of Political Science and Sociology in Florence. Her research interests are social inequalities and social citizenship, labor and welfare, social theory, the impact of globalization and Europeanization on local societies. Among her most recent publications: Ralf Dahrendorf. Between Social Theory and Political Praxis, Palgrave (2020 with Olaf Kühne); Social citizenship and inequalities. Is there a future for a social Europe? In Social challenges for Europe: Addressing failures and perspectives or the European project, pp. 189–204, Bologna: Il Mulino (2019 with G Scalise, eds.); Reddito di base e diseguaglianze sociali. Una questione di chances di vita. Iride, vol. XXXII, pp. 325–337, 2019.; Ipotesi di quadratura del cerchio. Diseguaglianze, chances di vita e politica sociale in Ralf Dahrendorf. SocietàMutamentoPolitica, vol. 10, pp. 127–139, 2019.